D1751343

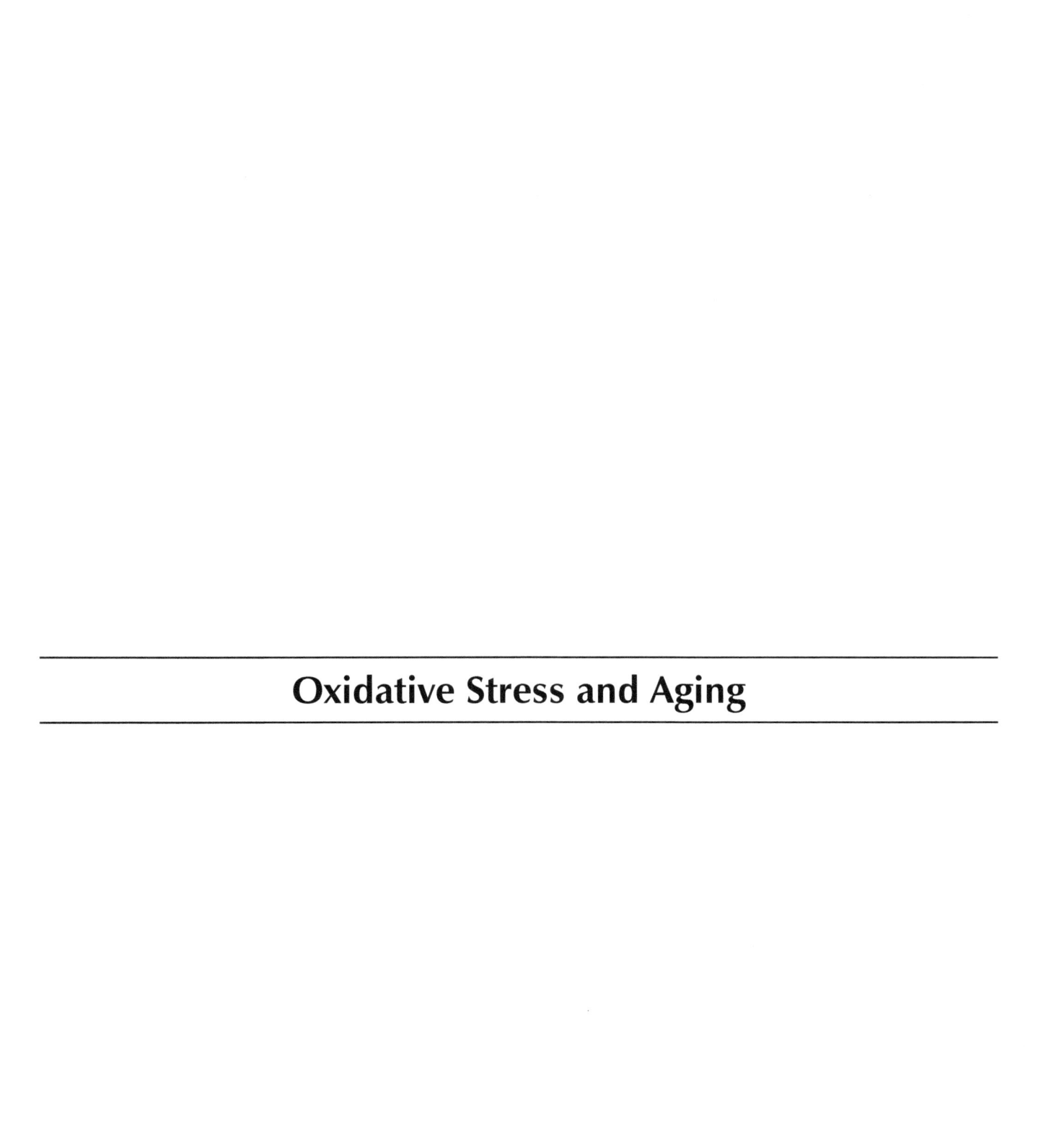

# Oxidative Stress and Aging

# Oxidative Stress and Aging

Editor: Nick Rees

FOSTER ACADEMICS

www.fosteracademics.com

www.fosteracademics.com

**FA**
FOSTER
ACADEMICS

Cataloging-in-Publication Data

Oxidative stress and aging / edited by Nick Rees.
    p. cm.
Includes bibliographical references and index.
ISBN 978-1-63242-906-3
1. Oxidative stress. 2. Aging. 3. Stress (Physiology). I. Rees, Nick.
RB170 .O95 2020
616.07--dc23

© Foster Academics, 2020

Foster Academics,
118-35 Queens Blvd., Suite 400,
Forest Hills, NY 11375, USA

ISBN 978-1-63242-906-3 (Hardback)

This book contains information obtained from authentic and highly regarded sources. Copyright for all individual chapters remain with the respective authors as indicated. All chapters are published with permission under the Creative Commons Attribution License or equivalent. A wide variety of references are listed. Permission and sources are indicated; for detailed attributions, please refer to the permissions page and list of contributors. Reasonable efforts have been made to publish reliable data and information, but the authors, editors and publisher cannot assume any responsibility for the validity of all materials or the consequences of their use.

**Trademark Notice:** Registered trademark of products or corporate names are used only for explanation and identification without intent to infringe.

# Contents

Preface .................................................................................................................................. VII

Chapter 1 **Beneficial Effects Exerted by Paeonol in the Management of Atherosclerosis** ........................ 1
Li Lu, Yating Qin, Chen Chen and Xiaomei Guo

Chapter 2 **c-Met Signaling Protects from Nonalcoholic Steatohepatitis- (NASH-) Induced Fibrosis in Different Liver Cell Types** ........................ 12
Hannah K. Drescher, Fabienne Schumacher, Teresa Schenker, Maike Baues, Twan Lammers, Thomas Hieronymus, Christian Trautwein, Konrad L. Streetz and Daniela C. Kroy

Chapter 3 **Biological Effects of Tetrahydroxystilbene Glucoside: An Active Component of a Rhizome Extracted from *Polygonum multiflorum*** ........................ 26
Lingling Zhang and Jianzong Chen

Chapter 4 **The Potential Roles of Extracellular Vesicles in Cigarette Smoke-Associated Diseases** ........................ 41
A-Reum Ryu, Do Hyun Kim, Eunjoo Kim and Mi Young Lee

Chapter 5 **Dietary DHA/EPA Ratio Changes Fatty Acid Composition and Attenuates Diet-Induced Accumulation of Lipid in the Liver of ApoE$^{-/-}$ Mice** ........................ 49
Liang Liu, Qinling Hu, Huihui Wu, Xiujing Wang, Chao Gao, Guoxun Chen, Ping Yao and Zhiyong Gong

Chapter 6 **Hydrogen Sulfide Ameliorates Developmental Impairments of Rat Offspring with Prenatal Hyperhomocysteinemia** ........................ 61
O. V. Yakovleva, A. R. Ziganshina, S. A. Dmitrieva, A. N. Arslanova, A. V. Yakovlev, F. V. Minibayeva, N. N. Khaertdinov, G. K. Ziyatdinova, R. A. Giniatullin and G. F. Sitdikova

Chapter 7 **Neuroprotective Mechanisms of Resveratrol in Alzheimer's Disease: Role of SIRT1** ........................ 74
Bruno Alexandre Quadros Gomes, João Paulo Bastos Silva, Camila Fernanda Rodrigues Romeiro, Sávio Monteiro dos Santos, Caroline Azulay Rodrigues, Pricila Rodrigues Gonçalves, Joni Tetsuo Sakai, Paulo Fernando Santos Mendes, Everton Luiz Pompeu Varela and Marta Chagas Monteiro

Chapter 8 **Nox2 Activity is Required in Obesity-Mediated Alteration of Bone Remodeling** ........................ 89
Md Mizanur Rahman, Amina El Jamali, Ganesh V. Halade, Allal Ouhtit, Haissam Abou-Saleh and Gianfranco Pintus

Chapter 9 **Upregulation of Heme Oxygenase-1 by Hemin Alleviates Sepsis-Induced Muscle Wasting in Mice** ........................ 99
Xiongwei Yu, Wenjun Han, Changli Wang, Daming Sui, Jinjun Bian, Lulong Bo and Xiaoming Deng

| | | |
|---|---|---|
| Chapter 10 | **Oxidative Stress, Maternal Diabetes and Autism Spectrum Disorders** | **109** |
| | Barbara Carpita, Dario Muti and Liliana Dell'Osso | |
| Chapter 11 | **Mitoproteomics: Tackling Mitochondrial Dysfunction in Human Disease** | **118** |
| | María Gómez-Serrano, Emilio Camafeita, Marta Loureiro and Belén Peral | |
| Chapter 12 | **Resveratrol Decreases Oxidative Stress by Restoring Mitophagy and Improves the Pathophysiology of Dystrophin-Deficient *mdx* Mice** | **144** |
| | Rio Sebori, Atsushi Kuno, Ryusuke Hosoda, Takashi Hayashi and Yoshiyuki Horio | |
| Chapter 13 | **Protective Effects of Aqueous Extracts of *Flos lonicerae Japonicae* against Hydroquinone-Induced Toxicity in Hepatic L02 Cells** | **157** |
| | Yanfang Gao, Huanwen Tang, Liang Xiong, Lijun Zou, Wenjuan Dai, Hailong Liu and Gonghua Hu | |
| Chapter 14 | **Multifunctional Phytocompounds in *Cotoneaster* Fruits: Phytochemical Profiling, Cellular Safety, Anti-Inflammatory and Antioxidant Effects in Chemical and Human Plasma Models *In Vitro*** | **167** |
| | Agnieszka Kicel, Joanna Kolodziejczyk-Czepas, Aleksandra Owczarek, Magdalena Rutkowska, Anna Wajs-Bonikowska, Sebastian Granica, Pawel Nowak and Monika A. Olszewska | |
| Chapter 15 | **Photobiomodulation at Multiple Wavelengths Differentially Modulates Oxidative Stress *In Vitro* and *In Vivo*** | **183** |
| | Katia Rupel, Luisa Zupin, Andrea Colliva, Anselmo Kamada, Augusto Poropat, Giulia Ottaviani, Margherita Gobbo, Lidia Fanfoni, Rossella Gratton, Massimo Santoro, Roberto Di Lenarda, Matteo Biasotto and Serena Zacchigna | |

**Permissions**

**List of Contributors**

**Index**

# Preface

Over the recent decade, advancements and applications have progressed exponentially. This has led to the increased interest in this field and projects are being conducted to enhance knowledge. The main objective of this book is to present some of the critical challenges and provide insights into possible solutions. This book will answer the varied questions that arise in the field and also provide an increased scope for furthering studies.

Oxidative stress arises when there exists an imbalance between the biological system's inability to detoxify reactive intermediates and thereby repair the damage, and the systemic manifestation of reactive oxygen species. Aging is a physiological state in which there is a progressive decline of organ function along with the incidence of age-related diseases. The oxidative stress theory of aging suggests that free radicals or reactive oxygen species create damage which gives rise to symptoms of aging. It is claimed that the oxidative stress within mitochondria may lead to a progression of events in which damaged mitochondria produces increased number of reactive oxygen species, which leads to further damage. This book covers in detail some existing theories and innovative concepts revolving around oxidative stress and aging. Different approaches, evaluations, methodologies and advanced studies on the role of oxidative stress in aging have been included herein. This book, with its detailed analyses and data, will prove immensely beneficial to professionals and students involved in this area at various levels.

I hope that this book, with its visionary approach, will be a valuable addition and will promote interest among readers. Each of the authors has provided their extraordinary competence in their specific fields by providing different perspectives as they come from diverse nations and regions. I thank them for their contributions.

Editor

# Beneficial Effects Exerted by Paeonol in the Management of Atherosclerosis

### Li Lu, Yating Qin, Chen Chen, and Xiaomei Guo

*Department of Cardiology, Tongji Hospital, Tongji Medical College, Huazhong University of Science and Technology, Wuhan 430030, China*

Correspondence should be addressed to Xiaomei Guo; xiaomguo@yeah.net

Academic Editor: Ryuichi Morishita

Atherosclerosis, a chronic luminal stenosis disorder occurred in large and medium arteries, is the principle pathological basis of cardiovascular diseases with the highest morbidity and mortality worldwide. In oriental countries, traditional Chinese medicine Cortex Moutan has been widely used for the treatment of atherosclerosis-related illnesses for thousands of years. Paeonol, a bioactive monomer extracted from Cortex Moutan, is an important pharmacological component responsible for the antiatherosclerotic effects. Numerous lines of findings have established that paeonol offers beneficial roles against the initiation and progression of atherosclerotic lesions through inhibiting proatherogenic processes, such as endothelium damage, chronic inflammation, disturbance of lipid metabolism, uncontrolled oxidative stress, excessive growth, and mobilization of vascular smooth muscle cells as well as abnormality of platelet activation. Investigations identifying the atheroprotective effects of paeonol present substantial evidence for potential clinical application of paeonol as a therapeutic agent in atherosclerosis management. In this review, we summarize the antiatherosclerotic actions by which paeonol suppresses atherogenesis and provide newly insights into its atheroprotective mechanisms and the future clinical practice.

## 1. Introduction

Exposed to numerous social and health problems such as aging of population, progressive urbanization, elevated energy intake, reduced physical exercise, and air pollution, more and more people are insulted by cardiovascular diseases (CVDs) [1]. According to the statistics of the World Health Organization, more than 40% of deaths of noncommunicable diseases are attributable to CVDs annually [2]. Atherosclerosis, characterized by multifactor-induced vascular stenosis occurred in large- and medium-sized arteries, is a crucial predisposed pathogenic process toward CVDs [3]. It has been demonstrated that statins are effectively used for the treatment of atherosclerosis. However, 5%–20% of patients with indications for statin therapy show inability to tolerate routine dosages due to muscle symptoms caused by statins [4].

Cortex Moutan, the root bark of *Paeonia suffruticosa* Andrews, has been widely applied as a traditional Chinese medicine (TCM) in the prevention and management of various diseases for thousands of years, such as CVD, diabetes, arthritis, and cancer [5]. Paeonol ($2'$-hydroxy-$4'$-methoxyacetophenone) is a bioactive constituent extracted from Cortex Moutan and has been reported to possess extensive pharmacological properties for alleviating atherosclerotic lesions, which is associated with improvement of endothelial injury, repression of vascular smooth muscle cell (VSMC) proliferation and migration, amelioration of inflammation and oxidative stress, inhibition of platelet activation and aggregation and decrease of blood lipids, etc. [6–11]. In oriental countries, paeonol has been employed alone or in combination with other TCMs to effectively protect the cardiovascular system, suggesting that paeonol is potentially to act as an alternative or complementary agent for compensating for the limited efficiency and uncertain safety of modern drugs regarding atherosclerosis treatment [4, 12–14]. Considering the pharmacological activities and therapeutic potentials of paeonol in dealing with atherosclerosis, we put forward an overview concerning the atheroprotective roles of paeonol and the underlying mechanisms identified in preclinical studies.

## 2. Pharmacological Features of Paeonol

*2.1. Bioactive Components in Cortex Moutan.* Since multiple lines of evidence have clarified the cardioprotective effects of Cortex Moutan, the molecular mechanisms are difficult to be recognized and accepted because of its complex mixture nature. Increasing attention paid by the cardiovascular research community is focused on the bioactive chemical monomers comprised in Cortex Moutan responsible for the pharmacological abilities [5, 15]. Phytochemical studies indicate that there are more than 80 compounds isolated from Cortex Moutan, mainly divided into the following categories with different structural formulas: monoterpenoid glycosides, flavonoids, tannins, phenols, and paeonols. Among them, paeoniflorin, catechin, 1,2,3,4,6-penta-O-galloyl-$\beta$-D-glucose, gallic acid, and paeonol are the representative extracts in the above groups, respectively (Figure 1) [15, 16]. In terms of the involvement of these chemicals in atherosclerosis development, paeonol is the main bioactive component which is extensively investigated.

*2.2. Pharmacokinetics of Paeonol.* It is universally established that pharmacokinetic detection is beneficial to evaluate the efficacy and possible toxicity of herbal products and explore the interactions between medicinal herbs among the preparations. With some advanced detection methods, the pharmacokinetic researches of paeonol have been broadly performed in the past years (Table 1) [17–21]. Paeonol is rapidly absorbed into the circulation from the intestinal tract after oral administration and quickly distributed in multiple organs including the heart, liver, kidney, and brain without long-term accumulation, as explained by short $T_{max}$ and $T_{1/2}$ [17, 18]. This rapid clearance of paeonol from the body appears to guarantee its safety. Furthermore, the level and duration of paeonol in the heart and brain could be significantly increased via coadministration with danshensu, which might be helpful to interpret the synergistic actions of combination remedy amalgamating paeonol and danshensu in treating cardiovascular disorders [13, 18, 22]. However, fast and complete first-pass metabolism of paeonol, along with the features of low aqueous solubility and high volatility, determines its poor bioavailability.

*2.3. Drug Delivery System.* A number of drug delivery systems connected with paeonol have been developed to enhance the dissolution rate and unsatisfactory bioavailability, because the hydrophobicity of paeonol hinders its clinical application as a promising therapeutic agent [23–25]. Microemulsion gel, transethosome, porous microsphere, liquid crystalline nanoparticle, and microsponge formulations have been designed for transdermal delivery of paeonol. Results show that these carriers, with high encapsulation efficiency and stability, are biocompatible with paeonol and dramatically raise skin permeability, control release of drug, extend time of drug residence in local tissues, and lower irritation to covered areas, implying superior effects of these complexes in treating skin diseases [23, 26–29]. Additionally, paeonol-loaded nanoparticles are more effective in cancer treatment when compared to paeonol alone [30]. Nonetheless, there is no paeonol-related drug delivery system which has been designed and prepared for atherosclerosis therapy up to now. Given that cancer and atherosclerosis share some common pathogenic mechanisms, vehicles loading paeonol used for cancer management might be suitable for alleviating atherosclerosis progression, which is needed to be further elucidated [31]. Moreover, nanoemulsions prepared by Chen et al. are likely to strengthen therapeutic effects of paeonol in the cardiovascular system, for which augment the bioavailability of paeonol by enhancing its oral absorption and transport in the digestive tract through blocking p-glycoprotein regulated efflux [32]. In addition, stents carrying paeonol-laden microparticles or poly (butyl-2-cyanoacrylate) nanocapsules are likely to have high efficiency in preventing restenosis and stent thrombosis occurrence after percutaneous coronary intervention, by the fact that the formulations could slow and control the release of paeonol, which probably sustainedly inhibited VSMC proliferation and platelet activation in the local environment [23, 24].

## 3. Mechanism of Action Underlying Paeonol Alleviates Atherosclerosis

Accumulating studies support the notion that atherosclerosis is a multifaceted vascular impairment involving functional abnormalities of diverse cell types like endothelial cells (ECs), macrophages, VSMCs, and platelets. Endothelium damage, chronic inflammation, disturbance of lipid metabolism, uncontrolled oxidative stress, excessive growth, and mobilization of VSMCs and abnormality of platelet activation are important contributors to atherogenesis [33, 34]. Targeting these proatherogenic processes is indicated to be the pivotal mechanisms underlying paeonol mitigates atherosclerotic lesions and subsequent cardiovascular events (Figure 2).

*3.1. Amelioration of Endothelial Injury.* The intact vascular endothelium, known as nature's container for circulating blood in vivo, has been delineated to be deeply associated with diverse biological processes. When incited by proatherogenic factors, apoptotic signaling in ECs is amplified and the barrier of arterial vasculature is deranged, which causes increased permeability of the endothelial lining of lesion-prone areas, followed by trapping and epigenetic modification of blood lipoproteins as well as macrophage deposition and succedent foam cell formation in the subendothelial region, thereby favoring atherosclerosis initiation [33]. Independent research teams have uncovered that paeonol could improve endothelial damage by enhancing endothelial nitric oxide synthase- (eNOS-) induced production of nitric oxide (NO) in ECs in response to diverse stimuli, owing that NO is an EC-protecting factor capable of repressing activities of apoptosis-related pathways and elevating cellular survival rates [9, 35–37]. Through suppression activation of phosphatidylinositol 3-kinbase (PI3K)/Akt/nuclear factor kappa

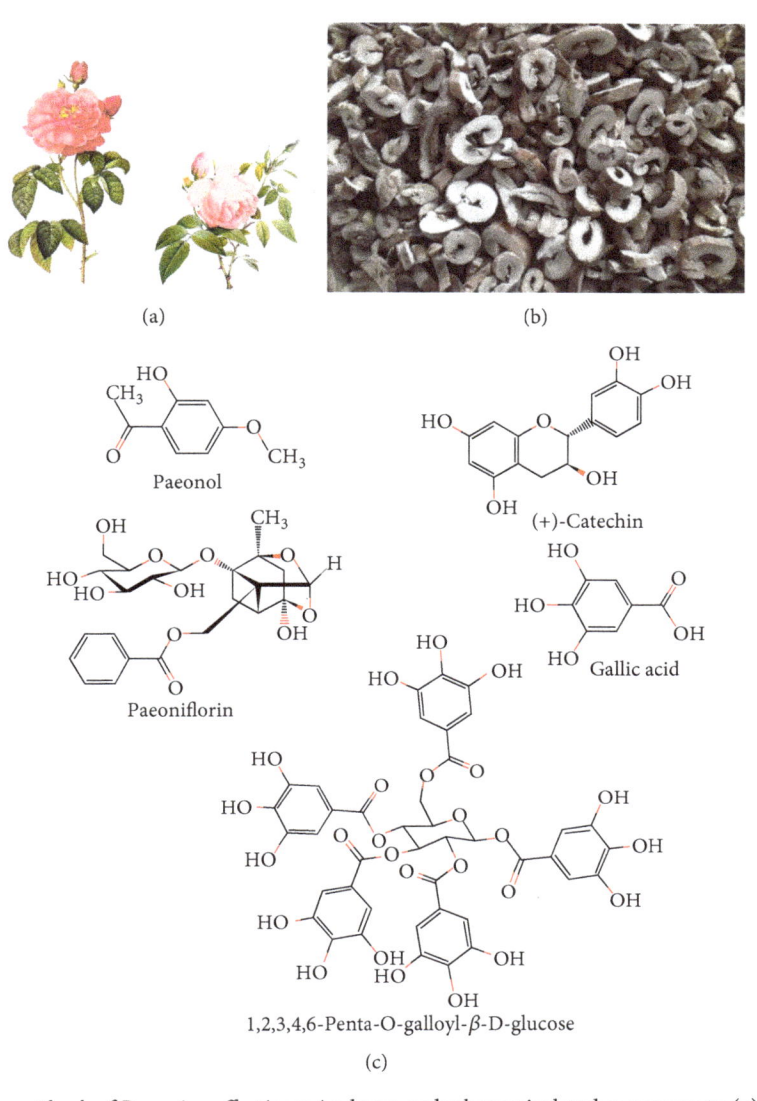

Figure 1: The whole plant and root bark of *Paeonia suffruticosa* Andrews and relevant isolated components. (a) *Paeonia suffruticosa* Andrews is a kind of elegant ornamental plant with great medicinal value. (b) Cortex Moutan, the root bark of *Paeonia suffruticosa* Andrews, contains a variety of bioactive pharmacological compounds including paeonol. (c) The chemical structural formula of the main components extracted from Cortex Moutan.

B (NF-$\kappa$B) and lectin-like low-density lipoprotein receptor-1 (LOX-1)/p38/NF-$\kappa$B cascade, paeonol inhibits apoptosis and increases viability of EC damaged by lipopolysaccharides (LPS) and ox-LDL, as seen by downregulation of caspase-3 level and LDH leakage and upmodulation of Bcl-2 expression and OD value of MTT test [38, 39]. Other aspects associated with endothelium dysfunction encompass premature senescence, aberrant autophagy and microRNA (miRNA) mediation, reactive oxygen species (ROS), and inflammation stimulation [33]. As untimely aging of ECs deteriorates their actions of growth and antioxidation, pretreatment with paeonol is showed to reduce the number of senescent cells, propel cell cycle, and DNA synthesis and then restore abilities resistant to dysfunction in the model of endothelial senescence, which is linked to mediation of sirtuin 1 (Sirt1)/p53 axis [40, 41]. Emerging findings depict that autophagy is an evolutionarily conserved process degrading own damaged proteins and macromolecule substances and uncontrolled autophagy results in atherosclerosis-related vascular ECs death [42]. Paeonol has been demonstrated to ameliorate cell injury induced by excessive autophagy in ox-LDL-triggered ECs through raising the level of miR-30a which downmodulates expression of Beclin-1 and LC3II [6]. Moreover, another mechanism by which paeonol recovers the proliferation of ECs damaged by ox-LDL is ascribe to paeonol-induced decrement of proapoptotic miR-21 expression and following TNF-$\alpha$ release [43]. Taken together, it is rational to discern that paeonol possesses great beneficial potentials in the treatment of endothelium dysfunction.

*3.2. Inhibition of Oxidative Stress.* Under physiological conditions, generation and elimination of ROS is in a dynamic equilibrium. In diseases, the overproduction of oxidants or shortage of antioxidants leads to the imbalance of the redox

TABLE 1: The pharmacokinetic parameters of paeonol.

| Object | Agent | Route | Dose of paeonol (mg/kg) | $C_{max}$ ($\mu$g/mL) | $T_{max}$ (min) | $T_{1/2}$ (min) | MRT (min) | AUC ($\mu$g·min/mL) | CL/F (L/kg·min) |
|---|---|---|---|---|---|---|---|---|---|
| Wistar rat plasma | Cortex Moutan | Oral | 20 | 2.69 ± 0.26 | 19.26 ± 4.4 | 80.93 ± 16.26 | — | 172.7 ± 48.86 | 0.12 ± 0.03 |
| SD rat plasma | Paeonol | Oral | 40 | 3.04 ± 0.61 | 17.5 ± 5 | 62.48 ± 17.41 | 91.25 ± 15.59 | 334 ± 81.29 | 0.13 ± 0.03 |
|  | Paeonol plus danshensu | Oral | 40 | 0.87 ± 0.08 | 12.5 ± 5 | 159.45 ± 56.38 | 250.85 ± 42.45 | 186 ± 9.88 | 0.16 ± 0.04 |
| Wistar rat plasma | Paeonol | Intramuscular | 10 | 0.71 ± 0.13 | 7.5 ± 2.73 | 59.85 ± 10.23 | 77.67 ± 10.48 | 43.06 ± 6.1 | 0.24 ± 0.03 |
| SD rat plasma | DA-9805 | Oral | 58 | 5.23 ± 3.9 | 60 | 90.13 ± 35.97 | — | 846.82 ± 347.58 | — |
| SD rat plasma | Qingfu Guanjieshu capsule | Oral | 70 | 8.54 ± 1.36 | 5 ± 0 | 43.62 ± 3.01 | 47.97 ± 3.91 | 265.47 ± 46.71 | 0.32 ± 0.054 |
|  | Qingfu Guanjieshu capsule | Oral | 17.75 | 2.16 ± 0.27 | 5 ± 0 | 27.31 ± 1.73 | 75.5 ± 32 | 70.78 ± 11.49 | 0.3 ± 0.06 |

$C_{max}$: the maximum plasma concentration; $T_{max}$: the time to reach $C_{max}$; $T_{1/2}$: half-time of elimination; MRT: mean residence time; AUC: area under the concentration-time curve; CL/F: total clearance; $V_d$: volume of distribution; DA-9805: a formulation comprising extracts from root cortex of *Paeonia suffruticosa* Andrews, root of *Bupleurum falcatum* L., and root of *Angelica dahurica* Benth et Hook; Qingfu Guanjieshu: a formulation containing Caulis Sinomenii, Radix Aconiti Lateralis Preparata, Rhizoma Curcumae Longae, Radix Paeoniae Alba, and Cortex Moutan.

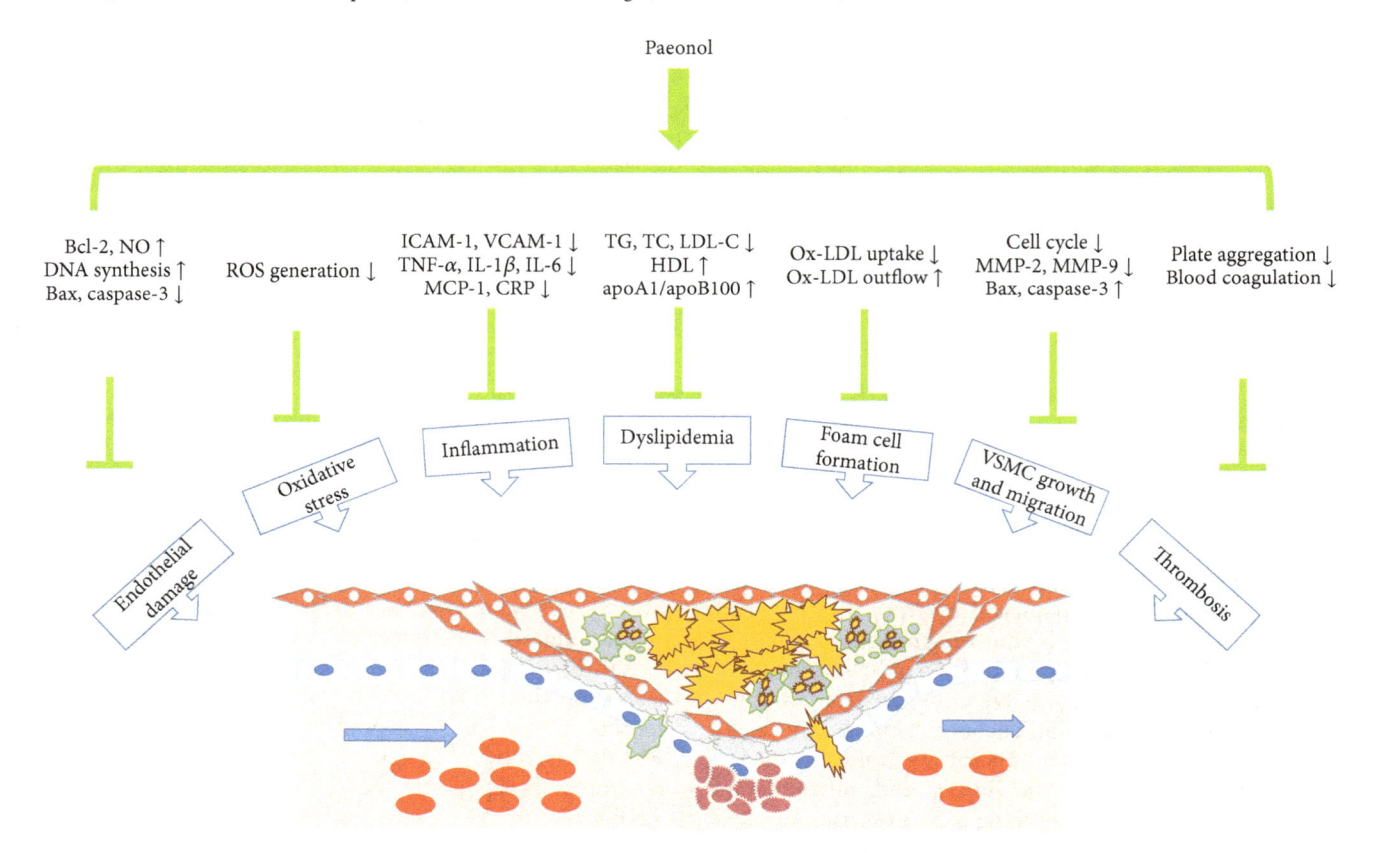

FIGURE 2: The antiatherosclerotic effects by which paeonol alleviates the development of AS.

status, causing ROS overload and then proximal and distal impairments called oxidative stress. Considerable documentations reveal that oxidative stress exerts positive roles in the pathogenesis of atherosclerosis [37, 44]. Experimental data have manifested that paeonol obviously lowers ROS content, abrogates the upregulation of MDA and oxidized low-density lipoprotein (ox-LDL), restores the level of Bcl-2/Bax and caspase-3, and decreases expression of tumor necrosis factor (TNF)-α, interleukin (IL)-1β, IL-6, and monocyte chemotactic protein (MCP)-1 in the oxidative stress environment [45–47]. These findings show that encumbering oxidative stress-evoked acceleration of lipid peroxidation, induction of vascular endothelial injury, and activation of inflammatory pathways are imperative components in the antiatherogenic effects of paeonol. Investigations on the molecular mechanisms suggest that paeonol could induce ROS decline through activating AMP-activated protein kinase α (AMPKα)/peroxisome proliferator-activated receptor δ (PPARδ) cascade and blocking endoplasmic reticulum (ER) stress signaling, followed by reduction of NADPH oxidase (NOX) which is the main enzyme catalyzing ROS generation, indicating that synthesis inhibition of ROS is an important action of paeonol to improve oxidative stress [9, 44, 48]. In terms of the effects of paeonol on the antioxidative system, previous evidences had uncovered that paeonol was capable of increasing contents of antioxidants and scavenging ROS-evoked cardiac and cerebral injury via inducing nuclear factor E2-related factor 2 signaling and downstream expression of heme oxygenase-1, superoxide dismutase, catalase, and glutathione peroxidase [13, 49]. Then, the potent abilities of elevating antioxidant enzymes of paeonol might directly eliminate ROS and then ameliorate oxidative stress-elicited vascular wall damage.

### 3.3. Mitigation of Inflammatory Response. 
Tremendous basic studies have elaborated the essential atheroprone impacts of inflammation in all stages of atherosclerosis from fatty streak formation to luminal occlusion. Activation of inflammatory cascades in vascular ECs stimulates the biosynthesis of adhesion molecules (vascular cell adhesion molecule-1 (VCAM-1), intercellular adhesion molecule-1 (ICAM-1), and E-selection) and chemokines (MCP-1) which promote recruitment and retention of circulating monocytes in the intima where they differentiate into macrophages and aggravate atherosclerotic lesions via changing into foam cells and secreting vast proinflammatory factors [3, 50]. In vitro experiments confirm that paeonol forcefully retards the detainment of monocytes by ECs via weakening the expression of VCAM-1 and ICAM-1 in ECs upon stimulation of TNF-α, and the inner mechanism is due to abolishment of extracellular signal-regulated kinase 1/2 (ERK1/2) and p38 signaling and then NF-κB inactivation [51, 52]. Similarly, results from Wang et al. and Zhou et al. show that paeonol extenuates contents of ICAM-1, VCAM-1, and E-selection through blocking mitogen-activated protein kinase (MAPKs) and NF-κB cascade, diminishing ECs ability to capture monocytes in the inflammation circumstance [53, 54]. Furthermore, the elevated adhesion of monocytes to ox-LDL-injured vascular ECs is normalized in the presence of paeonol, which is attributed to drug-mediated level promotion of miR-126 delaying the activity of downstream PI3K/Akt/NF-κB axis [55]. Ample literatures have documented that inflammatory factors are cytotoxic that undermine the endothelial barrier and boost the release of proteolytic enzymes, consequently contributing to atherosclerosis onset and destabilization of atheroma plaques [34, 56]. In macrophages, level ascent of IL-1, inducible nitric oxide synthase (iNOS), COX-2, and TNF-α elicited by LPS is attenuated after paeonol diminishes activation of Toll-like receptor 4 (TLR4)/NF-κB and ERK1/2 cascade [57, 58]. Moreover, paeonol has been shown to remit inflammation responses via modulating signal flow of other MAPKs such as p38 and c-jun N-terminal kinase (JNK) [59]. According to Choy et al., LPS triggered inflammatory reactions and caused EC apoptosis through stimulating the NADPH/ROS/MAPK cascade and relevant upstream mediator TLR4 and bone morphogenic protein 4 (BMP4), while coadministration with paeonol significantly reversed these events [8]. With microarray analysis, Huang and colleagues proved that paeonol served as an anti-inflammatory agent by repressing actions of signal pathways concerning Toll receptor, interleukin, interferon-γ, Janus kinase/signal transducers and activators of transcription, etc. [60]. Apart from directly influencing activities of signal molecules, paeonol has been discovered to prohibit proinflammatory signaling and cytokine generation via affecting specific regulators, as exemplified by paeonol-impeded expression of miR-21 followed by inactivation of Ras/MKK3/6/p38 pathway in ox-LDL-damaged ECs [61]. Additionally, in vivo studies depict that paeonol exerts markedly atheroprotective effects by the way of its proficiency in reducing inflammatory mediators including CRP, TNF-α, and IL-1β [62].

### 3.4. Improvement of Lipid Profiles and Foam Cell Formation. 
Compelling evidence indicates that dyslipidemia is one of the crucial activators of atherosclerosis occurrence and progression. Hyperlipidemia perturbs the permeability of vascular wall to promote the proatherogenic sedimentation and oxidation of lipoproteins in the subendothelium. Medications targeting lipid dysbolism have been proved to be effective in controlling atherogenesis [14, 63]. There is evidence establishing that paeonol is able to lower the contents of blood triglyceride (TG), total cholesterol (TC), and low-density lipoprotein cholesterol (LDL-C) and ameliorate atherosclerosis development in rats fed with high-fat diet [64]. Furthermore, in a quail model of atherosclerosis, the decrement of TC, LDL, VLDL, and apolipoprotein (apo) B100 and increment of high-density lipoprotein (HDL), HDL/TC, and apoA1/apoB100 are seen after gavage with paeonol [65]. The lipid-lowering and antiatherogenic effects of paeonol are also identified in a study reported by Qian et al., as assessed by decrease of concentration of TC, TG, and LDL-C and extent of atheroma lesions [66]. In terms of the molecular mechanisms of paeonol-modulated lipid metabolism, Chen and Kang unraveled that paeonol lowered TG level via delaying the de novo synthesis and favoring lipid oxidation through blocking sterol regulatory element-binding protein 1c (SREBP-1c)/fatty acid synthetase (FAS) and SREBP-1c/acetyl CoA carboxylase α (ACCα) pathway and inducing PPAR-α/carnitine palmitoyltransferase I (CPT-1) cascade, respectively. And the decrement of TC and LDL-C was linked to paeonol-evoked depression of 3-hydroxy-3-methylglutaryl-coenzyme A reductase (HMGCR) and ascent of LDL receptor (LDLR), separately [11].

It is widely held that foam cell formation is a hallmark of the early phase of atherosclerosis. Scavenger receptors like

CD36, SR-A, and LOX-1 facilitate foam cell formation by internalizing cholesterol, while ATP-binding cassette transporter A1 (ABCA1) and ATP-binding cassette transporter G1 repress macrophage conversion via favoring cholesterol efflux [14, 63]. Recent results have demonstrated that paeonol could activate the liver X receptor $\alpha$ (LXR$\alpha$)/ABCA1 pathway to accelerate ox-LDL outflow in macrophages, accompanied by reduction of foam cell formation and pathogenic changes of atherosclerosis [67]. In addition, another study found that paeonol abated macrophages switching into foam cells not only through promoting the efflux of ox-LDL by maintaining stabilization of ABCA1 but also via blocking the cholesterol uptake by abolishing c-Jun-mediated CD36 synthesis and then leading to attenuation of atherosclerosis burden in apoE$^{-/-}$ mice [68]. HDL is reported to be responsible for reverse transport of cholesterol from peripheral organs to the liver, and paeonol is capable of upregulating the circulating level of HDL, implying that increasing HDL might be a mechanism of action for paeonol to expedite cholesterol ejection from macrophages and weaken their conversion [69–71]. Thus, the therapeutic utility of paeonol in atherosclerosis has been at least partly ascribed to regulation of lipid metabolism and suppression of macrophages turn into lipid-laden foam cells.

*3.5. Suppression of VSMC Growth and Mobilization.* Pharmacological efforts with antiproliferatory and antimigratory properties on VSMCs are beneficial for treating atherosclerosis, given that unlimited proliferation and movement of VSMCs within the arterial wall contribute to plaque expansion and vascular narrowing. Once irritated by mitogens, VSMCs in a resting state turn into the synthetic phenotype and begin to proliferate and move from tunica media to intima [72]. Paeonol has been found to restrain phenotype change to suppress VSMC proliferation induced by hyperlipemic serum, suggesting its favorable roles against intima thickening [73]. Furthermore, paeonol decreases platelet-derived growth factor- (PDGF-) triggered VSMC growth by arresting cell cycle at $G_0/G_1$ phase through inactivating mitogenic signal ERK1/2/c-fos [74]. High glucose, one of the predominant contributors to atherosclerosis progression, is capable of promoting EC damage and VSMC growth. It is reported that pretreatment with paeonol markedly reverses high glucose-elicited proliferation of VSMCs in the cell coculture model, and this effect is due to decrease of vascular endothelial growth factor and PDGF release and following blockade of Ras/Raf/ERK1/2 pathway transduction [75]. There is evidence showing that paeonol could decrease the level of blood glucose in hyperglycemic state, hinting that paeonol probably reduces hyperglycemia to indirectly retard VSMC growth [76]. Additionally, TNF-$\alpha$-stimulated enhancement of cellular proliferatory and migratory abilities is restored by paeonol which activates mitochondria-related apoptotic cascade and diminishes extracellular matrix degradation, as explained by increase of Bax and cleaved caspase-9 and caspase-3 and decline of Bcl-2 and matrix metalloprotein (MMP)-2 and MMP-9 [77]. Owing that emerging evidence has determined the implication of autophagy in weakening VSMC growth, Wu and colleagues investigated whether paeonol regulated cellular proliferatory activities via mediating autophagic processes. They discovered that paeonol produced cell cycle arrest in ox-LDL-affected VSMCs and reduced the number of VSMCs in tunica media of apoE$^{-/-}$ mice, both of which were ascribe to the mechanism that paeonol induced enhancement of autophagy via upregulating LC3II expression, p62 degradation, and autophagosome formation through stimulating the AMPK/mammalian target of rapamycin (mTOR) signaling axis [7]. With respect to the roles of paeonol in vascular restenosis, Zhang et al. clarified that local administration of paeonol mitigated early neointimal thickening of graft veins by abrogating mitogenic cytokine-triggered proliferation of VSMCs and apoptosis of ECs, implying the potential application of paeonol for preventing occurrence of in-stent restenosis, a severe complication of angioplasty [78]. According to the above findings, it is apparent that blockade of VSMC proliferation and migration is an important constituent of atheroprotective effects of paeonol.

*3.6. Repression of Platelet Activity and Thrombosis.* Because of circulating hemorheology abnormality or procoagulant material upmodulation in atherosclerotic lesion areas, platelets are extensively activated and recruited to the damaged endothelium, which is an initiation of coagulation cascade, thereby inducing artery thrombosis and vascular occlusion, a life-threatening acute coronary event [79]. Previous studies had deciphered that paeonol and its analogues offered advantageous roles against thrombus formation via directly restraining platelet aggregation and blood coagulation [10, 80]. With improvement of hemorheological parameters, paeonol is considered as a protective agent lessening thrombogenic incidents, by the fact that aberrant whole blood viscosity, plasma viscosity, and fibrinogen participate in coagulation processes [65, 71, 81]. Another antiatherothrombotic effects of paeonol might be associated with the increase of NO and $PGI_2$ acting as antagonists of platelet activity while the reduction of ET-1 and $TXA_2$ that are agonists of platelet activation and aggregation [82].

## 4. Other Potential Therapeutic Targets

Cumulative findings have demonstrated the potential of paeonol in the control of atherosclerotic lesions and held promise for clinical use of paeonol in atherosclerosis treatment [5]. According to the published papers, proatherogenic actions of cells in vascular wall have been effectively inhibited by paeonol, and the molecular mechanisms well investigated have been delineated in Figure 3. Other worthy possible therapeutic targets involved in atheroprotective effects of paeonol are as follows: (1) miRNAs in foam cell formation: given that some miRNAs are related with the processes of macrophages turn into foam cell, such as miR-155 and miR-21, and paeonol is illustrated to be a modulator of this two miRNAs, affecting their expression is likely be a potential target for paeonol to attenuate foam cell formation [61, 83, 84]; (2) autophagy in endothelial impairment: it is recognized that autophagy is required for diverse biological activities including cellular apoptosis, and paeonol offers antiapoptotic effects against ROS-induced myocardial death via abolishing

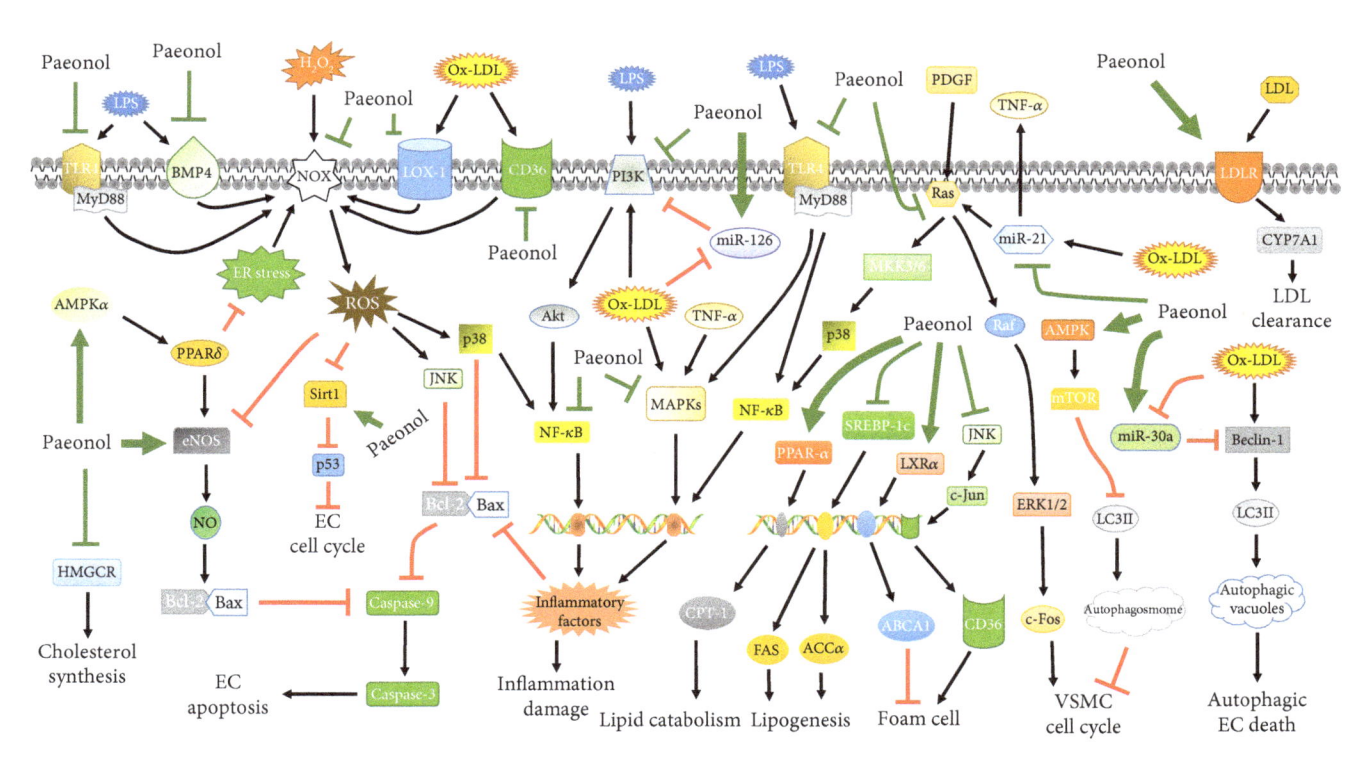

FIGURE 3: Schematic diagram of molecular mechanism underlying paeonol protects against atherogenesis. Paeonol-induced ROS elimination is associated with inhibition of NOX/ROS pathway. Paeonol mainly suppresses the MAPKs and NF-κB cascade to weaken inflammatory responses and EC apoptosis. In addition, paeonol activates PPAR-α/CPT-1 pathway and represses SREBP-1c signaling to accelerate TG catabolism and block TG synthesis, respectively. Then, paeonol weakens foam cell formation by increasing level of reverse transport axis LXRα/ABCA1 and reducing activities of JNK signaling involved in CD36 production. Moreover, paeonol mediates autophagic factors and cell cycle-related single molecules to block the VSMC proliferation via the AMPK/mTOR pathway. TLR4: Toll-like receptor 4; MyD88: myeloid differentiation primary response protein 88; AMPKα: AMP-activated protein kinase α; HMGCR: 3-hydroxy-3-methylglutaryl-coenzyme A reductase; BMP4: bone morphogenic protein 4; PPARδ: peroxisome proliferator-activated receptor δ; eNOS: endothelial nitric oxide synthase; NOX: NADPH oxidase; Sirt1: sirtuin 1; LOX-1: lectin-like low-density lipoprotein receptor-1; JNK: c-jun N-terminal kinase; NF-κB: nuclear factor kappa B; PI3K: phosphatidylinositol 3-kinbase; MAPKs: mitogen-activated protein kinases; LPS: lipopolysaccharides; CPT-1: carnitine palmitoyltransferase I; FAS: fatty acid synthetase; ACCα: acetyl CoA carboxylase α; SREBP-1c: sterol regulatory element-binding protein 1c; MKK: mitogen-activated protein kinase kinase; ABCA1: ATP-binding cassette transporter A1; LXRα: liver X receptor α; PDGF: platelet-derived growth factor; ERK1/2: extracellular signal-regulated kinase 1/2; mTOR: mammalian target of rapamycin; LC3II: microtubule-associated protein 1 light chain 3 II; LDLR: low-density lipoprotein receptor; CYP7A1: cholesterol 7 alpha-hydroxylase.

antiautophagic PI3K/Akt/mTOR pathway [42, 85]. Exploring the implication of autophagy in paeonol-mediated ECs protection is rewarding; (3) angiogenesis: paeonol has been viewed as an angiogenesis inhibitor repressing tumor growth and metastasis. As angiogenesis in the atheroma lesions accelerates plaque rupture, it is worth figuring out whether paeonol improves the vulnerability of plaques by regulating angiogenic events in atherosclerotic areas [86, 87]; (4) phenotype switching: paeonol is reported to delay the transformation of VSMCs from quiescent state to proliferatory status, but the mechanisms are poorly defined. Moreover, it is indicated that macrophage polarization from proinflammatory M1 phenotype to anti-inflammatory M2 state is an important contributor to atherosclerosis development. Several molecules regulating the phenotype change have been shown to serve as targets of paeonol, such as miR-21, MAPKs, NF-κB, TNF-α, and IL-10 [88, 89]. It can be speculated that improvement of macrophage polarization might be involved in paeonol-reduced atherosclerosis progression;

(5) promotion of vasodilation: emerging evidence suggests that paeonol dramatically facilitates arterial dilation through decreasing intercellular calcium content via repressing $Ca^{2+}$ influx and $Ca^{2+}$ release. This vasodilation-promoting feature is promising to explain paeonol-alleviated coronary noreflow, which is remained to be further elucidated [90, 91]; (6) vascular remodeling: considering that paeonol potently weakens tissue pathological remodeling by abating extracellular matrix production and fibrosis via blocking the transforming growth factor-β/Smads cascade, another antiatherogenic target of paeonol is probably linked to restraint of vascular remodeling, a key event favoring expansion of atherosclerotic lesions [3, 92]; and (7) gut microbiota and immunity regulation: evidence has begun to emerge that dysfunction of gut microbiota and autoimmune responses plays encouraging roles in atherogenesis [93, 94]. Seeking effects of paeonol in mediating functions of microbiota and immune cells would provide newly insights into the understanding of antiatherogenic mechanisms of paeonol.

## 5. Conclusion

In summary, considerable research evidence has pointed to the fact that paeonol, a naturally occurring bioactive compound in Cortex Moutan, is a promising therapeutic agent for atherosclerosis management. The antiatherosclerotic roles of paeonol are attributed to its multifactorial actions involving restoring endothelial integrity, repressing oxidative stress, alleviating inflammation, regulating lipid metabolism, inhibiting VSMC proliferation, and ameliorating platelet activation. These pleiotropic pharmacological activities of paeonol suggest great potential of its clinical application in atherosclerosis prevention and treatment. With respect to the undesirable physical characteristics of paeonol, there are reports indicating that several paeonol-loaded carriers have overcome the shortcomings of poor solubility and stability and improving the bioavailability and residence of paeonol in vivo, providing reliable technical support for paeonol in practice. However, the clinical trials monitoring the therapeutic effects of paeonol are scarce in recent years. So, large randomized, controlled, and double-blind trials are urgently needed to evaluate the efficacy and safety of paeonol in atherosclerosis treatment from the perspective of clinical practice.

## Acknowledgments

This study was supported by the grant from the National Natural Science Foundation of China (no. 81270353).

## References

[1] C. Shen and J. Ge, "Epidemic of cardiovascular disease in China," *Circulation*, vol. 138, no. 4, pp. 342–344, 2018.

[2] S. Mendis, S. Davis, and B. Norrving, "Organizational update: the world health organization global status report on noncommunicable diseases 2014; one more landmark step in the combat against stroke and vascular disease," *Stroke*, vol. 46, no. 5, pp. e121–e122, 2015.

[3] A. Solanki, L. K. Bhatt, and T. P. Johnston, "Evolving targets for the treatment of atherosclerosis," *Pharmacology & Therapeutics*, vol. 187, pp. 1–12, 2018.

[4] S. E. Nissen, E. Stroes, R. E. Dent-Acosta et al., "Efficacy and tolerability of evolocumab vs ezetimibe in patients with muscle-related statin intolerance: the GAUSS-3 randomized clinical trial," *JAMA*, vol. 315, no. 15, pp. 1580–1590, 2016.

[5] Y. Hu and G. Xu, "Pharmacological activities of cortex moutan radicis and its main components of paeonol," *Anhui Medical and Pharmaceutical Journal*, vol. 18, no. 4, pp. 589–592, 2014.

[6] C. Li, L. Yang, H. Wu, and M. Dai, "Paeonol inhibits oxidized low-density lipoprotein-induced vascular endothelial cells autophagy by upregulating the expression of miRNA-30a," *Frontiers in Pharmacology*, vol. 9, 2018.

[7] H. Wu, A. Song, W. Hu, and M. Dai, "The anti-atherosclerotic effect of paeonol against vascular smooth muscle cell proliferation by up-regulation of autophagy via the AMPK/mTOR signaling pathway," *Frontiers in Pharmacology*, vol. 8, 2018.

[8] K. W. Choy, Y. S. Lau, D. Murugan, P. M. Vanhoutte, and M. R. Mustafa, "Paeonol attenuates LPS-induced endothelial dysfunction and apoptosis by inhibiting BMP4 and TLR4 signaling simultaneously but independently," *The Journal of Pharmacology and Experimental Therapeutics*, vol. 364, no. 3, pp. 420–432, 2018.

[9] K.-W. Choy, M. R. Mustafa, Y. S. Lau et al., "Paeonol protects against endoplasmic reticulum stress-induced endothelial dysfunction via AMPK/PPARδ signaling pathway," *Biochemical Pharmacology*, vol. 116, pp. 51–62, 2016.

[10] Y. K. Koo, J. M. Kim, J. Y. Koo et al., "Platelet anti-aggregatory and blood anti-coagulant effects of compounds isolated from Paeonia lactiflora and Paeonia suffruticosa," *Pharmazie*, vol. 65, no. 8, pp. 624–628, 2010.

[11] Y. Chen and L. Kang, "Protection effect of paeonol on regulation of lipid metabolism in hyperlipidemia mice and its mechanisms," *The Chinese Journal of Clinical Pharmacology*, vol. 33, no. 22, pp. 2273–2277, 2017.

[12] H. Li, S. Wang, Y. Xie et al., "Simultaneous determination of danshensu, salvianolic acid B, and paeonol in ShuangDan oral liquid by HPLC," *Journal of AOAC International*, vol. 96, no. 1, pp. 20–23, 2013.

[13] H. Li, Y.-H. Xie, Q. Yang et al., "Cardioprotective effect of paeonol and danshensu combination on isoproterenol-induced myocardial injury in rats," *PLoS One*, vol. 7, no. 11, article e48872, 2012.

[14] X. L. Lin, L. L. Xiao, Z. H. Tang, Z. S. Jiang, and M. H. Liu, "Role of PCSK9 in lipid metabolism and atherosclerosis," *Biomedicine & Pharmacotherapy*, vol. 104, pp. 36–44, 2018.

[15] C. N. He, Y. Peng, Y. C. Zhang, L. J. Xu, J. Gu, and P. G. Xiao, "Phytochemical and biological studies of paeoniaceae," *Chemistry & Biodiversity*, vol. 7, no. 4, pp. 805–838, 2010.

[16] S. S. Li, Q. Wu, D. D. Yin, C. Y. Feng, Z. A. Liu, and L. S. Wang, "Phytochemical variation among the traditional Chinese medicine Mu Dan Pi from Paeonia suffruticosa (tree peony)," *Phytochemistry*, vol. 146, pp. 16–24, 2018.

[17] X. Wu, H. Chen, X. Chen, and Z. Hu, "Determination of paeonol in rat plasma by high-performance liquid chromatography and its application to pharmacokinetic studies following oral administration of Moutan cortex decoction," *Biomedical Chromatography*, vol. 17, no. 8, pp. 504–508, 2003.

[18] H. Li, S. Wang, B. Zhang et al., "Influence of co-administered danshensu on pharmacokinetic fate and tissue distribution of paeonol in rats," *Planta Medica*, vol. 78, no. 2, pp. 135–140, 2012.

[19] L. Y. Ma, X. D. Xu, Q. Zhang, J. H. Miao, and B. L. Tang, "Paeonol pharmacokinetics in the rat following i. m. administration," *European Journal of Drug Metabolism and Pharmacokinetics*, vol. 33, no. 3, pp. 133–136, 2008.

[20] M. H. Kwon, J. S. Jeong, J. Ryu, Y. W. Cho, and H. E. Kang, "Simultaneous determination of saikosaponin a, paeonol, and imperatorin, components of DA-9805, in rat plasma by LC-MS/MS and application to a pharmacokinetic study," *Journal of Chromatography*, vol. 1068-1069, pp. 289–296, 2017.

[21] Y. Xie, Z. H. Jiang, H. Zhou et al., "The pharmacokinetic study of sinomenine, paeoniflorin and paeonol in rats after oral administration of a herbal product Qingfu Guanjiesu capsule by HPLC," *Biomedical Chromatography*, vol. 28, no. 9, pp. 1294–1302, 2014.

[22] H. Li, F. Song, L. R. Duan et al., "Paeonol and danshensu combination attenuates apoptosis in myocardial infarcted rats by inhibiting oxidative stress: roles of Nrf 2/HO-1 and PI3K/Akt pathway," *Scientific Reports*, vol. 6, no. 1, article 23693, 2016.

[23] S. S. Li, G. F. Li, L. Liu et al., "Optimization of paeonol-loaded microparticle formulation by response surface methodology," *Journal of Microencapsulation*, vol. 32, no. 7, pp. 677–686, 2015.

[24] J. Yao, Y. Zhang, Q. Hu et al., "Optimization of paeonol-loaded poly (butyl-2-cyanoacrylate) nanocapsules by central composite design with response surface methodology together with the antibacterial properties," *European Journal of Pharmaceutical Sciences*, vol. 101, pp. 189–199, 2017.

[25] H. Ma, D. Guo, Y. Fan, J. Wang, J. Cheng, and X. Zhang, "Paeonol-loaded ethosomes as transdermal delivery carriers: design, preparation and evaluation," *Molecules*, vol. 23, no. 7, 2018.

[26] M. Luo, Q. Shen, and J. Chen, "Transdermal delivery of paeonol using cubic gel and microemulsion gel," *International Journal of Nanomedicine*, vol. 6, pp. 1603–1610, 2011.

[27] Z. X. Chen, B. Li, T. Liu et al., "Evaluation of paeonol-loaded transethosomes as transdermal delivery carriers," *European Journal of Pharmaceutical Sciences*, vol. 99, pp. 240–245, 2017.

[28] S. S. Li, G. F. Li, L. Liu et al., "Evaluation of paeonol skin-target delivery from its microsponge formulation: in vitro skin permeation and in vivo microdialysis," *PLoS One*, vol. 8, no. 11, article e79881, 2013.

[29] J.-C. Li, N. Zhu, J.-X. Zhu et al., "Self-assembled cubic liquid crystalline nanoparticles for transdermal delivery of paeonol," *Medical Science Monitor*, vol. 21, pp. 3298–3310, 2015.

[30] C. Chen, F. Jia, Z. Hou, S. Ruan, and Q. Lu, "Delivery of paeonol by nanoparticles enhances its in vitro and in vivo antitumor effects," *International Journal of Nanomedicine*, vol. 12, pp. 6605–6616, 2017.

[31] J. V. Tapia-Vieyra, B. Delgado-Coello, and J. Mas-Oliva, "Atherosclerosis and cancer; a resemblance with far-reaching implications," *Archives of Medical Research*, vol. 48, no. 1, pp. 12–26, 2017.

[32] S. Chen, J. Zhang, L. Wu, H. Wu, and M. Dai, "Paeonol nanoemulsion for enhanced oral bioavailability: optimization and mechanism," *Nanomedicine*, vol. 13, no. 3, pp. 269–282, 2018.

[33] M. A. Gimbrone Jr. and G. Garcia-Cardena, "Endothelial cell dysfunction and the pathobiology of atherosclerosis," *Circulation Research*, vol. 118, no. 4, pp. 620–636, 2016.

[34] C. Weber and H. Noels, "Atherosclerosis: current pathogenesis and therapeutic options," *Nature Medicine*, vol. 17, no. 11, pp. 1410–1422, 2011.

[35] C. Niu, X. Zhou, J. Zhang, Q. Xu, and K. Cao, "Protective effect of paeonol on human umbilical vein endothelial cells injured by hyperlipidemic serum," *Chinese Journal of Pharmacology and Toxicology*, vol. 25, no. 5, pp. 413–418, 2011.

[36] Q. Xu, K. Cao, X. H. Zhou, X. J. Wang, Y. F. Wang, and D. U. Chao, "Effects of paeonol on expression of eNOS and level of NO in HUVECs injured by homocysteine," *Chinese Traditional Patent Medicine*, vol. 34, no. 12, pp. 2286–2289, 2012.

[37] U. Forstermann, N. Xia, and H. Li, "Roles of vascular oxidative stress and nitric oxide in the pathogenesis of atherosclerosis," *Circulation Research*, vol. 120, no. 4, pp. 713–735, 2017.

[38] W. Hu, Z. Zhang, and M. Dai, "Paeonol affects proliferation activity of rat vasular endothelial cells induced by lipopolysaccharide and co-cultured with smooth muscle cells via inhibiting pathway of PI3K/AKT-NF-κB signaling," *China Journal of Chinese Materia Medica*, vol. 41, no. 12, pp. 2298–2302, 2016.

[39] M. H. Bao, Y. W. Zhang, and H. H. Zhou, "Paeonol suppresses oxidized low-density lipoprotein induced endothelial cell apoptosis via activation of LOX-1/p38MAPK/NF-κB pathway," *Journal of Ethnopharmacology*, vol. 146, no. 2, pp. 543–551, 2013.

[40] B. G. Childs, M. Gluscevic, D. J. Baker et al., "Senescent cells: an emerging target for diseases of ageing," *Nature Reviews Drug Discovery*, vol. 16, no. 10, pp. 718–735, 2017.

[41] J. Jamal, M. R. Mustafa, and P. F. Wong, "Paeonol protects against premature senescence in endothelial cells by modulating sirtuin 1 pathway," *Journal of Ethnopharmacology*, vol. 154, no. 2, pp. 428–436, 2014.

[42] M. O. J. Grootaert, L. Roth, D. M. Schrijvers, G. R. Y. De Meyer, and W. Martinet, "Defective autophagy in atherosclerosis: to die or to senesce?," *Oxidative Medicine and Cellular Longevity*, vol. 2018, Article ID 7687083, 12 pages, 2018.

[43] Y.-r. Liu, J.-j. Chen, and M. Dai, "Paeonol protects rat vascular endothelial cells from ox-LDL-induced injury in vitro via downregulating microRNA-21 expression and TNF-α release," *Acta Pharmacologica Sinica*, vol. 35, no. 4, pp. 483–488, 2014.

[44] Y. C. Cheng, J. M. Sheen, W. L. Hu, and Y. C. Hung, "Polyphenols and oxidative stress in atherosclerosis-related ischemic heart disease and stroke," *Oxidative Medicine and Cellular Longevity*, vol. 2017, Article ID 8526438, 16 pages, 2017.

[45] Q. Yang, Y. Xie, J. Sun, L. Bi, X. Zhou, and S. Wang, "Effects of salvianolic acid B with paeonol on the bcl-2, bax, caspase-3 mRNA and protein expression in human vascular endothelial cell oxidative damage induced by $H_2O_2$," *China Medical Herald*, vol. 10, no. 2, pp. 26–28, 2013.

[46] M. Dai, Q. Liu, C. Gu, and H. Zhang, "Inhibitory effect of paeonol on lipid peroxidational reaction and oxidational decorate of low density lipoprotein," *China Journal of Chinese Materia Medica*, vol. 25, no. 10, pp. 625–627, 2000.

[47] M. Ping, W. Xiao, L. Mo et al., "Paeonol attenuates advanced oxidation protein product-induced oxidative stress injury in THP-1 macrophages," *Pharmacology*, vol. 93, no. 5-6, pp. 286–295, 2014.

[48] K. W. Choy, Y. S. Lau, D. Murugan, and M. R. Mustafa, "Chronic treatment with paeonol improves endothelial function in mice through inhibition of endoplasmic reticulum stress-mediated oxidative stress," *PLoS One*, vol. 12, no. 5, article e0178365, 2017.

[49] Y. Zhao, B. Fu, X. Zhang et al., "Paeonol pretreatment attenuates cerebral ischemic injury via upregulating expression of pAkt, Nrf 2, HO-1 and ameliorating BBB permeability in mice," *Brain Research Bulletin*, vol. 109, pp. 61–67, 2014.

[50] D. Tousoulis, E. Oikonomou, E. K. Economou, F. Crea, and J. C. Kaski, "Inflammatory cytokines in atherosclerosis: current therapeutic approaches," *European Heart Journal*, vol. 37, no. 22, pp. 1723–1732, 2016.

[51] L. L. Pan and M. Dai, "Paeonol from Paeonia suffruticosa prevents TNF-α-induced monocytic cell adhesion to rat aortic endothelial cells by suppression of VCAM-1 expression," *Phytomedicine*, vol. 16, no. 11, pp. 1027–1032, 2009.

[52] I. T. Nizamutdinova, H. M. Oh, Y. N. Min et al., "Paeonol suppresses intercellular adhesion molecule-1 expression in tumor necrosis factor-α-stimulated human umbilical vein endothelial cells by blocking p 38, ERK and nuclear factor-κB signaling pathways," *International Immunopharmacology*, vol. 7, no. 3, pp. 343–350, 2007.

[53] Y.-Q. Wang, M. Dai, J.-C. Zhong, and D.-K. Yin, "Paeonol

inhibits oxidized low density lipoprotein-induced monocyte adhesion to vascular endothelial cells by inhibiting the mitogen activated protein kinase pathway," *Biological & Pharmaceutical Bulletin*, vol. 35, no. 5, pp. 767–772, 2012.

[54] X. Zhou, C. Niu, K. Cao, and Q. Xu, "Paeonol reduced expression of adhesion molecules in HUVECs induced by hyperlipidemic serum via inhibiting the pathway of NF-κB signaling," *Chinese Journal of Pathophysiology*, vol. 27, no. 2, pp. 249–253, 2011.

[55] X. Yuan, J. Chen, and M. Dai, "Paeonol promotes microRNA-126 expression to inhibit monocyte adhesion to ox-LDL-injured vascular endothelial cells and block the activation of the PI3K/Akt/NF-κB pathway," *International Journal of Molecular Medicine*, vol. 38, no. 6, pp. 1871–1878, 2016.

[56] J. P. Kolb, T. H. Oguin III, A. Oberst, and J. Martinez, "Programmed cell death and inflammation: winter is coming," *Trends in Immunology*, vol. 38, no. 10, pp. 705–718, 2017.

[57] L. Yang, M. Dai, and P. Chen, "Effects of paeonol on lipopolysaccharide-induced release of vascular endothelial cell adhesion molecule-1 and tumor necrosis factor-alpha and TLR4/NF-κB signaling pathway in rat vascular endothelial cells," *Journal of Anhui University of Chinese Medicine*, vol. 34, no. 1, pp. 46–50, 2015.

[58] H. S. Chae, O. H. Kang, Y. S. Lee et al., "Inhibition of LPS-induced iNOS, COX-2 and inflammatory mediator expression by paeonol through the MAPKs inactivation in RAW 264.7 cells," *The American Journal of Chinese Medicine*, vol. 37, no. 1, pp. 181–194, 2009.

[59] N. Chen, D. Liu, L. W. Soromou et al., "Paeonol suppresses lipopolysaccharide-induced inflammatory cytokines in macrophage cells and protects mice from lethal endotoxin shock," *Fundamental & Clinical Pharmacology*, vol. 28, no. 3, pp. 268–276, 2014.

[60] H. Huang, E. J. Chang, Y. Lee, J. S. Kim, S. S. Kang, and H. H. Kim, "A genome-wide microarray analysis reveals anti-inflammatory target genes of paeonol in macrophages," *Inflammation Research*, vol. 57, no. 4, pp. 189–198, 2008.

[61] Y. Liu, J. Chen, M. Dai, X. Su, and Y. Ye, "Paeonol reduces release of TNF-α in vascular endothelial cells induced by oxidized low-density lipoprotein through micro RNA-21-mediated p 38 MAPK pathway," *Journal of Anhui University of Chinese Medicine*, vol. 33, no. 1, pp. 51–56, 2014.

[62] H. Li, M. Dai, and W. Jia, "Paeonol attenuates high-fat-diet-induced atherosclerosis in rabbits by anti-inflammatory activity," *Planta Medica*, vol. 75, no. 1, pp. 7–11, 2009.

[63] M. S. Rahman, A. J. Murphy, and K. J. Woollard, "Effects of dyslipidaemia on monocyte production and function in cardiovascular disease," *Nature Reviews. Cardiology*, vol. 14, no. 7, pp. 387–400, 2017.

[64] Q. Li, X. Zheng, M. Gong, Y. Fan, J. Qu, and X. Zhou, "Influence of paeonol on coronary atherosclerosis," *Chinese Journal of Evidence-Based Cardiovascular Medicine*, vol. 8, no. 9, pp. 1037–1039, 2016.

[65] M. Dai, X. Zhi, D. Peng, and Q. Liu, "Inhibitory effect of paeonol on experimental atherosclerosis in quails," *China Journal of Chinese Materia Medica*, vol. 24, no. 8, pp. 488–90, 512, 1999.

[66] X. Qian, C. Hu, D. Liu, L. Qiang, and X. Shi, "Effects of paeonol on aortic NF-κB activity and blood lipid levels in rats with atherosclerosis," *China Pharmacist*, vol. 17, no. 9, pp. 1441–1443, 2014.

[67] J. F. Zhao, S. J. Jim Leu, S. K. Shyue, K. H. Su, J. Wei, and T. S. Lee, "Novel effect of paeonol on the formation of foam cells: promotion of LXRα-ABCA1-dependent cholesterol efflux in macrophages," *The American Journal of Chinese Medicine*, vol. 41, no. 5, pp. 1079–1096, 2013.

[68] X. Li, Y. Zhou, C. Yu et al., "Paeonol suppresses lipid accumulation in macrophages via upregulation of the ATP-binding cassette transporter A1 and downregulation of the cluster of differentiation 36," *International Journal of Oncology*, vol. 46, no. 2, pp. 764–774, 2015.

[69] R. S. Rosenson, H. Bryan Brewer, B. J. Ansell et al., "Dysfunctional HDL and atherosclerotic cardiovascular disease," *Nature Reviews Cardiology*, vol. 13, no. 1, pp. 48–60, 2016.

[70] Y. Dong, S. Zhang, S. Dong et al., "Study on the antilipemic effect and protective effect of thoracic aorta of paeonol in hyperlipidemic rats," *Journal of Clinical and Experimental Medicine*, vol. 13, no. 14, pp. 1132–1134, 2014.

[71] Q. Yang, S. Wang, Y. Xie et al., "Effect of salvianolic acid B and paeonol on blood lipid metabolism and hemorrheology in myocardial ischemia rabbits induced by pituitruin," *International Journal of Molecular Sciences*, vol. 11, no. 10, pp. 3696–3704, 2010.

[72] M. R. Bennett, S. Sinha, and G. K. Owens, "Vascular smooth muscle cells in atherosclerosis," *Circulation Research*, vol. 118, no. 4, pp. 692–702, 2016.

[73] X. Zhou, X. Zhou, Q. Xu, and L. Wang, "Effect of paeonol on proliferation of cultured smooth muscle cells from rats by hyperlipidemia," *Hebei Journal of Traditional Chinese Medicine*, vol. 22, no. 6, pp. 477–478, 2000.

[74] H. Lin, J. Zeng, L. Bai, and C. Chen, "Effect of paeonol on cell cycle of rat vascular smooth muscle cell proliferation induced by platelet-derived growth factor-BB," *South China Journal of Cardiovascular Diseases*, vol. 19, no. 5, pp. 621–625, 2013.

[75] J. Chen, M. Dai, and Y. Wang, «Paeonol inhibits proliferation of vascular smooth muscle cells stimulated by high glucose via Ras-Raf-ERK1/2 signaling pathway in coculture model," *Evidence-based Complementary and Alternative Medicine*, vol. 2014, Article ID 484269, 9 pages, 2014.

[76] Z. Jia, K. Shi, Y. Fan, Q. Li, and X. Zhou, "Effect of paeonol combined with Panax notoginseng saponins on diabetic myocardial fibrosis in rats," *Chinese Journal of Experimental Traditional Medical Formulae*, vol. 24, no. 6, pp. 133–138, 2018.

[77] L. Meng, W. Xu, L. Guo, W. Ning, and X. Zeng, "Paeonol inhibits the proliferation, invasion, and inflammatory reaction induced by TNF-α in vascular smooth muscle cells," *Cell Biochemistry and Biophysics*, vol. 73, no. 2, pp. 495–503, 2015.

[78] J.-y. Zhang, L. Lei, J. Shang et al., "Local application of paeonol prevents early restenosis: a study with a rabbit vein graft model," *Journal of Surgical Research*, vol. 212, pp. 278–287, 2017.

[79] J. Yeung, W. Li, and M. Holinstat, "Platelet signaling and disease: targeted therapy for thrombosis and other related diseases," *Pharmacological Reviews*, vol. 70, no. 3, pp. 526–548, 2018.

[80] M. Doble, S. Karthikeyan, P. A. Padmaswar, and K. G. Akamanchi, "QSAR studies of paeonol analogues for inhibition of platelet aggregation," *Bioorganic & Medicinal Chemistry*, vol. 13, no. 21, pp. 5996–6001, 2005.

[81] Q. Bai, "Influence of endothelial function and blood rheology in diabetic rats by paeonol and tanshinol," *Chinese Journal of Experimental Traditional Medical Formulae*, vol. 21, no. 3, pp. 110–113, 2015.

[82] C. Min, H. Liu, F. Zhan, and W. Qiu, "Effect of Cortex Moutan on PGI2, TXA2, ET and NO in diabetic rats," *Journal of Chinese Medicinal Materials*, vol. 30, no. 6, pp. 687–690, 2007.

[83] I. Andreou, X. Sun, P. H. Stone, E. R. Edelman, and M. W. Feinberg, "miRNAs in atherosclerotic plaque initiation, progression, and rupture," *Trends in Molecular Medicine*, vol. 21, no. 5, pp. 307–318, 2015.

[84] N. Liu, X. Feng, W. Wang, X. Zhao, and X. Li, "Paeonol protects against TNF-α-induced proliferation and cytokine release of rheumatoid arthritis fibroblast-like synoviocytes by upregulating FOXO3 through inhibition of miR-155 expression," *Inflammation Research*, vol. 66, no. 7, pp. 603–610, 2017.

[85] J. Wu, C. Sun, R. Wang et al., "Cardioprotective effect of paeonol against epirubicin-induced heart injury via regulating miR-1 and PI3K/AKT pathway," *Chemico-Biological Interactions*, vol. 286, pp. 17–25, 2018.

[86] A. S. Jaipersad, G. Y. H. Lip, S. Silverman, and E. Shantsila, "The role of monocytes in angiogenesis and atherosclerosis," *Journal of the American College of Cardiology*, vol. 63, no. 1, pp. 1–11, 2014.

[87] H. J. Lee, S. A. Kim, H. J. Lee et al., "Paeonol oxime inhibits bFGF-induced angiogenesis and reduces VEGF levels in fibrosarcoma cells," *PLoS One*, vol. 5, no. 8, article e12358, 2010.

[88] Y. C. Koh, G. Yang, C. S. Lai, M. Weerawatanakorn, and M. H. Pan, "Chemopreventive effects of phytochemicals and medicines on M1/M2 polarized macrophage role in inflammation-related diseases," *International Journal of Molecular Sciences*, vol. 19, no. 8, 2018.

[89] J. B. Self-Fordham, A. R. Naqvi, J. R. Uttamani, V. Kulkarni, and S. Nares, "MicroRNA: dynamic regulators of macrophage polarization and plasticity," *Frontiers in Immunology*, vol. 8, 2017.

[90] Y. J. Li, J. X. Bao, J. W. Xu, F. Murad, and K. Bian, "Vascular dilation by paeonol–a mechanism study," *Vascular Pharmacology*, vol. 53, no. 3-4, pp. 169–176, 2010.

[91] L. Ma, C. C. Chuang, W. Weng et al., "Paeonol protects rat heart by improving regional blood perfusion during no-reflow," *Frontiers in Physiology*, vol. 7, 2016.

[92] Z. Shi, X. Zhou, Q. Xu, and K. Cao, "Effect of paeonol on Smad 2, Smad 3, Smad 7 mRNA expression in acute myocardial infarction rat heart," *Chinese Journal of Experimental Traditional Medical Formulae*, vol. 21, no. 2, pp. 146–150, 2015.

[93] A. L. Jonsson and F. Backhed, "Role of gut microbiota in atherosclerosis," *Nature Reviews. Cardiology*, vol. 14, no. 2, pp. 79–87, 2017.

[94] K. Miteva, R. Madonna, R. De Caterina, and S. Van Linthout, "Innate and adaptive immunity in atherosclerosis," *Vascular Pharmacology*, vol. 107, pp. 67–77, 2018.

# c-Met Signaling Protects from Nonalcoholic Steatohepatitis- (NASH-) Induced Fibrosis in Different Liver Cell Types

Hannah K. Drescher,[1] Fabienne Schumacher,[1] Teresa Schenker,[1] Maike Baues,[2] Twan Lammers,[2] Thomas Hieronymus,[3] Christian Trautwein,[1] Konrad L. Streetz,[1] and Daniela C. Kroy[1]

[1]Department of Internal Medicine III, University Hospital, RWTH, Aachen, Germany
[2]Institute for Experimental Molecular Imaging, University Hospital, RWTH, Aachen, Germany
[3]Institute for Biomedical Engineering-Cell Biology, University Hospital, RWTH, Aachen, Germany

Correspondence should be addressed to Daniela C. Kroy; dkroy@ukaachen.de

Academic Editor: Ryuichi Morishita

Nonalcoholic steatohepatitis (NASH) is the most common chronic, progressive liver disease in Western countries. The significance of cellular interactions of the HGF/c-Met axis in different liver cell subtypes and its relation to the oxidative stress response remains unclear so far. Hence, the present study is aimed at investigating the role of c-Met and the interaction with the oxidative stress response during NASH development in mice and humans. Conditional c-Met knockout (KO) lines (LysCre for Kupffer cells/macrophages, GFAPCre for $\alpha$-SMA$^+$ and CK19$^+$ cells and MxCre for bone marrow-derived immune cells) were fed chow and either methionine-choline-deficient diet (MCD) for 4 weeks or high-fat diet (HFD) for 24 weeks. Mice lacking c-Met either in Kupffer cells, $\alpha$-SMA$^+$ and CK19$^+$ cells, or bone marrow-derived immune cells displayed earlier and faster progressing steatohepatitis during dietary treatments. Severe fatty liver degeneration and histomorphological changes were accompanied by an increased infiltration of immune cells and a significant upregulation of inflammatory cytokine expression reflecting an earlier initiation of steatohepatitis development. In addition, animals with a cell-type-specific deletion of c-Met exhibited a strong generation of reactive oxygen species (ROS) by dihydroethidium (hydroethidine) (DHE) staining showing a significant increase in the oxidative stress response especially in LysCre/c-Met$^{mut}$ and MxCre/c-Met$^{mut}$ animals. All these changes finally lead to earlier and stronger fibrosis progression with strong accumulation of collagen within liver tissue of mice deficient for c-Met in different liver cell types. The HGF/c-Met signaling pathway prevents from steatosis development and has a protective function in the progression to steatohepatitis and fibrosis. It conveys an antifibrotic role independent on which cell type c-Met is missing (Kupffer cells/macrophages, $\alpha$-SMA$^+$ and CK19$^+$ cells, or bone marrow-derived immune cells). These results highlight a global protective capacity of c-Met in NASH development and progression.

## 1. Introduction

Nonalcoholic fatty liver disease (NAFLD) is a chronic and progressive liver disease which is evolving to be one of the leading causes for liver transplantation [1–3]. As the most common chronic liver disease in Western countries, NAFLD is accompanied by a specific sequence of hepatic alterations that begins with simple steatosis which can easily progress to more advanced stages like nonalcoholic steatohepatitis (NASH), liver fibrosis, and cirrhosis and finally the development of hepatocellular carcinoma (HCC) leading to an increased mortality [4, 5]. When considering that more than 20% of the general population in industrialized nations have a fatty liver with the risk to progress to NASH [6], the dramatic need for the development of therapeutic treatment options gets evident. A hallmark of NASH is its correlation to the metabolic syndrome and its close relationship to type II diabetes mellitus which is predominantly characterized by fat accumulation, insulin resistance, and dyslipidemia [7]. Also, on the molecular level, NASH is associated with a consecutive impairment of the metabolism and an intense inflammatory immune response. Alterations in the lipid metabolism lead to an increased secretion of free fatty acids from skeletal muscle and adipose tissue which in turn leads to the accumulation of triglycerides and other lipids in hepatocytes [8, 9]. Toxic lipids can provoke intrahepatic cell death inducing the oxidative stress and proinflammatory immune response. Histologically, the cause of NASH disease development can be observed by ballooning of hepatocytes, infiltration of immune cells, and finally the accumulation of collagen fibers within liver tissue. The progression to fibrosis can in the end provide a loss of organ function.

Met, the cellular receptor tyrosine kinase for the hepatocyte growth factor (HGF), plays a pivotal role in mediating processes like stem cell growth, wound healing, motility, and morphogenesis [10]. Previous studies could further show its importance in the growth of hepatocytes, fibrogenesis, and immune modulation [11, 12]. On the molecular level, the direct binding of HGF to its receptor c-Met leads to the homodimerization followed by phosphorylation of this receptor at the tyrosine residues Y1234 and Y1235 within the catalytic domain of the tyrosine kinase domain [13]. Phosphorylation leads to the activation of several downstream signaling pathways conveying multiple intracellular functions including the PI3K/AKT [14], JAK/STAT [15, 16], and Ras/MAPK cascades [17–19] controlling, e.g., proliferation and apoptosis [20].

The importance of c-Met gets more obvious when considering that mice with a deletion of this receptor exhibit serious developmental defects of several muscle groups [21], placenta, and liver between weeks 13 and 16 of embryogenesis up to death of these animals in utero [22–25].

c-Met is predominantly expressed on cells with endothelial and epithelial origins and has initially been described as a protooncogene [10]. In the recent review summarizing the role of c-Met as a therapeutic option in HCC, it was shown that only the selective inhibition of c-Met shows antitumoral potential in HCC, whereas the nonselective kinase inhibition with c-Met activity failed in clinical trials [26].

During chronic liver injury, especially in NASH, c-Met has been shown to play a benefitial, hepatoprotective role by suppressing chronic inflammation and progression to fibrosis [27]. The generation of conditional c-Met knockout mice first described by Borowiak et al. [28] highlighted the role of this receptor in liver regeneration. Mice with a MxCre-specific deletion for c-Met displayed reduced liver regeneration based on a cell cycle defect together with reduced ERK activity while Akt phosphorylation still occurred.

Based on the current knowledge, the aim of this study was to elucidate the particular role of c-Met on the different liver cell types during the development and progression of diet-induced steatohepatitis in two distinct murine feeding models. Mice with a deficiency of c-Met in either Kupffer cells/myeloid cells, cells under the glial fibrillary acid protein promotor, or bone marrow-derived immune cells were treated with high-fat diet (HFD) or methionine-choline-deficient diet (MCD). This study is the first to publish that the deletion of c-Met, insignificant on which liver cell types (Kupffer cells/macrophages, $\alpha$-SMA$^+$/CK19$^+$ cells, or bone marrow-derived immune cells), accelerated the disease outcome in diet-induced steatohepatitis and fibrosis.

## 2. Material and Methods

*2.1. Animal Studies.* The study was carried out in accordance to the law of the regional authorities for nature, environment, and consumer protection of North Rhine-Westfalia (Landesamt für Natur, Umwelt und Verbraucherschutz NRW (LANUV), Recklinghausen, Germany) and approved by the LANUV Committee (permit number: TV11018G). All experiments were performed in accordance with the German guidelines for animal housing and husbandry.

*2.2. Human Samples.* 52 liver specimens were obtained according to the local ethics committee rules from liver explantation or liver resection recruited at the RWTH Aachen University Hospital (local IRB permit number EK 166-12). When intrahepatic lesions were present, the tissue for this study was collected from the most distant section of the specimen. Overall, we analyzed 12 samples from patients with HBV, 17 HCV, 6 PBC, 8 PSC, 5 alcoholic cirrhosis, and 4 NASH patient samples.

*2.3. Housing and Generation of Mice.* Male C57BL/6J wild-type mice, conditional c-Met$^{fl/fl}$ and LysCre c-Met$^{mut}$, conditional c-Met$^{fl/fl}$ and GFAPCre c-Met$^{mut}$, and conditional c-Met$^{fl/fl}$ and MxCre c-Met$^{mut}$ mice all in C57BL/6J background were used as experimental animals. At least five animals per group and time point were treated in parallel. Respective c-Met$^{fl/fl}$ animals were always littermates to the corresponding c-Met$^{mut}$ mice. All experiments were repeated at least twice. Experimental animals were housed in the animal facility of the RWTH Aachen University Hospital with 12-hour light/dark cycles and water and food *ad libitum* available. Treatments were in accordance with the criteria of the German administrative panel on laboratory animal care.

*2.4. Dietary Treatments.* All dietary treatments were performed with 8–12-week-old male mice weighing at least 25 g. Mice were fed either with chow diet, methionine-choline-deficient diet (MCD) (Sniff, Soest, Germany), or high-fat diet (HFD) (40 kcal% fat (vegetable fats), 20 kcal% fructose, and 2% cholesterol) (Brogaarden, Lynge, Denmark). Wild-type and knockout mice showed a food intake of about 6 g per day without any differences between chow- and diet-treated animals.

Further material and methods' data are provided in Supplementary Materials (available here).

## 3. Results

*3.1. Effective c-Met Deletion in the Different Cre Lines.* To determine the specificity of the c-Met deletion in the different used Cre lines (LysCre, GFAPCre, and MxCre), we first conducted genomic PCR genotyping, detecting the recombinant and nonrecombinant c-Met allele. Analysis was performed for hepatocytes, Kupffer cells/macrophages, α-SMA$^+$/CK19$^+$ cells, and bone marrow-derived immune cells in all three tissue-specific genetic settings. These controls show the presence of the c-Met WT allele in the varying liver cell types not targeted by the specific Cre promotor (Supplementary Figure 1A). The next experiment was supposed to show the efficient deletion of c-Met in Kupffer cells/macrophages under the LysCre promotor (Supplementary Figure 1B), on α-SMA$^+$/CK19$^+$ cells under the GFAPCre promotor (Supplementary Figure 1C), and on bone marrow-derived immune cells under the MxCre promotor (Supplementary Figure 1D) at baseline conditions. All results show a deletion of c-Met on the specific cell compartment.

*3.2. The Loss of c-Met in Kupffer Cells/Myeloid Cells Increases Steatosis Development and Inflammation in the MCD and HFD Model of Steatohepatitis.* To evaluate the significance of c-Met in Kupffer cells/myeloid cells for steatohepatitis development, 8–12-week-old c-Met$^{fl/fl}$ and LysCre/c-Met$^{mut}$ littermates were fed either chow, methionine-choline-deficient diet (MCD) (4 weeks), or high-fat diet (HFD) (24 weeks) *ad libitum*. A significant increase in serum transaminase levels in LysCre/c-Met$^{mut}$ animals as well as in the MCD and HFD models points to a severe disease progression of nonalcoholic steatohepatitis (NASH) (Figures 1(a) and 1(b)). This invasive disease development was further reflected in faster progressing steatosis directly associated with detrimental changes in liver architecture and fatty liver degeneration which is shown in the histological analysis (H&E Figure 1(c), Oil Red O Figure 1(d)). To investigate metabolic changes over 24 weeks of treatment with steatosis-induced diets, we next performed a glucose tolerance test to detect typical features of the metabolic syndrome. Significantly elevated serum glucose levels in LysCre/c-Met$^{mut}$ mice compared to equally fed c-Met$^{fl/fl}$ animals indicated the existence of insulin resistance (Figure 1(e)). To directly measure the hepatic triglyceride accumulation as a hallmark of nonalcoholic fatty liver disease, homogenates of whole liver extracts of c-Met$^{fl/fl}$ and LysCre/c-Met$^{mut}$ animals were assessed regarding their triglyceride content after chow, MCD (4 weeks), and HFD (24 weeks) feeding. After treatment with both steatohepatitis-induced diets, LysCre/c-Met$^{mut}$ mice displayed a significantly higher accumulation of hepatic triglycerides compared to c-Met$^{fl/fl}$ controls (Figure 1(f)). The data suggests a protective role of c-Met in Kupffer cells in the development of diet-induced steatosis.

To further assess the progression from simple steatosis to a more advanced disease state of steatohepatitis, we next investigated whether the imbalance in systemic glucose and lipid metabolism resulted in alterations in the immune cell response.

Flow cytometric analysis of the intrahepatic immune cell infiltration revealed a decrease in the ratio of CD4$^+$/CD8$^+$ T cells (Figure 2(a), Supplementary Figure 2) which reflects a more dominant CD8$^+$ lymphocyte-driven immune cell response in LysCre/c-Met$^{mut}$ animals after MCD and HFD feeding compared to c-Met$^{fl/fl}$ mice. CD8$^+$ T cells exert several effector functions including the production of inflammatory cytokines and cytolysis. To investigate this in more detail real-time PCR analysis showed an increase in the mRNA expression of the proinflammatory mediators TNF-α and IL-6 and fibrosis markers, such as TGF-β and Collagen1α in LysCre/c-Met$^{mut}$ after both dietary treatments (Figure 2(b)). TNF-α is strongly expressed in animals treated with MCD compared to HFD feeding where it shows only a slight trend to be upregulated. This difference potentially occurs because the MCD model is a non-obesity-related steatohepatitis mouse model with strong inflammatory changes within the liver tissue. TGF-β on the contrary is strongly expressed in HFD-treated animals compared to MCD fed LysCre/c-Met$^{mut}$ mice. TGF-β is known to be involved in lipid accumulation in hepatocytes in the course of the metabolic syndrome which is more pronounced in the chronic HFD model of murine steatohepatitis compared to MCD feeding [29, 30].

To unravel a potential mechanism responsible for the observed differences in disease development and progression of NASH in mice lacking c-Met in Kupffer cells, we next investigated the amount of apoptotic cell death and the intrahepatic oxidative stress environment by TUNEL and DHE (dihydroethidium (hydroethidine)) staining (Figure 2(c), Supplementary Figures 3A and 3B). LysCre/c-Met$^{mut}$ mice display a trend to an increase in TUNEL-positive and DHE-positive cells after MCD and HFD feeding. An increase in oxidative stress is known to be present in the hepatic microenvironment of livers with NAFLD and NASH which is in turn known to induce the recruitment and infiltration of proinflammatory cells. In particular, T cells are critical in this regard for influencing steatohepatitis development and progression.

Our results thus indicate that the proinflammatory immune response provoked by oxidative stress may be mechanistically at least in parts responsible for the more invasive phenotype in mice with a c-Met deficiency in Kupffer cells.

*3.3. c-Met Deletion in Kupffer Cells/Myeloid Cells Promotes Fibrosis Progression in Two Dietary Models of Steatohepatitis (MCD, HFD).* A crucial part in the cause of NASH is the progression to fibrosis. To assess this increase in extracellular matrix accumulation, we first performed an immunofluorescence staining for α-SMA which revealed a significant increase in LysCre/c-Met$^{mut}$ mice after MCD and HFD feeding compared to equally treated c-Met$^{fl/fl}$ animals (Figure 2(d), Supplementary Figure 3C). An additional Sirius Red staining and the direct measurement of the intrahepatic hydroxyproline content support these results both showing a significant increase in matrix deposition in diet-treated LysCre/c-Met$^{mut}$ mice (Figures 2(e) and 2(f), Supplementary Figure 3D). These results indicate strong hepatic stellate cell activation reflecting fibrosis initiation in animals deficient of c-Met in Kupffer cells.

*3.4. c-Met Deficiency under the Glial Fibrillary Acid Protein (GFAP) Promoter Leads to Steatosis and Fibrosis.* In the next

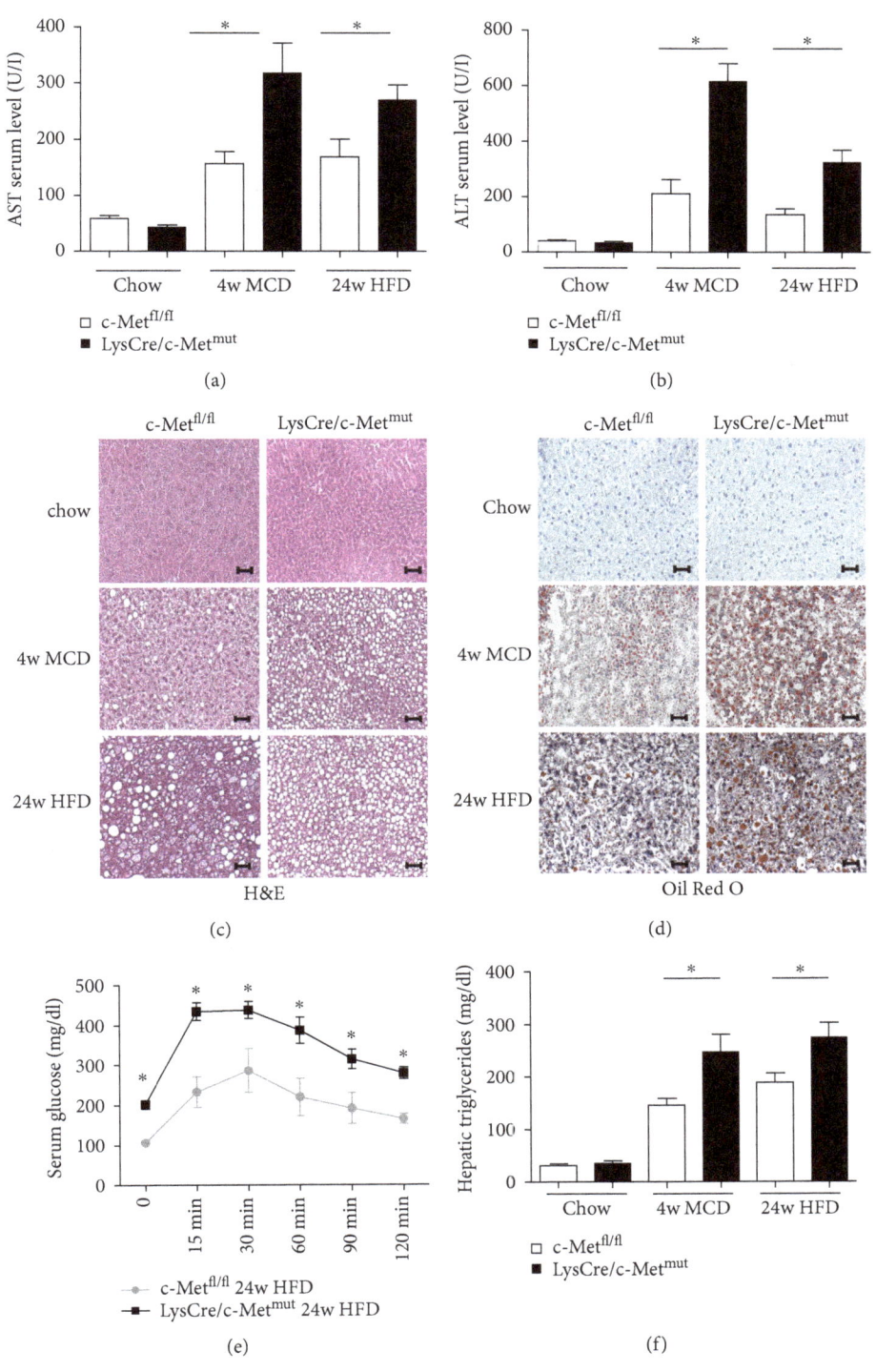

FIGURE 1: Increased steatosis in LysCre/c-Met$^{mut}$ mice after MCD and HFD feeding (a) AST levels in serum of c-Met$^{fl/fl}$ and LysCre/c-Met$^{mut}$ mice after chow, MCD (4 weeks), and HFD (24 weeks) feeding. Serum transaminases increase after treatment with steatosis-induced diets ($n = 8$) ($^*p < 0.05$). (b) ALT levels in serum of c-Met$^{fl/fl}$ and LysCre/c-Met$^{mut}$ mice after chow, MCD (4 weeks), and HFD (24 weeks) feeding. Serum transaminases increase after treatment with steatosis-induced diets ($n = 8$) ($^*p < 0.05$). (c) Representative H&E-stained liver sections of c-Met$^{fl/fl}$ and LysCre/c-Met$^{mut}$ animals (chow, MCD (4 w), and HFD (24 w)) show increased steatosis development in LysCre/c-Met$^{mut}$ mice. Magnification: 200x; scale bars: 100 $\mu$m. (d) Representative images of Oil Red O-stained liver sections from c-Met$^{fl/fl}$ and LysCre/c-Met$^{mut}$ mice after chow, MCD (4 weeks), and HFD (24 weeks) treatment are depicted. Magnification: 200x; scale bars: 100 $\mu$m. (e) After 24 weeks of HFD feeding, LysCre/c-Met$^{mut}$ animals displayed a significantly impaired glucose tolerance shown by a stronger increase in serum glucose levels compared to equally treated c-Met$^{fl/fl}$ mice ($n = 8$) ($^*p < 0.05$). (f) Intrahepatic triglyceride levels were determined in livers of chow, MCD (4 weeks), or HFD (24 weeks) fed c-Met$^{fl/fl}$ and LysCre/c-Met$^{mut}$ mice. At least 5 animals per group were included ($^*p < 0.05$).

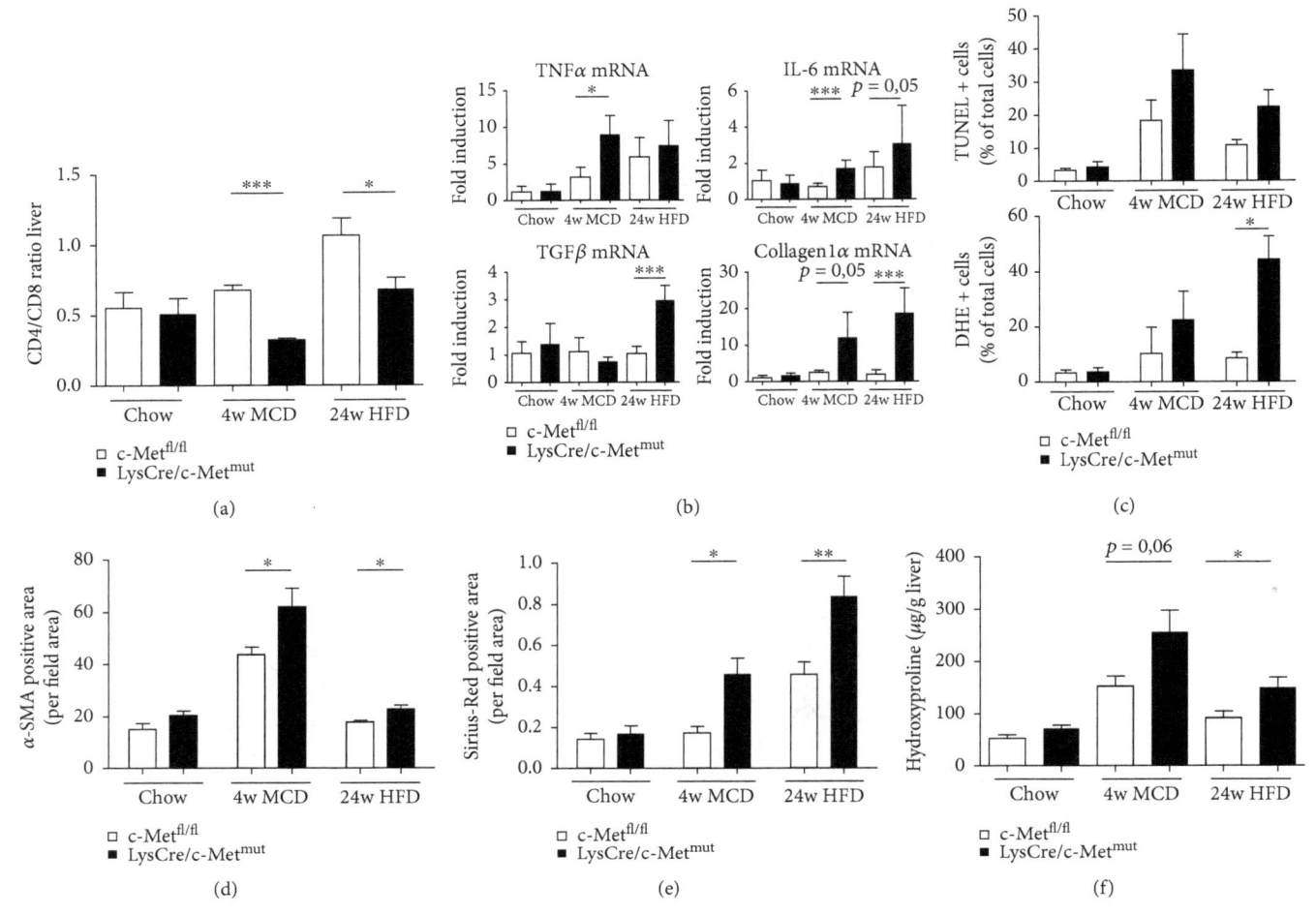

FIGURE 2: Stronger proinflammatory immune response and fibrosis development in livers of LysCre/c-Met$^{mut}$ animals. (a) Intrahepatic CD4$^+$ and CD8$^+$ T cells were analyzed by flow cytometry after chow, 4 weeks of MCD, or 24 weeks of HFD feeding of c-Met$^{fl/fl}$ and LysCre/c-Met$^{mut}$ mice. CD4$^+$ and CD8$^+$ T cells were gated by FSC/SSC (duplets were excluded), live/CD45$^+$, CD4$^+$, or CD8$^+$. A statistical analysis of the ratio of CD4$^+$/CD8$^+$ T cells of recorded cells was performed ($n = 5$) (*$p < 0.05$, ***$p < 0.001$). (b) mRNA expression levels of TNF-$\alpha$, IL-6, TGF-$\beta$, and Collagen1$\alpha$. Whole liver homogenates of c-Met$^{fl/fl}$ or LysCre/c-Met$^{mut}$ mice were analyzed via real-time PCR. The quantification is expressed as fold induction over the mean values obtained for chow fed c-Met$^{fl/fl}$ livers. At least 6 animals per group were included (*$p < 0.05$, ***$p < 0.001$). (c) Statistical analysis of the percentage of TUNEL$^+$ and DHE$^+$ cells referred to the number of total cells on stained liver sections of c-Met$^{fl/fl}$ and LysCre/c-Met$^{mut}$ mice treated either with chow or steatohepatitis-induced diets. 10 view fields/liver of at least $n = 4$ animals per genotype and time point were included (scale bars: 100 $\mu$m; magnification: 200x) (*$p < 0.05$). (d) Quantitative analysis of $\alpha$-SMA-stained liver cryosections of c-Met$^{fl/fl}$ and LysCre/c-Met$^{mut}$ mice after chow, MCD (4 weeks), or HFD (24 weeks) feeding. Scale bars: 100 $\mu$m; magnification: 200x. Quantitative analysis of $\alpha$-SMA staining was calculated by evaluating the $\alpha$-SMA$^+$ area per field area of 10 view fields/liver by ImageJ© (*$p < 0.05$). $n = 4$ animals per group and genotype were included. (e) For quantitative analysis of Sirius Red staining, the Sirius Red-positive area per view field of 10 view fields/liver of chow, MCD (4 weeks), or HFD (24 weeks) fed c-Met$^{fl/fl}$ or LysCre/c-Met$^{mut}$ animals was analyzed and recorded under polarized light by ImageJ© (*$p < 0.05$, **$p < 0.01$). Included were at least $n = 4$ animals per genotype and time point. (f) Displayed are hydroxyproline levels of chow, MCD (4 weeks), or HFD (24 weeks) fed c-Met$^{fl/fl}$ and LysCre/c-Met$^{mut}$ mice (*$p < 0.05$) ($n = 4$).

experiment, it should be investigated which cell types are affected by the deletion of c-Met under the GFAP promoter. In the literature, GFAP is described to be primarily expressed on neural glia cells, especially astrocytes, and on hepatic stellate cells (HSCs) [31]. A multiplex immunofluorescence staining approach for DAPI, c-Met, $\alpha$-SMA (activated hepatic stellate cells), and CK19 (cells of the hepatobiliary tract [32]) in c-Met$^{fl/fl}$ and GFAPCre/c-Met$^{mut}$ animals revealed detailed insight on the targeted cell types and the knockout efficiency (Supplementary Figure 4). Multispectral image analysis with a pattern recognition learning algorithm combined with a trained cell classifier tool identified a c-Met deficiency in GFAPCre/c-Met$^{mut}$ mice, especially in $\alpha$-SMA$^+$ and CK19$^+$ cells (Figure 3(a), Supplementary Figures 5 and 6). Further analysis of the inflammatory status of these mice under steatohepatitis-induced diets (MCD, HFD) showed a protective effect of c-Met in the targeted liver cells ($\alpha$-SMA$^+$/CK19$^+$ cells) which gets evident by significantly increased serum transaminase levels in GFAPCre/c-Met$^{mut}$ animals compared to equally treated c-Met$^{fl/fl}$ controls (Figure 3(b)). Histological analysis (H&E, Oil Red O staining) further demonstrate ballooned hepatocytes and fine and

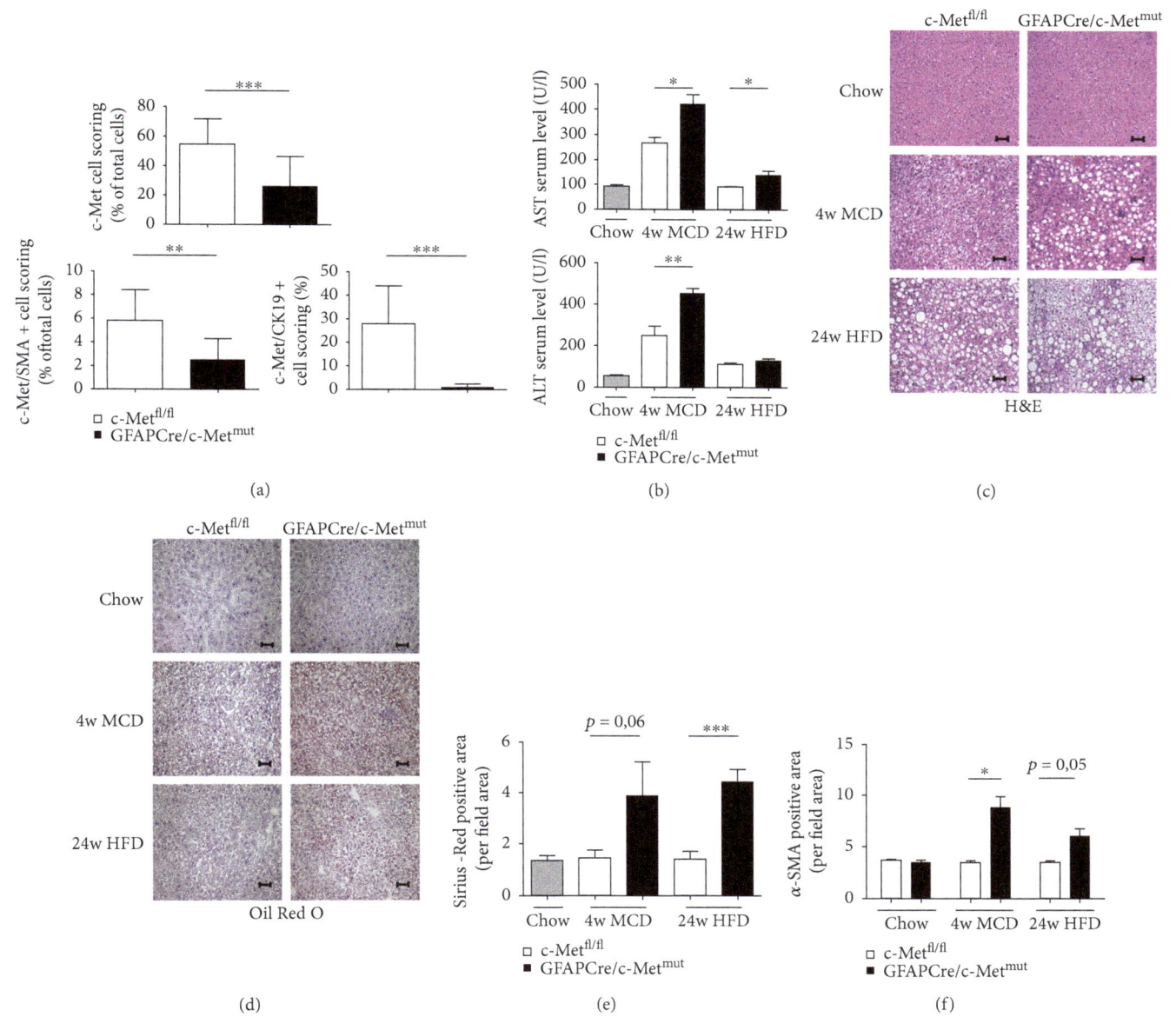

Figure 3: More pronounced disease progression of diet-induced steatohepatitis in GFAPCre/c-Met$^{mut}$ animals. (a) Quantitative evaluation of a multiplex staining approach with simultaneous staining of DAPI, c-Met, α-SMA, and CK19 of untreated c-Met$^{fl/fl}$ and GFAPCre/c-Met$^{mut}$ animals. Statistical analysis of the percentage of c-Met single positive cells, c-Met/α-SMA double positive cells, and c-Met/CK19 double positive cells referred to the number of total cells is depicted. At least 10 view fields/liver were included at a magnification of 200x (**$p < 0.01$, ***$p < 0.001$). (b) Serum transaminase levels (AST and ALT) of c-Met$^{fl/fl}$ and GFAPCre/c-Met$^{mut}$ mice were analyzed after chow, MCD (4 weeks), and HFD (24 weeks) feeding. Serum transaminases increase after treatment with steatohepatitis-induced diets ($n = 5$) (*$p < 0.05$, **$p < 0.01$). (c) H&E-stained liver sections of c-Met$^{fl/fl}$ and GFAPCre/c-Met$^{mut}$ animals (chow, MCD (4 w), and HFD (24 w)) show an increase in lipid droplets in GFAPCre/c-Met$^{mut}$ mice. Magnification: 200x; scale bars: 100 μm. (d) Representative images of Oil Red O-stained liver sections from c-Met$^{fl/fl}$ or GFAPCre/c-Met$^{mut}$ mice after chow, MCD (4 weeks), and HFD (24 weeks) treatment are depicted. Magnification: 200x; scale bars: 100 μm. (e) Quantitative analysis of Sirius Red staining of c-Met$^{fl/fl}$ and GFAPCre/c-Met$^{mut}$ mice after chow, 4 weeks of MCD, or 24 weeks of HFD diet. The Sirius Red-positive area was calculated per view field of 10 view fields/liver by ImageJ© (***$p < 0.001$). At least $n = 4$ animals per group and genotype were included. (f) Quantitative analysis of α-SMA-stained liver cryosections of c-Met$^{fl/fl}$ and GFAPCre/c-Met$^{mut}$ mice after chow, MCD (4 weeks), or HFD (24 weeks) feeding. Scale bars: 100 μm; magnification: 200x. Quantitative analysis of α-SMA staining was calculated by evaluating the α-SMA$^+$ area per field area of 10 view fields/liver by ImageJ© (*$p < 0.05$). $n = 4$ animals per group and genotype were included.

coarse fat droplet formation in c-Met$^{fl/fl}$ animals which is more pronounced in the GFAPCre/c-Met$^{mut}$ after MCD (4 weeks) and HFD (24 weeks) feeding (Figures 3(c) and 3(d)). To assess the effect of a c-Met knockout under the GFAP promoter especially in α-SMA$^+$ cells, we next performed a Sirius Red staining showing a stronger progressing fibrosis development in GFAPCre/c-Met$^{mut}$ mice compared to c-Met$^{fl/fl}$ controls in both dietary models (Figure 3(e), Supplementary Figure 7A). In line with these results, immunofluorescence staining for α-SMA reflected stronger fibrosis initiation in GFAPCre/c-Met$^{mut}$ animals after MCD treatment (Figure 3(f), Supplementary Figure 7B).

The data suggest that a deficiency of c-Met in different liver cell types expressing the GFAP promoter leads to a worsened disease outcome in two independent mouse models of steatohepatitis.

*3.5. Deletion of c-Met in Bone Marrow-Derived Immune Cells Leads to Severe Development and Progression of Steatohepatitis.* Next, we investigated a targeted c-Met deletion in hematopoietic cells under the inducible (Mx dynamin-like GTPase Cre) MxCre promoter. After bone marrow transplantation, the efficiency of the c-Met deletion was investigated by a knockout PCR approach. The 372 bp PCR product shows the effective deletion of c-Met either in the donor or in the associated recipient (Figure 4(a)). Surprisingly, c-Met deficiency in hematopoietic cells (MxCre/c-Met$^{mut}$) resulted in severe maintenance of survival compared to controls (c-Met$^{fl/fl}$) (Figure 4(b)). In support of the idea that c-Met is protective in different liver cell types, histological analysis of H&E and Oil Red O staining again displayed massive fatty liver degeneration and earlier and stronger signs of steatosis in MxCre/c-Met$^{mut}$ animals after MCD and HFD treatment compared to floxed controls (c-Met$^{fl/fl}$) (Figures 4(c) and 4(d)). Representative images illustrated more ballooned hepatocytes and fine and coarse dropping steatosis in mice deficient of c-Met in hematopoietic cells whereas c-Met$^{fl/fl}$ mice showed fewer hepatocyte ballooning and just fine fat droplet formation. The increase in steatosis development was further reflected by slightly increased serum aspartate aminotransferase (AST) and alanine aminotransferase (ALT) levels (Figure 4(e)) and higher hepatic triglyceride levels especially after 24-week HFD treatment (Figure 4(f)) in MxCre/c-Met$^{mut}$ compared to c-Met$^{fl/fl}$ mice. To specifically investigate whether the worsened disease outcome was associated with stronger immune cell infiltration, flow cytometric analysis was performed. As hepatic inflammation is one of the main factors differentiating simple steatosis from progressive steatohepatitis, the investigation of the hepatic infiltration of CD4$^+$ and CD8$^+$ T cells showed significant differences in the ratio of these cell types. After 4 weeks of MCD treatment, a significant shift to a more dominant CD8$^+$ T cell response was evident in MxCre/c-Met$^{mut}$ animals. On the contrary, after 24-week HFD feeding, the T cell response reverses to a CD4$^+$ cell-prone immune response in the MxCre/c-Met$^{mut}$ (Figure 5(a)). To test whether this contributes to a pro- or anti-inflammatory cytokine profile, we next determined the mRNA expression level of IL-6 which was significantly upregulated in MxCre/c-Met$^{mut}$ in both treatment groups compared to c-Met$^{fl/fl}$ controls (Figure 5(b)).

To elucidate the disparity of steatohepatitis development and manifestation between controls and mice lacking c-Met on hematopoietic cells, we tried to further characterize potential underlying mechanisms in more detail. In animals with a specific deletion of c-Met in Kupffer cells, we previously observed differences in the oxidative stress milieu which may trigger hepatic inflammation in the two dietary mouse models of steatohepatitis (MCD, HFD). To investigate whether this is a general phenomenon when c-Met signaling is blocked, we also performed TUNEL and DHE staining in c-Met$^{fl/fl}$ and MxCre/c-Met$^{mut}$ mice after chow, MCD, or HFD feeding. Comparable to what was found in c-Met$^{fl/fl}$ and LysCre/c-Met$^{mut}$ animals, mice deficient of c-Met in hematopoietic cells displayed a significantly greater amount of TUNEL- and DHE-positive cells whereas no difference in the expression of the antioxidative regulator Nrf2 could be detected after MCD and HFD treatment (Figure 5(c), Supplementary Figures 8A, 8B, and 8C). Collectively, these data suggest a general modulatory effect of c-Met on any intrahepatic cell on the oxidative microenvironment associated with exacerbated steatohepatitis.

*3.6. Stronger Fibrosis Progression in MxCre/c-Met$^{mut}$ Mice after HFD and MCD Feeding.* The progression from fatty liver degeneration to the excessive accumulation of extracellular matrix within liver tissue reflects the typical course of NASH disease development. To assess the progression towards fibrosis, the mRNA expression of TGF-β, Collagen1a1, and α-SMA was measured via real-time PCR. All investigated markers showed increases in MxCre/c-Met$^{mut}$ animals compared to c-Met$^{fl/fl}$ controls after dietary treatments whereas the increase was clearly more pronounced in MCD fed mice compared to HFD fed mice. HFD-treated animals displayed only a minor but visible upregulation (Figure 5(b)). In addition, we found a significant increase in intrahepatic α-SMA levels (Figure 5(d), Supplementary Figure 9A), Sirius Red-positive collagen fibers (Figure 5(e), Supplementary Figure 9B), and a moderately upregulated hydroxyproline content (Figure 5(f)) in MxCre/c-Met$^{mut}$ mice after MCD and HFD.

Taken together, the data clearly shows that c-Met in Kupffer cells/macrophages, α-SMA$^+$/CK19$^+$ cells, and bone marrow-derived immune cells is involved in pushing the intrahepatic microenvironment to more oxidative stress finally leading to severe development of steatohepatitis and the progression to fibrosis.

*3.7. c-Met Is Increased in Human Cirrhotic Patients but Drops in NASH Patients.* In two different mouse models of diet-induced steatohepatitis, we found strong development of nonalcoholic steatohepatitis and progression to fibrosis when c-Met is missing in any liver cell type. To confirm the dominant role of c-Met on nonclassical monocytes during fibrosis development, we correlated the expression of c-Met on CD16$^+$ immune cells with the NAFLD activity score (NAS) of patient samples. The correlation clearly demonstrates a decrease in c-Met expression with less severe disease development (Figure 6(a)). Supporting this data, immunohistochemistry analysis of human patient samples showed strong c-Met expression in patients suffering from HBV (hepatitis B virus), HCV (hepatitis C virus), PBC (primary biliary cholangitis), and PSC (primary sclerosing cholangitis). However, patients with alcoholic cirrhosis and NASH patients display

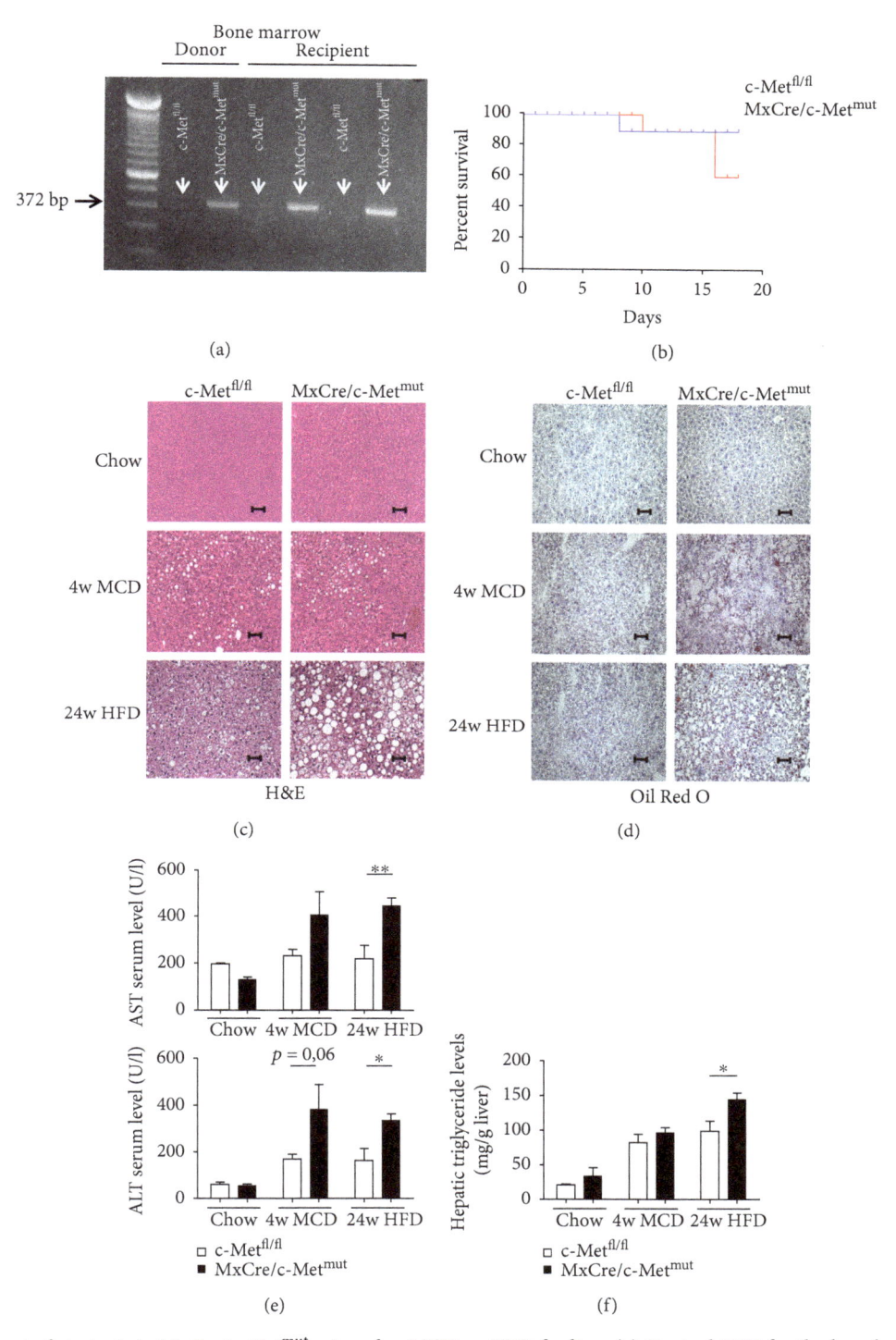

FIGURE 4: Earlier onset of steatosis in MxCre/c-Met$^{mut}$ mice after MCD or HFD feeding. (a) Control PCR for the knockout PCR product of c-Met$^{fl/fl}$ and MxCre/c-Met$^{mut}$ mice. DNA of isolated bone marrow after bone marrow transplantation was used as template. The 372 bp product represents the MxCre/c-Met$^{mut}$ allele. (b) After bone marrow transplantation, it was particularly noticed that nearly 50% MxCre/c-Met$^{mut}$ mice died within 16 days. Therefore, survival curves were calculated for c-Met$^{fl/fl}$ and MxCre/c-Met$^{mut}$ animals after bone marrow transplantation. (c) Representative H&E-stained liver sections of c-Met$^{fl/fl}$ and MxCre/c-Met$^{mut}$ animals (chow, MCD (4 w), and HFD (24 w)) show an increase in steatosis development in MxCre/c-Met$^{mut}$ mice after treatment with steatohepatitis-induced diets. Magnification: 200x; scale bars: 100 μm. (d) Oil Red O-stained liver sections from c-Met$^{fl/fl}$ and MxCre/c-Met$^{mut}$ mice after chow, MCD (4 weeks), and HFD (24 weeks) treatment are depicted. Magnification: 200x; scale bars: 100 μm. (e) AST and ALT serum levels of c-Met$^{fl/fl}$ and MxCre/c-Met$^{mut}$ mice after chow, 4 weeks of MCD, and 24 weeks of HFD feeding. Serum transaminases increase after treatment with chronic liver injury-induced diets ($n = 6$) ($^*p < 0.05$, $^{**}p < 0.01$). (f) Intrahepatic triglyceride levels were determined in livers of chow, MCD (4 weeks), or HFD (24 weeks) fed c-Met$^{fl/fl}$ and MxCre/c-Met$^{mut}$ mice. At least 6 animals per group were included ($^*p < 0.05$).

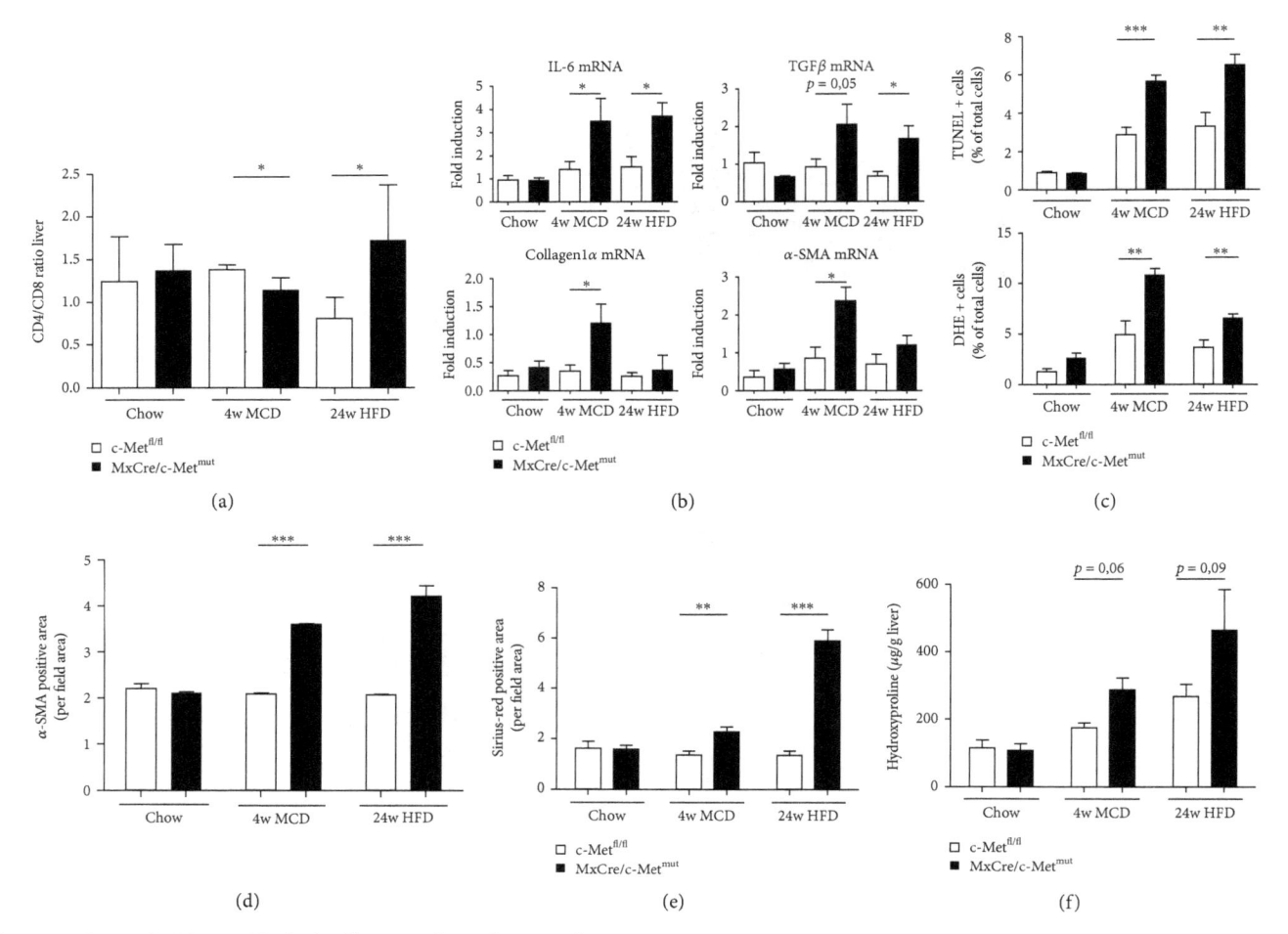

FIGURE 5: Loss of c-Met on MxCre+ cells strengthens the proinflammatory immune response after chronic liver injury. (a) Intrahepatic CD4+ and CD8+ T cells were analyzed by flow cytometry after chow, 4 weeks of MCD, or 24 weeks of HFD feeding of c-Met$^{fl/fl}$ and MxCre/c-Met$^{mut}$ mice. CD4+ and CD8+ T cells were gated by FSC/SSC (duplets were excluded), live/CD45+, CD4+, or CD8+. A statistical analysis of the ratio of CD4+/CD8+ T cells of recorded cells was performed ($n = 5$) (*$p < 0.05$). (b) mRNA expression levels of IL-6, TGF-$\beta$, Collagen1$\alpha$, and $\alpha$-SMA. Whole liver homogenates of c-Met$^{fl/fl}$ or MxCre/c-Met$^{mut}$ mice were analyzed via real-time PCR. The quantification is expressed as fold induction over the mean values obtained for chow fed c-Met$^{fl/fl}$ livers. At least 5 animals per group were included (*$p < 0.05$). (c) Statistical analysis of the percentage of TUNEL+ and DHE+ cells referred to the total number of cells on stained liver sections of c-Met$^{fl/fl}$ and MxCre/c-Met$^{mut}$ mice treated either with chow or steatohepatitis-induced diets. 10 view fields/liver of at least $n = 5$ animals per genotype and time point were included (scale bars: 100 $\mu$m; magnification: 200x) (**$p < 0.01$, ***$p < 0.001$). (d) Quantitative analysis of $\alpha$-SMA-stained liver cryosections of c-Met$^{fl/fl}$ and MxCre/c-Met$^{mut}$ mice after chow, MCD (4 weeks), or HFD (24 weeks) feeding. Scale bars: 100 $\mu$m; magnification: 200x. Quantitative analysis of $\alpha$-SMA staining was calculated by evaluating the $\alpha$-SMA+ area per field area of 10 view fields/liver by ImageJ© (*$p < 0.05$). $n = 4$ animals per group and genotype were included. (e) For quantitative analysis of Sirius Red staining, the Sirius Red-positive area per view field of 10 view fields/liver of chow, MCD (4 weeks), or HFD (24 weeks) fed c-Met$^{fl/fl}$ or MxCre/c-Met$^{mut}$ animals was analyzed and recorded under polarized light by ImageJ© (**$p < 0.01$, ***$p < 0.001$). Included were at least $n = 5$ animals per genotype and time point. (f) Hydroxyproline levels of chow, MCD (4 weeks), or HFD (24 weeks) fed c-Met$^{fl/fl}$ and MxCre/c-Met$^{mut}$ mice are displayed ($n = 5$).

decreased c-Met levels (Figure 6(b)). In addition to this, immunohistochemical staining of c-Met in human samples reflected a strong c-Met expression in patients with HVC-related cirrhosis (Figure 6(c)) and nearly no c-Met expression in NASH patients (Figure 6(d)).

## 4. Discussion

The first transgenic mouse experiments with c-Met were performed in 1996 by Liang and colleagues [33]. In this study, mice exhibited a constitutive activation of the c-Met protein caused by constant dimerization. This overactivation leads to the spontaneous development of mammary hyperplasia and tumors, as a direct effect of the transgene at 6–9 months of age. Also, in primary tumors of human patient samples (liver metastasis from colon carcinoma [34], non-small-cell lung carcinoma [35, 36]), c-Met overexpression and constitutive kinase activity were present. Mechanistically, it could be shown that the overexpression of c-Met is driven by hypoxia in the center of the growing tumor [37]. On the contrary,

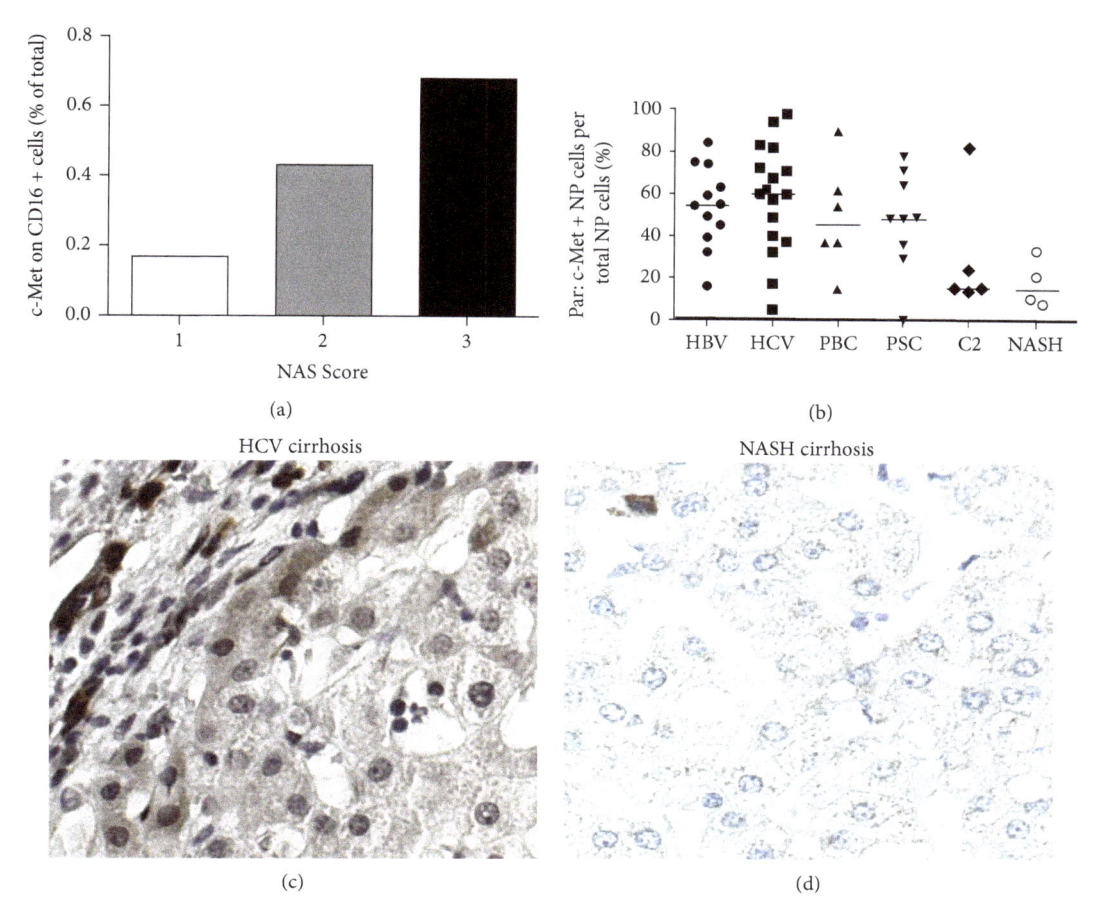

FIGURE 6: c-Met expression is decreased in NASH patients. (a) c-Met expression of intrahepatic nonclassical human monocytes analyzed by flow cytometry correlated with the NAFLD activity score (NAS). (b) c-Met expression on nonparenchymal (NP) cells of patients with HBV (hepatitis B virus), HCV (hepatitis C virus), PBC (primary biliary cholangitis), PSC (primary sclerosing cholangitis), C2 (alcoholic) cirrhosis, and NASH was analyzed by immunohistochemistry. Immunohistochemical staining of c-Met on liver sections of patients with (c) HCV-related cirrhosis and (d) NASH cirrhosis.

it could be shown that HGF, the only known ligand for c-Met, conveys protective effects in different chronic diseases such as liver cirrhosis and lung fibrosis [38–40].

In the previous study, our group could show that c-Met plays a pivotal role during the development of NASH in a hepatocyte-specific knockout mouse model [41]. Nevertheless, the contribution of c-Met on other liver cell types and on fibrosis initiation besides hepatocytes is only poorly understood.

Therefore, the aim of the present study was to investigate the significance of c-Met on different intrahepatic and infiltrating cell types in the initiation and progression of NASH-related fibrosis. Ueki et al. could show in 1999 that the activation of the HGF/c-Met signaling pathway decreases TGF-$\beta$1 levels. Based on these findings, they proposed HGF/c-Met gene therapy for the treatment of patients with liver cirrhosis.

We here hypothesized that c-Met prevents from NASH-related fibrosis and wanted to clarify in detail on which cell type (LysCre for Kupffer cells/macrophages, GFAPCre for $\alpha$-SMA$^+$ and CK19$^+$ cells, and MxCre for bone marrow-derived immune cells) c-Met is essentially involved in this process.

A large number of experiments in different inflammatory diseases show controversial effects of the activation of the HGF/c-Met axis. In atherosclerosis, HGF was found to be associated with disease progression [42], whereas in a pancreatitis model in mice and in drug-induced oxidative liver damage [43], results displayed protective effects of the administration of recombinant human HGF.

As hepatic manifestation of the metabolic syndrome, the development of NASH is associated with an impaired insulin metabolism. Varkaris et al. could show in 2013 that the insulin-like growth factor-1 (IGF-1) signaling through the IGF-1 receptor is able to delay the induction of c-Met activation, thereby showing a direct interaction between the HGF/c-Met signaling axis with metabolic changes as they can occur in steatohepatitis [44]. Thus, strengthening the hypothesis of a direct involvement of c-Met in diet-induced NASH fibrosis. In this study, the loss of c-Met on different liver cell types (Kupffer cells/macrophages, $\alpha$-SMA$^+$/CK19$^+$ cells, and bone marrow-derived immune cells) led to severe steatosis, steatohepatitis, and fibrosis progression in two different mouse models of diet-induced steatohepatitis in all three used Cre lines (Figures 1–5). Further, human patients with NASH show nearly no c-Met expression in liver tissue

(Figure 6). These results support the idea that c-Met protects from fibrosis. Strong fat droplet accumulation and the more pronounced inflammatory response are further associated with an increased number of apoptotic cells together with a strong activation of the oxidative stress response in livers of mice with a cell-specific c-Met deficiency after treatment with NASH-induced diets. Xiao et al. could show in 2001 that c-Met signals via AKT/protein kinase B are primarily responsible for cell survival [45]. Additionally, previous work of our group further demonstrated that HGF/c-Met actively regulates hepatocyte differentiation [46]. Both studies support the finding that the loss of c-Met on different liver cells (Kupffer cells/macrophages, $\alpha$-SMA$^+$/CK19$^+$ cells, and bone marrow-derived immune cells) leads to an increase in the oxidative stress response and apoptotic cell death under dietary treatment.

It is well described that the infiltration of proinflammatory monocytes contributes to NASH progression [47]. The role of HGF/c-Met signaling on liver resident and infiltrating monocytes during NASH development however remains unclear. The loss of c-Met on monocytes expressing the LysMCre promoter leads to a deterioration in the course of liver disease in MCD- as well as in HFD-induced steatohepatitis.

LysCre/c-Met$^{mut}$ animals display earlier fat droplet accumulation, a stronger onset of insulin resistance, and an increase in the proinflammatory immune response finally leading to massive extracellular matrix deposition (Figures 2 and 3). In 2012, Ishikawa et al. could show that c-Met deficiency leads to defects in the mobilization of Kupffer cells in the DDC model of chronic liver injury [48]. Based on this, we hypothesize less stimulation of the anti-inflammatory and antioxidative stress response leading to an increase in apoptosis, oxidative stress, and in the end fibrosis progression.

Previous studies could show that hepatic stellate cells (HSCs) that express the GFAPCre promotor play a major role in the development of liver fibrosis [49]. Under inflammatory conditions, they transdifferentiate into myofibroblast-like cells producing extracellular matrix proteins which accumulate within liver tissue [50]. To which extent the activation of the HGF/c-Met axis is involved in this process is not well understood yet. GFAPCre/c-Met$^{mut}$ mice exhibit severe steatohepatitis development after dietary treatment with increased transaminase levels, histomorphological changes, and the accumulation of extracellular matrix components (Figure 4). Previous studies in our group could show that the loss of c-Met on hepatocytes leads to chronic tissue damage and fibrosis progression in the bile duct ligation model of chronic liver injury [11]. This effect is based on an imbalance between antiapoptotic functions of c-Met and its proproliferative capacity. In c-Met$^{\Delta Hepa}$ mice, this lead to an increase in the proinflammatory immune response and strong activation of hepatic stellate cells. The same mechanism seems to be active when c-Met is missing on $\alpha$-SMA$^+$/CK19$^+$ cells directly. Strong activation of the proinflammatory immune response drives an increased activation of hepatic stellate cells and thereby the accumulation of extracellular matrix.

The infiltration of immune cells is one of the characteristic differences between simple steatosis and the more severe steatohepatitis. Recent investigations describe an immunoregulatory function of c-Met on hematopoietic progenitor cells [51, 52]. Further studies additionally show the expression of c-Met also on neutrophils and CD8$^+$ T cells upon proinflammatory cytokine exposure during tissue injury [53, 54]. In this study, we examined the role of c-Met on infiltrating bone marrow-derived immune cells to the liver in two dietary mouse models of steatohepatitis. Our experiments clearly show that the loss of c-Met on migrating immune cells carrying the MxCre promotor leads to severe steatosis, inflammation, and fibrosis development (Figures 5 and 6). The recent study of Wang et al. shows that the overexpression of c-Met on bone marrow-derived mesenchymal stem cells leads to an improvement of their homing capacity, their mobilization potential, and their repair function in acute liver failure in rats [55]. Defective migration for its part leads to the absence of the activation of anti-inflammatory immune responses resulting in the observed proinflammatory phenotype. This is further reflected by a significant increase in apoptotic cell death within liver tissue together with massive production of reactive oxygen species (ROS) causing tissue damage and thus fibrosis.

Taken together, our findings surprisingly show that we could find no difference in the degree and outcome of diet-induced steatohepatitis initiation and progression independent on which cell type c-Met is lacking. This indicates a global role of c-Met in a cross talk between intrahepatic cells involved in steatohepatitis development and a protective effect of c-Met in fibrosis generation. Mechanistically, the knockout of c-Met leads to an imbalance of different pro- and anti-inflammatory pathways with an increase in apoptotic cell death together with a strong activation of ROS production which may in part explain the increase in liver injury leading to severe fibrosis progression.

On the contrary, there is a growing quantity of clinical trials, e.g., ficlatuzumab, which is a humanized HGF inhibitory monoclonal antibody preventing HGF/c-Met signaling by blocking the ligand-mediated activation which is actually in a phase 1b clinical trial for the treatment of non-small-cell lung cancer [56]. Further, there is an increasing body of interest in testing different nonselective and selective c-Met inhibitors in the treatment of HCC [26]. Therefore, the role of c-Met signaling in the chronically injured liver seems to be contextual and also tissue specific.

## Abbreviations

ALT: Alanine aminotransferase
AST: Aspartate aminotransferase
DHE: Dihydroethidium (hydroethidine)
GFAP: Glial fibrillary acid protein
HBV: Hepatitis B virus
HCC: Hepatocellular carcinoma
HCV: Hepatitis C virus
HFD: High-fat diet
HGF: Hepatocyte growth factor
HSC: Hepatic stellate cell

KO: Knockout
MCD: Methionine-choline-deficient diet
NAFLD: Nonalcoholic fatty liver disease
NAS: NAFLD activity score
NASH: Nonalcoholic steatohepatitis
PBC: Primary biliary cholangitis
PSC: Primary sclerosing cholangitis
ROS: Reactive oxygen species
TGF-$\beta$: Transforming growth factor beta.

## Authors' Contributions

HKD did the study design, data acquisition, data analysis, and drafting of the manuscript. DCK and KLS performed the study design, fundraising, drafting of the manuscript, and study supervision. TS, FS, and MB did the data acquisition, data analysis, and manuscript revision. TL and TH were assigned for the technical and material support and manuscript revision. CT performed the critical revision of the manuscript and intellectual content.

## Acknowledgments

This work was supported by the SFB/TRR 57 of the Deutsche Forschungsgemeinschaft (DFG).

## Supplementary Materials

Supplementary Figure 1: (A) genomic PCR genotyping using oligos for recombined and nonrecombined alleles. Comparison of knockout PCR products using c-Met$^{fl/fl}$ and LysCre/c-Met$^{mut}$ mouse DNA of isolated primary hepatocytes, $\alpha$-SMA+/CK19+ cells and bone marrow-derived immune cells. GFAPCre/c-Met$^{mut}$ mouse DNA of isolated primary hepatocytes, Kupffer cells/macrophages and bone marrow-derived immune cells. MxCre/c-Met$^{mut}$ mouse DNA of isolated primary hepatocytes, $\alpha$-SMA+/CK19+ cells and Kupffer cells/macrophages. The 500 bp product represents the c-Met$^{fl/fl}$ allele; the 800 bp product shows the c-Met$^{mut}$ genotype. (B) Comparison of knockout PCR products using c-Met$^{fl/fl}$ and LysCre/c-Met$^{mut}$ mouse DNA of isolated primary Kupffer cells/macrophages. The 500 bp product represents the c-Met$^{fl/fl}$ allele; the 800 bp product shows the c-Met$^{mut}$ genotype. (C) Comparison of knockout PCR products using c-Met$^{fl/fl}$ and GFAP/c-Met$^{mut}$ mouse DNA of isolated primary $\alpha$-SMA+/CK19+ cells. The 500 bp product represents the c-Met$^{fl/fl}$ allele; the 800 bp product shows the c-Met$^{mut}$ genotype. (D) Comparison of knockout PCR products using c-Met$^{fl/fl}$ and Mx/c-Met$^{mut}$ mouse DNA of isolated primary bone marrow-derived immune cells. The 500 bp product represents the c-Met$^{fl/fl}$ allele; the 800 bp product shows the c-Met$^{mut}$ genotype. Supplementary Figure 2: representative gating of the analysis of intrahepatic CD4+ and CD8+ T cells in c-Met$^{fl/fl}$ animals after HFD treatment for 24 weeks. Cells are gated FSC/SSC, FSC-W/FSC-A, live dead/CD45+, CD4+, or CD8+. Supplementary Figure 3: representative photographs of (A) TUNEL-, (B) DHE-, (C) $\alpha$-SMA-, and (D) Sirius Red-stained liver sections of c-Met$^{fl/fl}$ and LysCre/c-Met$^{mut}$ mice after chow, MCD (4 weeks), or HFD (24 weeks) feeding. Scale bars: 100 $\mu$m; magnification: 200x. Supplementary Figure 4: images of multiplexed immunofluorescence-stained liver sections (DAPI, c-Met, alpha-SMA, and CK19) of c-Met$^{fl/fl}$ and GFAPCre/c-Met$^{mut}$ animals. Magnification: 200x and 400x. Together with the corresponding images of the single channels of the multiplexed immunofluorescence-stained liver sections (DAPI, c-Met, alpha-SMA, and CK19) of c-Met$^{fl/fl}$ and GFAPCre/c-Met$^{mut}$ mice. Magnification: 400x. Supplementary Figure 5: scans of the multispectral image acquisition and analysis showing the raw image, cell segmentation, phenotyping, and scoring of a c-Met$^{fl/fl}$ and GFAPCre/c-Met$^{mut}$ mouse. Supplementary Figure 6: (A) 2D and (B) 3D scattered plots based on the analysis of the mean fluorescent intensity of c-Met$^{fl/fl}$ and GFAPCre/c-Met$^{mut}$ animals with MATLAB R2016b (MathWorks, Natick, USA). Supplementary Figure 7: representative images of (A) Sirius Red-stained liver sections and (B) of $\alpha$-SMA-stained liver sections of c-Met$^{fl/fl}$ and GFAPCre/c-Met$^{mut}$ mice after dietary treatment (chow, MCD 6 weeks, and HFD 24 weeks). Scale bars: 100 $\mu$m; magnification: 200x. Supplementary Figure 8: representative photographs of (A) TUNEL-, (B) DHE-, and (C) Nrf2-stained liver sections of c-Met$^{fl/fl}$ and MxCre/c-Met$^{mut}$ mice after chow, MCD (4 weeks), or HFD (24 weeks) feeding. Scale bars: 100 $\mu$m; magnification: 200x. Supplementary Figure 9: representative photographs of (A) $\alpha$-SMA- and (B) Sirius Red-stained liver sections of c-Met$^{fl/fl}$ and MxCre/c-Met$^{mut}$ mice after chow, MCD (4 weeks), or HFD (24 weeks) feeding. Scale bars: 100 $\mu$m; magnification: 200x. *(Supplementary Materials)*

## References

[1] C. Estes, H. Razavi, R. Loomba, Z. Younossi, and A. J. Sanyal, "Modeling the epidemic of nonalcoholic fatty liver disease demonstrates an exponential increase in burden of disease," *Hepatology*, vol. 67, no. 1, pp. 123–133, 2018.

[2] L. Vigano, A. Lleo, and A. Aghemo, "Non-alcoholic fatty liver disease, non-alcoholic steatohepatitis, metabolic syndrome and hepatocellular carcinoma-a composite scenario," *HepatoBiliary Surgery and Nutrition*, vol. 7, no. 2, pp. 130–133, 2018.

[3] N. H. Bzowej, "Nonalcoholic steatohepatitis: the new frontier for liver transplantation," *Current Opinion in Organ Transplantation*, vol. 23, no. 2, pp. 169–174, 2018.

[4] J. M. Clark, "The epidemiology of nonalcoholic fatty liver disease in adults," *Journal of Clinical Gastroenterology*, vol. 40, Supplement 1, pp. S5–10, 2006.

[5] J. D. Browning and J. D. Horton, "Molecular mediators of hepatic steatosis and liver injury," *The Journal of Clinical Investigation*, vol. 114, no. 2, pp. 147–152, 2004.

[6] R. J. Wong, R. Cheung, and A. Ahmed, "Nonalcoholic steatohepatitis is the most rapidly growing indication for liver transplantation in patients with hepatocellular carcinoma in the U.S," *Hepatology*, vol. 59, no. 6, pp. 2188–2195, 2014.

[7] R. S. Khan and P. N. Newsome, "Non-alcoholic fatty liver disease and liver transplantation," *Metabolism*, vol. 65, no. 8, pp. 1208–1223, 2016.

[8] P. Angulo, "GI epidemiology: nonalcoholic fatty liver disease," *Alimentary Pharmacology & Therapeutics*, vol. 25, no. 8, pp. 883–889, 2007.

[9] S. S. Choi and A. M. Diehl, "Hepatic triglyceride synthesis and nonalcoholic fatty liver disease," *Current Opinion in Lipidology*, vol. 19, no. 3, pp. 295–300, 2008.

[10] S. L. Organ and M. S. Tsao, "An overview of the c-MET signaling pathway," *Therapeutic Advances in Medical Oncology*, vol. 3, 1_suppl, pp. S7–S19, 2011.

[11] A. Giebeler, M. V. Boekschoten, C. Klein et al., "c-Met confers protection against chronic liver tissue damage and fibrosis progression after bile duct ligation in mice," *Gastroenterology*, vol. 137, no. 1, pp. 297–308.e4, 2009.

[12] K. Fujiwara, A. Matsukawa, S. Ohkawara, K. Takagi, and M. Yoshinaga, "Functional distinction between CXC chemokines, interleukin-8 (IL-8), and growth related oncogene (GRO) alpha in neutrophil infiltration," *Laboratory Investigation*, vol. 82, no. 1, pp. 15–23, 2002.

[13] G. A. Rodrigues and M. Park, "Autophosphorylation modulates the kinase activity and oncogenic potential of the Met receptor tyrosine kinase," *Oncogene*, vol. 9, no. 7, pp. 2019–2027, 1994.

[14] C. Ponzetto, A. Bardelli, Z. Zhen et al., "A multifunctional docking site mediates signaling and transformation by the hepatocyte growth factor/scatter factor receptor family," *Cell*, vol. 77, no. 2, pp. 261–271, 1994.

[15] Y. W. Zhang, L. M. Wang, R. Jove, and G. F. Vande Woude, "Requirement of Stat 3 signaling for HGF/SF-Met mediated tumorigenesis," *Oncogene*, vol. 21, no. 2, pp. 217–226, 2002.

[16] C. Boccaccio, M. Andò, L. Tamagnone et al., "Induction of epithelial tubules by growth factor HGF depends on the STAT pathway," *Nature*, vol. 391, no. 6664, pp. 285–288, 1998.

[17] L. Trusolino, A. Bertotti, and P. M. Comoglio, "MET signalling: principles and functions in development, organ regeneration and cancer," *Nature Reviews. Molecular Cell Biology*, vol. 11, no. 12, pp. 834–848, 2010.

[18] R. Paumelle, D. Tulashe, Z. Kherrouche et al., "Hepatocyte growth factor/scatter factor activates the ETS1 transcription factor by a RAS-RAF-MEK-ERK signaling pathway," *Oncogene*, vol. 21, no. 15, pp. 2309–2319, 2002.

[19] E. D. Fixman, T. M. Fournier, D. M. Kamikura, M. A. Naujokas, and M. Park, "Pathways downstream of Shc and Grb 2 are required for cell transformation by the tpr-Met oncoprotein," *The Journal of Biological Chemistry*, vol. 271, no. 22, pp. 13116–13122, 1996.

[20] Y. Zhang, M. Xia, K. Jin et al., "Function of the c-Met receptor tyrosine kinase in carcinogenesis and associated therapeutic opportunities," *Molecular Cancer*, vol. 17, no. 1, p. 45, 2018.

[21] F. Bladt, D. Riethmacher, S. Isenmann, A. Aguzzi, and C. Birchmeier, "Essential role for the c-met receptor in the migration of myogenic precursor cells into the limb bud," *Nature*, vol. 376, no. 6543, pp. 768–771, 1995.

[22] C. Schmidt, F. Bladt, S. Goedecke et al., "Scatter factor/hepatocyte growth factor is essential for liver development," *Nature*, vol. 373, no. 6516, pp. 699–702, 1995.

[23] Y. Uehara, O. Minowa, C. Mori et al., "Placental defect and embryonic lethality in mice lacking hepatocyte growth factor/scatter factor," *Nature*, vol. 373, no. 6516, pp. 702–705, 1995.

[24] C. Kellendonk, C. Opherk, K. Anlag, G. Schutz, and F. Tronche, "Hepatocyte-specific expression of Cre recombinase," *Genesis*, vol. 26, no. 2, pp. 151–153, 2000.

[25] C. Postic, M. Shiota, K. D. Niswender et al., "Dual roles for glucokinase in glucose homeostasis as determined by liver and pancreatic beta cell-specific gene knock-outs using Cre recombinase," *The Journal of Biological Chemistry*, vol. 274, no. 1, pp. 305–315, 1999.

[26] M. Bouattour, E. Raymond, S. Qin et al., "Recent developments of c-Met as a therapeutic target in hepatocellular carcinoma," *Hepatology*, vol. 67, no. 3, pp. 1132–1149, 2018.

[27] T. Nakamura, K. Sakai, T. Nakamura, and K. Matsumoto, "Hepatocyte growth factor twenty years on: much more than a growth factor," *Journal of Gastroenterology and Hepatology*, vol. 26, Suppl 1, pp. 188–202, 2011.

[28] M. Borowiak, A. N. Garratt, T. Wustefeld, M. Strehle, C. Trautwein, and C. Birchmeier, "Met provides essential signals for liver regeneration," *Proceedings of the National Academy of Sciences of the United States of America*, vol. 101, no. 29, pp. 10608–10613, 2004.

[29] S. Dooley and P. ten Dijke, "TGF-β in progression of liver disease," *Cell and Tissue Research*, vol. 347, no. 1, pp. 245–256, 2012.

[30] L. Yang, Y. S. Roh, J. Song et al., "Transforming growth factor beta signaling in hepatocytes participates in steatohepatitis through regulation of cell death and lipid metabolism in mice," *Hepatology*, vol. 59, no. 2, pp. 483–495, 2014.

[31] G. Maubach, M. C. Lim, C. Y. Zhang, and L. Zhuo, "GFAP promoter directs lacZ expression specifically in a rat hepatic stellate cell line," *World Journal of Gastroenterology*, vol. 12, no. 5, pp. 723–730, 2006.

[32] R. Jain, S. Fischer, S. Serra, and R. Chetty, "The use of cytokeratin 19 (CK19) immunohistochemistry in lesions of the pancreas, gastrointestinal tract, and liver," *Applied Immunohistochemistry & Molecular Morphology*, vol. 18, no. 1, pp. 9–15, 2010.

[33] T. J. Liang, A. E. Reid, R. Xavier, R. D. Cardiff, and T. C. Wang, "Transgenic expression of tpr-met oncogene leads to development of mammary hyperplasia and tumors," *The Journal of Clinical Investigation*, vol. 97, no. 12, pp. 2872–2877, 1996.

[34] M. F. Di Renzo, R. Poulsom, M. Olivero, P. M. Comoglio, and N. R. Lemoine, "Expression of the Met/hepatocyte growth factor receptor in human pancreatic cancer," *Cancer Research*, vol. 55, no. 5, pp. 1129–1138, 1995.

[35] J. Bean, C. Brennan, J. Y. Shih et al., "MET amplification occurs with or without T790M mutations in EGFR mutant lung tumors with acquired resistance to gefitinib or erlotinib," *Proceedings of the National Academy of Sciences of the United States of America*, vol. 104, no. 52, pp. 20932–20937, 2007.

[36] J. A. Engelman, K. Zejnullahu, T. Mitsudomi et al., "MET amplification leads to gefitinib resistance in lung cancer by activating ERBB3 signaling," *Science*, vol. 316, no. 5827, pp. 1039–1043, 2007.

[37] S. Pennacchietti, P. Michieli, M. Galluzzo, M. Mazzone, S. Giordano, and P. M. Comoglio, "Hypoxia promotes invasive growth by transcriptional activation of the met protooncogene," *Cancer Cell*, vol. 3, no. 4, pp. 347–361, 2003.

[38] T. Ueki, Y. Kaneda, H. Tsutsui et al., "Hepatocyte growth factor gene therapy of liver cirrhosis in rats," *Nature Medicine*, vol. 5, no. 2, pp. 226–230, 1999.

[39] M. Watanabe, M. Ebina, F. M. Orson et al., "Hepatocyte growth factor gene transfer to alveolar septa for effective suppression of lung fibrosis," *Molecular Therapy*, vol. 12, no. 1, pp. 58–67, 2005.

[40] M. Yaekashiwa, S. Nakayama, K. Ohnuma et al., "Simultaneous or delayed administration of hepatocyte growth factor equally represses the fibrotic changes in murine lung injury induced by bleomycin. A morphologic study," *American Journal of Respiratory and Critical Care Medicine*, vol. 156, no. 6, pp. 1937–1944, 1997.

[41] D. C. Kroy, F. Schumacher, P. Ramadori et al., "Hepatocyte specific deletion of c-Met leads to the development of severe non-alcoholic steatohepatitis in mice," *Journal of Hepatology*, vol. 61, no. 4, pp. 883–890, 2014.

[42] E. J. Bell, P. A. Decker, M. Y. Tsai et al., "Hepatocyte growth factor is associated with progression of atherosclerosis: the Multi-Ethnic Study of Atherosclerosis (MESA)," *Atherosclerosis*, vol. 272, pp. 162–167, 2018.

[43] M. Palestino-Dominguez, M. Pelaez-Luna, R. Lazzarini-Lechuga et al., "Recombinant human hepatocyte growth factor provides protective effects in cerulein-induced acute pancreatitis in mice," *Journal of Cellular Physiology*, vol. 233, no. 12, pp. 9354–9364, 2018.

[44] A. Varkaris, S. Gaur, N. U. Parikh et al., "Ligand-independent activation of MET through IGF-1/IGF-1R signaling," *International Journal of Cancer*, vol. 133, no. 7, pp. 1536–1546, 2013.

[45] G. H. Xiao, M. Jeffers, A. Bellacosa, Y. Mitsuuchi, G. F. Vande Woude, and J. R. Testa, "Anti-apoptotic signaling by hepatocyte growth factor/Met via the phosphatidylinositol 3-kinase/Akt and mitogen-activated protein kinase pathways," *Proceedings of the National Academy of Sciences of the United States of America*, vol. 98, no. 1, pp. 247–252, 2001.

[46] M. Kaldenbach, A. Giebeler, D. F. Tschaharganeh et al., "Hepatocyte growth factor/c-Met signalling is important for the selection of transplanted hepatocytes," *Gut*, vol. 61, no. 8, pp. 1209–1218, 2012.

[47] O. Krenkel, T. Puengel, O. Govaere et al., "Therapeutic inhibition of inflammatory monocyte recruitment reduces steatohepatitis and liver fibrosis," *Hepatology*, vol. 67, no. 4, pp. 1270–1283, 2018.

[48] T. Ishikawa, V. M. Factor, J. U. Marquardt et al., "Hepatocyte growth factor/c-met signaling is required for stem-cell-mediated liver regeneration in mice," *Hepatology*, vol. 55, no. 4, pp. 1215–1226, 2012.

[49] J. Lin, S. Zheng, A. D. Attie et al., "Perilipin 5 and liver fatty acid binding protein function to restore quiescence in mouse hepatic stellate cells," *Journal of Lipid Research*, vol. 59, no. 3, pp. 416–428, 2018.

[50] D. Benten, J. Kluwe, J. W. Wirth et al., "A humanized mouse model of liver fibrosis following expansion of transplanted hepatic stellate cells," *Laboratory Investigation*, vol. 98, no. 4, pp. 525–536, 2018.

[51] T. Hieronymus, M. Zenke, J. H. Baek, and K. Sere, "The clash of Langerhans cell homeostasis in skin: should I stay or should I go?," *Seminars in Cell & Developmental Biology*, vol. 41, pp. 30–38, 2015.

[52] J. Hubel and T. Hieronymus, "HGF/met-signaling contributes to immune regulation by modulating tolerogenic and motogenic properties of dendritic cells," *Biomedicine*, vol. 3, no. 1, pp. 138–148, 2015.

[53] V. Finisguerra, G. Di Conza, M. Di Matteo et al., "MET is required for the recruitment of anti-tumoural neutrophils," *Nature*, vol. 522, no. 7556, pp. 349–353, 2015.

[54] M. Benkhoucha, N. Molnarfi, G. Kaya et al., "Identification of a novel population of highly cytotoxic c-Met-expressing CD8+ T lymphocytes," *EMBO Reports*, vol. 18, no. 9, pp. 1545–1558, 2017.

[55] K. Wang, Y. Li, T. Zhu et al., "Overexpression of c-Met in bone marrow mesenchymal stem cells improves their effectiveness in homing and repair of acute liver failure," *Stem Cell Research & Therapy*, vol. 8, no. 1, p. 162, 2017.

[56] E. H. Tan, W. T. Lim, M. J. Ahn et al., "Phase 1b trial of ficlatuzumab, a humanized hepatocyte growth factor inhibitory monoclonal antibody, in combination with gefitinib in Asian patients with NSCLC," *Clinical Pharmacology in Drug Development*, vol. 7, no. 5, pp. 532–542, 2018.

# Biological Effects of Tetrahydroxystilbene Glucoside: An Active Component of a Rhizome Extracted from *Polygonum multiflorum*

**Lingling Zhang [1] and Jianzong Chen [2]**

[1]*Translational Medicine Center, Honghui Hospital, Xi'an Jiaotong University, Xi'an 710054, China*
[2]*Traditional Chinese Medicine Department, Xijing Hospital, Fourth Military Medical University, Xi'an 710032, China*

Correspondence should be addressed to Lingling Zhang; sxjxzhangll2018@126.com and Jianzong Chen; chenjz2016@126.com

Academic Editor: Daniele Vergara

*Polygonum multiflorum* Thunb. (PM), a traditional Chinese medicinal herb, has been widely used in the Orient as a tonic and antiaging agent. 2,3,5,4'-Tetrahydroxystilbene-2-O-$\beta$-D-glucoside (TSG, $C_{20}H_{22}O_9$, FW = 406.38928) is one of the active components extracted from PM. TSG is an antioxidant agent, which exhibits remarkable antioxidative activities in vivo and in vitro. The antioxidant effect of TSG is achieved by its radical-scavenging effects. TSG can inhibit apoptosis and protect neuronal cells against injury through multifunctional cytoprotective pathways. TSG performs prophylactic and therapeutic activities against Alzheimer's disease, Parkinson's disease, and cerebral ischemia/reperfusion injury. It is also antiatherosclerotic and anti-inflammatory. However, the mechanisms underlying these pharmacological activities are unclear. This study aimed at reviewing experimental studies and describing the effectiveness and possible mechanisms of TSG.

## 1. Introduction

The root of *Polygonum multiflorum* Thunb. (PM), which is also known as heshouwu, is a famous traditional Chinese medicinal herb that has been used as a tonic and antiaging agent in the Orient. PM and its extract can be used to treat age-related diseases [1–3]. The medicinal effects of PM in the treatment of these age-related diseases are possibly mediated by the antioxidant capacity of this plant [4], because free radical-induced oxidative stress has been implicated in the aging process. PM consists of anthraquinone, stilbene, phospholipid, and other compounds, and modern chromatographic separation studies have demonstrated that many bioactive compounds (e.g., stilbene glycosides) in PM are responsible for its medicinal activities [5]. One of the main components that can be extracted from the root of PM is 2,3,5,4'-tetrahydroxystilbene-2-O-$\beta$-D-glucoside (TSG), which is a stilbene monomer [6]. Since the discovery of TSG by Hata et al. [7], who were Japanese scientists, the chemical components of PM have been widely explored in academic medicine. The pharmacological characteristics of the compound have also been investigated. TSG possesses special biological actions and considerable values in scientific research and clinical medicine. The structure of TSG is similar to that of resveratrol (3,4',5-trihydroxy-*trans*-stilbene), which is well known for its numerous biological activities, especially in cardiovascular protection and neuroprotection. TSG possesses strong antioxidant and free radical-scavenging activities, which are much stronger than those of resveratrol in superoxide anion radical scavenging, hydroxyl radical scavenging, and DPPH radical scavenging [8]. Parkinson's disease (PD) is among the most common age-related neurodegenerative disease. Current pharmacological treatments of PD remain largely symptomatic, and the development of new therapeutic strategies may provide effective alternative treatment options. Vergara et al. [9] provided experimental evidence supporting the beneficial effects of resveratrol in preserving cellular homeostasis in parkin-mutant fibroblasts, which may be relevant for PD treatment. The polyphenolic structure of TSG is similar to that of resveratrol. TSG also elicits neuroprotective effects in various neurodegenerative diseases and cerebral ischemia. In aged mice, TSG treatment rescues synapses and suppresses α-synuclein overexpression in the brain, subsequently improving their memory and movement functions [10]. TSG protects dopamine (DA) neurons against lipopolysaccharide- (LPS-) induced neurotoxicity through dual modulation on glial cells by attenuating microglial-mediated neuroinflammation and enhancing astroglial-derived neurotrophic effects [11]. In an ischemia-reperfusion-injured rat model, TSG promotes the postoperative recovery of rats by minimizing the volume of cerebral infarcts and improving neurological dysfunction, thereby upregulating the expression levels of CD31, angiopoietin 1, and angiopoietin receptor-2 [12]. TSG possesses antiatherosclerosis, anti-inflammatory, and anticardiac fibrotic effects. Although these effects are widely known, the mechanisms of action have yet to be established. Numerous mechanisms, such as activation of adenosine 5′-monophosphate-activated protein kinase [13, 14], protein kinase B (Akt) [15, 16], peroxisome proliferator-activated receptor gamma (PPAR-γ) [17], and silent mating type information regulation 2 homolog 1 [18] and inhibition of the classical nuclear factor-κB (NF-κB) signal [13], mediate the therapeutic effects of TSG. This study aimed at reviewing various experimental studies and describing the effectiveness and possible mechanisms of TSG.

## 2. Antioxidative Effect

Oxidative stress is a phenomenon that induces various health disturbances and diseases by enhancing the oxidation of biologically important molecules in vivo. Oxidation reactions by reactive oxygen species (ROS) are regarded as factors that trigger oxidative stress [19]. They are a group of very reactive short-lived chemicals produced during normal metabolism or after an oxidative reaction. ROS include free radicals, such as superoxide anion ($O_2^{\cdot-}$) and hydroxyl radical ($\cdot OH$), and nonradical molecules, such as hydrogen peroxide ($H_2O_2$) and singlet oxygen ($^1O_2$) [20]. Under physiological conditions, an appropriate level of intracellular ROS should be maintained to achieve redox balance and cell proliferation [21]. However, excessive ROS accumulation is highly cytotoxic because these chemicals induce DNA damage, lipid peroxidation, and protein degradation [20, 22]. Oxidative damages caused by ROS may cause various neurodegenerative and chronic diseases, such as coronary heart diseases, atherosclerosis, cancer, and aging [23]. DPPH is a free radical widely used to test the free radical-scavenging ability of various chemicals. TSG possesses a strong DPPH radical-scavenging activity in vitro with an $IC_{50}$ of 33.24 μM [24]. TSG also exhibits a strong capacity to scavenge $\cdot OH$ and superoxide anion radical with $IC_{50}$ of 2.75 and 0.57 μg/ml, respectively [8]. The crucial components of an antioxidant defense system in the body are cellular antioxidant enzymes (e.g., superoxide dismutase (SOD) and glutathione peroxidase (GSH-Px)), which are involved in the reduction of ROS and peroxides produced in living organisms and in the detoxification of certain exogenous compounds. In human brain microvascular endothelial cells, TSG exhibits cytoprotective effects against $H_2O_2$-induced oxidative stress by inhibiting malondialdehyde (MDA) and ROS and upregulating SOD and GSH [25]. To further evaluate the in vivo antioxidative effects of TSG on rats, Lv et al. [26] administered D-galactose (100 mg/kg/day body weight; single hypodermic injection) to senile rats and applied TSG (20 and 40 mg/kg/day body weight). They found an increase in the activities of SOD and GSH-Px and a decrease in the content of thiobarbituric acid reactive species in the serum and heart, brain, and liver tissues of rats. TSG also reduces serum MDA and tissue 8-OHDG levels and increases serum GSH-Px activity in hypertensive rats aged spontaneously [27]. Hemeoxygenase-1 (HO-1) is a highly inducible and stress-responsive protein (also called heat shock protein 32) that catalyzes the first and rate-limiting step in the degradation of heme, which is a potent oxidant [28]. NADPH-quinone oxidoreductase 1 (NQO1) is a predominantly cytosolic enzyme that provides cells with multiple layers of protection against oxidative stress, including the direct detoxification of highly reactive quinones [29]. The pretreatment of cells with TSG (5 and 10 μM) for 24 h can attenuate $H_2O_2$-induced ROS production and upregulate the expression of antioxidant enzymes HO-1 and NQO1 in a dose-dependent manner [30]. These results suggest that TSG is a potent antioxidant.

## 3. Neuroprotective Effect

*3.1. Protective Effect against Alzheimer's Disease.* Alzheimer's disease (AD) is a neurodegenerative disorder characterized by a progressive decline in cognitive abilities and associated with neuropathological features, including extensive loss of neurons (particularly cholinergic neurons), intracellular neurofibrillary tangles composed of the tau protein, and extracellular deposition of plaques composed of β-amyloid (Aβ), which is a cleavage product of amyloid precursor protein (APP). These two insoluble protein aggregates are accompanied by chronic inflammatory responses and extensive oxidative damage [31]. Learning and memory abilities are important cognition aspects that reflect the advanced integrative functions of the brain. Modern research has shown that an impaired memory function may be associated with cholinergic system synthases, such as acetylcholine transferase (ChAT) and hydrolytic enzymes, such as acetylcholinesterase (AChE). The levels of ChAT and acetylcholine (Ach) decrease in patients with AD, whereas the concentration of AChE increases [32]. Learning and memory abilities are closely related to central nervous system function, and their regulation involves monoamine neurotransmission; NE, DA, and 5-HT levels change with age [33]. Senescence-accelerated-prone 8 (SAMP8) mice exhibit early onset impaired learning and memory [34]. TSG significantly improves memory ability ($P < 0.01$) compared with that of the control group and prolongs the life span of SAMP8 mice by 17% ($P < 0.01$) compared with that of the control group [3]. TSG increases the protein level of neural klotho and reduces the levels of neural insulin, insulin receptor, insulin-like growth factor-1 (IGF-1), and IGF-1 receptor in the brain of SAMP8 mice ($P < 0.01$) compared with that of the control group [3]. In a hidden platform test and a spatial probe test, age-related learning and memory impairment in SAMP8 mice is prevented by daily treatment with TSG (33, 100, and 300 mg/kg/day) for 50 days [35]. In the same manner, TSG induces a dose-dependent enhancement of the activity of ChAT, decreases the activity of AChE, promotes the synthesis of monoamine neurotransmitter (NE, DA, and 5-HT), downregulates glutamic acid and aspartic acid, and retrieves the metabolic disorder of amino acid neurotransmitters in the brain tissues of SAMP8 mice compared with those of the SAMP8 group ($P < 0.05$ or $0.01$) [36, 37].

A$\beta$, a major protein component of senile plaques, is considered as a critical cause of the pathogenesis of AD [31]. The alternative splicing of APP exon 7 generates isoforms containing a Kunitz protease inhibitor (KPI) domain. APP-KPI levels in the brain are correlated with A$\beta$ production [38]. TSG protects nerve cells against A$\beta$-induced cell damage and improves learning-memory deficit in an AD model of APP transgenic mice and in a rat model of cholinergic damage induced by injecting ibotenic acid into the basal forebrain [39, 40]. Moreover, intragastric TSG (100 mg/kg/day) administration for 1 month can significantly improve learning, memory, spatial orientation, and other behavioral functions in transgenic mice and attenuate A$\beta$ neurotoxicity-induced injury to endoplasmic reticulum functions [41]. In cell lines and APP$_{SWE}$/PSEN$_{dE9}$ (APP/PS1) transgenic mice, TSG suppresses APP-KPI$^+$ and amyloid plaque formation. The mechanism of the neuroprotective effect of TSG may involve the activation of the AKT-GSK3$\beta$ signaling pathway, the attenuation of the splicing activity of ASF, and the decrease in APP-KPI$^+$ levels, leading to the decline in A$\beta$ deposition [15]. In addition, the intragastrical administration of 60 and 120 mg/kg TSG increases mitochondrial COX activities; decreases A$\beta_{1-42}$ contents; reduces APP, beta-site APP cleaving enzyme 1, and presenilin-1 expression; and enhances nerve growth factor, brain-derived neurotrophic factor, and tropomyosin-related kinase B expression in the hippocampus of NaN$_3$-infused rats [42]. A$\beta$ deposition-induced microglial activation is a crucial event in the pathology of AD [43]. Jiao et al. [44] found that TSG attenuates A$\beta$-induced microglial activation and inhibits the production of inflammatory molecules, such as inducible nitric oxide synthase (iNOS), nitric oxide (NO), cyclooxygenase 2 (COX-2), and prostaglandin E2. Furthermore, A$\beta$ exposure increases the levels of microglial M1 markers, interleukin- (IL-) 1$\beta$, IL-6, and tumor necrosis factor $\alpha$ (TNF-$\alpha$). TSG pretreatment suppresses the increase in M1 markers and enhances the levels of M2 markers, such as IL-10, brain-derived neurotrophic factor, glial cell-derived neurotrophic factor, and arginase-1, in N9 and BV2 cells. PU.1 overexpression partly eliminates the anti-inflammatory effects of TSG, suggesting that the roles of TSG in A$\beta$-induced microglial cells are mediated by PU.1 expression.

Oxidative damage in macromolecules is associated with the accumulation of A$\beta$ in the progressive development of AD [45] and is a critical event in A$\beta$-induced neuronal cell death [46]. In vitro models, TSG prevents A$\beta_{1-42}$-induced PC12 cell injury by improving the cell survival rate, reduces LDH release and MDA content, and increases SOD activity and Bcl-2 expression [47]. In vivo, the intragastric administration of TSG (100 mg/kg/day) in drinking water for 8 weeks provides protection against memory impairment induced by intracerebroventricular treatment with A$\beta_{1-40}$ (3 $\mu$g/mouse, i.c.v.) in mice. In addition, TSG reduces interleukin-6 (IL-6) content ($P < 0.05$), MDA content ($P < 0.01$), and MAO-B activity ($P < 0.05$) and increases T-AOC activity ($P < 0.01$) compared with those of the model group [48, 49]. Other studies have shown that TSG potentially reverses alterations in cognitive behavioral, biochemical changes, and oxidative damage induced by A$\beta_{1-42}$ in mice. These beneficial effects of TSG can be attributed partly by inhibiting the expression of the Keap1/Nrf2 pathway in hippocampus and cerebral cortex tissues [50]. Some investigations have also suggested that TSG protects neuronal HT-22 cells against A$\beta$-induced neurocytotoxicity by ameliorating mitochondrial-dependent oxidative stress and apoptotic pathway via the activation of Nrf2-HO-1 signaling. These data may elucidate a new mechanism about how TSG attenuates the pathologic process of AD by repairing A$\beta$-induced hippocampal neuron impairment [51].

A$\beta$ enhances the oligomerization, accumulation, and toxicity of $\alpha$-synuclein [52]. $\alpha$-Synuclein is a highly abundant protein at presynaptic terminals, is associated with the distal reserve pool of synaptic vesicles, and has a role in the regulation of neurotransmitter release, synaptic function, and plasticity [53]. Synaptic plasticity in the hippocampus has been considered the key phenomenon of learning and memory processes [54]. A previous study demonstrated that 120 and 240 $\mu$mol/kg/day TSG improves learning-memory impairment, decreases A$\beta$ content, and inhibits $\alpha$-synuclein overexpression and aggregation in the hippocampus of APPV717I transgenic mice in an AD model [40, 55].

The intragastric administration of 60 mg/kg/day TSG to 21-month-old rats for 3 months remarkably improves their learning-memory abilities in water maze tests, increases the number of synapses and synaptic vesicles, and increases the expression of synaptophysin in the hippocampal CA1 region of aged rats [56]. Another study has demonstrated that the oral administration of TSG for 3 months enhances memory and movement functions, protects the synaptic ultrastructure, and increases the synaptic connections and the levels of synapse-related proteins and p-CaMKII in the hippocampus, striatum, and cerebral cortex of aged mice [10]. The long-term potentiation (LTP) of synaptic transmission triggered by high-frequency stimulation (HFS) in the hippocampal CA1 area requires postsynaptic molecular mechanisms, such as the activation of N-methyl-D-aspartate (NMDA) receptors, calcium-calmodulin-dependent protein kinase II (CaMKII), and extracellular signal-regulated kinases (ERKs) of the mitogen-activated protein family [57-60]. In vitro, 1 and 5 $\mu$M TSG induce neurite outgrowth, promote PC12 cell differentiation, and increase intracellular calcium levels in hippocampal neurons in a concentration-dependent manner. Moreover, TSG facilitates HFS-induced hippocampal LTP in a bell-shaped manner. The facilitation of LTP induction by TSG requires CaMKII and ERK activation [61]. Therefore, TSG can be used to treat Alzheimer's disease.

*3.2. Protective Effect against Parkinson's Disease.* Parkinson's disease (PD) is the second-most common neurodegenerative disorder caused by the loss of dopaminergic (DA) neurons in the substantia nigra pars compacta (SNpc) and the presence of ubiquitinated alpha-synuclein- ($\alpha$-syn-) containing cytoplasmic inclusions called Lewy bodies in surviving SNpc neurons [62]. Several factors, including oxidative stress, mitochondrial dysfunction, neuroinflammation, and dysregulated kinase signaling, likely operate in the mechanism of death of nigrostriatal DA neurons in PD [63–65]. Many studies have used 1-methyl-4-phenyl-1,2,3,6-tetrahydropyridine (MPTP) or its active metabolite 1-methyl-4-phenylpyridinium ($MPP^+$), 6-hydroxydopamine (6-OHDA) and paraquat (PQ) as neurotoxins or Parkinsonism mimetics in cell cultures or animal PD models. These toxins induce oxidative stress and lead to cell death of DA neurons to mimic the situation in PD [66].

Oxidative stress is an important factor that can modulate intracellular signaling and lead to neuronal death by apoptosis or necrosis [67, 68]. These include JNK signaling, p38 activation, PI3K/Akt signaling inactivation, and signaling through bcl-2 family proteins [65]. MTT assay, flow cytometry, and DNA fragmentation have confirmed that TSG elicits protective effects against $MPP^+$-induced PC12 cell apoptosis by inhibiting ROS generation and modulating the activation of JNK and the PI3K/Akt pathway [69, 70]. TSG remarkably enhances the antioxidant enzyme activities of SOD, catalase (CAT), and GSH-Px and efficiently reduces the MDA content in PC12 cells [71]. $MPP^+$ transports into DA neurons and accumulates in the mitochondria, resulting in ATP depletion and mitochondrial membrane potential alteration [72]. TSG inhibits the increase in intracellular reactive oxygen species levels and the $MPP^+$-induced disruption of mitochondrial membrane potential. TSG markedly upregulates the Bcl-2/Bax ratio, reverses the cytochrome c release, and inhibits the caspase-3 activation in $MPP^+$-induced PC12 cells or SH-SY5Y cells [71, 73]. In vitro studies have shown that $MPP^+$ increases $\alpha$-Syn expression [74], and the overexpression of mutant human $\alpha$-Syn aggravates $MPP^+$-induced neurotoxicity [75]. TSG (3.125–50 $\mu$M) protects A53T AS cells against $MPP^+$-induced cell damage, and the mechanisms of TSG's neuroprotective effects are mediated by inhibiting $\alpha$-Syn overexpression and aggregation, enhancing mitochondrial function, reducing ROS levels, and inhibiting apoptosis in A53T AS cells exposed to $MPP^+$ [76]. NO is involved in the pathogenesis of PD [77]. NOS has four known isoforms: nNOS, iNOS, eNOS, and mitochondrial NO synthase. nNOS and iNOS are closely related to the pathogenesis of PD [78]. Our study showed that the exposure of PC12 cells to 75 mM 6-OHDA for 24 h significantly increases the level of intracellular ROS and NO, induces the overexpression of iNOS and nNOS, and elevates the level of 3-NT. However, these changes are markedly reversed in a dose-dependent manner after PC12 cells are pretreated with different TSG concentrations for 24 h. These results suggest that TSG may protect PC12 cells from 6-OHDA-induced apoptosis by regulating the ROS–NO pathway [79].

To further evaluate the in vivo neuroprotective effect of TSG, Zhang et al. and He et al. [16, 80] treated mice with TSG (20 or 40 mg/kg, i.g.) for 14 days, conducted pole and open field tests, and found that this treatment significantly reverses the MPTP-induced behavioral deficits compared with those of the MPTP treatment group ($P < 0.05$). Behavioral disorders occur when the number of TH-positive neurons in the SNpc of MPTP-treated mice decreases to 57.04% of the normal amount [81]. In another study, mice injected with MPTP show an approximately 63% decrease in TH neurons and manifest typical symptoms and behavioral disorders, such as tremor, piloerection, and bradykinesia. Conversely, mice treated with TSG exhibit a significant improvement in behavior and TH-positive neurons, which recover to 66% of that in the control group [80]. DA and its metabolites (DOPAC and HVA) decrease after MPTP is injected in striatal neurons. TSG dose-dependently counteracts the MPTP-induced loss of striatal DA [80], prevents the loss of striatal DA transporter protein [16], and provides protection against MPTP lesions partly by controlling ROS-mediated JNK, P38, and mitochondrial pathways and PI3K/Akt-mediated signaling mechanism. These in vivo effects extend previous in vitro observations [70, 71].

Neuroinflammation is an important contributor to PD pathogenesis with the hallmark of microglial activation [82]. Daily intraperitoneal injection of TSG for 14 consecutive days significantly protects DA neurons from 6-OHDA-induced neurotoxicity and suppresses microglial activation. A similar neuroprotection is shown in primary neuron–glial cocultures. In vitro studies have further demonstrated that TSG inhibits the activation of microglia and the subsequent release of proinflammatory factors. Moreover, TSG-mediated neuroprotection is closely related to the inactivation of mitogen-activated protein kinase signaling pathway [83]. Astroglia also plays an important role in PD development and becomes the prime target of PD treatment [84]. TSG protects DA neurons against LPS-induced neurotoxicity through dual modulation on glial cells by attenuating microglial-mediated neuroinflammation and enhancing astroglial-derived neurotrophic effects. These findings may serve as a basis for developing new alternative PD treatments [11].

*3.3. Protective Effects against Cerebral Ischemia Injury.* Multiple pathogenic mechanisms are involved in ischemia/reperfusion- (I/R-) related injury, including oxidative stress, inflammation, excitotoxicity, calcium overload, and apoptosis [85, 86]. ROS have been considered important mediators of brain damage after I/R injury [86, 87]. In vitro, various TSG concentrations (25, 50, and 100 $\mu$M) have been reported to protect the primary culture of cortical neurons against cytotoxicity induced by oxygen–glucose deprivation followed by reperfusion (OGD-R). Moreover, TSG (25 $\mu$M) reverses OGD-R-induced neuronal injury, intracellular ROS generation, and mitochondrial membrane potential dissipation and attenuates $H_2O_2$-induced increase in $[Ca^{2+}]_i$ [18]. Xu et al. [88] showed an ameliorating effect of TSG against focal cerebral I/R injury-induced apoptosis that can be attributed to its antioxidative actions. Doses of 0.038, 0.114, and 0.342 g/kg/day of TSG administered intragastrically at the onset of I/R in rats for 7 days result in a significant increase in GSH-Px and SOD activities ($P < 0.05$) but a decrease in the MDA content ($P < 0.01$) [88]. In another study that involves TUNEL staining, the intraperitoneal administration of TSG at dosages of 15 and 40 mg/kg at the onset of reperfusion after 90 min of middle cerebral artery occlusion (MCAO) in mice causes significant reductions in the brain infarct volume and the number of positive cells in the cerebral cortex compared with those of the MCAO group [18]. In addition, the continuous intragastric administration of TSG to mice for 6 days before I/R relieves the increase in the binding force of NMDA receptor ($P < 0.05$). Optical section results have shown that the calcium concentration in the group treated with TSG is remarkably lower than that in the ischemia model group ($P < 0.05$ or 0.01) [89]. A decrease in cellular oxygen tension during ischemic stroke also promotes an endogenous adaptive response accompanied by the upregulation of various cytoprotective factors to attenuate ischemic injury. The cytokines erythropoietin (EPO) and insulin-like growth factor I (IGF-I) are individually found to be effective [90]. The administration of TSG (60 mg/kg/d) can improve the neurological function of rats after reperfusion and increase the protein expression of HIF-1$\alpha$ and EPO after reperfusion compared with those of the model group [91]. In a related paper by Zhao et al. [92], TSG (30, 60, and 120 mg/kg/d) taken orally for 7 days can inhibit cell apoptosis caused by focal cerebral ischemia in rats, and its mechanism may be involved in the increase in the expression of Bcl-2 protein, which can inhibit cell apoptosis, and in the decrease in the expression of Bax protein, which can induce cell apoptosis and decrease the ratio of Bcl-2/Bax.

Neuronal death that follows 10 min of ischemia is associated with a late increase in AChE activity [93]. Microtubule-associated protein-2 (MAP-2) is depleted in the early hours after an in vivo ischemia insult in a rat hippocampal slice [94]. Experiments involving rats with chronic cerebral ischemia have demonstrated that the administration of TSG (30, 60, and 120 mg/kg) for 11 weeks causes a dose-dependent decrease in the AChE activity and increases the expression of protein phosphatase-2A (PP-2A) and MAP-2 in the hippocampus [95]. Angiogenesis is a prognostic marker of the survival and functional improvement of patients with cerebral stroke. Intraperitoneal injection of TSG (30, 60, and 120 mg/kg) promotes postoperative recovery in rats by minimizing the volume of cerebral infarcts and improving neurological dysfunction in a dose- and time-dependent manner. Additionally, TSG significantly increases the microvessel density in the brain and upregulates the CD31 expression in the ischemic penumbra relative to that in the control. Finally, treatment with TSG significantly upregulates the relative levels of vascular endothelial growth factor, angiopoietin 1, and angiopoietin receptor-2 expression in the brain lesions of rats [12]. Thus, TSG elicits neuroprotective effects against cerebral ischemia injury by modulating various pathways.

## 4. Antiatherosclerosis Effect

Atherosclerosis is a chronic and progressive disease in which plaques consisting of deposits of cholesterol and other lipids, calcium, and large inflammatory cells called macrophages build up in arterial walls. Although the pathophysiological mechanisms underlying atherosclerosis are poorly understood, inflammation and oxidative stress play important roles in all the phases of atherosclerosis evolution [96]. ROS can promote inflammation, alter vasomotion, induce cell death, cause platelet aggregation, and stimulate vascular smooth muscle cell (VSMC) proliferation [97]. One of its main risk factors is low-density lipoprotein (LDL) cholesterol [98]. When LDL penetrates endothelial cells, more ROS are generated. In turn, ROS oxidize LDL to oxidative low-density lipoprotein (OX-LDL), which can cause endothelial injury and inflammatory reaction [99]. All these events contribute to cardiovascular lesion formation.

TSG inhibits atherosclerosis by regulating blood fat, antioxidant and anti-inflammatory effects, suppressing matrix metalloproteinase expression, and relaxing blood vessels [100–103]. Gao et al. [100] reported that the administration of TSG (90 and 180 mg/kg/day) to hyperlipidemic rats for 1 week significantly reduces atherosclerosis index (AI, LDL-C/HDL-C) and serum total cholesterol (TC) and low-density lipoprotein cholesterol (LDL-C) levels ($P < 0.01$) but increases the mRNA expression of LDL receptor (LDLR) ($P < 0.05$) in liver cells. In another study, after rats are orally administered with 12 and 24 g/kg of TSG and a high-fat diet for 4 weeks, the contents of TC, TG, LDL-c, apoB, and MDA decrease significantly. The energy of serum SOD, CAT, GSH-Px, and T-AOC and the ratios of HDL-c/TC and apoAI/apoB also significantly increase in TSG groups, suggesting that TSG can regulate lipid metabolism and elicit an antioxidant effect [101]. ApoE$^{-/-}$ mouse is the first mouse model to develop lesions similar to those of humans and mimic the initiation and progression of human atherosclerosis. After 8 weeks of treatment, TSG ameliorates the serum levels of TC, TG, and LDL-C and increases the serum level of high-density lipoprotein cholesterol in ApoE$^{-/-}$ mice. TSG suppresses hepatic steatosis, atherosclerotic lesion formation, and macrophage foam cell formation in ApoE$^{-/-}$ mice. Moreover, TSG improves the expression levels of hepatic scavenger receptor class B type I (SR-BI), ABCG5, and CYP7A1 and upregulates the protein expression levels of aortic ATP-binding cassette transporters G1 and A1 (ABCG1 and ABCA1). An in vitro study has shown that TSG promotes the cholesterol efflux of macrophages and increases the protein expression levels of ABCA1 and ABCG1 [104].

Endothelial dysfunction is a key early event in atherosclerotic plaque formation and characterized by inflammatory processes, reduced NO bioavailability, and increased oxidative stress [105, 106]. TSG can prevent the development of atherosclerosis by influencing endothelial function in atherogenic diet-fed rats [103]. The administration of TSG (30, 60, and 120 mg/kg/day) after atherosclerosis is induced for 6 weeks results in a dose-dependent increase in the NO levels in the serum and aorta, the NOS content in the aorta, and the expression of eNOS but reduces the expression of iNOS in the aorta of atherosclerotic rats [107]. In the same manner, TSG improves ACh-induced endothelium-dependent relaxation, prevents intimal remodeling, and inhibits the decreased NOx content in the serum and aorta of atherogenic diet-fed rats after 12 weeks of treatment. The decreased mRNA and protein expression of eNOS and the increased mRNA and protein expression of iNOS in atherogenic diet-fed rats are attenuated by TSG treatment. These results suggest that TSG can restore vascular endothelial function, which may be related to its ability to prevent changes in eNOS and iNOS expression, leading to the preservation of NO bioactivity [103].

Inflammatory processes also play important roles in the onset, development, and remodeling of atherosclerotic lesions [108]. C-reactive protein (CRP) is probably a mediator of atherosclerosis and may increase the vulnerability of an atherosclerotic plaque to rupture [109]. After adhering to and migrating into the vascular wall at atherosclerotic sites, leukocytes secrete proinflammatory cytokines, such as IL-6, which likely play a major role in the pathogenesis of atherosclerosis [110]. Tumor necrosis factor- (TNF-) $\alpha$, another cytokine with proinflammatory effects, is upregulated in atherosclerotic plaques and may contribute to the pathogenesis of atherosclerosis [111]. Different doses of TSG (120, 60, or 30 mg/kg/day) administered to Sprague–Dawley rats for 12 weeks significantly and dose-dependently attenuate hyperlipidemic diet-induced alterations in serum lipid profile and increase CRP, IL-6, and TNF-$\alpha$ levels [102]. Zhao et al. [112, 113] showed that TSG inhibits lysophosphatidylcholine- (LPC-) induced apoptosis of human umbilical vein endothelial cells (HUVECs) by blocking the mitochondrial apoptotic pathway, and this process is accompanied with the activation of superoxide dismutase and glutathione peroxidase, the clearance of intracellular reactive oxygen species, and the reduction of lipid peroxidation. In addition, TSG provides protection against LPC-induced endothelial inflammatory damage. This protective effect is indicated by improved cell viability, adhesive ability, and migratory ability. Moreover, TSG reduces the expression and prevents the LPC-enhanced expression of Notch1, Hes1, and MCP-1. Therefore, the protective effects of TSG against inflammatory damage partly depends on Notch1 inhibition.

VSMCs are the main cellular components in the blood vessel wall, and their excessive proliferation plays an important role in the pathogenesis of atherosclerosis [114]. Although several growth factors and cytokines are involved in the development of atherosclerotic lesions, one of the principal regulators of mitogenesis in VSMCs is platelet-derived growth factor- (PDGF-) BB whose expression is increased in atherosclerotic lesions. PDGF-BB has been shown to activate key extracellular signaling transducers, including ERK 1/2, which are associated with cell growth and movement and critical for the initiation and progression of vascular lesions [115]. Xu et al. [116] demonstrated that TSG (10 and 100 $\mu$mol/l) significantly inhibits VSMC proliferation induced in the serum, cell cycle transition from the $G_0/G_1$ phase to the S phase, and proliferating cell nuclear antigen expression in the nucleus of VSMCs. Vimentin is a cytoskeletal protein involved in VSMC proliferation, monocyte migration across the endothelium walls, and foam cell formation [117]. Yao et al. [118, 119] reported that vimentin is a key protein in the TSG treatment for atherosclerosis in rats, and TSG attenuates vimentin mRNA and protein levels in oxLDL-induced HUVECs. This protective effect may be mediated partly by TSG-induced inhibition of vimentin expression via the interruption of the TGF$\beta$/Smad signaling pathway and caspase-3 activation. These findings help elucidate the molecular mechanism underlying the beneficial effects of TSG on atherosclerosis and suggest that TSG may be an effective agent for cardiovascular disease.

## 5. Effects on Inflammatory Disease

NF-$\kappa$B is a transcription factor that promotes the transcription of genes involved in proinflammatory responses [120]. The activation of NF-$\kappa$B leads to the formation of NF-$\kappa$B dimers (p65 and p50) that translocate to the nucleus to promote the transcription of the proinflammatory mediators TNF-$\alpha$, IL-6, and COX-2, resulting in a series of inflammatory cascade responses [121]. PPAR-$\gamma$, a member of the nuclear hormone receptor superfamily, can inhibit the activation of NF-$\kappa$B through several mechanisms and repress the NF-$\kappa$B-mediated transcription of proinflammatory cytokines [122]. The intragastric administration of TSG (10, 30, and 60 mg/kg/day) for 7 days after colitis is induced by acetic acid irrigation dramatically attenuates acetic acid-induced colon lesions in mice, reverses body weight loss, and improves histopathological changes. TSG apparently decreases the content of MDA, which is a marker of lipid peroxidation. TSG appears to exert its beneficial effects on acetic acid-induced experimental colitis through the upregulation of the mRNA and protein levels of PPAR-$\gamma$ and the inhibition of the NF-$\kappa$B pathway, which in turn decreases the protein overexpression of the downstream inflammatory mediators TNF-$\alpha$, IL-6, and COX-2 [123].

Neuroinflammation is closely implicated in the pathogenesis of neurological diseases [124]. The hallmark of neuroinflammation is microglial activation. The activation of microglial cells and the consequent release of proinflammatory and cytotoxic factors, such as TNF-$\alpha$, IL-1$\beta$, NO, and prostaglandin E synthase 2, possibly contribute to the onset of neurodegenerative diseases [125]. TSG suppresses matrix metalloproteinase expression and inflammation in rats with diet-induced atherosclerosis [102] and inhibits cyclooxygenase-2 activity and expression in RAW264.7 macrophage cells [126]. As a member of the intracellular phase II enzyme family, HO-1 is necessary to maintain cellular redox homeostasis. The overexpression of HO-1 markedly suppresses TNF-$\alpha$, thereby inducing airway inflammation by the inhibition of oxidative stress [127]. TSG treatment strongly induces the expression of HO-1 in an NRF2-dependent manner. Furthermore, TSG attenuates the LPS-mediated activation of RAW264.7 cells and the secretion of proinflammatory cytokines, including IL-6 and TNF-$\alpha$ [128]. NADPH oxidase is recognized as a key ROS-producing enzyme during inflammation and widely expressed in various immune cells, such as macrophages, eosinophils, microglia, and neutrophils [129]. TSG attenuates LPS-induced NADPH oxidase activation and subsequent ROS production [130]. Therefore, TSG may be used

TABLE 1: Antioxidative effect.

| Experimental model | IC$_{50}$/dose | Effects and possible mechanism | Reference number |
|---|---|---|---|
| | 33.24 $\mu$M | Scavenge DPPH radical | [24] |
| | 2.75 and 0.57 $\mu$g/ml | Scavenge ˙OH and superoxide anion radical | [8] |
| HBMECs | 50 and 100 $\mu$M | ↓ MDA and ROS; ↑ SOD and GSH | [25] |
| Rats | 20 and 40 mg/kg | ↑ SOD, GSH-Px activities; ↓ thiobarbituric acid reactive species content | [26] |
| Hypertensive rats | 50 mg/kg | ↓ Serum MDA and tissue 8-OHDG levels; ↑ serum GSH-Px activity | [27] |
| GES-1 cells and SGC-7901 cells | 5 and 10 $\mu$M | ↓ ROS production; ↑ the expression of HO-1 and NQO1 | [30] |

TABLE 2: Protective effect against Alzheimer's disease.

| Experimental model | Dose | Effects and possible mechanism | Reference number |
|---|---|---|---|
| SAMP8 mice | 2, 20, and 50 $\mu$M | Improves the memory ability and prolonged the life span of mice; ↑ the protein level of neural klotho; ↓ the levels of neural insulin, insulin-receptor, IGF-1, and IGF-1 receptor | [3] |
| SAMP8 mice | 33, 100, and 300 mg/kg | Prevention of age-related learning and memory impairment; ↑ ChAT activity; ↓ AChE activity; ↑the synthesis of NE, DA, and 5-HT; ↓ glutamic acid and aspartic acid contents | [35–37] |
| APP695V717I transgenic mice | 100 mg/kg | Improve learning, memory, and spatial orientation behavioral functions in mice; ↓ A$\beta$ neurotoxicity-induced injury to endoplasmic reticulum functions | [41] |
| HEK-293FT cells; SH-SY5Y cells; APP/PS1 transgenic mice NaN$_3$-induced rats | 10–200 $\mu$M; 50 mg/kg | Activation of AKT-GSK3$\beta$ signaling pathway; ↓ APP–KPI$^+$ and amyloid plaque formation; ↓ the splicing activity of ASF, APP–KPI$^+$ levels, A$\beta$ deposition | [15] |
| | 60 and 120 mg/kg | ↑ Mitochondrial COX activity; NGF, BDNF, TrkB expression; ↓ A$\beta_{1-42}$ content; APP, BACE1, and PS1 expression | [42] |
| A$\beta$-induced N9 and BV2 cells | 90 $\mu$M | ↓ iNOS, NO, COX-2, and PGE2, IL-1$\beta$, IL-6, and TNF-$\alpha$; ↑ IL-10, BDNF, and GDNF, Arg-1; ↓ PU.1 overexpression | [44] |
| A$\beta_{1-42}$-induced PC12 cells | 0.1, 1, and 10 $\mu$mol/l | ↑ Cell survival rate; ↓LDH release, MDA content; ↑SOD activity and Bcl-2 expression | [47] |
| A$\beta_{1-40}$-induced mice | 100 mg/kg | Prevention of learning and memory impairment; ↓ IL-6 content, MDA content, and MAO-B activity; ↑ T-AOC activity | [48, 49] |
| A$\beta_{1-42}$-induced mice | 30, 60, and 120 mg/kg | ↓ Oxidative damage; inhibited the expression of Keap1/Nrf2 pathway | [50] |
| A$\beta$-induced HT-22 cells | 60 $\mu$mol/l | ↓ Oxidative stress; activation of Nrf2-HO-1 signaling | [51] |
| APPV717I transgenic mice | 120 and 240 $\mu$mol/kg | Improves learning–memory impairment; ↓ A$\beta$ content; ↓ $\alpha$-synuclein overexpression and aggregation | [40, 55] |
| Aged rats | 30 and 60 mg/kg | ↑ Synapses and synaptic vesicles, SYP expression in the hippocampal CA1 region | [56] |
| Aged mice | 50, 100, and 200 mg/kg | Enhances the memory and movement functions, protected the synaptic ultrastructure; ↑ synapse-related proteins and p-CaMKII | [10] |
| PC12 cells | 1 $\mu$M and 5 $\mu$M | Induces neurite outgrowth, promotes PC12 cell differentiation; ↑ intracellular calcium levels, facilitates HFS-induced hippocampal LTP, activates CaMKII and extracellular ERK | [61] |

TABLE 3: Protective effect against Parkinson's disease.

| Experimental model | Dose | Effects and possible mechanism | Reference number |
| --- | --- | --- | --- |
| $MPP^+$-induced PC12 cells | 1–10 μM | ↑ Cell viability; ↓ cell apoptosis; ↓ intracellular ROS and the phosphorylated JNK; ↑ SOD, CAT, and GSH-Px activities, ↑ Bcl-2/Bax ratio, MMP; ↓ MDA content, cytochrome c, caspase-3; activation of PI3K/Akt pathway | [69–71] |
| $MPP^+$-induced A53T AS cells | 3.125–50 μM | ↓ α-Syn overexpression and aggregation; enhancing mitochondrial function; ↓ ROS level; ↓ cell apoptosis | [76] |
| 6-OHDA-induced PC12 cells | 10–50 μM | ↓ Intracellular ROS and NO; ↓ overexpression of iNOS, nNOS; ↓ 3-NT level | [79] |
| MPTP-induced mice | 20 or 40 mg/kg | Reverses the MPTP-induced behavioral deficits; ↑ TH-positive neurons in SNpc ↑DA and its metabolites contents, and DAT protein in striatum; activation of PI3K/Akt pathway; inhibition of the ROS-mediated JNK, P38, and mitochondrial pathways | [16, 80] |
| 6-OHDA-induced rats; primary rat midbrain neuron-glia cocultures | 10 and 50 mg/kg; 20–80 μM | ↓ Neurotoxicity; suppressed microglia activation and proinflammatory factors; inactivation of MAPK signaling pathway | [83] |
| Primary rat microglia- and astroglia-enriched cultures; LPS-induced rats | 20–80 μM; 10 and 50 mg/kg | ↓ Microglia-mediated neuroinflammation; enhancing astroglia-derived neurotrophic effects | [11] |

TABLE 4: Protective effects against cerebral ischemia injury.

| Experimental model | Dose | Effects and possible mechanism | Reference number |
| --- | --- | --- | --- |
| OGD-R-induced cell; MCAO mice | 25 μM; 15 and 40 mg/kg | ↓ Neuronal injury; ↓ intracellular ROS; ↓ $[Ca^{2+}]_i$, ↑ MMP; ↓ brain infarct volume; ↓ cell apoptosis | [18] |
| Rat | 0.038, 0.114, and 0.342 g/kg | ↑ GSH-$P_X$ and SOD activities; ↓ MDA content | [88] |
| Gerbils | 0.038, 0.114, and 0.342 g/kg | ↓ Binding force of NMDA receptor; ↓ intracellular $[Ca^{2+}]_i$ | [89] |
| Rats | 60 mg/kg | ↑ The protein expression of HIF-1α and EPO | [91] |
| Rats | 30, 60, and 120 mg/kg | ↓ Cell apoptosis; ↑ Bcl-2; ↓ Bax | [92] |
| Rats | 30, 60, and 120 mg/kg | ↓ AChE activity; ↑ the expression of protein PP-2A and MAP-2 | [95] |
| Rats | 30, 60, and 120 mg/kg | Promoted postoperative recovery in rats; ↓ volume of cerebral infarcts; improving neurological dysfunction; ↑ microvessel density in the brain; ↑ CD31 expression; ↑ levels of VEGF, Ang 1, and Tie-2 | [12] |

to treat inflammatory diseases. However, further research should be performed.

## 6. Other Effects

Oxidative stress is believed to play a role in physiological and pathological aging processes, such as age-related neurodegenerative diseases. Various studies have been performed about the antiaging effects of TSG [2]. Klotho, a serum secretory protein, is closely related to age. Klotho has many physiological functions, such as regulating calcium and phosphorus levels in vivo, delaying senescence, improving cognition, reducing oxidative stress, and protecting vascular endothelial cells [131]. Klotho and insulin/IGF-1 signaling pathways are closely associated with ROS accumulation, and the protein levels of klotho and insulin signaling pathway are critical for antiaging [131]. The long-term administration of TSG improves the memory ability and regulates the body weight of mice with D-galactose-induced aging, reduces the levels of IGF-1, and increases the levels of klotho in serum. TSG upregulates the expression of klotho in the cerebrum, heart, kidney, testis, and epididymis tissues of mice with D-galactose-induced aging [132]. Further studies have shown that TSG extends the life span of mice by upregulating neural klotho and downregulating neural insulin or

TABLE 5: Effects on inflammatory disease.

| Experimental model | Dose | Effects and possible mechanism | Reference number |
|---|---|---|---|
| Mice | 10, 30, and 60 mg/kg | Attenuates acetic acid-induced colon lesions; reverses body weight loss, and improves histopathological changes; ↓ the content of MDA; ↑ the mRNA and protein levels of PPAR-γ; inhibition of the NF-κB pathway; ↓ the protein overexpression of the TNF-α, IL-6, and COX-2 | [123] |
| RAW264.7 macrophage cells | 1, 10, and 100 μmol/l | ↓ COX-2 enzyme activity and expression | [126] |
| Microglia BV2 cell lines | 20–80 μM | ↓ NADPH oxidase activation; ↓ ROS production | [130] |

TABLE 6: Antiatherosclerosis effect.

| Experimental model | Dose | Effects and possible mechanism | Reference number |
|---|---|---|---|
| Hyperlipidemic rats | 90 and 180 mg/kg | ↓ Serum TC, LDL-C, and AI levels; ↑ the mRNA expression of LDLR in the liver cells | [100] |
| Hyperlipidemia rats | 12 and 24 g/kg | ↓ Serum TC, TG, LDL-c, apoB, and MDA; ↑ serum SOD, CAT, GSH-$P_X$, and T-AOC; ↑ the ratios of HDL-c/TC and apoAI/apoB | [101] |
| ApoE$^{-/-}$ mice; RAW264.7 cells | 50 and 100 mg/kg; 10 and 100 μM | ↓ TC, TG, and LDL-C; ↑ HDL-c, SR-BI, ABCG5, and CYP7A1 expression; ↑ the protein expressions of ABCA1 and ABCG1 | [104] |
| Atherosclerotic rats | 30, 60, and 120 mg/kg | ↑ NO levels in the serum and aorta; ↑ NOS content and the expression of eNOS, eNOS mRNA; ↓ the expression of iNOS, iNOS mRNA | [103, 107] |
| Atherosclerotic rats | 120, 60, or 30 mg/kg | ↓ CRP, IL-6, and TNF-α levels | [102] |
| HUVECs | 0.1, 1, or 10 μmol/l | ↓ Mitochondrial apoptotic pathway, lipid peroxidation, ROS, and MDA; ↑ SOD and GSH-Px; ↓ the expression of Notch1, Hes1, and MCP-1 | [112, 113] |
| VSMCs | 10 and 100 μmol/l | ↓ VSMC proliferation; ↓ cell cycle transition from G0/G1 phase to S phase; ↓ PCNA expression in the nucleus of VSMCs | [116] |
| HUVECs | 50 and 100 μM | ↓ Vimentin mRNA and protein levels; ↓ TGFβ/Smad signaling pathway and caspase-3 activation | [119] |

TABLE 7: Other effects.

| Experimental model | Dose | Effects and possible mechanism | Reference number |
|---|---|---|---|
| D-galactose-induced aging mice | 42, 84, and 168 mg/kg | Improves the memory ability and regulates the body weight of mice; ↓ the levels of IGF-1; ↑ the expression of klotho | [132] |
| C. elegans | 50 and 100 μM | Enhances the stress resistance; ↑ the life span of the nematode C. elegans | [1] |
| Rat cardiac fibroblasts | 3–100 μmol/l | ↓ ERK1/2 activation; ↓ overall production of ECM components | [133] |
| Pressure overload rats | 60 and 120 mg/kg | ↓ Angiotensin II level; ↓ transforming growth factor-β1 expression; ↓ ERK1/2 and p38 MAPK activation | [134] |
| Pressure overload rats | 120 mg/kg | ↑ Endogenous PPAR-γ expression | [17] |
| MC3T3-E1 cells | 0.1–10 μM | ↓ Osteoblastic differentiation; ↓ oxidative damage | [136] |
| SD rats | 150, 300, and 600 mg/kg | ↑ The density, content, and size of minerals in bone tissues; enhances the resistance to exogenic action, structural toughness, and strength of bone tissues | [137] |

insulin-like growth factor 1 [3]. TSG enhances the stress resistance and increases the life span of the nematode Caenorhabditis elegans [1]. These results strongly confirm the potential of TSG as a pharmaceutical antiaging drug.

Cardiac fibroblasts play an important role in regulating normal myocardial function and adverse myocardial remodeling. Angiotensin (Ang) II, the effector peptide of the renin-angiotensin system, is a key pathogenic factor in the development of hypertension and heart failure. TSG can inhibit Ang II-induced cardiac fibroblast proliferation by suppressing the ERK1/2 pathway and reducing the overall production of extracellular matrix components [133]. TSG can prevent cardiac remodeling induced by pressure overload in rats. The underlying mechanisms may be related to a decreasing angiotensin II level, an antioxidant effect of the tested compound, the suppression of transforming growth factor-$\beta$1 expression, and the inhibition of ERK1/2 and p38 mitogen-activated protein kinase activation [134]. Further studies have suggested that the upregulation of endogenous PPAR-$\gamma$ expression by TSG may be involved in the beneficial effect of TSG on pressure overload-induced cardiac fibrosis [17].

ROS can enhance bone resorption by directly or indirectly promoting osteoclast formation and activity [135]. TSG protects MC3T3-E1 cells from $H_2O_2$-induced cell damage and inhibition of osteoblastic differentiation. The protective effect of TSG on osteoblastic MC3T3-E1 cells may be mediated partly by its antioxidant ability [136]. In addition, TSG increases the density, content, and size of minerals in bone tissues and enhances the resistance to exogenic action, structural toughness, and strength of bone tissues [137]. These results suggest that TSG can be used as a good candidate for the protection of osteoblasts against oxidative stress-induced dysfunction and may have a potential therapeutic value for osteoporosis.

## 7. Conclusion

TSG has broad biological actions, including radical-scavenging effects and beneficial effects, for the treatment of various conditions, such as neuronal disease, cardiovascular disease, inflammatory disease, and osteoporosis. Several studies have elucidated the underlying mechanism of TSG action.

The effects of TSG against oxidative stress have helped elucidate many aspects of its mechanism of action. TSG is a strong free radical scavenger and potent antioxidant [24] (Table 1). TSG can enhance cognitive performance and prevent or treat AD in multistages and multitargets (Table 2). TSG may also act as an effective neuroprotective agent against PD through the modulation of the PI3K/Akt signaling pathway and ROS-mediated JNK, p38, and mitochondrial pathways [69–71, 80] (Table 3). TSG elicits anti-inflammatory effects and provides protection against cerebral I/R injury through multifunctional cytoprotective pathways (Tables 4 and 5). Previous studies demonstrated that TSG reduces the blood lipid content and inhibits the atherosclerotic process [102] (Table 6). TSG can prevent pressure overload-induced cardiac remodeling and Ang II-induced cardiac fibrosis in vitro [17] and may function against $H_2O_2$-induced dysfunction and oxidative stress in osteoblastic MC3T3-E1 cells (Table 7), but these abilities should be further investigated.

TSG has a wide variety of pharmaceutical properties. Its ameliorating effects and possible mechanisms should be summarized and determined. This review may provide a foundation for further studies to assess the mechanisms underlying the effects and clinical applications of TSG.

## Abbreviations

| | |
|---|---|
| PM: | *Polygonum multiflorum* Thunb |
| TSG: | 2,3,5,4′-Tetrahydroxystilbene-2-O-$\beta$-D-glucoside |
| ROS: | Reactive oxygen species |
| $H_2O_2$: | Hydrogen peroxide |
| SOD: | Superoxide dismutase |
| GSH-Px: | Glutathione peroxidase |
| TBARS: | Thiobarbituric acid reactive species |
| HUVECs: | Human umbilical vein endothelial cells |
| HO-1: | Hemeoxygenase-1 |
| NQO1: | NADPH-quinone oxidoreductase 1 |
| A$\beta$: | Beta-amyloid peptide |
| APP: | Amyloid precursor protein |
| ChAT: | Acetylcholine transferase |
| ACH: | Acetylcholine |
| SAMP8: | Senescence-accelerated-prone 8 |
| IGF-1: | Insulin-like growth factor-1 |
| KPI: | Kunitz protease inhibitor |
| BACE1: | Beta-site APP cleaving enzyme 1 |
| PS1: | Presenilin 1 |
| TrkB: | Tropomyosin-related kinase B |
| SYP: | Synaptophysin |
| NMDA: | N-Methyl-D-aspartate |
| CaMKII: | Calcium–calmodulin-dependent protein kinase II |
| ERK: | Extracellular signaling-regulated kinase |
| IL-6: | Interleukin-6 |
| MPTP: | 1-Methyl-4-phenyl-1,2,3,6-tetrahydropyridine |
| MPP$^+$: | 1-Methyl-4-phenylpyridinium |
| MDA: | Malondialdehyde |
| Bax: | Bcl2-associated protein |
| Bcl-2: | B-cell lymphoma protein-2 |
| MCAO: | Middle cerebral artery occlusion |
| MAP-2: | Microtubule-associated protein-2 |
| PP-2A: | Protein phosphatase-2A |
| VSMC: | Vascular smooth muscle cell |
| LDL: | Low-density lipoprotein |
| LDL-C: | Low-density lipoprotein cholesterol |
| NO: | Nitric oxide |
| iNOS: | Inducible NO synthase |
| TNF-$\alpha$: | Tumor necrosis factor |
| CRP: | C-reactive protein |
| PCNA: | Proliferating cell nuclear antigen |
| COX: | Cyclooxygenase. |

## Acknowledgments

This work was supported the Research Foundation of Xi'an Honghui Hospital (No. YJ2017010).

# References

[1] C. Büchter, L. Zhao, S. Havermann et al., "TSG (2,3,5,4′-tetrahydroxystilbene-2-O- β-D-glucoside) from the Chinese herb *Polygonum multiflorum* increases life span and stress resistance of *Caenorhabditis elegans*," *Oxidative Medicine and Cellular Longevity*, vol. 2015, Article ID 124357, 12 pages, 2015.

[2] S. Ling and J. W. Xu, "Biological activities of 2,3,5,4′-tetrahydroxystilbene-2-O-β-D-glucoside in antiaging and antiaging-related disease treatments," *Oxidative Medicine and Cellular Longevity*, vol. 2016, Article ID 4973239, 14 pages, 2016.

[3] X. Zhou, Q. Yang, Y. Xie et al., "Tetrahydroxystilbene glucoside extends mouse life span via upregulating neural klotho and downregulating neural insulin or insulin-like growth factor 1," *Neurobiology of Aging*, vol. 36, no. 3, pp. 1462–1470, 2015.

[4] S. P. Ip, A. S. M. Tse, M. K. T. Poon, K. M. Ko, and C. Y. Ma, "Antioxidant activities of *Polygonum multiflorum* Thunb., in vivo and in vitro," *Phytotherapy Research*, vol. 11, no. 1, pp. 42–44, 1997.

[5] L. Han, B. Wu, G. Pan, Y. Wang, X. Song, and X. Gao, "UPLC-PDA analysis for simultaneous quantification of four active compounds in crude and processed rhizome of *Polygonum multiflorum* Thunb," *Chromatographia*, vol. 70, no. 3-4, pp. 657–659, 2009.

[6] X. Qiu, J. Zhang, Z. Huang, D. Zhu, and W. Xu, "Profiling of phenolic constituents in *Polygonum multiflorum* Thunb. by combination of ultra-high-pressure liquid chromatography with linear ion trap-Orbitrap mass spectrometry," *Journal of Chromatography A*, vol. 1292, no. 5, pp. 121–131, 2013.

[7] K. Hata, M. Kozawa, and K. Baba, "A new stilbene glucoside from Chinese crude drug "Heshouwu," the roots of *Polygonum multiflorum* Thunb," *Yakugaku Zasshi*, vol. 95, no. 2, pp. 211–213, 1975.

[8] L. S. Lv, "Study on stilbene from roots of Polygonum multiglorum Thunb antioxidant activities in vitro," *Food Science*, vol. 28, no. 1, pp. 313–317, 2007.

[9] D. Vergara, A. Gaballo, A. Signorile et al., "Resveratrol modulation of protein expression in *parkin*-mutant human skin fibroblasts: a proteomic approach," *Oxidative Medicine & Cellular Longevity*, vol. 2017, Article ID 2198243, 22 pages, 2017.

[10] C. Shen, F. L. Sun, R. Y. Zhang et al., "Tetrahydroxystilbene glucoside ameliorates memory and movement functions, protects synapses and inhibits α-synuclein aggregation in hippocampus and striatum in aged mice," *Restorative Neurology and Neuroscience*, vol. 33, no. 4, pp. 531–541, 2015.

[11] Y. Zhou, G. Wang, D. Li et al., "Dual modulation on glial cells by tetrahydroxystilbene glucoside protects against dopamine neuronal loss," *Journal of Neuroinflammation*, vol. 15, no. 1, p. 161, 2018.

[12] Y. Mu, Z. Xu, X. Zhou et al., "2,3,5,4′-Tetrahydroxystilbene-2-O-β-D-glucoside attenuates ischemia/reperfusion-induced brain injury in rats by promoting angiogenesis," *Planta Medica*, vol. 83, no. 8, pp. 676–683, 2017.

[13] S. Y. Park, M. L. Jin, Z. Wang, G. Park, and Y. W. Choi, "2,3,4′,5-Tetrahydroxystilbene-2-O-β-d-glucoside exerts anti-inflammatory effects on lipopolysaccharide-stimulated microglia by inhibiting NF-κB and activating AMPK/Nrf 2 pathways," *Food and Chemical Toxicology*, vol. 97, pp. 159–167, 2016.

[14] Z. Ning, Y. Li, D. Liu et al., "Tetrahydroxystilbene glucoside delayed senile symptoms in old mice via regulation of the AMPK/SIRT1/PGC-1α signaling cascade," *Gerontology*, vol. 64, no. 5, pp. 457–465, 2018.

[15] X. Yin, C. Chen, T. Xu, L. Li, and L. Zhang, "Tetrahydroxystilbene glucoside modulates amyloid precursor protein processing via activation of AKT-GSK3β pathway in cells and in APP/PS1 transgenic mice," *Biochemical and Biophysical Research Communications*, vol. 495, no. 1, pp. 672–678, 2018.

[16] L. Zhang, L. Huang, L. Chen, D. Hao, and J. Chen, "Neuroprotection by tetrahydroxystilbene glucoside in the MPTP mouse model of Parkinson's disease," *Toxicology Letters*, vol. 222, no. 2, pp. 155–163, 2013.

[17] Y. Peng, Y. Zeng, J. Xu, X. L. Huang, W. Zhang, and X. L. Xu, "PPAR-γ is involved in the protective effect of 2,3,4′,5-tetrahydroxystilbene-2-O-β-D-glucoside against cardiac fibrosis in pressure-overloaded rats," *European Journal of Pharmacology*, vol. 791, pp. 105–114, 2016.

[18] T. Wang, J. Gu, P. F. Wu et al., "Protection by tetrahydroxystilbene glucoside against cerebral ischemia: involvement of JNK, SIRT1, and NF-κB pathways and inhibition of intracellular ROS/RNS generation," *Free Radical Biology & Medicine*, vol. 47, no. 3, pp. 229–240, 2009.

[19] M. Kuwabara, T. Asanuma, K. Niwa, and O. Inanami, "Regulation of cell survival and death signals induced by oxidative stress," *Journal of Clinical Biochemistry and Nutrition*, vol. 43, no. 2, pp. 51–57, 2008.

[20] Y. Sun, "Free radicals, antioxidant enzymes, and carcinogenesis," *Free Radical Biology & Medicine*, vol. 8, no. 6, pp. 583–599, 1990.

[21] K. R. Martin and J. C. Barrett, "Reactive oxygen species as double-edged swords in cellular processes: low-dose cell signaling versus high-dose toxicity," *Human & Experimental Toxicology*, vol. 21, no. 2, pp. 71–75, 2002.

[22] R. J. Mallis, J. E. Buss, and J. A. Thomas, "Oxidative modification of H-ras: S-thiolation and S-nitrosylation of reactive cysteines," *Biochemical Journal*, vol. 355, no. 1, pp. 145–153, 2001.

[23] O. I. Aruoma, "Free radicals, oxidative stress, and antioxidants in human health and disease," *Journal of the American Oil Chemists' Society*, vol. 75, no. 2, pp. 199–212, 1998.

[24] Y. Chen, M. Wang, R. T. Rosen, and C. T. Ho, "2,2-Diphenyl-1-picrylhydrazyl radical-scavenging active components from *Polygonum multiflorum* Thunb," *Journal of Agricultural and Food Chemistry*, vol. 47, no. 6, pp. 2226–2228, 1999.

[25] Z. Jiang, W. Wang, and C. Guo, "Tetrahydroxy stilbene glucoside ameliorates $H_2O_2$-induced human brain microvascular endothelial cell dysfunction *in vitro* by inhibiting oxidative stress and inflammatory responses," *Molecular Medicine Reports*, vol. 16, no. 4, pp. 5219–5224, 2017.

[26] L. Lv, X. Gu, J. Tang, and C.-T. Ho, "Antioxidant activity of stilbene glycoside from *Polygonum multiflorum* Thunb in vivo," *Food Chemistry*, vol. 104, no. 4, pp. 1678–1681, 2007.

[27] S. Ling, J. Duan, R. Ni, and J. W. Xu, "2,3,5,4′-Tetrahydroxystilbene-2-O-β-D-glucoside promotes expression of the longevity gene klotho," *Oxidative Medicine and Cellular Longevity*, vol. 2016, Article ID 3128235, 11 pages, 2016.

[28] L. E. Otterbein and A. M. K. Choi, "Heme oxygenase: colors of defense against cellular stress," *American Journal of Physiology-Lung Cellular and Molecular Physiology*, vol. 279, no. 6, pp. L1029–L1037, 2000.

[29] X. Cui, L. Li, G. Yan et al., "High expression of NQO1 is associated with poor prognosis in serous ovarian carcinoma," *BMC Cancer*, vol. 15, no. 1, p. 244, 2015.

[30] S. S. Tian, P. Song, L. Y. Liu, R. Zhu, and H. J. Zhao, "Tetrahydroxystilbene glucoside exerts cytoprotective effect against hydrogen peroxide-induced cell death involving ROS production and antioxidant enzyme activation," *Indian Journal of Pharmaceutical Sciences*, vol. 78, no. 5, pp. 602–607, 2017.

[31] M. N. Rossor, "Molecular pathology of Alzheimer's disease," *Journal of Neurology, Neurosurgery & Psychiatry*, vol. 56, no. 6, pp. 583–586, 1993.

[32] M. Grothe, L. Zaborszky, M. Atienza et al., "Reduction of basal forebrain cholinergic system parallels cognitive impairment in patients at high risk of developing Alzheimer's disease," *Cerebral Cortex*, vol. 20, no. 7, pp. 1685–1695, 2010.

[33] J. T. Coyle, D. L. Price, and M. R. Delong, "Alzheimer's disease: a disorder of cortical cholinergic innervation," *Science*, vol. 219, no. 4589, pp. 1184–1190, 1983.

[34] Y. Nomura and Y. Okuma, "Age-related defects in lifespan and learning ability in SAMP8 mice," *Neurobiology of Aging*, vol. 20, no. 2, pp. 111–115, 1999.

[35] R. C. Hunag, Z. S. Huang, Y. Li, S. Lai, and J. Huang, "Effects of TSG on learning and memory abilities and free radical scavenging in SAMP8 mice," *Chinese Journal of New Clinical Medicine*, vol. 2, no. 9, pp. 893–896, 2009.

[36] Z. S. Huang, Y. Li, J. Huang, S. B. Li, and S. Nong, "Effects of TSG on cholinergic neurotransmitter and monoamine neurotransmitter in SAMP8 mice's brain," *Youjiang Medical Journal*, vol. 38, no. 4, pp. 381–383, 2010.

[37] Z. S. Huang, Y. Li, J. Huang, S. B. Li, T. Li, and S. Nong, "The effects of tetrahydroxystilbene glucoside on amino acids neurotransmitter of SAMP8 mice brain tissues," *Journal of Youjiang Medical University for Nationalities*, vol. 32, no. 4, pp. 469–471, 2010.

[38] R. Sandbrink, C. L. Masters, and K. Beyreuther, "APP gene family alternative splicing generates functionally related isoforms," *Annals of the New York Academy of Sciences*, vol. 777, no. 1, pp. 281–287, 1996.

[39] L. Zhang, C. F. Ye, Y. Q. Chu, B. Li, and L. Li, "Effects of tetrahyroxystilbene glucoside on cholinergic system in dementia rats model induced by ibotenic acid," *Chinese Pharmaceutical Journal*, vol. 40, no. 10, pp. 749–752, 2005.

[40] L. Zhang, Y. Xing, C.-F. Ye, H.-X. Ai, H.-F. Wei, and L. Li, "Learning-memory deficit with aging in APP transgenic mice of Alzheimer's disease and intervention by using tetrahydroxystilbene glucoside," *Behavioural Brain Research*, vol. 173, no. 2, pp. 246–254, 2006.

[41] H. Luo, Y. Li, J. Guo et al., "Tetrahydroxy stilbene glucoside improved the behavioral disorders of APP695V717I transgenic mice by inhibiting the expression of Beclin-1 and LC3-II," *Journal of Traditional Chinese Medicine*, vol. 35, no. 3, pp. 295–300, 2015.

[42] R. Y. Zhang, L. Zhang, L. Zhang, Y. L. Wang, and L. Li, "Antiamyloidgenic and neurotrophic effects of tetrahydroxystilbene glucoside on a chronic mitochondrial dysfunction rat model induced by sodium azide," *Journal of Natural Medicines*, vol. 72, no. 3, pp. 596–606, 2018.

[43] R. E. Mrak, "Microglia in Alzheimer brain: a neuropathological perspective," *International Journal of Alzheimer's Disease*, vol. 2012, Article ID 165021, 6 pages, 2012.

[44] C. Jiao, F. Gao, L. Ou et al., "Tetrahydroxystilbene glycoside antagonizes β-amyloid-induced inflammatory injury in microglia cells by regulating PU.1 expression," *NeuroReport*, vol. 29, no. 10, pp. 787–793, 2018.

[45] B. Frank and S. Gupta, "A review of antioxidants and Alzheimer's disease," *Annals of Clinical Psychiatry*, vol. 17, no. 4, pp. 269–286, 2005.

[46] D. A. Butterfield and C. M. Lauderback, "Lipid peroxidation and protein oxidation in Alzheimer's disease brain: potential causes and consequences involving amyloid β-peptide-associated free radical oxidative stress," *Free Radical Biology & Medicine*, vol. 32, no. 11, pp. 1050–1060, 2002.

[47] H. H. Ma, Y. Shao, L. X. Chen, and Q. D. Yang, "The protective effects of stilbene glycoside on the PC12 cells injury induced by $A\beta_{(1-42)}$," *Chinese Journal of Gerontology*, vol. 31, no. 1, pp. 55–57, 2011.

[48] J. Chu, C. F. Ye, and L. Li, "Effects of stilbene-glycoside on learning and memory function and free radicals metabolism in dementia model mice," *Chinese Journal of Rehabilitation Theory & Practice*, vol. 9, no. 11, pp. 643–645, 2003.

[49] J. Chu, C. F. Ye, and L. Li, "Effects of stilbene-glycoside on learning and memory and inflammatory reaction of brain in dementia mice," *Traditional Chinese Drug Research & Clinical Pharmacology*, vol. 15, no. 4, pp. 235–237, 2004.

[50] M. Xie, G. Zhang, W. Yin, X. X. Hei, and T. Liu, "Cognitive enhancing and antioxidant effects of tetrahydroxystilbene glucoside in Aβ1-42-induced neurodegeneration in mice," *Journal of Integrative Neuroscience*, vol. 17, no. 3-4, pp. 355–365, 2018.

[51] C. Jiao, F. Gao, L. Ou et al., "Tetrahydroxy stilbene glycoside (TSG) antagonizes Aβ-induced hippocampal neuron injury by suppressing mitochondrial dysfunction via Nrf2-dependent HO-1 pathway," *Biomedicine & Pharmacotherapy*, vol. 96, pp. 222–228, 2017.

[52] E. Masliah, E. Rockenstein, I. Veinbergs et al., "β-amyloid peptides enhance α-synuclein accumulation and neuronal deficits in a transgenic mouse model linking Alzheimer's disease and Parkinson's disease," *Proceedings of the National Academy of Sciences of the United States of America*, vol. 98, no. 21, pp. 12245–12250, 2001.

[53] K. J. Vargas, S. Makani, T. Davis, C. H. Westphal, P. E. Castillo, and S. S. Chandra, "Synucleins regulate the kinetics of synaptic vesicle endocytosis," *Journal of Neuroscience*, vol. 34, no. 28, pp. 9364–9376, 2014.

[54] T. V. P. Bliss and G. L. Collingridge, "A synaptic model of memory: long-term potentiation in the hippocampus," *Nature*, vol. 361, no. 6407, pp. 31–39, 1993.

[55] L. Zhang, S. Yu, R. Zhang, Y. Xing, Y. Li, and L. Li, "Tetrahydroxystilbene glucoside antagonizes age-related α-synuclein overexpression in the hippocampus of APP transgenic mouse model of Alzheimer's disease," *Restorative Neurology and Neuroscience*, vol. 31, no. 1, pp. 41–52, 2013.

[56] R. Wang, Y. Tang, B. Feng et al., "Changes in hippocampal synapses and learning-memory abilities in age-increasing rats and effects of tetrahydroxystilbene glucoside in aged rats," *Neuroscience*, vol. 149, no. 4, pp. 739–746, 2007.

[57] S. Berberich, V. Jensen, Ø. Hvalby, P. H. Seeburg, and G. Köhr, "The role of NMDAR subtypes and charge transfer during hippocampal LTP induction," *Neuropharmacology*, vol. 52, no. 1, pp. 77–86, 2007.

[58] K. Fukunaga, D. Muller, and E. Miyamoto, "Increased phosphorylation of $Ca^{2+}$/calmodulin-dependent protein kinase II and its endogenous substrates in the induction of long-term potentiation," *Journal of Biological Chemistry*, vol. 270, no. 11, pp. 6119–6124, 1995.

[59] J. D. Sweatt, "Mitogen-activated protein kinases in synaptic plasticity and memory," *Current Opinion in Neurobiology*, vol. 14, no. 3, pp. 311–317, 2004.

[60] G. M. Thomas and R. L. Huganir, "MAPK cascade signalling and synaptic plasticity," *Nature Reviews Neuroscience*, vol. 5, no. 3, pp. 173–183, 2004.

[61] T. Wang, Y. J. Yang, P. F. Wu et al., "Tetrahydroxystilbene glucoside, a plant-derived cognitive enhancer, promotes hippocampal synaptic plasticity," *European Journal of Pharmacology*, vol. 650, no. 1, pp. 206–214, 2011.

[62] A. J. Lees, J. Hardy, and T. Revesz, "Parkinson's disease," *The Lancet*, vol. 373, no. 9680, pp. 2055–2066, 2009.

[63] M. T. Lin and M. F. Beal, "Mitochondrial dysfunction and oxidative stress in neurodegenerative diseases," *Nature*, vol. 443, no. 7113, pp. 787–795, 2006.

[64] E. C. Hirsch, S. Vyas, and S. Hunot, "Neuroinflammation in Parkinson's disease," *Parkinsonism & Related Disorders*, vol. 18, no. 4, pp. S210–S212, 2012.

[65] O. A. Levy, C. Malagelada, and L. A. Greene, "Cell death pathways in Parkinson's disease: proximal triggers, distal effectors, and final steps," *Apoptosis*, vol. 14, no. 4, pp. 478–500, 2009.

[66] W. Dauer and S. Przedborski, "Parkinson's disease: mechanisms and models," *Neuron*, vol. 39, no. 6, pp. 889–909, 2003.

[67] K. A. Jellinger, "Recent advances in our understanding of neurodegeneration," *Journal of Neural Transmission*, vol. 116, no. 9, pp. 1111–1162, 2009.

[68] H. U. Simon, A. Haj-Yehia, and F. Levi-Schaffer, "Role of reactive oxygen species (ROS) in apoptosis induction," *Apoptosis*, vol. 5, no. 5, pp. 415–418, 2000.

[69] X. Li, Y. Li, J. Chen et al., "Tetrahydroxystilbene glucoside attenuates $MPP^+$-induced apoptosis in PC12 cells by inhibiting ROS generation and modulating JNK activation," *Neuroscience Letters*, vol. 483, no. 1, pp. 1–5, 2010.

[70] R. Qin, X. Li, G. Li et al., "Protection by tetrahydroxystilbene glucoside against neurotoxicity induced by $MPP^+$: the involvement of PI3K/Akt pathway activation," *Toxicology Letters*, vol. 202, no. 1, pp. 1–7, 2011.

[71] L. Zhang, L. Huang, X. Li et al., "Potential molecular mechanisms mediating the protective effects of tetrahydroxystilbene glucoside on $MPP^+$-induced PC12 cell apoptosis," *Molecular and Cellular Biochemistry*, vol. 436, no. 1-2, pp. 203–213, 2017.

[72] A. Ghosh, K. Chandran, S. V. Kalivendi et al., "Neuroprotection by a mitochondria-targeted drug in a Parkinson's disease model," *Free Radical Biology & Medicine*, vol. 49, no. 11, pp. 1674–1684, 2010.

[73] F. L. Sun, L. Zhang, R. Y. Zhang, and L. Li, "Tetrahydroxystilbene glucoside protects human neuroblastoma SH-SY5Y cells against $MPP^+$-induced cytotoxicity," *European Journal of Pharmacology*, vol. 660, no. 2-3, pp. 283–290, 2011.

[74] J. Xu, C. Wei, C. Xu et al., "Rifampicin protects PC12 cells against $MPP^+$-induced apoptosis and inhibits the expression of an α-Synuclein multimer," *Brain Research*, vol. 1139, no. 1, pp. 220–225, 2007.

[75] J. J. Qian, Y. B. Cheng, Y. P. Yang et al., "Differential effects of overexpression of wild-type and mutant human α-synuclein on $MPP^+$-induced neurotoxicity in PC12 cells," *Neuroscience Letters*, vol. 435, no. 2, pp. 142–146, 2008.

[76] R. Zhang, F. Sun, L. Zhang, X. Sun, and L. Li, "Tetrahydroxystilbene glucoside inhibits α-synuclein aggregation and apoptosis in A53T α-synuclein-transfected cells exposed to $MPP^+$," *Canadian Journal of Physiology and Pharmacology*, vol. 95, no. 6, pp. 750–758, 2017.

[77] K. M. K. Boje, "Nitric oxide neurotoxicity in neurodegenerative diseases," *Frontiers in Bioscience*, vol. 9, no. 1-3, pp. 763–776, 2004.

[78] T. Dehmer, J. Lindenau, S. Haid, J. Dichgans, and J. B. Schulz, "Deficiency of inducible nitric oxide synthase protects against MPTP toxicity in vivo," *Journal of Neurochemistry*, vol. 74, no. 5, pp. 2213–2216, 2000.

[79] L. Tao, X. Li, L. Zhang et al., "Protective effect of tetrahydroxystilbene glucoside on 6-OHDA-induced apoptosis in PC12 Cells through the ROS-NO pathway," *PLoS One*, vol. 6, no. 10, article e26055, 2011.

[80] H. He, S. Wang, J. Tian et al., "Protective effects of 2,3,5,4′-tetrahydroxystilbene-2-O-β-D-glucoside in the MPTP-induced mouse model of Parkinson's disease: involvement of reactive oxygen species-mediated JNK, P 38 and mitochondrial pathways," *European Journal of Pharmacology*, vol. 767, pp. 175–182, 2015.

[81] X. H. Wang, G. Lu, X. Hu et al., "Quantitative assessment of gait and neurochemical correlation in a classical murine model of Parkinson's disease," *BMC Neuroscience*, vol. 13, no. 1, p. 142, 2012.

[82] V. Sanchez-Guajardo, C. J. Barnum, M. G. Tansey, and M. Romero-Ramos, "Neuroimmunological processes in Parkinson's disease and their relation to α-synuclein: microglia as the referee between neuronal processes and peripheral immunity," *ASN Neuro*, vol. 5, no. 2, pp. 113–139, 2013.

[83] C. Huang, F. Lin, G. Wang et al., "Tetrahydroxystilbene glucoside produces neuroprotection against 6-OHDA-induced dopamine neurotoxicity," *Oxidative Medicine and Cellular Longevity*, vol. 2018, Article ID 7927568, 9 pages, 2018.

[84] H.-H. Tsai, H. Li, L. C. Fuentealba et al., "Regional astrocyte allocation regulates CNS synaptogenesis and repair," *Science*, vol. 337, no. 6092, pp. 358–362, 2012.

[85] C. Sierra, A. Coca, and E. L. Schiffrin, "Vascular mechanisms in the pathogenesis of stroke," *Current Hypertension Reports*, vol. 13, no. 3, pp. 200–207, 2011.

[86] P. J. Crack and J. M. Taylor, "Reactive oxygen species and the modulation of stroke," *Free Radical Biology & Medicine*, vol. 38, no. 11, pp. 1433–1444, 2005.

[87] H. Chen, H. Yoshioka, G. S. Kim et al., "Oxidative stress in ischemic brain damage: mechanisms of cell death and potential molecular targets for neuroprotection," *Antioxidants & Redox Signaling*, vol. 14, no. 8, pp. 1505–1517, 2011.

[88] Y. M. Xu, M. H. Zhou, Y. P. Zheng, and Y. S. Fu, "Protective effects of stilbene glucoside preconditioning against cell apoptosis in rats subjected to cerebral ischemia-reperfusion," *Journal of Nanchang University (Medical Sciences)*, vol. 53, no. 5, pp. 13–16, 2013.

[89] Z. J. Liu, L. Li, C. F. Ye, and H. X. Ai, "Effects of tetrahydroxystilbene glucoside on NMDA receptors and intracellular calcium ions in brain of ischemia rodent," *Chinese Pharmacological Bulletin*, vol. 19, no. 10, pp. 1112–1115, 2003.

[90] M. Digicaylioglu, G. Garden, S. Timberlake, L. Fletcher, and S. A. Lipton, "Acute neuroprotective synergy of erythropoietin and insulin-like growth factor I," *Proceedings of the National Academy of Sciences of the United States of America*, vol. 101, no. 26, pp. 9855–9860, 2004.

[91] H. Q. Yuan, J. Yang, J. Zeng, and R. J. Guo, "Effects of tetrahydroxystilbene glucoside on hypoxia inducible factor-1 alpha and erythropoietin in old rat after cerebral ischemia reperfusion," *Journal of Apoplexy & Nervous Diseases*, vol. 28, no. 10, pp. 868–871, 2011.

[92] L. Zhao, C. Y. Li, L. Zhang, W. Cui, L. Zhang, and L. Li, "Effect of tetrahydroxy-stilbene glucoside on cell apoptosis in focal cerebral ischemia rats," *Chinese Traditional & Herbal Drugs*, vol. 39, no. 3, pp. 394–397, 2008.

[93] M. R. Schetinger, C. D. Bonan, S. S. Frassetto et al., "Pre-conditioning to global cerebral ischemia changes hippocampal acetylcholinesterase in the rat," *IUBMB Life*, vol. 47, no. 3, pp. 473–478, 1999.

[94] K. M. Raley-Susman and J. Murata, "Time course of protein changes following in vitro ischemia in the rat hippocampal slice," *Brain Research*, vol. 694, no. 1-2, pp. 94–102, 1995.

[95] L. Liu, L. Li, L. Zhao, L. Zhang, Y. L. Li, and C. F. Ye, "Effects of 2,3,5,4'-tetrahydroxystilbene-2-O-$\beta$-glucoside on learning and memory abilities of rats with chronic cerebral ischemia," *Chinese Journal of Pharmacology & Toxicology*, vol. 22, no. 2, pp. 108–115, 2008.

[96] F. Cipollone, M. L. Fazia, and A. Mezzetti, "Oxidative stress, inflammation and atherosclerotic plaque development," *International Congress Series*, vol. 1303, pp. 35–40, 2007.

[97] S. K. Yao, J. C. Ober, A. Gonenne et al., "Active oxygen species play a role in mediating platelet aggregation and cyclic flow variations in severely stenosed and endothelium-injured coronary arteries," *Circulation Research*, vol. 73, no. 5, pp. 952–967, 1993.

[98] R. Ross, "Atherosclerosis — an inflammatory disease," *The New England Journal of Medicine*, vol. 340, no. 2, pp. 115–126, 1999.

[99] M. H. Shishehbor and S. L. Hazen, "Inflammatory and oxidative markers in atherosclerosis: relationship to outcome," *Current Atherosclerosis Reports*, vol. 6, no. 3, pp. 243–250, 2004.

[100] X. Gao, Y. J. Hu, and L. C. Fu, "Blood lipid-regulation of stilbene glycoside from *Polygonum multiflorum*," *China Journal of Chinese Materia Medica*, vol. 32, no. 4, pp. 323–326, 2007.

[101] C. K. Xiang, R. Wang, and Z. F. Yuan, "Study on effect of Polygonum mutiflorum extract on lipid metabolism and its anti-oxidation in SD rats with hyperlipemia," *China Pharmaceuticals*, vol. 18, no. 24, pp. 19-20, 2009.

[102] W. Zhang, C. H. Wang, F. Li, and W. Z. Zhu, "2,3,4',5-Tetrahydroxystilbene-2-O-$\beta$-D- glucoside suppresses matrix metalloproteinase expression and inflammation in atherosclerotic rats," *Clinical and Experimental Pharmacology and Physiology*, vol. 35, no. 3, pp. 310–316, 2008.

[103] W. Zhang, X. L. Xu, Y. Q. Wang, C. H. Wang, and W. Z. Zhu, "Effects of 2,3,4',5-tetrahydroxystilbene 2-O-$\beta$-D-glucoside on vascular endothelial dysfunction in atherogenic-diet rats," *Planta Medica*, vol. 75, no. 11, pp. 1209–1214, 2009.

[104] X. Chen, K. Tang, Y. Peng, and X. Xu, "2,3,4',5-Tetrahydroxystilbene-2-O-$\beta$-d-glycoside attenuates atherosclerosis in apolipoprotein E-deficient mice: role of reverse cholesterol transport," *Canadian Journal of Physiology and Pharmacology*, vol. 96, no. 1, pp. 8–17, 2018.

[105] M. Simionescu, "Implications of early structural-functional changes in the endothelium for vascular disease," *Arteriosclerosis, Thrombosis, and Vascular Biology*, vol. 27, no. 2, pp. 266–274, 2007.

[106] C. S. Stancu, L. Toma, and A. V. Sima, "Dual role of lipoproteins in endothelial cell dysfunction in atherosclerosis," *Cell and Tissue Research*, vol. 349, no. 2, pp. 433–446, 2012.

[107] W. Zhang, Y. Shen, C. H. Wang, Y. Q. Wang, and F. Li, "Effects of 2,3,4',5-tetrahydroxystilbene-2-O-$\beta$-d- glucoside on conten of nitrc oxide and expression of nitric oxide synthase in aorta of atherosclerotic rats," *Chinese Journal of New Drugs*, vol. 17, no. 8, pp. 652–655, 2008.

[108] P. Libby, "Inflammation in atherosclerosis," *Nature*, vol. 420, no. 6917, pp. 868–874, 2002.

[109] P. M. Ridker, J. E. Buring, J. Shih, M. Matias, and C. H. Hennekens, "Prospective study of C-reactive protein and the risk of future cardiovascular events among apparently healthy women," *Circulation*, vol. 98, no. 8, pp. 731–733, 1998.

[110] U. Ikeda, M. Ikeda, T. Oohara et al., "Interleukin 6 stimulates growth of vascular smooth muscle cells in a PDGF-dependent manner," *American Journal of Physiology-Heart and Circulatory Physiology*, vol. 260, no. 5, pp. H1713–H1717, 1991.

[111] F. G. Rus, F. Niculescu, and R. Vlaicu, "Tumor necrosis factor-alpha in human arterial wall with atherosclerosis," *Atherosclerosis*, vol. 89, no. 2-3, pp. 247–254, 1991.

[112] J. Zhao, S. Xu, F. Song, L. Nian, X. Zhou, and S. Wang, "2,3,5,4'-tetrahydroxystilbene-2-O-$\beta$-D-glucoside protects human umbilical vein endothelial cells against lysophosphatidylcholine-induced apoptosis by upregulating superoxide dismutase and glutathione peroxidase," *IUBMB Life*, vol. 66, no. 10, pp. 711–722, 2014.

[113] J. Zhao, Y. Liang, F. Song et al., "TSG attenuates LPC-induced endothelial cells inflammatory damage through notch signaling inhibition," *IUBMB Life*, vol. 68, no. 1, pp. 37–50, 2016.

[114] V. J. Dzau, R. C. Braun-Dullaeus, and D. G. Sedding, "Vascular proliferation and atherosclerosis: new perspectives and therapeutic strategies," *Nature Medicine*, vol. 8, no. 11, pp. 1249–1256, 2002.

[115] E. Millette, B. H. Rauch, R. D. Kenagy, G. Daum, and A. W. Clowes, "Platelet-derived growth factor–BB transactivates the fibroblast growth factor receptor to induce proliferation in human smooth muscle cells," *Trends in Cardiovascular Medicine*, vol. 16, no. 1, pp. 25–28, 2006.

[116] X. L. Xu, W. Zhang, Y. J. Huang, and Y. Q. Wang, "Effect of 2,3,4',5-tetrahydroxystilbene-2-O-$\beta$-D glucoside on proliferation and antioxidation of vascular smooth muscle cell," *Chinese Pharmacological Bulletin*, vol. 26, no. 7, pp. 934–939, 2010.

[117] M. Nieminen, T. Henttinen, M. Merinen, F. Marttila-Ichihara, J. E. Eriksson, and S. Jalkanen, "Vimentin function in lymphocyte adhesion and transcellular migration," *Nature Cell Biology*, vol. 8, no. 2, pp. 156–162, 2006.

[118] W. Yao, W. Fan, C. Huang, H. Zhong, X. Chen, and W. Zhang, "Proteomic analysis for anti-atherosclerotic effect of tetrahydroxystilbene glucoside in rats," *Biomedicine & Pharmacotherapy*, vol. 67, no. 2, pp. 140–145, 2013.

[119] W. Yao, C. Huang, Q. Sun, X. Jing, H. Wang, and W. Zhang, "Tetrahydroxystilbene glucoside protects against oxidized LDL-induced endothelial dysfunction via regulating vimentin cytoskeleton and its colocalization with ICAM-1 and VCAM-1," *Cellular Physiology and Biochemistry*, vol. 34, no. 5, pp. 1442–1454, 2014.

[120] R. E. Simmonds and B. M. Foxwell, "Signalling, inflammation and arthritis: NF-κB and its relevance to arthritis and inflammation," *Rheumatology*, vol. 47, no. 5, pp. 584–590, 2008.

[121] J. M. Müller, H. W. L. Ziegler-Heitbrock, and P. A. Baeuerle, "Nuclear factor kappa B, a mediator of lipopolysaccharide effects," *Immunobiology*, vol. 187, no. 3–5, pp. 233–256, 1993.

[122] D. Kelly, J. I. Campbell, T. P. King et al., "Commensal anaerobic gut bacteria attenuate inflammation by regulating nuclear-cytoplasmic shuttling of PPAR-γ and RelA," *Nature Immunology*, vol. 5, no. 1, pp. 104–112, 2004.

[123] C. Zeng, J. H. Xiao, M. J. Chang, and J. L. Wang, "Beneficial effects of THSG on acetic acid-induced experimental colitis: involvement of upregulation of PPAR-γ and inhibition of the Nf-κb inflammatory pathway," *Molecules*, vol. 16, no. 10, pp. 8552–8568, 2011.

[124] A. J. Nimmo and R. Vink, "Recent patents in CNS drug discovery: the management of inflammation in the central nervous system," *Recent Patents on CNS Drug Discovery*, vol. 4, no. 2, pp. 86–95, 2009.

[125] H. González, D. Elgueta, A. Montoya, and R. Pacheco, "Neuroimmune regulation of microglial activity involved in neuroinflammation and neurodegenerative diseases," *Journal of Neuroimmunology*, vol. 274, no. 1-2, pp. 1–13, 2014.

[126] Y. Z. Zhang, J. F. Shen, J. Y. Xu, J. H. Xiao, and J. L. Wang, "Inhibitory effects of 2,3,5,4'-tetrahydroxystilbene-2-O-β-D-glucoside on experimental inflammation and cyclooxygenase 2 activity," *Journal of Asian Natural Products Research*, vol. 9, no. 4, pp. 355–363, 2007.

[127] I. T. Lee, S. F. Luo, C. W. Lee et al., "Overexpression of HO-1 protects against TNF-α-mediated airway inflammation by down-regulation of TNFR1-dependent oxidative stress," *The American Journal of Pathology*, vol. 175, no. 2, pp. 519–532, 2009.

[128] W. Yu, X. Zhang, H. Wu et al., "HO-1 is essential for tetrahydroxystilbene glucoside mediated mitochondrial biogenesis and anti-inflammation process in LPS-treated RAW264.7 macrophages," *Oxidative Medicine and Cellular Longevity*, vol. 2017, Article ID 1818575, 13 pages, 2017.

[129] M. S. Hernandes and L. R. G. Britto, "NADPH oxidase and neurodegeneration," *Current Neuropharmacology*, vol. 10, no. 4, pp. 321–327, 2012.

[130] F. Zhang, Y. Y. Wang, J. Yang, Y. F. Lu, J. Liu, and J. S. Shi, "Tetrahydroxystilbene glucoside attenuates neuroinflammation through the inhibition of microglia activation," *Oxidative Medicine and Cellular Longevity*, vol. 2013, Article ID 680545, 8 pages, 2013.

[131] M. Yamamoto, J. D. Clark, J. V. Pastor et al., "Regulation of oxidative stress by the anti-aging hormone klotho," *Journal of Biological Chemistry*, vol. 280, no. 45, pp. 38029–38034, 2005.

[132] X. X. Zhou, Q. Yang, Y. H. Xie et al., "Protective effect of tetrahydroxystilbene glucoside against d-galactose induced aging process in mice," *Phytochemistry Letters*, vol. 6, no. 3, pp. 372–378, 2013.

[133] W. Zhang, X. F. Chen, Y. J. Huang, Q. Q. Chen, Y. J. Bao, and W. Zhu, "2,3,4',5-Tetrahydroxystilbene-2-O-β-D-glucoside inhibits angiotensin II-induced cardiac fibroblast proliferation via suppression of the reactive oxygen species-extracellular signal-regulated kinase 1/2 pathway," *Clinical and Experimental Pharmacology and Physiology*, vol. 39, no. 5, pp. 429–437, 2012.

[134] X. L. Xu, Q. Y. Zhu, C. Zhao et al., "The effect of 2,3,4',5-tetrahydroxystilbene-2-O-β-D-glucoside on pressure overload-induced cardiac remodeling in rats and its possible mechanism," *Planta Medica*, vol. 80, no. 2-3, pp. 130–138, 2014.

[135] N. K. Lee, Y. G. Choi, J. Y. Baik et al., "A crucial role for reactive oxygen species in RANKL-induced osteoclast differentiation," *Blood*, vol. 106, no. 3, pp. 852–859, 2005.

[136] J. K. Zhang, L. Yang, G. L. Meng et al., "Protective effect of tetrahydroxystilbene glucoside against hydrogen peroxide-induced dysfunction and oxidative stress in osteoblastic MC3T3-E1 cells," *European Journal of Pharmacology*, vol. 689, no. 1–3, pp. 31–37, 2012.

[137] X. Q. Hu, J. Zhou, G. Z. Liu, and R. M. Xie, "Effect of stilbene glycoside on bone mineral density and bone strength of rats," *China Journal of Chinese Medicine*, vol. 26, no. 6, pp. 696–698, 2011.

# The Potential Roles of Extracellular Vesicles in Cigarette Smoke-Associated Diseases

A-Reum Ryu,[1] Do Hyun Kim,[2] Eunjoo Kim,[3] and Mi Young Lee[1,4]

[1]Department of Medical Science, Soonchunhyang University, 22 Soonchunhyang-ro, Asan, Chungnam, Republic of Korea
[2]Department of Biology, The College of Wooster, 1189 Beall Ave, Wooster, OH, USA
[3]Companion Diagnostics and Medical Technology Research Group, Daegu Gyeongbuk Institute of Science and Technology (DGIST), Daegu, Republic of Korea
[4]Department of Medical Biotechnology, Soonchunhyang University, 22 Soonchunhyang-ro, Asan, Chungnam, Republic of Korea

Correspondence should be addressed to Mi Young Lee; miyoung@sch.ac.kr

Academic Editor: Ji C. Bihl

Cigarette smoke contains more than 4,500 chemicals; most of which are highly reactive free radicals, which induce proinflammatory and carcinogenic reactions. Numerous efforts have focused extensively on the role of cigarette smoking as a cause of many diseases. Extracellular vesicles and exosomes have recently received increasing interest for their diagnostic and therapeutic roles in many diseases. However, research done on the role of extracellular vesicles and exosomes on cigarette smoke-induced chronic disease is still in its infancy. In this review, we summarize the recently addressed roles of extracellular vesicles and exosomes in the pathogenesis of cigarette smoke-related diseases, such as chronic obstructive pulmonary disease, cardiovascular disease, lung cancer, and oral cancer. Moreover, their potential utilization and future prospects as diagnostic biomarkers for cigarette smoke-related diseases are described.

## 1. Introduction

More than 6 million deaths are attributable to direct tobacco use per year globally, according to the World Health Organization (WHO) (2017). The death toll is expected to rise to 10 million per year by the 2020's or early 2030's. Cigarette smoking was the cause of 1 in 5 deaths in the United States and 1 in 3 deaths among men older than 30 years of age in Korea [1]. Cigarette smoke (CS) contains more than 4,500 chemicals; most of which are highly reactive free radicals, including peroxy radicals, nitrogen radicals, and other oxygen-derived species, which induce proinflammatory and carcinogenic reactions [2]. Thus, cigarette smoking causes serious health problems associated with oxidative and nitrosative stress and is the most relevant risk factor for chronic obstructive pulmonary disease (COPD), cardiovascular disease, and lung and oral cancers [3, 4].

Numerous efforts have focused extensively on the role of cigarette smoking as a cause of many diseases, studying the epidemiological and biomedical mechanisms. Moreover, despite the recent progress in research on extracellular vesicles (EVs) and exosomes as diagnostic and therapeutic targets in many diseases, the understanding of the role of EVs and exosomes associated with cigarette smoking is in its infancy. EVs are heterogeneous and include exosomes, microvesicles (microparticles or extosomes), and other vesicles still under controversy. Thus, we use the term "EV" in this review, except for instances where we clearly refer to exosomes.

EVs and exosomes are nanosized particles that participate in intercellular communication via delivery of crucial molecules to both distant and adjacent cells [5]. Exosomes are enriched with several molecular cargos, such as DNAs, mRNAs, miRNAs, and proteins, affecting diverse biological processes of the recipient cells [5–7]. Significantly enriched biomarkers via exosome shedding can be detected in certain disease states, whereas they are undetectable under normal conditions, suggesting that exosomes could play a role as diagnostic signatures reflecting the physiological condition of the parental cells [8, 9]. Cancer cell-derived exosomes might deliver miRNAs or proteins to recipient cells to trigger a favorable microenvironment for the proliferation and metastasis of tumors via horizontal exchange of tumorigenic information [9].

The research focusing on the role of EVs and exosomes in CS exposure-associated chronic diseases is limited; approximately 70 articles on this topic can be found in PubMed and Google Scholar. Exposure to cigarette smoke (CS)/extract (CSE) stimulated EVs and exosome release into the serum, saliva, urine of smokers/ex-smokers, and cancer patients, as well as from various cultured cells [10, 11]. Moreover, exosome secretion by CS led to several biochemical and cellular processes, including angiogenesis [12], endothelial dysfunction [8], and tissue remodeling [13], as well as proinflammatory [14] and prothrombotic effects [15, 16], promoting the pathogenesis of CS-related chronic diseases as categorized by Benedikter et al. [12, 17, 18].

Recently, a number of studies have suggested that oxidative stress affects the release of EVs in various cell types and human samples [15, 16, 19]. Although the mechanism underlying exosome release by cigarette smoke is not fully elucidated, redox-dependent thiol modification seems to be a plausible explanation for the exosome release under stress condition by CS [18]. The pathophysiological and cellular processes implicated in cigarette smoke-triggered exosome release are suggested to be the potential mechanisms that account for cigarette smoke-induced pathogenesis [6]. In this review, we briefly introduce and summarize the recently addressed roles of EVs and exosomes in the pathogenesis of cigarette smoke-related diseases, such as COPD, cardiovascular disease, lung cancer, and oral cancers. Moreover, the future prospects and potential utilization of EVs and exosomes as biomarkers for the diagnosis of cigarette smoke-related diseases are described.

## 2. The Release of EVs in Response to CS Exposure

EVs, induced under oxidative stress conditions, exhibit proinflammatory and prothrombotic effects which contribute to the pathogenesis of chronic diseases. The thiol-reactive compounds like acrolein, an endogenous inflammatory metabolite, contributed to exosome release by CS by reacting with cell surface thiols in the airway epithelial cells [18, 20, 21]. EV release was also enhanced with plasma membrane blebbing in response to thiol-reactive RCS (reactive carbonyl species) and ROS (reactive oxygen species). Moreover, EV shedding by thiol modifications protected the secreting cell against oxidative damage; however, at the same time, it induced a proinflammatory and prothrombotic state of the cells. However, both exosome release by CS [21] and the oxidation of protein thiols of the EV-secreting cells [18] were blocked by thiol antioxidants like N-acetyl cysteine (NAC) and glutathione S-transferase (GSH). These data clearly indicate that CS-induced EV release could be suppressed by thiol antioxidants. In other words, EV production under oxidative stress and a proinflammatory state was modulated by redox-dependent modification of protein thiols [12, 18]. In conclusion, thiol protection seems to be necessary for inhibiting the detrimental alterations caused by EV signaling under inflammatory and oxidative stress.

Recently, there is accumulating evidence that oxidative thiol modification promotes coagulation associated with EVs. The production of prothrombotic EVs was enhanced under oxidative and proinflammatory states with thiol depletion [22–24]. Phosphatidylserine (PS) [25] and tissue factor (TF) [13, 26] were implicated in the prothrombotic effect of EVs. Additionally, recent publications discuss in detail the targets of redox-dependent thiol modifications related with EV release, including exofacial thiols, protein disulfide isomerases, actin filaments, and redox-sensitive calcium channels. Taken together, thiols on the protein modulate EV release under thiol-depleting states via membrane fusion and blebbing, likely as an adaptation strategy against various oxidative stressors including CS.

## 3. Pathophysiological Processes Associated with CS-Induced Release of EVs

CS-triggered oxidative and nitrosative stress involved in the pathogenesis of several chronic diseases has been reported in diseases such as chronic obstructive pulmonary diseases (COPD), cardiovascular disease, lung cancer, and oral cancer. Biochemical and cellular processes, such as inflammation, prothrombosis, angiogenesis, endothelial dysfunction, and tissue remodeling, are ascribed to the pathogenesis of CS-related diseases [12]. Here, the recent literature, focusing on the most studied issues of inflammation and thrombophilia based on reported article numbers, is summarized in detail. They specifically imply a link between EV release by CS exposure and the biochemical processes at the cellular level.

*3.1. Inflammation.* The modulation of inflammatory responses associated with CS exposure-induced EVs was described in human peripheral blood mononuclear cells [25], human bronchial epithelial cells [17], macrophages [27, 28], and human epithelial cells [29], as well as in smokers with COPD [8] and lung cancer [14]. SOCS (suppressor of cytokine signaling proteins) is a gene family which regulates inflammation and Th cell differentiation via inhibiting the JAK-STAT signaling pathway. SOCS concentration in CS-exposed mice and BALF of human smokers were lower than those of the nonexposed groups [27]. Alveolar macrophages play a crucial role in host defense against respiratory tract infections via their phagocytic properties. In addition, they regulate host inflammatory response via cytokine production and anti-inflammatory microenvironment induction. Notably, the inflammatory response was suppressed by delivering alveolar macrophage-derived EVs with SOCS 1 and 3 to epithelial cells [27]. The internalization of alveolar macrophage-derived EVs by target cells was downregulated by CSE, showing loss of the EV-dependent anti-inflammatory state [30]. CSE-exposed mononuclear cells induced significant production of microparticles. They showed a proinflammatory potential with the expression of

ICAM-1, IL-8, and MCP-1. In conclusion, CS inhibits EV-dependent anti-inflammatory signaling, whereas it activates EV-dependent proinflammatory signaling [28].

According to the results of Héliot et al. [14], the exposure of BEAS-2B cells to smoker EVs elevated IL-6 and IL-8 levels in comparison with exposure to nonsmoker EVs. The production of CYFRA21-1, a prognostic factor in non-small cell lung cancer (NSCLC) was also upregulated in BEAS-2B cells exposed to smoker EVs. In addition, a comparison of EV miRNAs between smokers and nonsmokers demonstrated that four miRNAs, miR-21, miR-27a, let-7g, and let-7e, which are potentially attributed to lung carcinogenesis and might be used as lung cancer biomarkers, were differentially expressed [14]. The expression of miR-21 and miR-27a in BEAS-2B cells increased after treatment with smoker EVs, when compared with exposure to nonsmoker EVs, whereas let-7e and let-7g expression was decreased with smoker EV treatment. The reduced expression of NF-E2-related factor 2 (NRF2), a key regulator of antioxidants implicated in lung carcinogenesis, was significantly and inversely correlated with the expression of its target miRNA, miR-27a.

Other indirect data on the effect of CS-exposed EVs in inflammation were also obtained from transcriptomic (microRNA) and proteomic analyses of EV contents via bioinformatics approaches. CS exposure to human samples resulted in a parallel increase in EVs and proinflammatory molecules, yet a clear association between EVs and inflammation has not been confirmed *in vivo*. Moreover, there is conflicting data regarding the ability of CS exposure to affect the release of EVs and inflammatory molecules in human blood [15]. The discrepancy might be due to the differences in EVs and exosome phenotypes, depending on age, gender, hormone profile, and smoking status [15, 19, 26]. The level and function of microvesicles differed with age; higher levels of microvesicles with CD41, CD235, tissue factor, and phosphatidylserine were found in the younger and older groups, compared to those of middle age [26]. Moreover, AREG on plasma EVs was decreased with increase in age, in both males and females [19]. With respect to cigarette smoking, the levels of microparticles with CD41 and CD45 were enhanced during active smoking of one cigarette, while CD144 did not change [15]. In the case of gender, plasma EVs from male smokers had higher levels of CD171, PD-L1, and TSG101 than those from female smokers. The levels of AREG, MUC1, CD146, CD13, and TSG101 in EVs were notably reduced in female smokers but not reduced in male smokers [19]. Thus, more in-depth research will help elucidate the mechanism by which CS induces exosome shedding and how the exosomes modulate inflammatory responses, contributing to chronic disease development and aggravation.

*3.2. Thrombophilia.* Thrombophilia (or hypercoagulability) is the abnormal tendency to develop blood clots. CSE-induced EV release is thought to be a significant cardiovascular risk factor for smokers. Tissue factor (TF) and phosphatidylserine (PS) in the EV membrane promoted thrombin production via assembly of coagulation factors [16, 26]. *In vitro* experiments showed increased release of endothelial EVs with PS and TF associated with procoagulant activity [8]. EVs derived from mononuclear cells in response to CSE express active TF, thus potentially contributing to the pathogenesis of cardiovascular diseases.

At present, information on the involvement of procoagulant EVs with TF and PS exposed to CSE *in vivo* is limited, probably due to small sample size, limitation of correction for confounding factors, and low sensitivity of antibody-based detection. In the case of PS, induction of EVs with PS expression [8, 16, 25] has been clearly detected *in vitro*. However, there are conflicting data on the levels of EVs with PS in the blood of smokers compared to nonsmokers [26, 31, 32]. Further research should be carried out to build on the current findings and understand whether EVs with PS could play a key role in thrombophilia caused by CS exposure *in vitro* and *in vivo*.

## 4. Chronic Diseases Associated with CS-Induced Release of EVs

*4.1. Cardiovascular Disease (CVD).* Smoking is a major risk factor of cardiovascular disease (CVD), which is the leading cause of death worldwide. Smokers have a higher risk of developing vascular diseases such as atherosclerosis and venous thromboembolism than nonsmokers [33]. CS-induced alteration of lipid profile via oxidation by free radicals and oxidants originated from CS [34]. CS-induced CVD development occurs via several potential mechanisms such as inflammation, vascular dysfunction, thrombophilia, and atherosclerosis via lipid peroxidation, thrombus formation, and foam cell formation [35].

Microvesicles (MVs) released from human macrophages in response to CS possess proteolytic activity. The gelatinolytic and collagenolytic activities of CS-induced release of microvesicles from macrophages were predominantly ascribed to MMP14 production [13]. The MMP14-positive MVs may contribute to matrix damage, leading to instability of atherosclerotic plaques and emphysematous lung destruction [13]. CS exposure elevated the shedding of microparticles with proinflammatory and procoagulant potential in human mononuclear cells, via calcium-dependent mechanisms [25]. The roles of exosomes as biomarkers for acute coronary syndromes, myocardial infarction, and heart failure have been suggested. Cardiac-specific miRNAs, such as miRNA-1 and miRNA-133a, were released in injured cardiomyocytes and showed the highest plasma levels following the onset of myocardial infarction symptoms [36]. Moreover, circulating p53-responsive miRNAs, such as miRNA-192, miRNA-194, and miRNA-34a, have been reported to play as predictors of ischemic heart failure associated with acute myocardial infarction [37]. Circulating plasma microvesicles (PMVs) and their microRNAs are suspected to be major contributing factors for atherosclerosis, serving as biomarkers for CVD progression. However, little is known about how smoking affects PMV shedding and microRNA signatures *in vivo* [31]. Badrnya et al. provided evidence that smoking affects the PMV profile and microRNA cargo. Smokers showed increased levels of circulating leukocyte-derived PMVs (lPMV) and miR-29b and decreased levels of platelet-derived PMVs (pPMVs), the major PMV, and miR-

223. They suggested that alteration of the PMV profile and miRNA, contributable to atherogenesis, could serve as an early biomarker in smoking-related diseases, despite sample sizes being small [31]. CS-exposed neutrophils accelerated the production of membrane microvesicles (MVs) with enzymatically active transmembrane ADAM proteases, ADAM10 and ADAM17. Production of ADAM10- and ADAM17-positive MVs from neutrophils on exposure to CS indicates the molecular mechanism underlying the dramatically elevated risk of abdominal aortic aneurysm (AAA) development in smokers [38].

### 4.2. Chronic Obstructive Pulmonary Disease (COPD).

COPD is a representative disease mainly triggered by inhalation of CS, despite the genetic aspect of COPD [39, 40]. The pathogenesis of COPD is manifested by airway inflammation (chronic bronchitis), degradation of lung tissue (emphysema), and progressive airway limitation associated with fibrotic airway remodeling. A small but significant pool of literature cites the importance of exosomes in CS-related COPD. Fujita et al. reported that the myofibroblast differentiation for airway remodeling in COPD pathogenesis was driven by exosomal miR-210 [10]. CS induced miR-210 expression in exosomes from human bronchial epithelial cells (HBECs). The upregulated miR-210 promoted myofibroblast differentiation in primary lung fibroblasts, whereas it suppressed autophagy via downregulating ATG7, an essential component of autophagy. Therefore, miR-210 was suggested to be a potential therapeutic target for COPD. In addition, downregulation of the exosomal let-7 (tumor suppressive miRNA) family following cigarette exposure was reported in the COPD model and lung cancer development.

The roles of exosomes in COPD can be inferred from the report by Moon et al. using endothelium-derived exosomes after exposure to CS [17]. CS upregulated full-length CCN1 expression in exosomes from epithelial cells, allowing transfer of inflammatory signals to distant portions of the lung. However, prolonged exposure to CS cleaved the full-length CCN1 in exosomes into the truncated form. The cleaved CCN1 activates the secretion of MMP1 by interacting with integrin-$\alpha$7 in lung epithelial cells. In addition, CS exposure suppressed the transport of alpha-1 antitrypsin, which is involved in protecting the lung from degradation, inflammation, and apoptosis [41]. In addition, exosomes also played a role in COPD exacerbation. Exosomes with CD144, CD31, and CD62E were notably enriched in patients during COPD exacerbation than in stable patients [42].

Thus, EV and exosome shedding promoted by the exposure of CS appears to contribute to the development of CS-related COPD. Understanding how EVs and exosomes contribute in this process and changes in the EVs and exosome contents, which may be used as biomarkers during disease progression, in response to cigarette smoking might help us to inquire into pathogenesis and develop novel therapeutic strategies.

### 4.3. Lung Cancer.

EVs have been reported to be genetic cargo in lung cancer, which could be used for diagnostic, prognostic, and predictive biomarkers [43, 44]. Tumor-derived exosomal miR-1247-3p induces cancer-associated fibroblast activation to foster lung metastasis of liver cancer [45, 46]. Knockdown of TGF-$\beta$1 expression in human umbilical cord mesenchymal stem cells reverts their exosome-mediated EMT-promoting effect on lung cancer cells [47]. Exosomal contents were reported to be involved in anticancer drug resistance [48]. In addition, exosomes are involved in the crosstalk between lung cancer stem cells (CSC) and cancer-associated fibroblasts (CAF) [49], which is the primary component of the lung cancer microenvironment. CSCs interact with the CAFs via exosomes and could be activated by their crosstalk. In tumor-associated cells, alpha-smooth muscle actin ($\alpha$-SMA) was identified as a specific marker for myofibroblasts in lung cancer [50].

However, few studies reported the involvement of EV in CS-induced lung cancer disease. EVs released from airway epithelial cells show pathological properties when exposed to extracts obtained from CS. As CS extract causes oxidative stress, the oxidative components were proved to be responsible for inducing EV release and subsequent progress to lung cancer [21]. Lung cancer is more common in people with HIV than in the general population [51], and a major risk factor in this is smoking. A role of EV in CS-mediated HIV-1 pathogenesis has been reported. CS is known to exacerbate HIV-1 pathogenesis, especially in monocytes, through the oxidative stress pathway. EVs are known to alter HIV-1 pathogenesis through intercellular communication. The key factors were suggested by a potential role of antioxidant enzymes, which are differentially packaged into CSC-exposed HIV-1-infected cell-derived EVs, on HIV-1 replication of recipient cells [46].

In addition, there are reports on bronchial epithelial cell-derived EVs modulated by CS, which revealed a crucial pathogenic role for CS in the development of lung tumors. Six members of the let-7 family, namely, let-7a, let-7b, let-7c, let-7f, let-7g, and let-7i, were reported to be significantly reduced in human bronchial epithelial cell-derived EVs, after exposure to CS extract [10, 52]. The expression levels of the let-7 family members are lower in lung cancer tissues compared with normal lung tissues. Alterations in these tumor-suppressive miRNAs contribute to lung tumor development, suggesting a causal link between the reduction in the expression of the let-7 family in EVs and lung carcinogenesis in response to CS exposure, via EV transfer-mediated communication among these pathological cells [48, 53]. Understanding how exosomes contribute in this process and the alterations in EV contents derived from CS-induced lung tumors during disease progression may help us to understand the mechanisms and develop novel treatment strategies.

### 4.4. Oral Cancer.

Oral cancer, the sixth most common cancer worldwide, is a subtype of the head and neck cancer; most of these occurred in less developed regions. Oral squamous cell carcinoma (OSCC) constitutes over 90% of all cancers of the oral cavity [54, 55]. Cigarette smoking leads to oral cancer because of the exposure of oral epithelial cells to free radicals, reactive oxygen species, and reactive nitrogen species that contribute to oxidative damages [56]. In addition, CS caused a profound effect on salivary TNF-$\alpha$ and MMP-8, which are

TABLE 1: Potential biomarkers of EVs and exosomes in response to cigarette smoke in chronic diseases.

| Components | Species | Sources | Vesicle | Disease | Reference |
|---|---|---|---|---|---|
| miRNA | | | | | |
| miR-378a, miR-379, miR-139-5p, miR-200b-5p | Human | Plasma | Exosome | Lung cancer | [64] |
| miR-7, miR-20a, miR-28-3p, miR-34c-5p, miR-100 | Human | Serum | Exosome | COPD | [65] |
| let-7d, miR-191 miR-125a, miR-126 | Human Mouse | Cell supernatant Plasma | Endothelial microparticles (EMPs) | COPD | [8] |
| let-7e, let-7g, miR-26b | Human | Bronchoalveolar lavage (BAL) | EVs | COPD Lung cancer | [14] |
| miR-210 | Human | Cell supernatant | EVs | COPD | [10] |
| miR-223, miR-29b | Human | Plasma | Microvesicle | CVD | [31] |
| Protein | | | | | |
| LRG1 | Human | Urine | Exosome | NSCLC | [66] |
| $\mu$-Calpain, m-calpain | Human | Cell supernatant | Vesicles | Lung cancer | [11] |

markers of periodontal disease, in chronic periodontitis subjects in comparison to healthy controls [57]. As cigarette smoking is considered the most prevalent risk factor for oral cancer, saliva analysis serves as a crucial diagnostic tool. Seven mRNA biomarkers (DUSP1, H3F3A, IL-1B, IL-8, OAZ1, S100P, and SAT) in OSCC saliva [58] and two miRNAs (miR-125a and miR-200a) in saliva were characterized as potential biomarkers for oral cancer detection [59].

Recently, the field of salivary exosomics, which focuses on the nucleic acids and proteins in salivary exosomes as potential cancer biomarkers, has been rapidly expanding [60–62]. The use of salivary exosomes is promising due to their ready availability and noninvasiveness. Therefore, oral cancer-derived exosomes in saliva have great potential as biomarkers for cancer diagnosis. The exosomal miRNAs, miR-342-3p and miR-1246, were elevated in exosomes from oral cancer. Exosomal miRNA transfer from highly metastatic oral cancer cells to cells with low malignancy resulted in rapid cell proliferation, compared to the non-exosome-applied control cells by Sakha et al. [63].

However, the information on the saliva exosome research, targeting oral cancer in response to CS, is not available. Moreover, CS-induced salivary exosome shedding in oral cancers has not been reported yet. It is necessary to expand research on salivary exosomes, integrating current and past data and linking salivary exosomics with other biomedical research.

## 5. EVs as Potential Biomarkers in Exposure to CS for Chronic Diseases

Recently, diverse EVs and exosomal biomarkers have been reported to be potential diagnostic biomarkers for several diseases. However, CS-induced EV and exosomal biomarkers, miRNAs, and proteins, for CS-related diseases including CVD, COPD, and cancers, are limited. Table 1 summarizes the current data on potential EVs and exosomal biomarkers and their corresponding diseases.

## 6. Conclusions and Perspective

Chronic exposure to cigarette smoke is the major risk factor for numerous diseases. Several studies cited the gaps between acute *in vitro* CS exposure and chronic *in vivo* CS exposure in smokers [12, 17]. Therefore, development of a suitable *in vitro* exposure model that replicates/mimics *in vivo* exposure is required for intensive research to fully elucidate the functional properties of CS-induced EV shedding, which is implicated in disease pathogenesis.

It is generally accepted that chronic exposure to environmental toxicants, including CS, could stimulate EVs and exosome biogenesis and the subsequent release that is associated with disease onset. However, how CS exposure triggers intracellular signal transductions that affect EVs and exosomes and how the subsequent EV and exosome release modulates a series of cellular events responsible for disease development remain unclear so far. Therefore, further studies are needed for elucidating the network between EVs and exosome alteration, which is implicated in CS exposure and disease progression. A wide range of EV and exosome studies depicting the diagnostic or therapeutic feasibility of them are in clinical test [67, 68]. However, it is premature to find the links between EV and exosome alterations and CS-specific diseases. At present, *in vivo*, preclinical, and clinical data do not significantly correlate with the *in vitro* data. Moreover, development of exosome isolation technology for obtaining pure and abundant exosomes is needed to overcome limitations in integrating and validating current findings. Finally, functional studies of CS-induced EVs and exosomes open up promising research leading to the development of diagnostic and therapeutic EVs and exosomes in CS-associated diseases.

## Authors' Contributions

Eunjoo Kim and Mi Young Lee have contributed equally to this study as cocorrespondence.

## Acknowledgments

This work was supported by the Industrial Strategic Technology Development Program (10077559) funded by the Ministry of Trade, Industry and Energy (MOTIE, Korea). This work was also supported by Soonchunhyang University.

## References

[1] H. A. Lee, H. Park, H. Kim, and K. Jung-Choi, "The effect of community-level smoke-free ordinances on smoking rates in men based on community health surveys," *Epidemiology and Health*, vol. 36, article e2014037, 2014.

[2] A. Mossina, C. Lukas, J. Merl-Pham et al., "Cigarette smoke alters the secretome of lung epithelial cells," *Proteomics*, vol. 17, no. 1-2, article 1600243, 2017.

[3] A. Sharma, M. Bansal-Travers, P. Celestino et al., "Using a smoking cessation quitline to promote lung cancer screening," *American Journal of Health Behavior*, vol. 42, no. 6, pp. 85–100, 2018.

[4] S. Kundu, V. Ramshankar, A. K. Verma et al., "Association of DFNA5, SYK, and NELL1 variants along with HPV infection in oral cancer among the prolonged tobacco-chewers," *Tumour Biology*, vol. 40, no. 8, p. 101042831879302, 2018.

[5] D. S. Harischandra, S. Ghaisas, D. Rokad, and A. G. Kanthasamy, "Exosomes in toxicology: relevance to chemical exposure and pathogenesis of environmentally linked diseases," *Toxicological Sciences*, vol. 158, no. 1, pp. 3–13, 2017.

[6] Y. Fujita, N. Kosaka, J. Araya, K. Kuwano, and T. Ochiya, "Extracellular vesicles in lung microenvironment and pathogenesis," *Trends in Molecular Medicine*, vol. 21, no. 9, pp. 533–542, 2015.

[7] Q. Huang, J. Yang, J. Zheng, C. Hsueh, Y. Guo, and L. Zhou, "Characterization of selective exosomal microRNA expression profile derived from laryngeal squamous cell carcinoma detected by next generation sequencing," *Oncology Reports*, vol. 40, no. 5, pp. 2584–2594, 2018.

[8] K. A. Serban, S. Rezania, D. N. Petrusca et al., "Structural and functional characterization of endothelial microparticles released by cigarette smoke," *Scientific Reports*, vol. 6, no. 1, article 31596, 2016.

[9] J. H. Kim, E. Kim, and M. Y. Lee, "Exosomes as diagnostic biomarkers in cancer," *Molecular & Cellular Toxicology*, vol. 14, no. 2, pp. 113–122, 2018.

[10] Y. Fujita, J. Araya, S. Ito et al., "Suppression of autophagy by extracellular vesicles promotes myofibroblast differentiation in COPD pathogenesis," *Journal of Extracellular Vesicles*, vol. 4, no. 1, article 28388, 2015.

[11] L. Xu and X. Deng, "Tobacco-specific nitrosamine 4-(methylnitrosamino)-1-(3-pyridyl)-1-butanone induces phosphorylation of mu- and m-calpain in association with increased secretion, cell migration, and invasion," *The Journal of Biological Chemistry*, vol. 279, no. 51, pp. 53683–53690, 2004.

[12] B. J. Benedikter, E. F. M. Wouters, P. H. M. Savelkoul, G. G. U. Rohde, and F. R. M. Stassen, "Extracellular vesicles released in response to respiratory exposures: implications for chronic disease," *Journal of Toxicology and Environmental Health. Part B, Critical Reviews*, vol. 21, no. 3, pp. 142–160, 2018.

[13] C. J. Li, Y. Liu, Y. Chen, D. Yu, K. J. Williams, and M. L. Liu, "Novel proteolytic microvesicles released from human macrophages after exposure to tobacco smoke," *The American Journal of Pathology*, vol. 182, no. 5, pp. 1552–1562, 2013.

[14] A. Héliot, Y. Landkocz, F. Roy Saint-Georges et al., "Smoker extracellular vesicles influence status of human bronchial epithelial cells," *International Journal of Hygiene and Environmental Health*, vol. 220, no. 2, Part B, pp. 445–454, 2017.

[15] F. Mobarrez, L. Antoniewicz, J. A. Bosson, J. Kuhl, D. S. Pisetsky, and M. Lundbäck, "The effects of smoking on levels of endothelial progenitor cells and microparticles in the blood of healthy volunteers," *PLoS One*, vol. 9, no. 2, article e90314, 2014.

[16] M. Li, D. Yu, K. J. Williams, and M. L. Liu, "Tobacco smoke induces the generation of procoagulant microvesicles from human monocytes/macrophages," *Arteriosclerosis, Thrombosis, and Vascular Biology*, vol. 30, no. 9, pp. 1818–1824, 2010.

[17] H. G. Moon, S. H. Kim, J. Gao et al., "CCN1 secretion and cleavage regulate the lung epithelial cell functions after cigarette smoke," *American Journal of Physiology. Lung Cellular and Molecular Physiology*, vol. 307, no. 4, pp. L326–L337, 2014.

[18] B. J. Benedikter, A. R. Weseler, E. F. M. Wouters, P. H. M. Savelkoul, G. G. U. Rohde, and F. R. M. Stassen, "Redox-dependent thiol modifications: implications for the release of extracellular vesicles," *Cellular and Molecular Life Sciences*, vol. 75, no. 13, pp. 2321–2337, 2018.

[19] R. Bæk, K. Varming, and M. M. Jørgensen, "Does smoking, age or gender affect the protein phenotype of extracellular vesicles in plasma?," *Transfusion and Apheresis Science*, vol. 55, no. 1, pp. 44–52, 2016.

[20] K. Bein and G. D. Leikauf, "Acrolein - a pulmonary hazard," *Molecular Nutrition & Food Research*, vol. 55, no. 9, pp. 1342–1360, 2011.

[21] B. J. Benedikter, C. Volgers, P. H. van Eijck et al., "Cigarette smoke extract induced exosome release is mediated by depletion of exofacial thiols and can be inhibited by thiol-antioxidants," *Free Radical Biology and Medicine*, vol. 108, pp. 334–344, 2017.

[22] K. M. Lim, S. Kim, J. Y. Noh et al., "Low-level mercury can enhance procoagulant activity of erythrocytes: a new contributing factor for mercury-related thrombotic disease," *Environmental Health Perspectives*, vol. 118, no. 7, pp. 928–935, 2010.

[23] R. Vatsyayan, H. Kothari, U. R. Pendurthi, and L. V. M. Rao, "4-Hydroxy-2-nonenal enhances tissue factor activity in human monocytic cells via p38 mitogen-activated protein kinase activation-dependent phosphatidylserine exposure," *Arteriosclerosis, Thrombosis, and Vascular Biology*, vol. 33, no. 7, pp. 1601–1611, 2013.

[24] F. Novelli, T. Neri, L. Tavanti et al., "Procoagulant, tissue factor-bearing microparticles in bronchoalveolar lavage of interstitial lung disease patients: an observational study," *PLoS One*, vol. 9, no. 4, article e95013, 2014.

[25] C. Cordazzo, S. Petrini, T. Neri et al., "Rapid shedding of proinflammatory microparticles by human mononuclear cells exposed to cigarette smoke is dependent on $Ca^{2+}$ mobilization," *Inflammation Research*, vol. 63, no. 7, pp. 539–547, 2014.

[26] A. K. Enjeti, A. Ariyarajah, A. D'Crus, M. Seldon, and L. F. Lincz, "Circulating microvesicle number, function and small RNA content vary with age, gender, smoking status, lipid and hormone profiles," *Thrombosis Research*, vol. 156, pp. 65-72, 2017.

[27] E. Bourdonnay, Z. Zasłona, L. R. K. Penke et al., "Transcellular delivery of vesicular SOCS proteins from macrophages to epithelial cells blunts inflammatory signaling," *The Journal of Experimental Medicine*, vol. 212, no. 5, pp. 729-742, 2015.

[28] Y. Chen, G. Li, Y. Liu, V. P. Werth, K. J. Williams, and M. L. Liu, "Translocation of endogenous danger signal HMGB1 from nucleus to membrane microvesicles in macrophages," *Journal of Cellular Physiology*, vol. 231, no. 11, pp. 2319-2326, 2016.

[29] S. Sheller, J. Papaconstantinou, R. Urrabaz-Garza et al., "Amnion-epithelial-cell-derived exosomes demonstrate physiologic state of cell under oxidative stress," *PLoS One*, vol. 11, no. 6, article e0157614, 2016.

[30] D. J. Schneider, J. M. Speth, L. R. Penke, S. H. Wettlaufer, J. A. Swanson, and M. Peters-Golden, "Mechanisms and modulation of microvesicle uptake in a model of alveolar cell communication," *The Journal of Biological Chemistry*, vol. 292, no. 51, pp. 20897-20910, 2017.

[31] S. Badrnya, R. Baumgartner, and A. Assinger, "Smoking alters circulating plasma microvesicle pattern and microRNA signatures," *Thrombosis and Haemostasis*, vol. 112, no. 07, pp. 128-136, 2017.

[32] R. Grant, E. Ansa-Addo, D. Stratton et al., "A filtration-based protocol to isolate human plasma membrane-derived vesicles and exosomes from blood plasma," *Journal of Immunological Methods*, vol. 371, no. 1-2, pp. 143-151, 2011.

[33] M. T. Severinsen, S. R. Kristensin, S. P. Johnsen, C. Dethlefsen, A. Yjonneland, and K. Overvad, "Smoking and venous thromboembolism: a Danish follow-up study," *Journal of Thrombosis and Haemostasis*, vol. 7, no. 8, pp. 1297-1303, 2009.

[34] U. Garbin, A. Fratta Pasini, C. Stranieri et al., "Cigarette smoking blocks the protective expression of Nrf2/ARE pathway in peripheral mononuclear cells of young heavy smokers favouring inflammation," *PLoS One*, vol. 4, no. 12, article e8225, 2009.

[35] L. Erhardt, "Cigarette smoking: an undertreated risk factor for cardiovascular disease," *Atherosclerosis*, vol. 205, no. 1, pp. 23-32, 2009.

[36] Y. Kuwabara, K. Ono, T. Horie et al., "Increased microRNA-1 and microRNA-133a levels in serum of patients with cardiovascular disease indicate myocardial damage," *Circulation Cardiovascular Genetics*, vol. 4, no. 4, pp. 446-454, 2011.

[37] S. Matsumoto, Y. Sakata, S. Suna et al., "Circulating p53-responsive microRNAs are predictive indicators of heart failure after acute myocardial infarction," *Circulation Research*, vol. 113, no. 3, pp. 322-326, 2013.

[38] M. Folkesson, C. Li, S. Frebelius et al., "Proteolytically active ADAM10 and ADAM17 carried on membrane microvesicles in human abdominal aortic aneurysms," *Thrombosis and Haemostasis*, vol. 114, no. 6, pp. 1165-1174, 2015.

[39] W. J. Kim, J. H. Lim, Y. Hong et al., "Altered miRNA expression in lung tissues of patients with chronic obstructive pulmonary disease," *Molecular & Cellular Toxicology*, vol. 13, no. 2, pp. 207-212, 2017.

[40] W. J. Kim and C. Y. Lee, "Environmental exposures and chronic obstructive pulmonary disease," *Molecular & Cellular Toxicology*, vol. 13, no. 3, pp. 251-255, 2017.

[41] A. D. Lockett, M. B. Brown, N. Santos-Falcon et al., "Active trafficking of alpha 1 antitrypsin across the lung endothelium," *PLoS One*, vol. 9, no. 4, article e93979, 2014.

[42] T. Takahashi, S. Kobayashi, N. Fujino et al., "Increased circulating endothelial microparticles in COPD patients: a potential biomarker for COPD exacerbation susceptibility," *Thorax*, vol. 67, no. 12, pp. 1067-1074, 2012.

[43] P. Reclusa, R. Sirera, A. Araujo et al., "Exosomes genetic cargo in lung cancer: a truly Pandora's box," *Translational Lung Cancer Research*, vol. 5, no. 5, pp. 483-491, 2016.

[44] S. Cui, Z. Cheng, W. Qin, and L. Jiang, "Exosomes as a liquid biopsy for lung cancer," *Lung Cancer*, vol. 116, pp. 46-54, 2018.

[45] J. Guo and Y. Cheng, "MicroRNA-1247 inhibits lipopolysaccharides-induced acute pneumonia in A549 cells via targeting CC chemokine ligand 16," *Biomedicine & Pharmacotherapy*, vol. 104, pp. 60-68, 2018.

[46] T. Fang, H. Lv, G. Lv et al., "Tumor-derived exosomal miR-1247-3p induces cancer-associated fibroblast activation to foster lung metastasis of liver cancer," *Nature Communications*, vol. 9, no. 1, article 191, 2018.

[47] X. Zhao, X. Wu, M. Qian, Y. Song, D. Wu, and W. Zhang, "Knockdown of TGF-$\beta$1 expression in human umbilical cord mesenchymal stem cells reverts their exosome-mediated EMT promoting effect on lung cancer cells," *Cancer Letters*, vol. 428, pp. 34-44, 2018.

[48] C. Jing, H. Cao, X. Qin et al., "Exosome-mediated gefitinib resistance in lung cancer HCC827 cells via delivery of miR-21," *Oncology Letters*, vol. 15, no. 6, pp. 9811-9817, 2018.

[49] C. Alguacil-Núñez, I. Ferrer-Ortiz, E. García-Verdú, P. López-Pirez, I. M. Llorente-Cortijo, and B. Sainz Jr, "Current perspectives on the crosstalk between lung cancer stem cells and cancer-associated fibroblasts," *Critical Reviews in Oncology/Hematology*, vol. 125, pp. 102-110, 2018.

[50] H. Sugimoto, T. M. Mundel, M. W. Kieran, and R. Kalluri, "Identification of fibroblast heterogeneity in the tumor microenvironment," *Cancer Biology & Therapy*, vol. 5, no. 12, pp. 1640-1646, 2006.

[51] J. Cadranel, D. Garfield, A. Lavole, M. Wislez, B. Milleron, and C. Mayaud, "Lung cancer in HIV infected patients: facts, questions and challenges," *Thorax*, vol. 61, no. 11, pp. 1000-1008, 2006.

[52] Y. Fujita, J. Araya, and T. Ochiya, "Extracellular vesicles in smoking-related lung diseases," *Oncotarget*, vol. 6, no. 41, pp. 43144-43145, 2015.

[53] H. Osada and T. Takahashi, "Let-7 and miR-17-92: small-sized major players in lung cancer development," *Cancer Science*, vol. 102, no. 1, pp. 9-17, 2011.

[54] N. Suzuki, K. Nakanishi, M. Yoneda, T. Hirofuji, and T. Hanioka, "Relationship between salivary stress biomarker levels and cigarette smoking in healthy young adults: an exploratory analysis," *Tobacco Induced Diseases*, vol. 14, no. 1, p. 20, 2016.

[55] T. Kawakubo-Yasukochi, M. Morioka, M. Hazekawa et al., "miR-200c-3p spreads invasive capacity in human oral squamous cell carcinoma microenvironment," *Molecular Carcinogenesis*, vol. 57, no. 2, pp. 295-302, 2018.

[56] A. Valavanidis, T. Vlachogianni, and K. Fiotakis, "Tobacco smoke: involvement of reactive oxygen species and stable free radicals in mechanisms of oxidative damage, carcinogenesis and synergistic effects with other respirable particles," *International Journal of Environmental Research and Public Health*, vol. 6, no. 2, pp. 445–462, 2009.

[57] N. D. Gupta, N. Agrawal, N. Gupta, S. Khan, and P. Singh, "Effect of smoking on potential salivary markers of periodontal disease: a clinical and biochemical study," *Journal of Indian Association of Public Health Dentistry*, vol. 14, no. 4, pp. 377–382, 2016.

[58] Y. Li, M. A. St John, X. Zhou et al., "Salivary transcriptome diagnostics for oral cancer detection," *Clinical Cancer Research*, vol. 10, no. 24, pp. 8442–8450, 2004.

[59] N. J. Park, H. Zhou, D. Elashoff et al., "Salivary microRNA: discovery, characterization, and clinical utility for oral cancer detection," *Clinical Cancer Research*, vol. 15, no. 17, pp. 5473–5477, 2009.

[60] J. M. Yoshizawa and D. T. W. Wong, "Salivary microRNAs and oral cancer detection," *Methods in Molecular Biology*, vol. 936, pp. 313–324, 2013.

[61] L. Li, C. Li, S. Wang et al., "Exosomes derived from hypoxic oral squamous cell carcinoma cells deliver miR-21 to normoxic cells to elicit a prometastatic phenotype," *Cancer Research*, vol. 76, no. 7, pp. 1770–1780, 2016.

[62] S. Langevin, D. Kuhnell, T. Parry et al., "Comprehensive microRNA-sequencing of exosomes derived from head and neck carcinoma cells *in vitro* reveals common secretion profiles and potential utility as salivary biomarkers," *Oncotarget*, vol. 8, no. 47, pp. 82459–82474, 2017.

[63] S. Sakha, T. Muramatsu, K. Ueda, and J. Inazawa, "Exosomal microRNA miR-1246 induces cell motility and invasion through the regulation of DENND2D in oral squamous cell carcinoma," *Scientific Reports*, vol. 6, no. 1, article 38750, 2016.

[64] R. Cazzoli, F. Buttitta, M. di Nicola et al., "microRNAs derived from circulating exosomes as noninvasive biomarkers for screening and diagnosing lung cancer," *Journal of Thoracic Oncology*, vol. 8, no. 9, pp. 1156–1162, 2013.

[65] F. Akbas, E. Coskunpinar, E. Aynacı, Y. Müsteri Oltulu, and P. Yildiz, "Analysis of serum micro-RNAs as potential biomarker in chronic obstructive pulmonary disease," *Experimental Lung Research*, vol. 38, no. 6, pp. 286–294, 2012.

[66] Y. Li, Y. Zhang, F. Qiu, and Z. Qiu, "Proteomic identification of exosomal LRG1: a potential urinary biomarker for detecting NSCLC," *Electrophoresis*, vol. 32, no. 15, pp. 1976–1983, 2011.

[67] N. Suh, D. Subramanyam, and M. Y. Lee, "Molecular signatures of secretomes from mesenchymal stem cells: therapeutic benefits," *Molecular & Cellular Toxicology*, vol. 13, no. 2, pp. 133–141, 2017.

[68] V. Sundararajan, F. H. Sarkar, and T. S. Ramasamy, "The versatile role of exosomes in cancer progression: diagnostic and therapeutic implications," *Cellular Oncology*, vol. 41, no. 3, pp. 223–252, 2018.

# Dietary DHA/EPA Ratio Changes Fatty Acid Composition and Attenuates Diet-Induced Accumulation of Lipid in the Liver of ApoE$^{-/-}$ Mice

Liang Liu,[1,2,3] Qinling Hu,[1,2] Huihui Wu,[1,2] Xiujing Wang,[1,2] Chao Gao,[4] Guoxun Chen,[5] Ping Yao,[6] and Zhiyong Gong[1,2,3]

[1]*Key Laboratory for Deep Processing of Major Grain and Oil, Ministry of Education, Wuhan 430023, China*
[2]*College of Food Science and Engineering, Wuhan Polytechnic University, Wuhan 430023, China*
[3]*Hubei Key Laboratory for Processing and Transformation of Agricultural Products (Wuhan Polytechnic University), Wuhan 430023, China*
[4]*National Institute for Nutrition and Health, Chinese Center for Disease Control and Prevention, Beijing 100050, China*
[5]*Department of Nutrition, University of Tennessee at Knoxville, Knoxville 37996, USA*
[6]*Department of Nutrition and Food Hygiene, School of Public Health, Tongji Medical College, Huazhong University of Science and Technology, Wuhan 430030, China*

Correspondence should be addressed to Zhiyong Gong; gongwhpu@126.com

Academic Editor: Joan Roselló-Catafau

Diets containing various docosahexaenoic acid (DHA)/eicosapentaenoic acid (EPA) ratios protect against liver damage in mice fed with a high-fat diet (HFD). However, it is unclear whether these beneficial roles of DHA and EPA are associated with alterations of fatty acid (FA) composition in the liver. This study evaluated the positive impacts of n-6/n-3 polyunsaturated fatty acids (PUFAs) containing different DHA/EPA ratios on HFD-induced liver disease and alterations of the hepatic FA composition. ApoE$^{-/-}$ mice were fed with HFDs with various ratios of DHA/EPA (2 : 1, 1 : 1, and 1 : 2) and an n-6/n-3 ratio of 4 : 1 for 12 weeks. After treatment, the serum and hepatic FA compositions, serum biochemical parameters, liver injury, and hepatic lipid metabolism-related gene expression were determined. Our results demonstrated that dietary DHA/EPA changed serum and hepatic FA composition by increasing contents of n-6 and n-3 PUFAs and decreasing amounts of monounsaturated fatty acids (MUFAs) and the n-6/n-3 ratio. Among the three DHA/EPA groups, the DHA/EPA 2 : 1 group tended to raise n-3 PUFAs concentration and lower the n-6/n-3 ratio in the liver, whereas DHA/EPA 1 : 2 tended to raise n-6 PUFAs concentration and improve the n-6/n-3 ratio. DHA/EPA supplementation reduced the hepatic impairment of lipid homeostasis, oxidative stress, and the inflammatory responses in HFD-fed mice. The DHA/EPA 2 : 1 group had lower serum levels of total cholesterol, triglycerides, and low-density lipoprotein cholesterol and higher levels of adiponectin than HFD group. The DHA/EPA 1 : 2 group had elevated serum levels of aspartate aminotransferase, alanine aminotransferase, and alkaline phosphatase, without significant change the expression of genes for inflammation or hepatic lipid metabolism among the three DHA/EPA groups. The results suggest that DHA/EPA-enriched diet with an n-6/n-3 ratio of 4 : 1 may reverse HFD-induced nonalcoholic fatty liver disease to some extent by increasing n-6 and n-3 PUFAs and decreasing the amount of MUFAs and the n-6/n-3 ratio.

## 1. Introduction

Nonalcoholic fatty liver disease (NAFLD), the most common chronic liver disease worldwide, is one of the major causes of the fatty liver, occurring when fat is deposited in the liver in the absence of excessive alcohol intake [1, 2]. Currently, its prevalence in Asia is estimated to be 25%, similar to the incidence in many western countries (20–30%), and is even as high as 40% in westernized Asian populations [3]. The development of NAFLD is directly associated with enhancement in prooxidant status [4], proinflammatory status [5], and lipid content [4, 6] of the liver in mice fed with a high-fat diet (HFD) [7].

Lifestyle modification, including dietary changes, weight loss, and physical activity, is the initial treatment option for patients with NAFLD [8]. On the other hand, dietary modification may benefit the treatment of NAFLD without significant weight loss [9]. Accumulating clinical evidence has revealed that low levels of n-3 polyunsaturated fatty acids (n-3 PUFAs), including α-linolenic acid (ALA), in serum and liver tissue samples are common characteristics of patients with alcoholic disease and NAFLD [10, 11], which may be attributed to impaired bioavailability of liver n-6 and n-3 PUFAs [11–13]. Jump et al. [14] provided an in-depth rationale for the use of dietary n-3 PUFA supplements as a treatment option for NAFLD. Experimental and clinical data on n-3 PUFAs have also demonstrated that dietary supplementation with eicosapentaenoic acid (EPA, C20:5) and docosahexaenoic acid (DHA, C22:6) prevents or alleviates NAFLD [15]. Additionally, a recent transcriptomic study showed that fish oil protected against HFD- and high-cholesterol diet-induced NALFD by improving lipid metabolism and ameliorating hepatic inflammation in Sprague-Dawley rats [16]. We also reported that diet rich in DHA and/or EPA improved lipid metabolism and had anti-inflammatory effects in HFD-induced NALFD in C57BL/6J mice [17]. Thus, daily intake of DHA and EPA for healthy adults as well as those with coronary artery diseases and hypertriglyceridemia is strongly recommended by authority organizations. However, the precise requirement for marine n-3 PUFAs is not known [9].

Recently, the effects of different dietary n-6/n-3 ratio on health and disease have drawn close attention. A higher intake of n-6 FA and higher dietary n-6/n-3 FA ratio were reported in NAFLD subjects [18]. On the other hand, additional evidence also highlighted the role of ratios of DHA and EPA in the prevention and treatment of chronic disease in rat models [19–22], indicating the importance of both n-6/n-3 ratios and DHA/EPA ratios. It has been known that the intake of dietary fat alters the FA composition of plasma and various organs, including the liver [12, 18]. Lipidomics analysis has also revealed the role of different EPA/DHA ratios in the modulation of inflammation and oxidative markers in genetically obese hypertensive rats through the downregulation of the production of proinflammatory n-6 eicosanoids [23]. We previously showed that an oral administration of n-6/n-3 PUFAs with varying DHA/EPA ratios for 12 weeks ameliorated atherosclerosis lesions [24] and liver damage [17] in mice fed with an HFD. Data from aforementioned studies suggested the positive effects of supplementation with varying DHA/EPA ratios on the metabolic parameters of HFD-fed animals. However, there have been few studies on the protective role of n-6/n-3 PUFA supplementation with varying DHA/EPA ratios against HFD-induced liver damage and its correlation with hepatic FA composition.

Therefore, the focus of this study was to evaluate the positive effects of n-6/n-3 PUFA supplementation with varying DHA/EPA ratios on liver disease induced by an HFD as well as the associated alterations of FA composition of the liver.

## 2. Materials and Methods

*2.1. Animals and Diets.* Male apolipoprotein E knockout ($ApoE^{-/-}$) mice at weaning (C57/BL6 background, 6 weeks old, $20 \pm 2$ g) were obtained from Vital River Laboratories (Beijing, China). All of the mice were housed in a humidity and temperature controlled room (relative humidity, 65–75%; temperature, 20–24°C) with a 12 h : 12 h light/dark cycle and were given *ad libitum* access to their specific diets and water. After a 1-week acclimation, the mice were randomly divided into the following five groups: (1) normal diet (ND) group (control group received an ND of basic feed 86%, casein 4%, and yolk powder 10%), (2) HFD group received HFD I (basic feed 70%, 15% lard, 1% cholesterol, casein 4%, and yolk powder 10%), and (3–5) DHA/EPA groups (2 : 1, 1 : 1, and 1 : 2) received HFD II (basic feed 75%, 10% lard, 1% cholesterol, casein 4%, and yolk powder 10%) plus mixed oil. The mixed oil (including sunflower seed, perilla, fish, and algal oils) was formulated by the previous method [24] for partial replacement of 5% lard, with adjustment of the n-6/n-3 ratio to 4 : 1 and with variation in the DHA/EPA ratios (2 : 1, 1 : 1, and 1 : 2). The diets were prepared according to the previous method [17, 24]. The FA profiles of oils, basic feed, and HFDs were quantified by gas chromatography [24]. The FA compositions of oils, basic feed, control diet, and HFDs are shown in Table 1. The lipids were administered orally (1 g/kg body weight (BW)) for 12 weeks. The ND and HFD groups were given the same dose of physiological saline via intragastric administration. Their BWs were recorded once a week. The *Guide for the Care and Use of Laboratory Animals* by the National Institutes of Health (Bethesda, MD, USA) was followed during the experiments [25]. The animal protocol was approved by the Tongji Medical College Council on Animal Care Committee (Wuhan, China). At the end of the experiments, mice after 12 h of fasting were anesthetized with isoflurane before blood and tissue sample collections. Serum was collected from blood after agglutination and centrifugation at 4000 ×g at 4°C for 10 min and then stored at −80°C. Fresh tissue samples were fixed for histopathology determinations or were quick-frozen in liquid nitrogen for quantitative PCR (qPCR) and western blot analyses.

*2.2. Lipid Extraction and FA Analysis.* Total lipid from serum or liver tissue homogenates was extracted with ice-cold chloroform/methanol (2 : 1 *v/v*) with 0.01% butylated hydroxytoluene. After centrifugation, the phase interface was washed with chloroform/methanol/water (3 : 48 : 47 *v/v/v*). Methyl esterification of the lipids was conducted according to the previous report [26]. Fatty acid methyl esters (FAMEs) were quantified using the Agilent Technologies 6890 Gas Chromatograph (Agilent Technologies Inc., Savage, MD, USA) with a flame ionization detector. Separation of the FAMEs was performed on the HP-INNOWax capillary column (30 × 0.32, 0.25 μm; Agilent) using helium as carrier gas at a constant flow of 1.5 mL/min. The samples were injected at a

TABLE 1: Fatty acid composition of oils and feed supplemented to mice.

| Fatty acid | Sunflower seed oil | Perilla oil | Fish oil | Algal oil | Basic diet | HFD I | HFD II | DHA/EPA = 2:1 | DHA/EPA = 1:1 | DHA/EPA = 1:2 |
|---|---|---|---|---|---|---|---|---|---|---|
| | | | | | mg per 100 mg total fatty acid | | | | | |
| C14:0 | 0.1 | 0 | 0.3 | 8.7 | 0 | 0 | 0 | 1.4 | 0.8 | 0.1 |
| C14:1 | 0 | 0 | 0.2 | 0.3 | 0 | 0 | 0 | 0.1 | 0.1 | 0.1 |
| C15:0 | 0 | 0 | 0.5 | 0.2 | 0 | 0 | 0 | 0.1 | 0.1 | 0.1 |
| C16:0 | 6.5 | 0.4 | 10.1 | 15.8 | 27.0 | 42.0 | 43.7 | 8.6 | 8.1 | 8.0 |
| C16:1 | 0.1 | 0.1 | 4.7 | 3.5 | 0 | 1.8 | 1.7 | 1.1 | 1.2 | 1.5 |
| C17:0 | 0 | 0 | 0.1 | 0.3 | 0 | 0 | 0 | 0.1 | 0.1 | 0 |
| C17:1 | 0 | 0 | 0.1 | 0 | 0 | 0 | 0 | 0 | 0 | 0 |
| C18:0 | 5.2 | 0 | 2.8 | 0 | 4.7 | 0.2 | 0.3 | 4.4 | 4.6 | 4.9 |
| C18:1 | 24.9 | 10.1 | 7.9 | 13.5 | 21.2 | 16.5 | 15.6 | 22.2 | 22.4 | 22.2 |
| C18:2 | 60.7 | 15.4 | 1.9 | 1.6 | 36.9 | 21.5 | 21.7 | 48.4 | 49.3 | 49.9 |
| C18:3 | 0 | 73.3 | 2.0 | 0 | 0 | 16.0 | 15.3 | 0.2 | 0.4 | 0.2 |
| C20:1 | 0 | 0 | 0.1 | 0.3 | 0 | 0.13 | 0.12 | 0.1 | 0.1 | 0 |
| C20:2 | 0 | 0 | 0.2 | 0 | 0 | 0 | 0 | 0 | 0 | 0.1 |
| C20:4 | 0 | 0 | 1.5 | 0.5 | 0.5 | 0.19 | 0.20 | 0.2 | 0.3 | 0.4 |
| C20:5 | 0.2 | 0 | 35.1 | 0.0 | 0 | 0 | 0 | 4.3 | 6.2 | 8.1 |
| C22:1 | 0 | 0 | 0.3 | 3.5 | 0 | 0 | 0 | 0.5 | 0.3 | 0.1 |
| C22:5 | 0 | 0 | 2.4 | 0.0 | 0 | 0 | 0 | 0.3 | 0.4 | 0.6 |
| C22:6 | 0 | 0 | 18.2 | 40.1 | 0 | 0 | 0 | 8.0 | 6.0 | 4.3 |
| $\sum$SATs | 11.8 | 0.4 | 13.8 | 25 | 31.7 | 42.2 | 44 | 14.6 | 13.7 | 13.1 |
| $\sum$MUFAs | 25 | 10.2 | 13.3 | 21.1 | 21.2 | 18.43 | 17.42 | 24 | 24.1 | 23.9 |
| $\sum$PUFAs | 60.9 | 88.7 | 61.3 | 42.2 | 37.4 | 37.69 | 37.2 | 61.4 | 62.6 | 63.6 |
| $\sum$n-6 | 60.7 | 15.4 | 3.4 | 2.1 | 36.9 | 21.5 | 21.7 | 48.6 | 49.6 | 50.3 |
| $\sum$n-3 | 0.2 | 73.3 | 57.7 | 40.1 | 0 | 16.0 | 15.3 | 12.8 | 13.0 | 13.2 |
| n-6/n-3 | | | | | | 1.3 | 1.4 | 3.9 | 3.8 | 3.8 |
| EPA/DHA | 0 | 0 | 1.9 | 0 | 0 | 0 | 0 | 1.9 | 1.0 | 0.5 |

starting oven temperature of 50°C; the injector and detector temperatures were 250°C. The oven temperature was programmed as follows: 50°C, 1 min, 15°C/min to 175°C, 5 min, and 1°C/min to 250°C. The FAMEs were identified by comparing with authentic standards (Nu-Chek-Prep) and were calculated as the percent area of total FAs.

2.3. Histopathological Analysis. Fresh liver slices were processed by hematoxylin and eosin (H&E) staining. Briefly, liver tissues were cut into slices and fixed, and then, the samples were dehydrated and embedded with paraffin. Paraffin-embedded tissue sections (5 μm) were stained with H&E and observed under the Olympus BX50 light microscope (Olympus, Tokyo, Japan).

2.4. Measurements of Serum Parameters and Fat Liver Content. Serum total cholesterol (TC, mM), triglyceride (TG, mM), low-density lipoprotein cholesterol (LDL-C, mM), high-density lipoprotein cholesterol (HDL-C, mM) levels, and hepatic TC (mM/g protein) and TG (mM/g protein) were determined by spectrophotometric methods using the respective kits (Biosino Biotechnology Co. Ltd., Beijing, China) according to the manufacturer's instructions. Serum aspartate transaminase (AST, U/L), alanine transaminase (ALT, U/L), and alkaline phosphatase activities (AKP, U/L) were measured using specific diagnostic kits (Nanjing Jiancheng Corporation, Nanjing, China). Enzyme-linked immunoassay (ELISA) kits were used to assess the serum levels of tumor necrosis factor alpha (TNF-$\alpha$, pg/mL), interleukin-1$\beta$ (IL-1$\beta$, pg/mL), and adiponectin (Cloud-Clone Corp., Wuhan, China).

2.5. Analysis of Hepatic Malondialdehyde, Superoxide Dismutase, and Glutathione. Hepatic malondialdehyde (MDA, μM/g protein), glutathione (GSH, μM/g protein), and superoxide dismutase (SOD, U/mg protein) were determined using the respective kits (Nanjing Jiancheng Corporation, Nanjing, China).

2.6. qPCR Analysis. Total RNA of mouse liver samples was extracted using the TRIzol reagent (Ambion®, Life Technologies, Austin, TX, USA) according to the manufacturer's instructions. Messenger RNA (mRNA) expression levels of the target genes were quantified using the SYBR Green-based Kit (Takara Bio Inc., Dalian, China) with specific primers and a real-time PCR machine for qPCR (IQ5;

TABLE 2: Quantitative PCR primer sequences.

| Gene | Forward primer 5'-3' | Reverse primer 5'-3' |
|---|---|---|
| IL-6 | TCCAGTTGCCTTCTTGGGAC | AGTCTCCTCTCCGGACTTGT |
| IL-10 | GCTGCCTGCTCTTACTGACT | CTGGGAAGTGGGTGCAGTTA |
| IL-1$\beta$ | TGCCACCTTTTGACAGTGATG | TGATGTGCTGCTGCGAGATT |
| TNF-$\alpha$ | ATGGCCTCCCTCTCATCAGT | TTTGCTACGACGTGGGCTAC |
| MCP-1 | TATTGGCTGGACCAGATGCG | CCGGACGTGAATCTTCTGCT |
| VCAM-1 | CTGGGAAGCTGGAACGAAGT | GCCAAACACTTGACCGTGAC |
| ICAM-1 | TATGGCAACGACTCCTTCT | CATTCAGCGTCACCTTGG |
| CD36 | CGGGCCACGTAGAAAACACT | CAGCCAGGACTGCACCAATA |
| MSR-1 | GACTTCGTCATCCTGCTCAAT | GCTGTCGTTCTTCTCATCCTC |
| LOX-1 | TCACCTGCTCCCTGTCCTT | GGTTCTTTGCCTCAATGCC |
| ABCA-1 | CGACCATGAAAGTGACACGC | AGCACATAGGTCAGCTCGTG |
| ABCG-1 | AGAGCTGTGTGCTGTCAGTC | AGCAGGTCTCAGGGTCTAGG |
| LAL | CCCACCAAGTAGGTGTAGGC | GAGTTGCATCGGGAGTGGTC |
| ACAT-1 | CCAATGCCAGCACACTGAAC | TCTACGGCAGCATCAGCAAA |
| $\beta$-Actin | TTCGTTGCCGGTCCACACCC | GCTTTGCACATGCCGGAGCC |

TABLE 3: Effects of DHA/EPA supplementation on body and liver weights in each group.

|  | Initial weight (g) | Final weight (g) | Weight gain (g) | Liver weight (g) | Liver ratio to weight (%) |
|---|---|---|---|---|---|
| ND | 20.6 ± 1.2 | 25.4 ± 2.8 | 4.8 ± 3.0 | 1.03 ± 0.23 | 4.2 ± 0.4 |
| HFD | 20.5 ± 1.1 | 27.0 ± 2.5 | 6.5 ± 2.7 | 1.15 ± 0.17 | 4.5 ± 0.4 |
| DHA/EPA = 2 : 1 | 20.8 ± 1.0 | 26.8 ± 2.5 | 6.0 ± 2.5 | 1.10 ± 0.21 | 4.3 ± 0.4 |
| DHA/EPA = 1 : 1 | 20.7 ± 1.2 | 25.8 ± 2.6 | 4.9 ± 2.6 | 1.04 ± 0.20 | 4.2 ± 0.4 |
| DHA/EPA = 1 : 2 | 20.3 ± 1.5 | 25.6 ± 2.4 | 5.3 ± 2.5 | 1.06 ± 0.15 | 4.3 ± 0.3 |

Data are given as the mean ± SEM, $n = 10$.

Bio-Rad, Hercules, CA, USA). The mRNA level of $\beta$-actin was used as the invariable control for quantification, and the results were calculated by the comparative $2^{-\Delta\Delta Ct}$ method. The sequences of the forward and reverse primers used for the detection of the target genes are listed in Table 2.

2.7. Western Blot Analysis. The liver tissues were homogenized in radioimmunoprecipitation assay lysis buffer (1% Triton X-100, 1% deoxycholate, and 0.1% sodium dodecyl sulfate (SDS)), and protein concentration was measured. Equal amounts of protein extracts were mixed (3 : 1, v/v) and processed in loading buffer for electrophoresis in 10% acrylamide SDS gels and subsequently electroblotted to a nitrocellulose transfer membrane (Merck Millipore, Burlington, MA, USA) using a Trans-Blot SD semidry electrophoretic transfer cell (Bio-Rad). Target proteins were probed with specific primary antibodies, and then, the bound primary antibodies were recognized with species-specific secondary antibodies. The chemiluminescence intensity of the specific proteins on the membrane was subsequently detected using the SuperSignal West Pico Chemiluminescent Substrate (Thermo Fisher Scientific, Waltham, MA, USA) and a western blotting detection system (Bio-Rad). The optical densities (OD) of the bands were quantified using the Gel-Pro 3.0 software (Biometra, Goettingen, Germany). The density of the specific protein band was corrected to eliminate background noise and normalized to that of GAPDH (Boster Biological Technology Ltd., Wuhan, China) as OD/mm$^2$.

2.8. Statistical Analysis. Statistical analysis was performed with the GraphPad Prism 4.0.3 software (GraphPad Prism Software Inc., San Diego, USA). Data were presented as mean ± standard error of the mean (SEM). One-way analysis of variance was performed with Fisher's least significant difference multiple comparison post hoc test. A $P < 0.05$ was considered statistically significant.

## 3. Results

3.1. Dietary DHA/EPA Reduces HFD-Induced Liver Injury. Treatment with DHA/EPA did not change the BWs and liver weights in the study. The mice in the five dietary groups showed similar initial BWs, final BWs, and liver/BW ratio (Table 3). The hepatic histological changes were observed by light microscopy of tissue sections with stained H&E (Figure 1). The main change that occurred in the liver from the HFD group was macrovesicular steatosis, as determined by the observation of lipid vesicles in the cytosolic

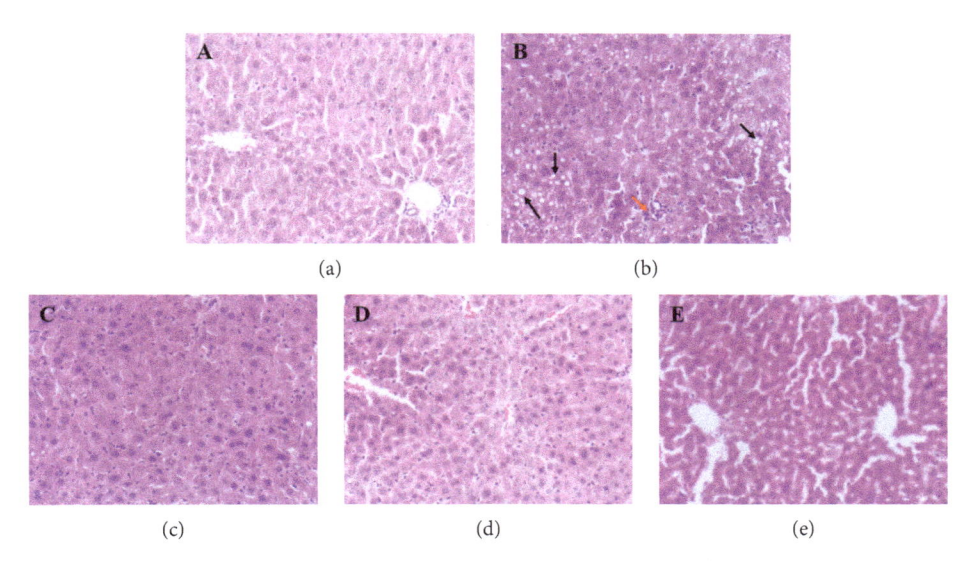

FIGURE 1: Effects of the supplementation of various DHA/EPA ratios on hepatic lipid metabolism. H&E staining of liver sections in each group, followed by observation under a light microscope (magnification 200x). Notice the fatty vesicles (black arrow) and lymphocyte infiltration (red arrow). (a) Normal diet (ND) group, (b) high-fat diet (HFD) group, (c) DHA/EPA 2:1 (DHA/EPA = 2:1) group, (d) DHA/EPA 1:1 (DHA/EPA = 1:1) group, and (e) DHA/EPA 1:2 (DHA/EPA = 1:2) group.

TABLE 4: General and biochemical parameters in serum and liver tissues.

|  | ND | HFD | DHA/EPA = 2:1 | DHA/EPA = 1:1 | DHA/EPA = 1:2 |
| --- | --- | --- | --- | --- | --- |
| Serum parameters |  |  |  |  |  |
| TC (mM) | $9.50 \pm 0.46$ | $19.38 \pm 0.66^a$ | $5.43 \pm 0.52^{a,b}$ | $7.82 \pm 0.84^{b,c}$ | $10.29 \pm 0.31^{b,c,d}$ |
| TG (mM) | $1.19 \pm 0.05$ | $2.38 \pm 0.24^a$ | $0.58 \pm 0.05^{a,b}$ | $0.75 \pm 0.08^{a,b}$ | $1.30 \pm 0.07^{b,c,d}$ |
| LDL (mM) | $3.51 \pm 0.19$ | $7.03 \pm 0.46^a$ | $2.55 \pm 0.43^b$ | $4.17 \pm 0.40^{b,c}$ | $4.34 \pm 0.17^{a,b,c}$ |
| HDL (mM) | $0.30 \pm 0.03$ | $0.13 \pm 0.03^a$ | $0.26 \pm 0.03^{a,b}$ | $0.35 \pm 0.05^b$ | $0.21 \pm 0.01^{a,b,d}$ |
| Adiponectin (pg/mg) | $159.76 \pm 23.19$ | $81.64 \pm 8.36^a$ | $196.77 \pm 18.68^b$ | $114.91 \pm 14.16^c$ | $103.97 \pm 7.43^{b,c}$ |
| OX-LDL ($\mu$g/L) | $223.46 \pm 25.32$ | $269.00 \pm 14.73$ | $171.58 \pm 8.58^b$ | $165.90 \pm 8.29^b$ | $165.89 \pm 10.76^{a,b}$ |
| AST (U/L) | $143.79 \pm 21.97$ | $487.5 \pm 95.19^a$ | $63.62 \pm 7.36^{a,b}$ | $110.23 \pm 13.31^{b,c}$ | $138.32 \pm 28.99^{b,c}$ |
| ALT (U/L) | $72.92 \pm 9.06$ | $210.82 \pm 23.72^a$ | $40.69 \pm 4.88^{a,b}$ | $51.67 \pm 3.05^{a,b}$ | $93.06 \pm 13.03^{b,c,d}$ |
| AKP (U/L) | $48.94 \pm 5.18$ | $90.66 \pm 7.22^a$ | $42.08 \pm 3.50^b$ | $58.93 \pm 5.92^{b,c}$ | $70.34 \pm 4.55^{a,b,c}$ |
| Liver parameters |  |  |  |  |  |
| TC (mM/g protein) | $74.94 \pm 3.62$ | $144.57 \pm 4.72^a$ | $73.81 \pm 4.22^b$ | $83.72 \pm 3.30^{b,c}$ | $77.59 \pm 6.03^b$ |
| TG (mM/g protein) | $204.01 \pm 25.74$ | $231.19 \pm 14.54$ | $125.21 \pm 14.26^{a,b}$ | $114.81 \pm 7.32^{a,b}$ | $160.34 \pm 15.76$ |
| MDA ($\mu$M/g protein) | $1.75 \pm 0.17$ | $2.31 \pm 0.18^a$ | $1.81 \pm 0.14^b$ | $1.73 \pm 0.07^b$ | $1.90 \pm 0.08$ |
| SOD (U/mg protein) | $6.2 \pm 0.32$ | $5.79 \pm 0.16$ | $6.9 \pm 0.14^b$ | $6.85 \pm 0.03^{a,b}$ | $6.86 \pm 0.26^b$ |
| GSH ($\mu$M/g protein) | $19.46 \pm 2.37$ | $10.8 \pm 1.57^a$ | $21.9 \pm 2.59^b$ | $23.81 \pm 1.86^b$ | $24.33 \pm 2.69^b$ |

Data are given as mean ± SEM, $n = 8$. $^aP < 0.05$ versus the ND group; $^bP < 0.05$ versus the HFD group; $^cP < 0.05$ versus the DHA/EPA = 2:1 group; $^dP < 0.05$ versus the DHA/EPA = 1:1 group.

compartment, along with neutrophil and lymphocyte infiltration. However, DHA/EPA-supplemented mice were much fewer and smaller hepatic fatty vesicles than the HFD group mice did.

As illustrated in Table 4, compared with ND-fed mice, serum levels of AST, ALT, and AKP levels were higher ($P < 0.05$) in HFD-fed mice. However, various ratios of DHA/EPA supplementation significantly alleviated HFD-induced liver injury by reducing serum levels of AST (ranging from 71.6% to 86.9%), ALT (ranging from 66.6% to 80.7%), and AKP (ranging from 22.4% to 53.6%). No significant change was observed in the activities of serum aminotransferases among the DHA/EPA groups; AST, ALT, and AKP levels were highest in the DHA/EPA 1:2 group.

Hepatic MDA was significantly boosted in HFD-fed mice compared to that in the ND-fed mice (Table 4). The MDA production was markedly decreased by DHA/EPA supplementation. However, the inhibitory effects of different DHA/EPA ratios on MDA production were not significantly different (Table 4). In contrast to that in the HFD group, serum levels of GSH (increased more than 2-fold) and SOD

Table 5: Fatty acid composition (%) of the serum of mice during the experimental period.

| Serum fatty acid | ND ($n=5$) | HFD ($n=5$) | DHA/EPA = 2 : 1 ($n=4$) | DHA/EPA = 1 : 1 ($n=5$) | DHA/EPA = 1 : 2 ($n=5$) |
| --- | --- | --- | --- | --- | --- |
| C16:0 | 22.881 ± 0.863 | 23.293 ± 0.271 | 22.759 ± 0.762 | 22.068 ± 0.763 | 20.838 ± 1.006 |
| C16:1 | 1.069 ± 0.117 | 1.473 ± 0.145[a] | 0.773 ± 0.287[b] | 0.65 ± 0.168[b] | 0.666 ± 0.174[b] |
| C18:0 | 8.151 ± 0.345 | 11.729 ± 0.440[a] | 9.163 ± 0.510[b] | 9.071 ± 0.336[b] | 8.871 ± 0.576[b] |
| C18:1 | 16.136 ± 0.603 | 26.315 ± 0.857[a] | 17.374 ± 1.279[b] | 16.886 ± 1.21[b] | 16.174 ± 1.967[b] |
| C18:2 | 30.819 ± 1.416 | 20.912 ± 0.605[a] | 30.538 ± 1.028[b] | 30.515 ± 1.039[b] | 28.062 ± 1.046[a,b] |
| C18:3 | 0.00 ± 0.00 | 0.00 ± 0.00 | 0.04 ± 0.08 | 0.00 ± 0.00 | 0.00 ± 0.00 |
| C19:0 | 0.836 ± 0.040 | 0.538 ± 0.137 | 0.500 ± 0.168 | 0.266 ± 0.1646[a] | 0.260 ± 0.162[a] |
| C20:0 | 0.112 ± 0.112 | 0.082 ± 0.0.082 | 0.133 ± 0.133 | 0.098 ± 0.098 | 0.00 ± 0.00 |
| C20:1 | 0.106 ± 0.106 | 0.228 ± 0.140 | 0.748 ± 0.329 | 0.416 ± 0.289 | 0.424 ± 0.309 |
| C20:4 | 4.265 ± 0.272 | 7.177 ± 0.458[a] | 3.67 ± 0.132[b] | 3.946 ± 0.126[b] | 4.242 ± 0.184[b] |
| C22:0 | 2.01 ± 0.059 | 0.660 ± 0.093 | 4.018 ± 1.264[a,b] | 4.383 ± 0.518[a,b] | 5.472 ± 0.775[a,b] |
| C20:5 | 0.752 ± 0.313 | 0.00 ± 0.00[a] | 0.275 ± 0.166 | 0.102 ± 0.102[a] | 0.130 ± 0.130[a] |
| C22:6 | 7.028 ± 0.305 | 3.555 ± 0.221[a] | 8.389 ± 0.371[a,b] | 8.347 ± 0.265[a,b] | 7.358 ± 0.429[b,c,d] |
| ∑SFAs | 33.987 ± 1.153 | 36.305 ± 0.321 | 36.575 ± 0.697 | 35.885 ± 0.986 | 35.439 ± 0.969 |
| ∑MUFAs | 17.314 ± 0.598 | 28.018 ± 0.805[a] | 18.895 ± 1.131[b] | 17.952 ± 1.090[b] | 17.266 ± 1.845[b] |
| ∑PUFAs | 42.866 ± 1.969 | 31.648 ± 0.510[a] | 42.878 ± 1.189[b] | 42.908 ± 0.721[b] | 39.794 ± 1.124[b] |
| ∑n-6 | 35.086 ± 1.659 | 28.092 ± 0.339[a] | 34.213 ± 1.021[b] | 34.460 ± 0.927[b] | 32.306 ± 1.134[b] |
| ∑n-3 | 7.780 ± 0.393 | 3.556 ± 0.222[a] | 8.665 ± 0.493[b] | 8.448 ± 0.253[b] | 7.488 ± 0.388[b,c] |
| n-6/n-3 | 4.522 ± 0.162 | 8.016 ± 0.469[a] | 3.983 ± 0.241[b] | 4.105 ± 0.224[b] | 4.372 ± 0.320[b] |

Data are given as the mean ± SEM. [a]$P < 0.05$ versus the ND group; [b]$P < 0.05$ versus the HFD group; [c]$P < 0.05$ versus the DHA/EPA = 2 : 1 group; [d]$P < 0.05$ versus the DHA/EPA = 1 : 1 group.

(increased by 18.5%) were notably elevated in DHA/EPA-treated mice (Table 4). However, no significant differences of MDA, SOD, and GSH among the three DHA/EPA ratios were observed.

### 3.2. Dietary DHA/EPA Changes FA Composition of the Serum and Liver.

FA compositions of the serum and liver samples in mice after the 12-week feeding of the HFD are shown in Tables 5 and 6, respectively. When the FA compositions of total liver lipids were compared, a significant decrease ($P < 0.05$) of total saturated fatty acids (SFAs) was observed in the HFD group compared with that in the ND group. This trend occurred in the abundance of total PUFAs (26.7% difference) ($P < 0.001$), including total n-6 and n-3 PUFAs (16.6% and 54.7% difference, respectively) with an 84.9% increase in the ratio of n-6/n-3. Also, the content of total MUFAs was significantly increased ($P < 0.01$) due to significant increases in 16:1 (palmitoleic acid) and C18:1 (oleic acid; 130% difference).

Among the varying ratios of DHA/EPA groups, we found an increase in SFAs (DHA/EPA 2 : 1 group, 19.6%; DHA/EPA 1 : 1 group, 14.5%), PUFAs n-6 series (DHA/EPA 2 : 1 group, 11.1%; DHA/EPA 1 : 1 group, 9.1%; and DHA/EPA 1 : 2 group, 17.9%), and PUFA n-3 series (DHA/EPA 2 : 1 group, 166.4%; DHA/EPA 1 : 1 group, 151.7%; and DHA/EPA 1 : 2 group, 126.3%) in the liver compared to the HFD group. Also, the amount of MUFAs (DHA/EPA 2 : 1 group, 49.7%; DHA/EPA 1 : 1 group, 41.8%; and DHA/EPA 1 : 2 group, 35.3%) and the ratio of n-6/n-3 (DHA/EPA 2 : 1 group, 58.3%; DHA/EPA 1 : 1 group, 55.9%; and DHA/EPA 1 : 2 group, 48.1%) showed a marked decrease after DHA/EPA supplementation. Among the three DHA/EPA groups, DHA/EPA 1 : 2 group had the lowest C18:0 and C20:1 concentration and the highest C18:2 and n-6 PUFA concentration. The DHA/EPA 2 : 1 group showed a tendency to raise n-3 PUFA concentration and lower SFAs, C20:5 and C22:0 concentrations, and n-6/n-3 ratio.

Concerning serum FA composition, the same trend was observed for the amount of MUFAs, PUFAs n-6 series, PUFAs n-3 series, and the ratio of n-6/n-3 in the three DHA/EPA groups compared with the HFD group. However, no significant difference among the three DHA/EPA ratios was found for the amount of SFAs, MUFAs, PUFAs n-6 series, and the ratio of n-6/n-3.

### 3.3. Dietary DHA/EPA Ameliorates HFD-Induced Hepatic Inflammation.

The serum concentrations of both IL-1$\beta$ and TNF-$\alpha$ were significantly lower in the three DHA/EPA-treated groups than those in the HFD group (Figure 2). In DHA/EPA-treated mice, the TNF-$\alpha$ level decreased by more than 30%. A similar trend was observed for serum levels of IL-1$\beta$. Consistent with findings for serum levels of proinflammatory cytokines, the data of qPCR analysis demonstrated significantly reduced hepatic expression levels of IL-6, IL-1$\beta$, TNF-$\alpha$, monocyte chemoattractant protein-1 (MCP-1), vascular cell adhesion molecule-1 (VCAM-1), and intercellular adhesion molecule-1 (ICAM-1) in DHA/EPA-treated mice compared to those in the HFD-treated mice (Figure 2). The mRNA expression levels of the anti-inflammatory cytokine IL-10 were increased by 51.0%, 47.8%, and 38.0% in mice treated with DHA/EPA ratios of 1 : 2, 1 : 1, and 2 : 1, respectively.

TABLE 6: Fatty acid composition (%) of the liver of mice during the experimental period.

| Hepatic fatty acid | ND ($n=5$) | HFD ($n=4$) | DHA/EPA = 2:1 ($n=5$) | DHA/EPA = 1:1 ($n=4$) | DHA/EPA = 1:2 ($n=3$) |
|---|---|---|---|---|---|
| C16:0 | 26.228 ± 0.799 | 20.211 ± 0.217[a] | 22.596 ± 0.689[a] | 21.261 ± 1.272[a] | 20.590 ± 0.116[a] |
| C16:1 | 0.00 ± 0.000 | 1.640 ± 0.247[a] | 0.48 ± 0.045[a,b] | 0.489 ± 0.104[a,b] | 0.577 ± 0.044[a,b] |
| C18:0 | 10.905 ± 0.797 | 8.772 ± 0.399 | 11.234 ± 0.922 | 10.545 ± 1.303 | 7.260 ± 0.297[a,c,d] |
| C18:1 | 13.451 ± 0.814 | 30.939 ± 0.911[a] | 15.386 ± 1.691[b] | 17.971 ± 3.113[b] | 20.368 ± 1.028[a,b] |
| C18:2 | 25.962 ± 0.722 | 19.456 ± 0.555[a] | 24.647 ± 0.708[b] | 23.941 ± 0.966[b] | 26.889 ± 0.404[b,d] |
| C18:3 | 0.452 ± 0.029 | 0.546 ± 0.059 | 0.099 ± 0.011[a,b] | 0.231 ± 0.049[a,b,c] | 0.133 ± 0.009[a,b] |
| C19:0 | 0.397 ± 0.049 | 0.293 ± 0.093 | 0.284 ± 0.057 | 0.188 ± 0.069[a] | 0.334 ± 0.024 |
| C20:0 | 0.358 ± 0.019 | 0.617 ± 0.208 | 0.333 ± 0.026[b] | 0.431 ± 0.043 | 0.391 ± 0.032 |
| C20:1 | 0.710 ± 0.074 | 0.689 ± 0.021 | 0.879 ± 0.099 | 0.894 ± 0.118 | 0.567 ± 0.041[c,d] |
| C20:4 | 6.371 ± 0.435 | 7.525 ± 0.381 | 5.332 ± 0.388[b] | 5.489 ± 0.738[b] | 4.932 ± 0.335[b] |
| C22:0 | 0.880 ± 0.041 | 0.156 ± 0.012[a] | 1.506 ± 0.046[a,b] | 1.98 ± 0.081[a,b,c] | 1.800 ± 0.138[a,b,c] |
| C20:5 | 0.804 ± 0.064 | 0.451 ± 0.076[a] | 1.086 ± 0.056[a,b] | 1.315 ± 0.074[a,b,c] | 1.527 ± 0.096[a,b,c] |
| C22:6 | 10.425 ± 0.223 | 4.289 ± 0.281[a] | 12.895 ± 0.530[a,b] | 11.761 ± 1.000[b] | 10.301 ± 0.483[b,c] |
| ∑SFAs | 38.770 ± 1.345 | 30.053 ± 0.432[a] | 35.948 ± 1.489[b] | 34.403 ± 2.263[a] | 30.377 ± 0.126[a,c] |
| ∑MUFAs | 14.166 ± 0.782 | 33.269 ± 1.066[a] | 16.746 ± 1.644[b] | 19.358 ± 3.044[a,b] | 21.517 ± 1.032[a,b] |
| ∑PUFAs | 44.016 ± 0.467 | 32.268 ± 0.757[a] | 44.064 ± 0.578[b] | 42.738 ± 0.826[b] | 43.787 ± 0.492[b] |
| ∑n-6 | 32.334 ± 0.643 | 26.983 ± 0.484[a] | 29.982 ± 0.383[a,b] | 29.43 ± 0.249[a,b] | 31.823 ± 0.143[b,c,d] |
| ∑n-3 | 11.682 ± 0.255 | 5.285 ± 0.293[a] | 14.082 ± 0.519[a,b] | 13.308 ± 1.069[b] | 11.963 ± 0.387[b,c] |
| n-6/n-3 | 2.778 ± 0.116 | 5.138 ± 0.202[a] | 2.141 ± 0.081[a,b] | 2.266 ± 0.228[a,b] | 2.665 ± 0.082[b,c] |

Data are given as mean ± SEM. [a]$P < 0.05$ versus the ND group; [b]$P < 0.05$ versus the HFD group; [c]$P < 0.05$ versus the DHA/EPA = 2:1 group; [d]$P < 0.05$ versus the DHA/EPA = 1:1 group.

### 3.4. Dietary DHA/EPA Improves HFD-Induced Lipid Dyshomeostasis in Liver Tissue.

DHA/EPA treatment for 12 weeks resulted in a significant reduction in serum levels of TC (reduced by 46.9–72%), TG (reduced by 45.4–75.6%), LDL-C (reduced by 38.3–63.7%), and ox-LDL (reduced by 36.2–38.3%) compared to the HFD group (Table 4). Although the reduction effects of DHA/EPA on the hepatic lipid level have no significant difference among the three DHA/EPA groups, daily DHA/EPA treatment alleviated hepatic fatty accumulation. Moreover, the three groups treated with DHA/EPA had higher serum levels of HDL-C (increased by 61.5–169.2%) and adiponectin (increased by 27.4–141%) than the HFD group did. In particular, DHA/EPA 1:2 group had the lowest serum TC, TG, and LDL levels and the highest adiponectin level among the three DHA/EPA groups.

As illustrated in Figure 3, 66.5%, 69.7%, and 58.0% increases in the mRNA expression of ATP-binding cassette transporter A1 (ABCA1) were, respectively, observed in the DHA/EPA 1:2, DHA/EPA 1:1, and DHA/EPA 2:1 groups, compared with that in the HFD-treated mice (Figure 3). No significant difference of the ABCA1 expression level was found among the DHA/EPA groups. Compared to that in the HFD group, the same trend was observed in ATP-binding cassette transporter G1 (ABCG1) and acyl-coenzyme A:cholesterol acyltransferase (ACAT-1) in the DHP/EPA groups, although only the DHA/EPA 1:1 group showed a significant increase in lysosomal acid lipase (LAL) ($P < 0.05$). In liver tissue, cluster of differentiation 36 (CD36), macrophage scavenger receptor 1 (MSR-1), and lectin-like oxidized low-density lipoprotein receptor 1 (LOX-1) expression levels were significantly downregulated at both the mRNA and protein levels in DHA/EPA-treated mice compared to that in the HFD-treated mice. Additionally, the feeding of the HFD significantly downregulated the protein levels of proliferator-activated receptor alpha (PPARα) and adenosine monophosphate-activated protein kinase (AMPK) and upregulated the protein levels of sterol regulatory element-binding protein 1c (SREBP-1c), compared with that of the ND, which were partially reversed with the supplementation of dietary DHA/EPA (Figure 3).

## 4. Discussion

Dietary n-3 PUFAs can reduce hepatic inflammation, fibrosis, and steatosis, decrease plasma TG concentrations, and regulate hepatic fatty acid and TG metabolism in NAFLD. We previously created a mouse model in which NAFLD, lipid disorder, oxidative stress, and inflammation were induced by an HFD in C57BL/6J mice [17]. Our findings showed that the consumption of diets with various ratios of DHA/EPA (2:1, 1:1, and 1:2) ameliorated liver steatosis in mice. This is probably due to the repletion of hepatic total n-3 PUFA content and decrease of the n-6/n-3 ratio, concomitant with a reduction of oxidative stress, proinflammatory cytokine secretion, and hepatic lipid content. ApoE is a class of proteins involved in the metabolism of fats in humans and mice. Its absence predisposes to metabolic syndrome (e.g., Alzheimer's disease, atherosclerosis, and obesity) and might be associated with NAFLD [27]. Therefore, ApoE$^{-/-}$ mice have been extensively employed as models for metabolic syndrome and NAFLD in recent years [28, 29].

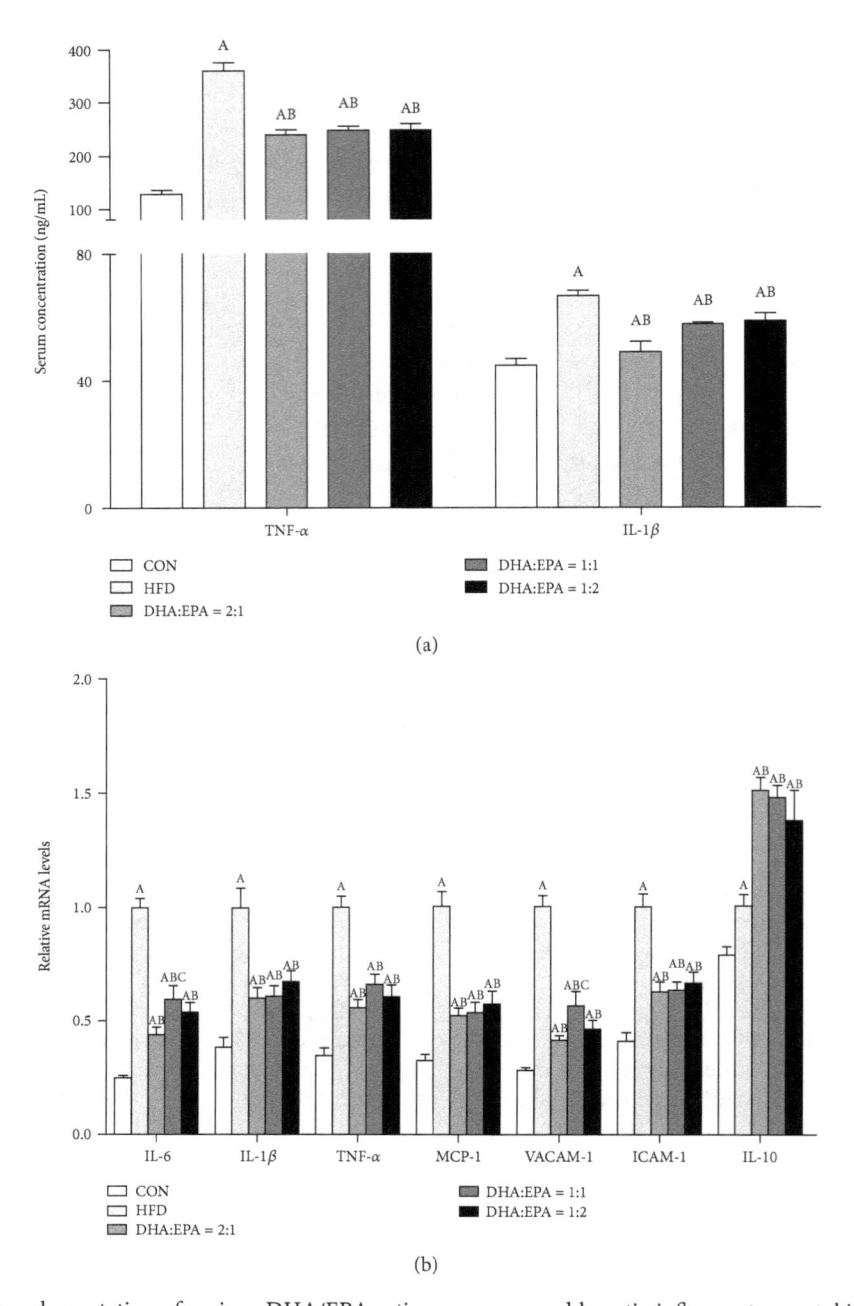

FIGURE 2: Effects of the supplementation of various DHA/EPA ratios on serum and hepatic inflammatory cytokine expression. (a) Serum inflammatory cytokines ($n = 8$). (b) Hepatic inflammatory cytokine expression ($n = 6$). The mRNA expression of $\beta$-actin was quantified as the endogenous control. (A) $P < 0.05$ versus the ND group; (B) $P < 0.05$ versus the HFD group; (C) $P < 0.05$ versus the DHA/EPA = 2 : 1 group; (D) $P < 0.05$ versus the DHA/EPA = 1 : 1 group.

It has been reported that consuming DHA and EPA directly from foods and/or dietary supplements is the only practical way to increase the levels of these FAs in the body. The contents of DHA and EPA in the serum and liver tissue of DHA/EPA-treated mice were notably increased in our study. It is also well known that dietary fat, including DHA and EPA, alters the FA composition of various organs [12, 18]. Our results showed that the increased MUFAs and decreased SFAs, n-6 PUFAs, and n-3 PUFAs with an increase of the n-6/n-3 ratio were observed in liver tissue of HFD-fed mice compared to that in the ND-fed mice. This phenomenon is most likely due to the increased activity of Δ-9 desaturase activity [30, 31] and the defective pathway for desaturation and elongation of essential precursors, linoleic acid, and ALA [32]. Our findings are in agreement with the observations of other authors [13, 33]. Interestingly, these changes were either reversed or normalized to the control levels in mice fed the diets supplemented with DHA/EPA (2 : 1, 1 : 1, and 1 : 2). Our study showed that the DHA/EPA 2 : 1 group showed a tendency to raise DHA and n-3 PUFA

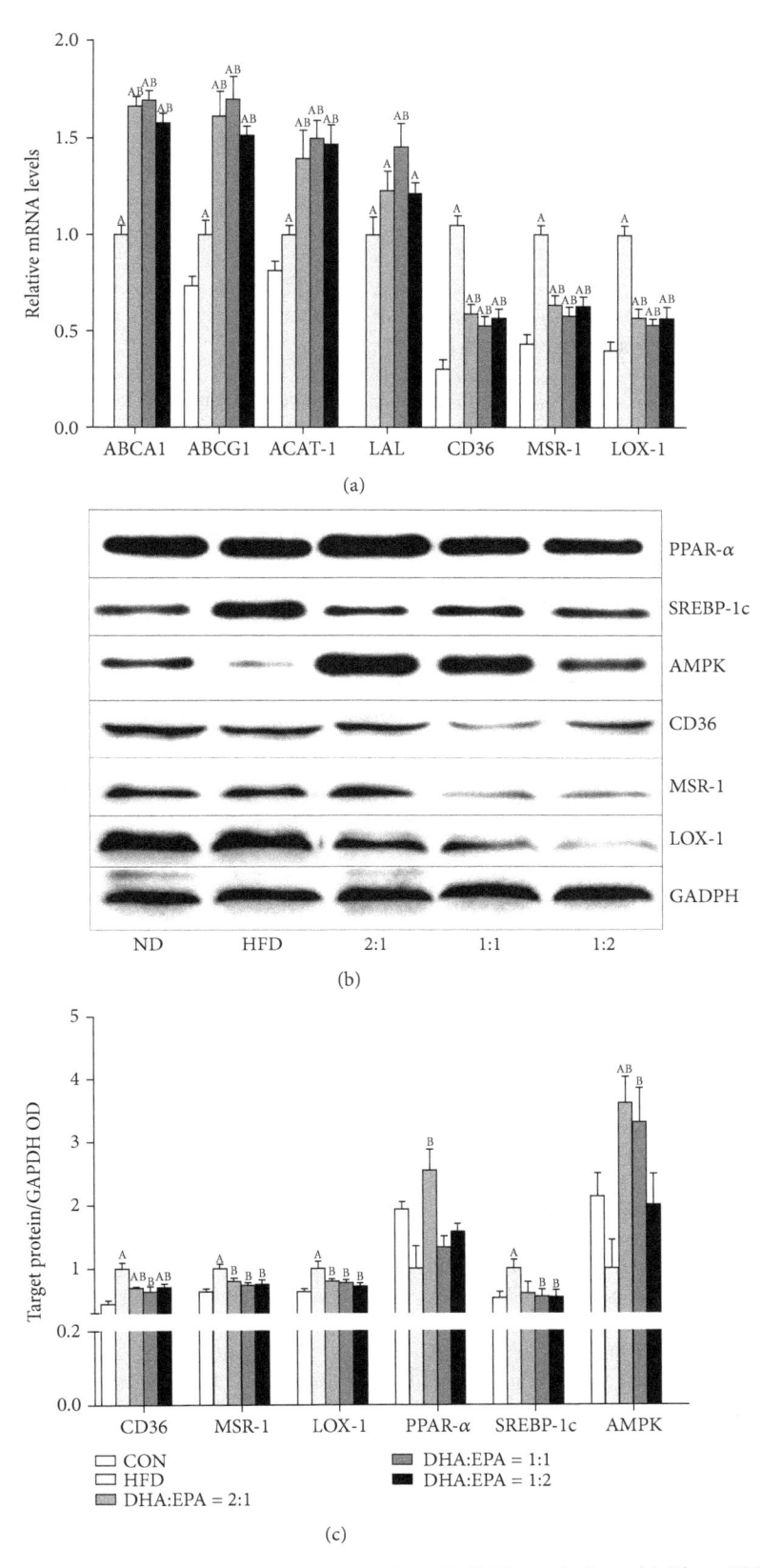

Figure 3: Effects of the supplementation of various DHA/EPA ratios on hepatic lipid metabolism. (a) The mRNA expression of ABCA1, ABCG1, ACAT-1, LAL, CD36, MSR-1, and LOX-1 in liver tissues, as measured by qPCR ($n = 6$). (b, c) Protein expression of PPARα, SREBP-1c, AMPK, CD36, MSR-1, and LOX-1 in liver tissues, as measured by western blotting ($n = 3 - 4$). (A) $P < 0.05$ versus the ND group; (B) $P < 0.05$ versus the HFD group; (C) $P < 0.05$ versus the DHA/EPA = 2 : 1 group; (D) $P < 0.05$ versus the DHA/EPA = 1 : 1 group.

concentration and lower the n-6/n-3 ratio in the liver. On the other hand, the DHA/EPA 1:2 group showed a tendency to raise EPA, n-6 PUFA concentration, and the n-6/n-3 ratio in the liver. The results suggest that DHA/EPA supplementation moderately attenuated the HFD-induced NAFLD, at least partly due to the alteration of FA composition of serum and liver tissue.

The impairment of normal redox homeostasis and the consequent accumulation of oxidized biomolecules have been linked to the onset and/or development of a large variety of diet-induced diseases. An established source of oxidative stress is reactive oxygen species (ROS), which are generated by free FA metabolism and can attack PUFAs and initiate lipid peroxidation within cells. The formation of aldehyde by-products during lipid peroxidation, including MDA, activates the inflammatory response, propagating tissue injury and activating cellular stress signaling pathways. We previously found that the supplementation of various DHA/EPA ratios with an n-6/n-3 ratio of 4:1 reversed HFD-induced oxidative stress, as evidenced by the lower content of MDA. These effects are correlated with the induction of serum SOD activity and enhancement in serum levels of GSH and serum total antioxidant capacity, although no significant differences were observed among the DHA/EPA groups (2:1, 1:1, and 1:2) [24]. However, Mendez et al. [21] revealed significant differences in the carbonylation status of albumin in plasma among the DHA/EPA dietary groups, and the EPA:DHA 1:1 ratio exhibited the lowest protein oxidation scores. In this study, the general changes in hepatic MDA, SOD, and GSH levels were similar to those observed in our previous report [24]. The difference between the results of our study and those of Mendez et al. may lie in the different FA compositions in the diets. HFD-induced liver oxidative stress is associated with progressively increasing availability and oxidation of FAs in the liver [34] and/or TNF-$\alpha$-induced enhancement in mitochondrial ROS production [35], while the DHA/EPA-reversed liver oxidative stress is possibly related to liver n-6 PUFAs and n-3 PUFA repletion with a decreased n-6/n-3 ratio [36].

Dysfunction of fat storage in adipose tissue may increase adipocyte lipolysis, subsequently causing excessive adipose-derived fatty acid influx into the liver, eventually resulting in hepatic steatosis [37]. By upregulating genes encoding proteins involved in FA oxidation and downregulating genes encoding proteins involved in lipid synthesis, n-3 PUFAs provide their protective effects on NAFLD. SREBP-1c, the key lipogenic transcription factor that is highly expressed in the liver, increases the expression of genes connected with fatty acid and TG synthesis. Our recent study showed that the treatment of C57BL/6J mice with various DHA/EPA ratios repressed SREBP-1c-mediated downregulation of FA synthase, stearoyl desaturase-1, and acetyl-CoA carboxylase with a concomitant reduction in *de novo* lipogenesis and activated PPAR$\alpha$-mediated upregulation of carnitine palmitoyl transferase-1 and acyl-CoA oxidase expression with a parallel enhancement in FA oxidation [17]. As one of the critical adipokines secreted by endocrine organs, adiponectin modulates hepatic lipid homeostasis towards a reduction of lipid content [10]. Activated adiponectin signaling leads to the activation of the AMPK pathway, which modulates hepatic lipid metabolism by simultaneously inhibiting *de novo* lipogenesis and stimulating FA $\beta$-oxidation [38]. In this study, the reduction of hepatic lipid accumulation in DHA/EPA-treated mice may be attributed to the elevated serum levels of adiponectin. Additionally, mice treated with DHA/EPA showed significant diminution in total liver fat content compared to untreated animals, a finding that may be related to changes in the pattern of lipid metabolism in the liver. To explain the potential mechanism causing the changes, proteins involved in cholesterol efflux (ABCA1 and ABCG1), cholesterol esterification (ACAT1), cholesterol lipolysis (LAL), and cholesterol uptake (CD36, MSR-1, and LOX-1) were examined. This is supported by the higher mRNA expression of the ABCA1, ABCG1, LAL, and ACAT-1 and the lower expression of CD36, MSR-1, and LOX-1. We demonstrated that diets lacking DHA and EPA have no effects on the expression of ABCA1, ABCG1, and LAL, which indicated that DHA and EPA are much more likely to regulate cholesterol homeostasis by increasing cholesterol efflux and lipolysis [24].

In both NAFLD patients and animals subjected to HFD, hepatic proinflammatory status is characterized by Kupffer cell activation, an increased number of hepatic neutrophils, and higher levels of serum transaminases, TNF-$\alpha$, IL-1$\beta$, and IL-6 [39]. Our recent study showed that serum levels of ALT, AST, TNF-$\alpha$, IL-1$\beta$, and IL-6 in C57BL/6J mice were all significantly lower in the DHA/EPA groups compared to those in the HFD group [17]. In agreement with these findings, the data presented here show that transaminase activity, TNF-$\alpha$, and IL-1 $\beta$ levels in serum and TNF-$\alpha$, IL-1 $\beta$, IL-6, MCP-1, VCAM-1, and ICAM-1 mRNA expression in the liver were higher in HFD-fed ApoE$^{-/-}$ mice compared to the controls, a condition that was reverted upon supplementation with various DHA/EPA ratios. Furthermore, mRNA expression of the anti-inflammatory cytokine IL-10 was significantly upregulated by DHA/EPA supplementation. Activating protein-1, including c-Jun and c-Fos, is an important signal transduction pathway component of proinflammatory mediator expression and is independent of NF-$\kappa$B. We previously found that the consumption of DHA/EPA significantly suppressed the expression of c-Jun and c-Fos protein and their respective genes. Additionally, the critical role of PPAR$\alpha$ in preventing fat-induced nonalcoholic steatohepatitis by alleviating liver steatosis, oxidative stress, and inflammation has been proven [40]. The underlying mechanisms by which n-3 PUFAs protected against HFD-induced liver steatosis are probably that n-3 PUFA-activated PPAR$\alpha$ interact with proinflammatory factor NF-$\kappa$B p65 with the formation of inactive PPAR$\alpha$/NF-$\kappa$B p65 complexes [41] and the suppression of proinflammatory cytokine formation and secretion [7]. Moreover, DHA had a greater suppressive effect than EPA on an alcohol/high-fat diet-induced hepatic inflammation and ROS generation by increasing adiponectin production and secretion [42, 43], which has strong cellular protective properties, acting through the AMPK-activated mechanism [44]. In this study, DHA/EPA supplementation reversed the decrease of hepatic PPAR$\alpha$ expression in HFD-fed mice.

Although only the DHA/EPA 2 : 1 group had significantly increased PPARα expression, the DHA/EPA 2 : 1 group had the highest serum levels of adiponectin, the lowest hepatic mRNA expression of proinflammatory cytokines, and the highest protein levels of PPARα and AMPK, which may be due to the higher ratio of DHA in this group. These results suggest that the alleviation of inflammatory responses in DHA/EPA-treated mice may correlate with an increase in serum levels of adiponectin and hepatic protein levels of PPARα and AMPK.

## 5. Conclusion

In addition to reducing oxidative stress, decreasing proinflammatory cytokine secretion, and improving hepatic lipid metabolism, a DHA/EPA-enriched diet with an n-6/n-3 ratio of 4 : 1 may reverse HFD-induced NALFD to some extent by increasing n-6 and n-3 PUFAs and decreasing the amount of MUFAs and the n-6/n-3 ratio. Although no significant difference was found in the expression of inflammation- and hepatic lipid metabolism-related genes in the three DHA/EPA groups, the DHA/EPA 2 : 1 group showed the highest DHA and n-3 PUFA concentration and the DHA/EPA 1 : 2 group showed the highest EPA, n-6 PUFA concentration, and n-6/n-3 ratio.

## Acknowledgments

This work was supported by the National Key Research and Development Program of China (no. 2017YFC1600500), the National High-Tech Research and Development Projects (no. 2010AA023003), the National Natural Science Foundation of China (no. 31201351), the Young Elite Scientists Sponsorship Program by CAST (China Association for Science and Technology) (no. YESS20160164), and the 2015 Chinese Nutrition Society DSM Research Fund. We would like to thank LetPub for English language editing.

## References

[1] M. E. Rinella, "Nonalcoholic fatty liver disease: a systematic review," *JAMA*, vol. 313, no. 22, pp. 2263–2273, 2015.

[2] M. Shaker, A. Tabbaa, M. Albeldawi, and N. Alkhouri, "Liver transplantation for nonalcoholic fatty liver disease: new challenges and new opportunities," *World Journal of Gastroenterology*, vol. 20, no. 18, pp. 5320–5330, 2014.

[3] J. G. Fan, S. U. Kim, and V. W. Wong, "New trends on obesity and NAFLD in Asia," *Journal of Hepatology*, vol. 67, no. 4, pp. 862–873, 2017.

[4] R. Valenzuela, A. Espinosa, D. González-Mañán et al., "N-3 long-chain polyunsaturated fatty acid supplementation significantly reduces liver oxidative stress in high fat induced steatosis," *PLoS One*, vol. 7, no. 10, article e46400, 2012.

[5] S. Sundaram, M. R. Bukowski, W.-R. Lie, M. J. Picklo, and L. Yan, "High-fat diets containing different amounts of n3 and n6 polyunsaturated fatty acids modulate inflammatory cytokine production in mice," *Lipids*, vol. 51, no. 5, pp. 571–582, 2016.

[6] J. Ruzickova, M. Rossmeisl, T. Prazak et al., "Omega-3 PUFA of marine origin limit diet-induced obesity in mice by reducing cellularity of adipose tissue," *Lipids*, vol. 39, no. 12, pp. 1177–1185, 2004.

[7] R. Valenzuela, A. Espinosa, P. Llanos et al., "Anti-steatotic effects of an n-3 LCPUFA and extra virgin olive oil mixture in the liver of mice subjected to high-fat diet," *Food & Function*, vol. 7, no. 1, pp. 140–150, 2015.

[8] K. Stavropoulos, K. Imprialos, A. Pittaras, C. Faselis, P. Narayan, and P. Kokkinos, "Lifestyle modifications in non-alcoholic fatty liver disease and non-alcoholic steatohepatitis," *Current Vascular Pharmacology*, vol. 16, no. 3, pp. 239–245, 2018.

[9] G. S. de Castro and P. C. Calder, "Non-alcoholic fatty liver disease and its treatment with n-3 polyunsaturated fatty acids," *Clinical Nutrition*, vol. 37, no. 1, pp. 37–55, 2018.

[10] M. Wang, X. J. Zhang, K. Feng et al., "Dietary α-linolenic acid-rich flaxseed oil prevents against alcoholic hepatic steatosis *via* ameliorating lipid homeostasis at adipose tissue-liver axis in mice," *Scientific Reports*, vol. 6, no. 1, article 26826, 2016.

[11] H. Shapiro, M. Tehilla, J. Attal-Singer, R. Bruck, R. Luzzatti, and P. Singer, "The therapeutic potential of long-chain omega-3 fatty acids in nonalcoholic fatty liver disease," *Clinical Nutrition*, vol. 30, no. 1, pp. 6–19, 2011.

[12] N. Ivanovic, R. Minic, I. Djuricic et al., "Active Lactobacillus rhamnosus LA68 or Lactobacillus plantarum WCFS1 administration positively influences liver fatty acid composition in mice on a HFD regime," *Food & Function*, vol. 7, no. 6, pp. 2840–2848, 2016.

[13] J. Araya, R. Rodrigo, P. Pettinelli, A. V. Araya, J. Poniachik, and L. A. Videla, "Decreased liver fatty acid Δ-6 and Δ-5 desaturase activity in obese patients," *Obesity*, vol. 18, no. 7, pp. 1460–1463, 2010.

[14] D. B. Jump, K. A. Lytle, C. M. Depner, and S. Tripathy, "Omega-3 polyunsaturated fatty acids as a treatment strategy for nonalcoholic fatty liver disease," *Pharmacology & Therapeutics*, vol. 181, pp. 108–125, 2018.

[15] J. Delarue and J. P. Lalles, "Nonalcoholic fatty liver disease: roles of the gut and the liver and metabolic modulation by some dietary factors and especially long-chain n-3 PUFA," *Molecular Nutrition & Food Research*, vol. 60, no. 1, pp. 147–159, 2016.

[16] F. Yuan, H. Wang, Y. Tian et al., "Fish oil alleviated high-fat diet-induced non-alcoholic fatty liver disease via regulating hepatic lipids metabolism and metaflammation: a transcriptomic study," *Lipids in Health and Disease*, vol. 15, no. 1, p. 20, 2016.

[17] T. Shang, L. Liu, J. Zhou et al., "Protective effects of various ratios of DHA/EPA supplementation on high-fat diet-induced liver damage in mice," *Lipids in Health and Disease*, vol. 16, no. 1, p. 65, 2017.

[18] S. Khadge, J. G. Sharp, G. M. Thiele et al., "Dietary omega-3 and omega-6 polyunsaturated fatty acids modulate hepatic pathology," *The Journal of Nutritional Biochemistry*, vol. 52, pp. 92–102, 2018.

[19] E. Molinar-Toribio, J. Pérez-Jiménez, S. Ramos-Romero et al., "Effect of *n*-3 PUFA supplementation at different EPA:DHA ratios on the spontaneously hypertensive obese rat model of the metabolic syndrome," *British Journal of Nutrition*, vol. 113, no. 6, pp. 878–887, 2015.

[20] N. Taltavull, M. Muñoz-Cortés, L. Lluís et al., "Eicosapentaenoic acid/docosahexaenoic acid 1:1 ratio improves histological alterations in obese rats with metabolic syndrome," *Lipids in Health and Disease*, vol. 13, no. 1, p. 31, 2014.

[21] L. Mendez, M. Pazos, J. M. Gallardo et al., "Reduced protein oxidation in Wistar rats supplemented with marine ω3 PUFAs," *Free Radical Biology and Medicine*, vol. 55, pp. 8–20, 2013.

[22] L. Lluis, N. Taltavull, M. Munoz-Cortés et al., "Protective effect of the omega-3 polyunsaturated fatty acids: eicosapentaenoic acid/docosahexaenoic acid 1:1 ratio on cardiovascular disease risk markers in rats," *Lipids in Health and Disease*, vol. 12, no. 1, p. 140, 2013.

[23] G. Dasilva, M. Pazos, E. Garcia-Egido et al., "Lipidomics to analyze the influence of diets with different EPA:DHA ratios in the progression of metabolic syndrome using SHROB rats as a model," *Food Chemistry*, vol. 205, pp. 196–203, 2016.

[24] L. Liu, Q. Hu, H. Wu et al., "Protective role of n6/n3 PUFA supplementation with varying DHA/EPA ratios against atherosclerosis in mice," *The Journal of Nutritional Biochemistry*, vol. 32, pp. 171–180, 2016.

[25] G. F. T. C. Ed, *Guide for the Care and Use of Laboratory Animals*, National Research Council, Washington, DC, USA, 8th edition, 2011.

[26] J. J. Agren, A. Julkunen, and I. Penttilä, "Rapid separation of serum lipids for fatty acid analysis by a single aminopropyl column," *Journal of Lipid Research*, vol. 33, no. 12, pp. 1871–1876, 1993.

[27] R. Schierwagen, L. Maybuchen, S. Zimmer et al., "Seven weeks of Western diet in apolipoprotein-E-deficient mice induce metabolic syndrome and non-alcoholic steatohepatitis with liver fibrosis," *Scientific Reports*, vol. 5, no. 1, article 12931, 2015.

[28] E. Catry, A. M. Neyrinck, I. Lobysheva et al., "Nutritional depletion in n-3 PUFA in apoE knock-out mice: a new model of endothelial dysfunction associated with fatty liver disease," *Molecular Nutrition & Food Research*, vol. 60, no. 10, pp. 2198–2207, 2016.

[29] C. Spanos, E. M. Maldonado, C. P. Fisher et al., "Proteomic identification and characterization of hepatic glyoxalase 1 dysregulation in non-alcoholic fatty liver disease," *Proteome Science*, vol. 16, no. 1, p. 4, 2018.

[30] P. G. Mavrelis, H. V. Ammon, J. J. Gleysteen, R. A. Komorowski, and U. K. Charaf, "Hepatic free fatty acids in alcoholic liver disease and morbid obesity," *Hepatology*, vol. 3, no. 2, pp. 226–231, 1983.

[31] D. A. Pan, A. J. Hulbert, and L. H. Storlien, "Dietary fats, membrane phospholipids and obesity," *Journal of Nutrition*, vol. 124, no. 9, pp. 1555–1565, 1994.

[32] J. Araya, R. Rodrigo, L. A. Videla et al., "Increase in long-chain polyunsaturated fatty acid n−6/n−3 ratio in relation to hepatic steatosis in patients with non-alcoholic fatty liver disease," *Clinical Science*, vol. 106, no. 6, pp. 635–643, 2004.

[33] W. Xin, Y. Cao, Y. Fu, G. Guo, and X. Zhang, "Liver fatty acid composition in mice with or without nonalcoholic fatty liver disease," *Lipids in Health and Disease*, vol. 10, no. 1, pp. 234–237, 2011.

[34] L. A. Videla, R. Rodrigo, J. Araya, and J. Poniachik, "Insulin resistance and oxidative stress interdependency in non-alcoholic fatty liver disease," *Trends in Molecular Medicine*, vol. 12, no. 12, pp. 555–558, 2006.

[35] A. Aronis, Z. Madar, and O. Tirosh, "Mechanism underlying oxidative stress-mediated lipotoxicity: exposure of J774.2 macrophages to triacylglycerols facilitates mitochondrial reactive oxygen species production and cellular necrosis," *Free Radical Biology & Medicine*, vol. 38, no. 9, pp. 1221–1230, 2005.

[36] J. P. Allard, E. Aghdassi, S. Mohammed et al., "Nutritional assessment and hepatic fatty acid composition in non-alcoholic fatty liver disease (NAFLD): a cross-sectional study," *Journal of Hepatology*, vol. 48, no. 2, pp. 300–307, 2008.

[37] W. Zhong, Y. Zhao, Y. Tang et al., "Chronic alcohol exposure stimulates adipose tissue lipolysis in mice: role of reverse triglyceride transport in the pathogenesis of alcoholic steatosis," *The American Journal of Pathology*, vol. 180, no. 3, pp. 998–1007, 2012.

[38] C. Q. Rogers, J. M. Ajmo, and Y. Min, "Adiponectin and alcoholic fatty liver disease," *IUBMB Life*, vol. 60, no. 12, pp. 790–797, 2008.

[39] J. Crespo, A. Cayón, P. Fernández-Gil et al., "Gene expression of tumor necrosis factor alpha and TNF-receptors, p55 and p75, in nonalcoholic steatohepatitis patients," *Hepatology*, vol. 34, no. 6, pp. 1158–1163, 2001.

[40] C. G. Dossi, G. S. Tapia, A. Espinosa, L. A. Videla, and A. D'Espessailles, "Reversal of high-fat diet-induced hepatic steatosis by n-3 LCPUFA: role of PPAR-α and SREBP-1c," *The Journal of Nutritional Biochemistry*, vol. 25, no. 9, pp. 977–984, 2014.

[41] G. Tapia, R. Valenzuela, A. Espinosa et al., "N-3long-chain PUFA supplementation prevents high fat diet induced mouse liver steatosis and inflammation in relation to PPAR-α upregulation and NF-κB DNA binding abrogation," *Molecular Nutrition & Food Research*, vol. 58, no. 6, pp. 1333–1341, 2014.

[42] J. Song, C. Li, Y. Lv, Y. Zhang, W. K. Amakye, and L. Mao, "DHA increases adiponectin expression more effectively than EPA at relative low concentrations by regulating PPARγ and its phosphorylation at Ser273 in 3T3-L1 adipocytes," *Nutrition & Metabolism*, vol. 14, no. 1, p. 52, 2017.

[43] J. M. Tishinsky, D. W. Ma, and L. E. Robinson, "Eicosapentaenoic acid and rosiglitazone increase adiponectin in an additive and PPARγ-dependent manner in human adipocytes," *Obesity*, vol. 19, no. 2, pp. 262–268, 2011.

[44] C. Johnson, R. Williams, J. Y. Wei, and G. Ranganathan, "Regulation of serum response factor and adiponectin by PPARgamma agonist docosahexaenoic acid," *Journal of Lipids*, vol. 2011, Article ID 670479, 8 pages, 2011.

# Hydrogen Sulfide Ameliorates Developmental Impairments of Rat Offspring with Prenatal Hyperhomocysteinemia

O. V. Yakovleva,[1] A. R. Ziganshina,[1] S. A. Dmitrieva,[2] A. N. Arslanova,[1] A. V. Yakovlev,[1] F. V. Minibayeva,[2] N. N. Khaertdinov,[1] G. K. Ziyatdinova,[1] R. A. Giniatullin,[1,3] and G. F. Sitdikova[1]

[1]Kazan Federal University, Kazan 420008, Russia
[2]Kazan Institute of Biochemistry and Biophysics, FRC Kazan Scientific Center of RAS, Kazan 420011, Russia
[3]A.I. Virtanen Institute, University of Eastern Finland, Kuopio 70211, Finland

Correspondence should be addressed to O. V. Yakovleva; a-olay@yandex.ru

Guest Editor: Mohamed M. Abdel-Daim

Maternal high levels of the redox active amino acid homocysteine—called hyperhomocysteinemia (hHCY)—can affect the health state of the progeny. The effects of hydrogen sulfide ($H_2S$) treatment on rats with maternal hHCY remain unknown. In the present study, we characterized the physical development, reflex ontogeny, locomotion and exploratory activity, muscle strength, motor coordination, and brain redox state of pups with maternal hHCY and tested potential beneficial action of the $H_2S$ donor—sodium hydrosulfide (NaHS)—on these parameters. Our results indicate a significant decrease in litter size and body weight of pups from dams fed with methionine-rich diet. In hHCY pups, a delay in the formation of sensory-motor reflexes was observed. Locomotor activity tested in the open field by head rearings, crossed squares, and rearings of hHCY pups at all studied ages (P8, P16, and P26) was diminished. Exploratory activity was decreased, and emotionality was higher in rats with hHCY. Prenatal hHCY resulted in reduced muscle strength and motor coordination assessed by the paw grip endurance test and rotarod test. Remarkably, administration of NaHS to pregnant rats with hHCY prevented the observed deleterious effects of high homocysteine on fetus development. In rats with prenatal hHCY, the endogenous generation of $H_2S$ brain tissues was lower compared to control and NaHS administration restored the $H_2S$ level to control values. Moreover, using redox signaling assays, we found an increased level of malondialdehyde (MDA), the end product of lipid peroxidation, and decreased activity of antioxidant enzymes such as superoxide dismutase (SOD) and glutathione peroxidase (GPx) in the brain tissues of rats of the hHCY group. Notably, NaHS treatment restored the level of MDA and the activity of SOD and GPx. Our data suggest that $H_2S$ has neuroprotective/antioxidant effects against homocysteine-induced neurotoxicity providing a potential strategy for the prevention of developmental impairments in newborns.

## 1. Introduction

Homocysteine, a sulfur-containing amino acid, is an intermediate product of the methionine metabolism. The concentration of homocysteine is regulated by remethylation back to methionine by methionine synthase, using 5-methyl tetrahydrofolate as cosubstrate that requires folic acid, or it can be catabolized by cystathionine β-synthase (CBS), a vitamin B6-dependent enzyme, to form cysteine and hydrogen sulfide ($H_2S$) [1]. In humans, an increase of total plasma homocysteine to a level more than 15 μM is defined as hyperhomocysteinemia (hHCY). According to the total plasma homocysteine level, it is classified as mild (15–25 μM), moderate (25–50 μM), or severe (50–500 μM) hHCY [2]. hHCY may be induced by an increase of methionine in the diet, vitamin deficiency (folate, B12, or B6), mutations of genes encoding methylene tetrahydrofolate reductase (MTHFR), limiting the cells methylating capacity, or CBS [3]. hHCY is a risk factor of cardiovascular diseases, associated with cognitive impairments, increased risk of Alzheimer's disease, vascular dementia, or cerebrovascular stroke [4]. An elevated level of homocysteine is associated with common pregnancy complications such as pregnancy-induced hypertension, placenta abruptio, thromboembolic events, neural tube defects, and intrauterine growth restriction. Infants born from mothers with hHCY exhibit mental and physical retardation [1, 5]. In animal models, maternal hHCY induced oxidative stress and apoptosis in the fetal brain, resulting in postnatal neurodevelopmental deficits [6–10].

$H_2S$ is a one of the metabolites of homocysteine produced by CBS and cystathionine $\gamma$-lyase (CSE), enzymes of the transulfuration pathway of methionine metabolism [11]. In addition to the role of $H_2S$ as an important neuromodulator [12–14], $H_2S$ elicits neuroprotection against oxidative stress, neuroinflammation, apoptosis, and neurodegeneration caused by several pathophysiological conditions [15–17]. $H_2S$ donors attenuated lipopolysaccharide- or stress-induced learning and memory impairments in rats and prevented hippocampal long-term depression (LTD) [18, 19].

Altered $H_2S$ signaling was suggested to contribute in homocysteine-induced neurotoxicity [20, 21]. Indeed, intracerebroventricular administration of homocysteine decreased CBS expression and endogenous $H_2S$ generation in the hippocampus of rats along with learning and memory dysfunctions [22, 23, 24]. The results indicate that $H_2S$ is effective in providing protection against neurodegeneration and cognitive dysfunctions in homocysteine exposed rats. Nevertheless, the effects of $H_2S$ treatment on rats with maternal hHCY remain unknown. Current therapies for hHCY are limited to vitamin supplements, which serve as cofactors in the pathways of homocysteine metabolism. These therapies lower the level of homocysteine but generally do not alter disease consequences [11]. In the present study, we (1) evaluated the developmental consequences of maternal hHCY in rats; (2) assessed the effects of treatment with the $H_2S$ donor during pregnancy on physical parameters, neurobehavioral reflexes, muscle strength, and motor balance of the offspring; (3) evaluated the level of $H_2S$ and the rate of $H_2S$ generation in brain tissues of rats from control, hHCY, and NaHS-treated groups; (4) compared the oxidative stress level in brain tissues of pups born from the dams of control, hHCY, and NaHS-treated groups by measuring the concentrations of malondialdehyde (MDA), the end product of lipid peroxidation, and the activity of the antioxidant enzymes—superoxide dismutase (SOD) and glutathione peroxidase (GPx).

## 2. Materials and Methods

*2.1. Experimental Animals and the Model of hHCY.* Experiments were carried out on Wistar rats in accordance with EU Directive 2010/63/EU for animal experiments and the Local Ethical committee KFU (protocol no. 8 from 5.05.2015). Animals were housed in polypropylene cages ($32 \times 40 \times 18$ cm) under controlled temperature (22–24°C), with a 12:12 L/D light schedule (lights on at 6:00 a.m.) and free access to food and water. Pregnant rats were divided into four groups as follows. One group was fed ad libitum with a control diet ($n = 7$); the second group ($n = 11$) received daily methionine (7.7 g/kg body weight) with food starting 3 weeks prior to and during pregnancy [10, 25]. The third group ($n = 4$) received NaHS three weeks before and throughout pregnancy according the following protocol: 7 days of injections alternated with 7 days of adaptation. Rats of the fourth group ($n = 4$) received daily methionine and injections of NaHS according the abovementioned protocols. NaHS was used as the $H_2S$ donor and was diluted in sterilized saline and injected subcutaneously (i.s.c.) at a dose 3 mg/kg.

The offspring was divided into the following groups according to maternal diet: (1) control diet group ($n = 61$ pups/7 dams/7 litters), (2) methionine diet group (Hcy, $n = 85$ pups/11 dams/11 litters), (3) control diet group receiving NaHS ($H_2S$, $n = 54$ pups/4 dams/4 litters), and (4) methionine diet group receiving NaHS (Hcy$H_2S$, $n = 54$ pups/4 dams/4 litters).

*2.2. Maturation of Physical Features.* After delivery, the litter size, total litter weight, and weight of each pup were assessed. Body weight was measured daily using an electronic balance (Vibra, model AJ-1200CE, Japan). Mortality was calculated as percent of dead pups against all pups in a litter during the observation period (P2–P28). The analysis of the physical development and reflex ontogeny was started at P2 and was carried out daily between 12 and 17 p.m. until P28 according to the previous studies [10, 26, 27]. The following physical features were observed: eye opening, ear unfolding, incisor eruption, and hair appearance. The maturation age of a particular feature was defined as the day on which that features were observed for the first time.

*2.3. Reflex Testing.* The time of appearance of each reflex was defined as the first day of its occurrence (Table 1) [27]. The following reflexes were scored: negative geotaxis, head shake, righting, cliff avoidance, acoustic startle reflex, cliff avoidance caused by visual stimulus, free-fall righting, and olfactory discrimination [10, 27].

*2.4. Open Field Test.* Rats were subjected to an open field test at P8, P16, and P26. The apparatus used to measure locomotion and exploratory activity was a round arena 0.3 m in diameter for P8 pups and 0.6 m for P16 and P26 pups with a floor divided into 36 parts and walls 0.1 and 0.2 m high, correspondingly (Open Science, Moscow, Russia). P8 animals were placed in the middle of the open field for 1 min and P16 and P26 animals for 3 min. The following parameters were evaluated: the number of crossings, head rearings, rearings, exploratory activity, grooming episodes, and defecation scores. After each experimental session, the arena was cleaned with a 0.5% ethanol solution.

*2.5. Muscle Endurance.* Muscle endurance was assessed by the paw grip endurance (PaGE) test [28] at P4, P14, and P26. Rats were placed on a wire grid and gently shaken to prompt the rat to grip the grid. The lid was turned upside down over a housing cage and held at ~0.45 m above an open cage bottom. The time (s) spent on the grid before falling was assessed. The largest value from three individual trials was used for analysis.

TABLE 1: The effects of NaHS treatment on the development of neurobehavioral reflexes of pups with prenatal hHCY.

| Parameters | Control | Hcy | H$_2$S | HcyH$_2$S |
|---|---|---|---|---|
| Negative geotaxis (day of appearance) | 6 (5–7)<br>$n = 55$ | 6 (6–8)*<br>$n = 65$ | 6 (6–6)#<br>$n = 51$ | 6 (5–7)#<br>$n = 47$ |
| Head shake reflex (number of the head rotations per min at P8) | 7 (5–11)<br>$n = 53$ | 2 (1–4)*<br>$n = 60$ | 9 (4–14)#<br>$n = 51$ | 4 (2–6)*,#<br>$n = 47$ |
| Righting reflex (day of appearance) | 6 (4–7)<br>$n = 55$ | 6 (6–8)*<br>$n = 65$ | 6 (3–6)*,#<br>$n = 51$ | 4 (4–5)*,#<br>$n = 49$ |
| Righting reflex (time (s) at P6) | 1 (1–2)<br>$n = 55$ | 2 (2–4)*<br>$n = 65$ | 1 (1–0.75)#<br>$n = 51$ | 1 (1–1.5)#<br>$n = 49$ |
| Cliff avoidance test (day of appearance) | 6 (5–7)<br>$n = 55$ | 7 (6–8)*<br>$n = 60$ | 6 (5–6)#<br>$n = 51$ | 4 (4–4)*,#<br>$n = 49$ |
| Acoustic startle reflex (day of appearance) | 10 (8–10)<br>$n = 53$ | 10 (9–11)*<br>$n = 55$ | 8 (6–10)*,#<br>$n = 51$ | 9 (8–12)#<br>$n = 47$ |
| Cliff avoidance caused by visual stimulus (day of appearance) | 14 (12–15)<br>$n = 53$ | 16 (16–17)*<br>$n = 53$ | 14 (13–16)#<br>$n = 50$ | 14 (14–15)#<br>$n = 46$ |
| Free-fall righting (day of appearance) | 12 (12–16)<br>$n = 53$ | 19 (16–19)*<br>$n = 53$ | 14 (13–14)#<br>$n = 50$ | 14 (13–14)#<br>$n = 46$ |
| Test olfactory discrimination (day of appearance) | 14 (12–15)<br>$n = 53$ | 16 (14–19)*<br>$n = 53$ | 14 (14–16)#<br>$n = 50$ | 14 (13–15)#<br>$n = 46$ |

Data are expressed as median (Q1–Q3). Statistical significance between medians was calculated using the nonparametric ANOVA Kruskal-Wallis test, Kolmogorov-Smirnov normality test, and Mann-Whitney. *,# $p < 0.05$. * compared to the control group, # compared to the Hcy group. $n$: number of animals.

### 2.6. The Rotarod Test.

The rotarod test was used to assess the motor coordination of fore and hind limbs and balance at P16, P21, and P26 [29] using rotarod (Neurobotix, Russia). Each rat was placed on the rotating rod with a rotation speed of 5 rotations per min (rpm), and the time to fall off and the running distance were measured. Animals are subjected to three consecutive test sessions (trials) with an interval of 20–30 min. The best of the latency to fall off the rotating rod was recorded [30].

### 2.7. Assay for Homocysteine Concentration and H$_2$S synthesis.

The total homocysteine level in plasma was determined by voltammetric measurements of products of the reaction with o-quinone [10, 31].

H$_2$S synthesis assay was carried out using the N,N-dimethyl-p-phenylenediamine sulphate (NNDPD) method [32]. Brain tissues of rats (P28) were homogenized in ice-cold 0.15 M NaCl with phosphate buffer. The homogenate (10%, 860 μl) was mixed with zinc acetate (1%, 500 μl) and saline (140 μl) at room temperature. Trichloroacetic acid (10%, 500 μl) was added to precipitate proteins and stop the reaction. NNDPD (20 mM, 266 μl) in 7.2 M HCl and FeCl$_3$ (30 mM, 266 μl) in 1.2 M HCl were added to the mixture, and absorbance of resulting solution (600 μl) was measured by a spectrophotometer at 670 nm (PE-5300VI, ECOHIM, Russia).

H$_2$S generation rate was measured in a mixture containing homogenate (10%, 860 μl), L-cysteine (10 mM, 40 μl), pyridoxal 5′-phosphate (2 mM, 40 μl), and saline (60 μl). After incubation at 37°C for 60 min, zinc acetate (1%, 500 μl) was injected to trap the produced H$_2$S followed by trichloroacetic acid (10%, 500 μl) addition. Then, NNDPD (20 mM, 266 μl) in 7.2 M HCl and FeCl$_3$ (30 mM, 266 μl) in 1.2 M HCl were added, and absorbance of aliquots of the resulting solution (600 μl) was measured at 670 nm by a spectrophotometer. H$_2$S concentration was calculated against a calibration curve of NaHS, and H$_2$S synthesizing activity is expressed as μM H$_2$S produced by 1 g tissue per minute (μM/min/g).

### 2.8. Lipid Peroxidation and the Activity of SOD and GPx.

Malondialdehyde (MDA) was measured using a spectrophotometer according to the method of Ohkawa et al. [33]. Samples of brain tissue were fixed in liquid nitrogen, then homogenized and mixed at a ratio 1 : 1 with 0.3% Triton X-100, 0.1 M HCl, and 0.03 M 2-thiobarbituric acid (TBA). The mixture was heated for 45 min at 95°C and centrifuged for 10 min at 10,000$g$. Under this condition, MDA readily participates in a nucleophilic addition reaction with 2-thiobarbituric acid (TBA), generating a red, fluorescent 1 : 2 MDA adduct. The absorbance of the supernatant was monitored at 532 nm and at 560 nm ($\varepsilon_{TBA-MDA} = 1.55$ mM$^{-1}$ cm$^{-1}$). MDA levels were expressed as μg/g of tissues.

The antioxidant potential was determined by measuring activities of glutathione peroxidase (GPx) and superoxide dismutase (SOD). Samples of brain tissue were fixed in liquid nitrogen, homogenized in cold buffer solution (0.1 M MES at pH 6.0, ratio 1 : 10), and centrifuged for 10 minutes at 10,000$g$. SOD activity (Cu/Zn superoxide dismutase) was determined according to Weyder and Cullen [34]. Applying this method, a xanthine/xanthine oxidase system was used to generate O$_2^-$ and nitroblue tetrazolium (NBT) reduction was used as an indicator of $O_2^{\cdot-}$ production. SOD competes with NBT for $O_2^{\cdot-}$. The percent inhibition of NBT reduction reflects the amount of SOD which is assayed using a spectrophotometer at 560 nm. The reaction mixture contained 100 mM $Na_2HPO_4$ buffer (pH 10.2), 0.1 mM EDTA, 1 M cytochrome $c$, 1 mM xanthine, 0.04 mM NBT, and 150 $\mu$l of the sample. The reaction was initiated by the addition of 0.05 unit of xanthine oxidase. The inhibition of the produced chromogen is proportional to the activity of the SOD present in the sample. A 50% inhibition is defined as 1 unit of SOD, and specific activity is expressed as units per milligram of protein ($U_{SOD}$/min/mg).

GPx activity was also determined according to Weyder and Cullen [34]. GPx catalyzes the oxidation of glutathione by cumene hydroperoxide. In the presence of glutathione reductase and NADPH, the oxidized glutathione (GSSG) is quickly converted to the reduced form with a concomitant oxidation of NADPH to $NADP^+$. The decrease in absorbance was monitored with a spectrophotometer at 340 nm. The reaction mixture consisted of 50 mM $Na_2HPO_4$ buffer (pH 7.2), 1 mM reduction glutathione (GSH), 0.5 unit of glutathione reductase, 0.15 mM NADPH, 1 mM EDTA, and 150 $\mu$l of the sample. One GPx unit is defined as 1 $\mu$mol of GSH consumed per minute, and the specific activity is reported as units per mg of protein ($U_{POX}$/min/mg).

Protein content was measured using Bradford's assay [35] employing bovine serum albumin as standard. A volume of 20 $\mu$l of the sample or standard was mixed with a 1 ml Bradford reagent, and the absorbance was assessed by a spectrophotometer at 595 nm after 5 min.

### 2.9. Statistical Analysis.
Normality of the sample data was evaluated with the Shapiro-Wilk test (sample size less than 25) or Kolmogorov-Smirnov test (sample size more than 25) for equal variances using $F$-test Origin Pro software (OriginLab Corp., Northampton, MA, USA). Data are expressed as median (Q1–Q3) or mean ± SEM. Statistical significance between medians was calculated using the nonparametric ANOVA Kruskal-Wallis test and Mann-Whitney test in Origin Pro 2015 (OriginLab Corp., USA). Statistical significance between means was calculated using parametric one-way ANOVA followed by the Bonferroni test in Origin Pro 2015 (OriginLab Corp., USA). Differences were considered as statistically significant at $p < 0.05$ (for parametric test) and Pu < 0.05 (for nonparametric test); $n$ indicates the number of animals.

## 3. Results

### 3.1. Maturation of Physical Features.
The average litter size of control and Hcy groups at P0 did not differ significantly (8.6 ± 1.2 vs. 8.4 ± 1.6 pups in the Hcy group, Pu > 0.05). However, in $H_2S$ and $HcyH_2S$ groups, the average litter size was significantly higher (13.3 ± 1.5 and 13.5 ± 1.0, correspondingly, Pu < 0.05) (Figure 1(a)). At the same time the total litter weight was significantly lower in the Hcy group due to the low body weight of the pups (Figure 1(a)). In $H_2S$ and $HcyH_2S$ groups, the total litter weight was higher compared to the control and Hcy groups due to larger litter sizes (Figure 1(a)). Substantial growth retardation of pups from the Hcy group was recorded during all observation periods (P0–P28) (Figure 1(c)).

At P28, body weight was reduced from 79.9 ± 0.8 g in controls to 66.1 ± 2.1 g ($n = 53$) in deficient animals ($n = 53$, $p < 0.05$). The average body weight of pups in the Hcy and $HcyH_2S$ groups at P2 was significantly lower than in the control and $H_2S$ groups. However, beginning from P8, the weight gain of Hcy pups was lower compared to all other groups during the observation period (Figure 1(c)). The mortality of pups in the Hcy group was higher (48%) compared to the control group (16%). The mortality of pups in the $H_2S$ and $HcyH_2S$ groups did not differ from the control (Figure 1(b)). Other parameters of physical maturation such as ear unfolding, the primary hair appearance, incisor eruption, and eye opening were not different in all experimental groups.

### 3.2. Reflex Testing.
We studied reflex ontogeny (righting reflex, negative geotaxis, cliff avoidance, head shake, acoustic startle reflex, free-fall righting, cliff avoidance caused by visual stimulus, and olfactory discrimination) reflecting brain maturation and integrity of sensorimotor development [26] (Table 1). Almost all reflexes were impaired in the Hcy group. Namely, negative geotaxis formation was delayed in the Hcy group (Table 1). The head shake reflex started at P8 in rat pups of all groups, but the number of head rotations per min was significantly lower in the Hcy group compared to the control, $H_2S$, and $HcyH_2S$ groups (Table 1). In the Hcy group, the onset of the righting reflex was delayed and the time necessary to come back to a quadruped position was significantly increased compared to other groups (Table 1). In the pups of the Hcy group, the cliff avoidance reflex was formed later (at P7) compared to the control, $H_2S$, and $HcyH_2S$ groups (Table 1). The delay of the reflex onset was also observed in pups of the Hcy group in other sensorimotor tests (Table 1).

### 3.3. Locomotion and Exploratory Activity in the Open Field Test.
The locomotor and exploratory activity was studied in the open field test at the ages P8, P16, and P26. Head rearing was analyzed in pups of P8 and P16. At P8 and P16, the number of head rearings in pups of the Hcy group was decreased compared to that in control, and in the $H_2S$ and $HcyH_2S$ groups, this parameter did not differ from the control (Figure 2(a)).

Horizontal activity was significantly lower in the Hcy group compared to the control group at all studied ages (Figure 2(b)). Administration of NaHS increased this parameter in pups with prenatal hHCY compared to the Hcy group. The number of crossed squares of pups from the $H_2S$ group was not different from the control group at all ages (Figure 2(b)). Rearings or vertical activity of pups from the Hcy group was significantly lower compared to the control. Activity of pups from the $H_2S$ and $HcyH_2S$ groups was higher compared to pups from the Hcy group (Figure 2(c)). Exploratory activity was assessed by the number of head dips at P26 (Figure 3(a)). The number of head dips from the Hcy group was significantly lower than in the control, the $H_2S$, and

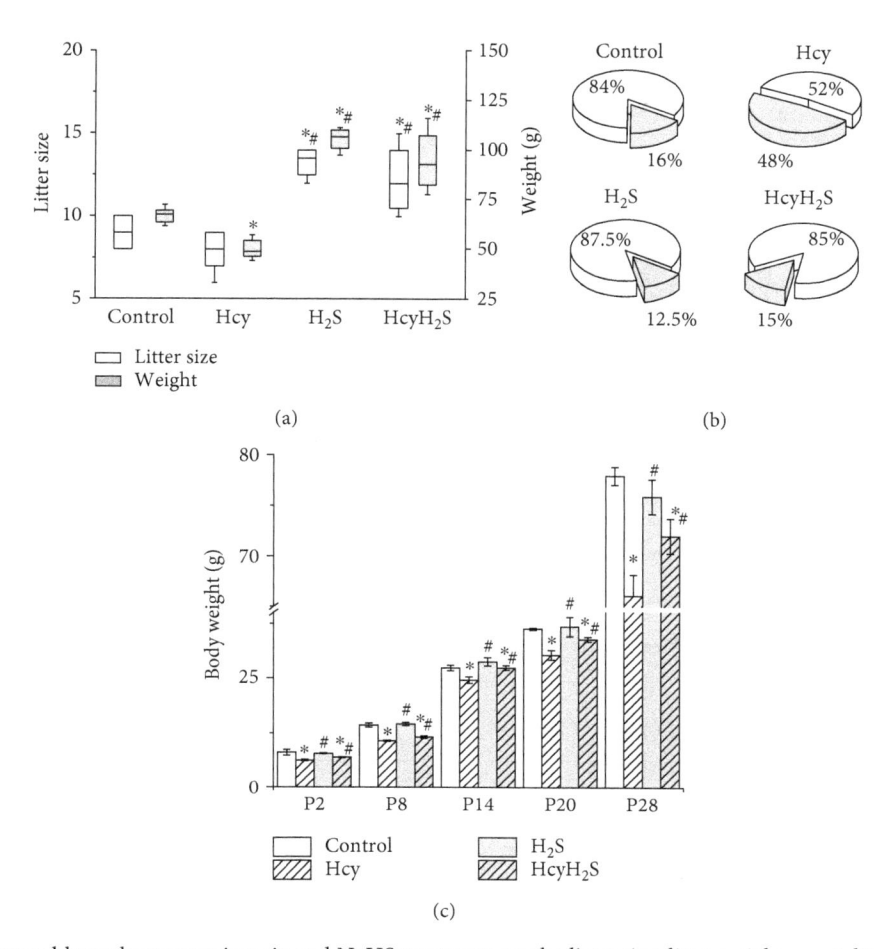

FIGURE 1: Effects of maternal hyperhomocysteinemia and NaHS treatment on the litter size, litter weight, mortality, and weight gain of the offspring. (a) Box plots reflecting the litter size (white boxes) and litter weight (grey boxes) in the control, Hcy, $H_2S$, and $HcyH_2S$ groups. (b) Mortality of pups during period P2–P28 in the control, $H_2S$, Hcy, and $HcyH_2S$ groups. White part—alive pups, grey part—dead pups in % relatively the litter size. (c) Body weights of rat dams during the period P2–P28 from the control, $H_2S$, Hcy, and $HcyH_2S$ groups. $*p < 0.05$ compared to the control group; $^{\#}p < 0.05$ compared to the Hcy group.

$HcyH_2S$ groups (Figure 3(a)). Grooming behavior and defecation scores were used as a measure of emotionality of animals [30, 36]. No significant intergroup difference was found in scores of defecation, but in animals of the Hcy group, higher numbers of grooming episodes were observed at P16 and P26 and were significantly decreased at P8, probably reflecting the deficit of motor coordination and locomotor activity (Figure 3(b)).

### 3.4. Rotarod Test and the Paw Grip Endurance (PaGE).
Motor coordination was assessed using the rotarod test, where the time to fall off and running distance were measured [29]. A significant reduction of the time spent on the rotarod was observed in the Hcy group at all age groups compared to the control (Figure 4(a)). Similar changes were also observed for the rotarod distance during experimental sessions for all studied ages (Figure 4(b)). NaHS treatment restored both parameters of the Hcy groups to control values.

In the control group, the time rats were able to stay on the grid increased with aging from $2.63 \pm 0.36$ s at P4 to $107.12 \pm 7.46$ s at P26 (Figure 4(c)). Rats from the Hcy group exhibited a deficit in the PaGE task as indicated by the reduction of time spent on the grid relatively to control rats (Figure 4(c)). NaHS treatment increased the time spent on the grid in pups of the Hcy group (Figure 4(c)).

### 3.5. Plasma Hcy Level.
The concentration of homocysteine in the plasma in control females was $8.16 \pm 0.29\,\mu M$ ($n = 7$) and in females fed with methionine-containing diet was $31.75 \pm 2.18\,\mu M$ ($n = 11$). The concentration of homocysteine in the plasma of pups born from control animals was $6.23 \pm 0.42\,\mu M$ ($n = 32$) and from females fed with methionine-containing diet was $22.07 \pm 2.60\,\mu M$ ($n = 32$). These results indicate the development of hHCY in dams and their offspring. NaHS treatment did not induce any changes of homocysteine levels in dams ($9.3 \pm 0.6\,\mu M$, $n = 4$) and pups ($6.5 \pm 0.3\,\mu M$, $n = 16$) of the control group, however, significantly reduced concentration of homocysteine in dams with hHCY ($17.4 \pm 1.4\,\mu M$, $n = 4$) and their offspring ($17.1 \pm 2.5\,\mu M$, $n = 16$).

### 3.6. $H_2S$ Generation in Brain Tissues.
It was shown previously that an exposure to homocysteine decreased the endogenous generation $H_2S$ in different tissues [21, 22, 24, 37]. In our

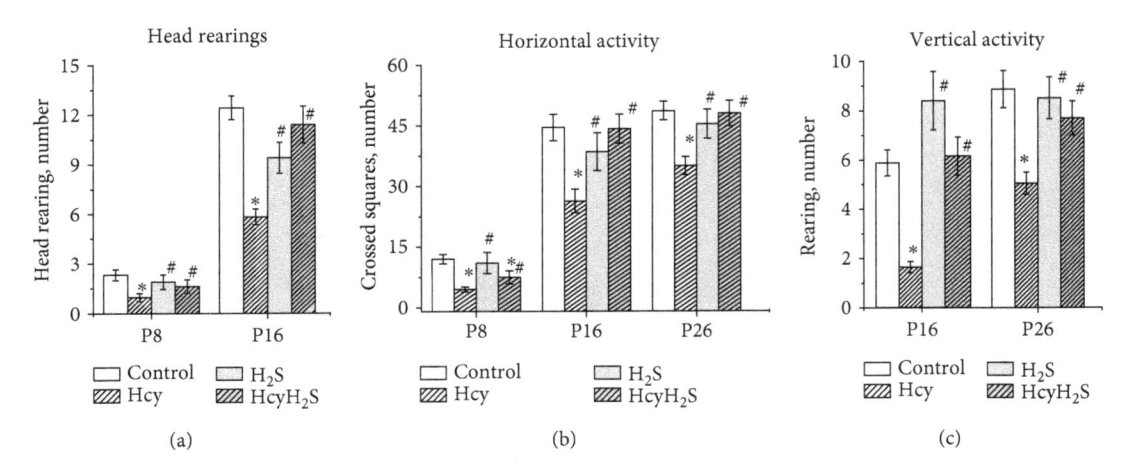

FIGURE 2: Effects of maternal hyperhomocysteinemia and NaHS treatment on locomotion in the open field test. Head rearings (a), the number of crossed squares (b), rearings (c) of pups from the control Hcy, $H_2S$, and $HcyH_2S$ groups. Data are expressed as mean ± SEM. $^*p < 0.05$ compared to the control group; $^\#p < 0.05$ compared to the Hcy group.

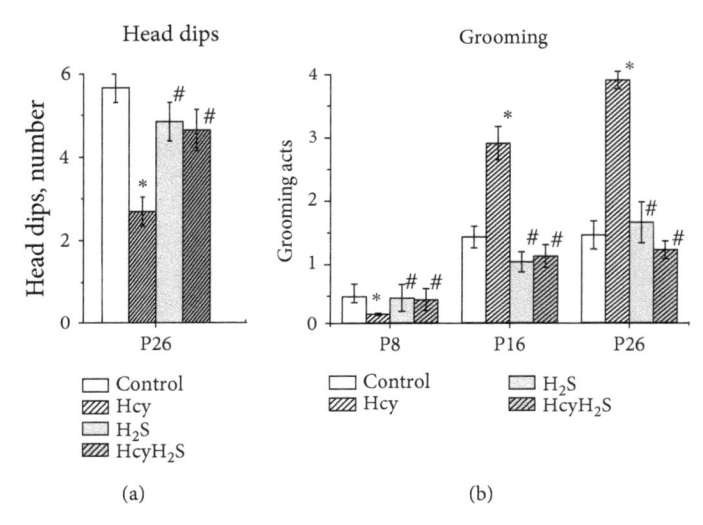

FIGURE 3: Effects of maternal hyperhomocysteinemia and NaHS treatment on the exploratory activity and emotionality in the open field test. Head dips (a) and grooming acts (b) of pups from the control Hcy, $H_2S$, and $HcyH_2S$ groups. Data are expressed as mean ± SEM. $^*p < 0.05$ compared to the control group; $^\#p < 0.05$ compared to the Hcy group.

experiments, $H_2S$ concentration, measured in brain tissues of control animals, was $12.76 \pm 0.72$ μM ($n = 7$). In rats of the Hcy group, we observed the decrease of $H_2S$ concentration to $7.97 \pm 0.87$ μM ($n = 7$, $p < 0.05$), which was elevated to $11.35 \pm 2.01$ μM by NaHS administration in the $HcyH_2S$ group ($n = 7$). The activity of $H_2S$-producing enzymes in the brain was measured as the rate of endogenous $H_2S$ generation when a high concentration of cysteine and pyridoxal 5′-phospate was added to brain homogenates. It was shown that the rate of $H_2S$ production decreased from $8.86 \pm 1.24$ μM/min/g in the control ($n = 7$) to $2.84 \pm 1.09$ μM/min/g in the Hcy group ($n = 7$, $p < 0.05$) and $2.25 \pm 0.98$ μM/min/g in the $HcyH_2S$ group ($n = 7$, $p < 0.05$). Our data indicate that in rats with prenatal hHCY, the rate of endogenous generation of $H_2S$ brain tissues was lower than in control conditions and administration of NaHS to dams with hHCY increased the $H_2S$ level to the control values but did not restore the activity of $H_2S$-producing enzymes.

### 3.7. Lipid Peroxidation and Antioxidant Enzymes Activity in Brain Tissues.
Severe oxidative stress during the prenatal period induces neuroinflammation and apoptosis followed by retardation of fetal growth and developmental impairments in postnatal life [8]. In order to estimate the extent of the oxidative stress in rats with prenatal hHCY, the level of MDA was measured in brain tissues of P13 and P28 animals from the control, Hcy, $H_2S$, and $HcyH_2S$ groups. At P13, the MDA level increased almost twice in the Hcy group which indicates a higher production of the reactive oxygen species (ROS) in rat brains with prenatal hHCY (Figure 5(a)). In rats of the $HcyH_2S$ group, the MDA level was significantly lower

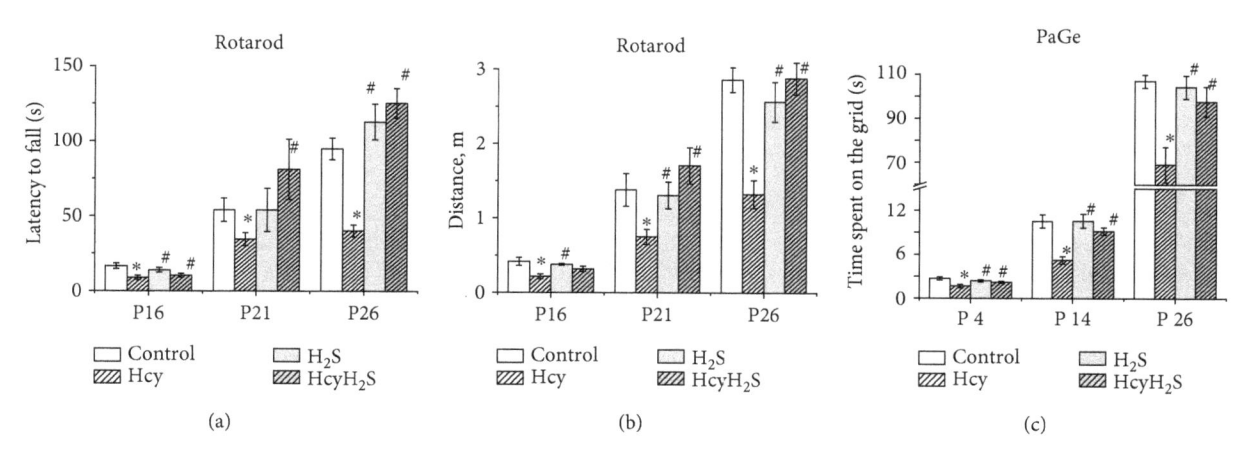

FIGURE 4: Effects of maternal hyperhomocysteinemia and NaHS treatment on muscle strength and motor coordination. Latency to fall (a) and running distance (b) in the rotarod test; the time spent on the grid (before falling) (c) in the paw grip endurance (PaGE) test of pups from the control Hcy, H$_2$S, and HcyH$_2$S groups. Data are expressed as mean ± SEM. *$p < 0.05$ compared to the control group; #$p < 0.05$ compared to the Hcy group.

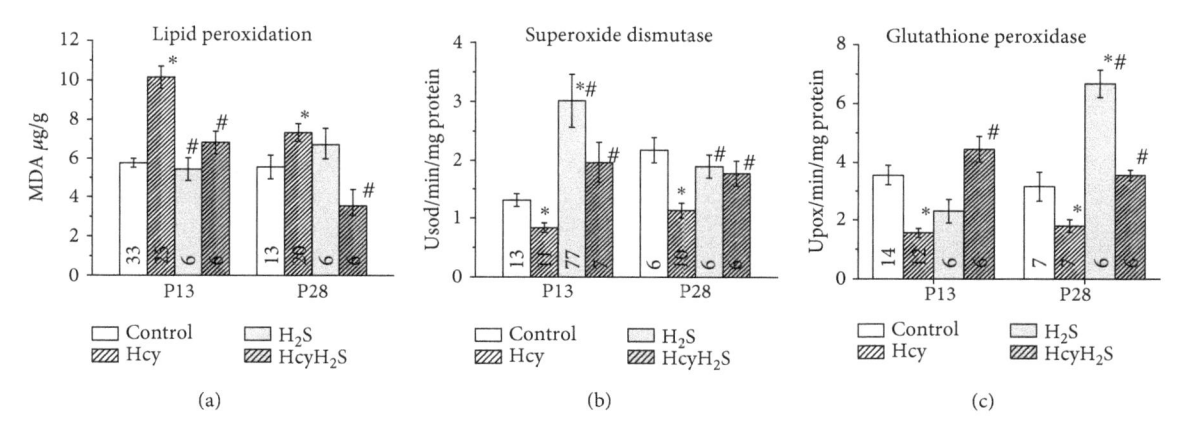

FIGURE 5: Effects of prenatal hHCY and NaHS treatment on lipid peroxidation and antioxidant enzyme activities measured in rat brain tissues. The level of MDA (an end product of lipid peroxidation) (a) and activities of antioxidant enzymes—superoxide dismutase 1 (b) and glutathione peroxidase 1 (c) measured in brain tissues of P13 and 28 rats from the control, Hcy, H$_2$S, and HcyH$_2$S groups. For each measurement, the number of samples is indicated inside the column. *$p < 0.05$ compared to the control group; #$p < 0.05$ compared to the Hcy group.

and did not differ from the control group (Figure 5(a)). In rats of the H$_2$S group, the MDA level was not different from the control level (Figure 5(a)). Similar values were observed in P28 rats (Figure 5(a)).

It is well known that homocysteine induces oxidative stress by the production of intracellular superoxide radicals but also impairs the activity of antioxidant enzymes [38, 39]. Therefore, we analyzed the enzymatic activities of SOD and GPx in brain tissues from the control, Hcy, H$_2$S, and HcyH$_2$S groups. We found that the activity of SOD that converts superoxide anions into H$_2$O$_2$ was significantly lower in the group of P13 and P28 Hcy rats (Figure 5(b)). Namely, at P13, the SOD activity decreased the Hcy groups. In rats from the HcyH$_2$S group, the SOD activity significantly increased and was not different from the control. Interestingly, in the H$_2$S group, SOD activity was higher than both in the control and Hcy groups ($n = 7$, $p < 0.05$). At P28, the level of SOD activity in the Hcy group was almost half of the control group and NaHS treatment restored its activity (Figure 5(b)). Similarly, decreased activity of GPx which reduces peroxides was observed in the Hcy group of P13 and P28 animals and NaHS treatment restored its activity to control values (Figure 5(c)). Evidently, the imbalance of prooxidant and antioxidant systems during chronic exposure of the fetus to high concentrations of homocysteine caused an oxidative stress and functional disability in the postnatal period. At the same time, low doses of NaHS during pregnancy provided antioxidant protection during prenatal and early postnatal development.

## 4. Discussion

During pregnancy, several complications have been associated with elevated homocysteine levels including preeclampsia, placental abruption, intrauterine growth retardation, or neural tube defects [40]. Several studies demonstrated that maternal hHCY resulted in a deficit of learning and memory in the offspring due to delayed brain maturation [6–9]. In most of the previous studies, the analysis of behavior was performed with offspring at almost adult level [7], whereas the present study focused on the detailed analysis of the physical development and reflex ontogeny, exploratory activity, and motor coordination of pups during the first 3 weeks of development. Our results indicate a significant decrease in litter size and body weight and delay in the formation of sensorimotor reflexes of pups with maternal hHCY. Locomotor and exploratory activity tested in the open field was diminished in the pups of the Hcy group. Prenatal hHCY also resulted in reduced muscle strength and motor coordination deficits assessed by the paw grip endurance test and the rotarod test. Simultaneously, we observed an increased level of oxidative stress and decreased activity of the antioxidant enzymes—SOD and GPx—in brain tissues of pups with hHCY. In rats with prenatal hHCY, the endogenous generation of $H_2S$ brain tissues was lower than in control conditions. Administration of the $H_2S$ donor—NaHS—to dams with hHCY during pregnancy prevented the deleterious effects of high homocysteine levels on fetus development, lowered oxidative stress, increased the $H_2S$ level in brain tissues, and restored the activity of SOD and GPx indicating its antioxidant potential.

*4.1. $H_2S$ Prevents Oxidative Stress and Decreases $H_2S$ Level in Brain Tissues of Rats with Prenatal hHCY.* In the model of prenatal hHCY used in our study, female rats received high methionine diet before and during pregnancy which induced an elevation of the plasma homocysteine level four times compared to control values. High blood plasma levels of homocysteine were not only observed in dams with hHCY but also in their offspring according to previous data [41]. Indeed, homocysteine can be transferred successfully through the placental exchange barrier and fetal cord homocysteine concentrations related to the maternal level [41, 42, 43]. In fetal brain, homocysteine can be produced from methionine or can be transported through the blood-brain barrier [44]. Under these circumstances, the fetal development occurs in hHCY conditions, which results in high mortality, low litter size, and low body weight of the offspring as was shown in our present and several previous studies [6–10].

Placental pathology due to endothelial dysfunctions, impaired NO synthesis, oxidative stress, and inflammation underlies adverse pregnancy outcome during hHCY conditions [45]. Oxidative stress is one of the main mechanisms of homocysteine-induced neurotoxicity as during prenatal period ROS highly affect embryo and fetus due to the lack of adequate antioxidant protection [46, 47]. Homocysteine itself can undergo autooxidation of its free thiol groups binding via a disulfide bridges with plasma proteins, low molecular thiols, or with a second homocysteine molecule [39]. Indirect oxidative effects of hHCY include the generation of superoxide from xanthine oxidase or uncoupled endothelial nitric oxide synthase, downregulation of antioxidant enzymes, or depletion of intracellular glutathione [39, 48, 49]. ROS, produced in these reactions, further oxidize various functionally important proteins, lipids, and nucleic acids [50]. Indeed, in our experiments, we observed an increased level of MDA, reflecting a higher level of oxidative stress in rats with prenatal hHCY similar to previous data [6–9]. Moreover, we found decreased activity of the antioxidant enzymes—SOD and GPx—in brain tissues of rats with prenatal hHCY which results in augmented accumulation of ROS during hHCY conditions. The altered activity/expression of SOD and GPx was also shown in vitro and in vivo studies [51, 52] including brain samples of rats with hHCY [6, 7, 9, 36, 39].

Recent data indicate the contribution of endogenous $H_2S$ for healthy placental vasculature which provides placental perfusion and optimal oxygen and nutrient diffusion [53, 54]. Moreover, inhibition of CSE reduced placental growth factor production, induced hypertension, promoted abnormal labyrinth vascularization in the placenta, and decreased fetal growth [53]. At the same time, $H_2S$ donor treatment prevented these changes and improved pregnancy outcome [54]. In addition, an insufficient $H_2S$ level has been suggested to be one of the potential causes of oxidative stress [55] which in turn results in the reduction of placental CSE activity, decreased $H_2S$ production, and intrauterine fetal growth restriction [54]. Worth noting, low level of $H_2S$ and diminished rate of endogenous $H_2S$ generation in brain tissues of rats with prenatal hHCY were shown in our experiments. Interestingly, that administration of the $H_2S$ donor before and during pregnancy increased the concentration of $H_2S$ without affecting the activity of $H_2S$-producing enzymes. $H_2S$ treatment not only restored the litter size and total litter weight of the offspring with maternal hHCY but even increased these parameters in control animals which appeared related to the improvement of placental blood supply and prevention of oxidative stress. Indeed, using spectrophotometric assays, we found that treatment with NaHS significantly lowered lipid peroxidation levels and restored the activity of SOD and GPx in brain tissues of rats with prenatal hHCY and even increased the activity of SOD and GPx in control animals. Positive effects of $H_2S$ were also shown in hHCY mice and rats where NaHS treatment attenuated oxidative stress, neurodegeneration, and neuroinflammation and restored the altered expression of synaptic proteins in hippocampal neurons and $H_2S$ level in brain tissues [16, 17, 23, 24]. Indeed, $H_2S$ with its reducing ability shows a high capacity to scavenge ROS [55]. $H_2S$ can react directly with superoxide anion ($O_2^-$), peroxynitrite, and other ROS [56]. Moreover, it was suggested that $H_2S$ can trigger antioxidant signaling pathways apart from its direct chemical reductant effect. Namely, $H_2S$ increases the level of two nonenzymatic antioxidants in animal cells, including intracellular reduced glutathione (GSH) and thioredoxin (Trx-1) [55, 57–59]. Mechanisms of $H_2S$ effects include the activation of the nuclear factor (erythroid-derived 2-) like 2 (Nrf2) and a transcription factor that regulates a wide variety of gene expression. Under oxidative stress conditions, Nrf2 is translocated into the nucleus and binds to promoters containing the antioxidant response element (ARE) sequence and inducing ARE-dependent gene expression such as Trx-1 and glutathione reductase [60–62].

H$_2$S also increases the activity of enzymatic antioxidants like SOD, catalase, and GPx which is likely mediated by an upregulation of NF-$\kappa$B transcription factor [55, 63-65] or Nrf2 signaling cascade [66]. Moreover, H$_2$S can directly bind at the catalytic Cu$^{2+}$ center of SOD as a substrate, increases the rate of superoxide anion scavenging [63], and directly stimulates the activity of GPx in vitro and in vivo studies [55, 67].

*4.2. H$_2$S Accelerates the Development of Neurobehavioral Maturation, Improves Exploratory Behavior, and Decreases Anxiety of Rats with Prenatal hHCY.* In rats, the period of two weeks after birth represents a critical phase in neurobehavioral maturation with rapid brain growth which corresponds to the last months of human fetal brain growth [26]. In our study, the development of the main parameters of physical maturation like eye opening, ear unfolding, incisor eruption, and hair appearance was not significantly different in all groups of animals. However, the development of sensorimotor reflexes important for the establishment of appropriate behavioral responses [68] was delayed in rats with prenatal hHCY. The day of appearance of negative geotaxis, righting reflex, cliff avoidance, and acoustic startle reflexes measured before P10 was slightly but significantly delayed in rats of the Hcy group. Reflexes which developed later and involved more complicated motor functions and different sensory systems were significantly delayed compared to the control group. Free-fall righting reflexes mediated by the visual, vestibular system, surface body senses, and proprioceptive senses appeared only at P19 (in control, at P12). The same delay was observed for cliff avoidance caused by visual stimuli and test olfactory discrimination, indicating variable development of different sensory systems. Similar observations were found in pups with gestational vitamin B deficiency where the implementation time of the negative geotaxis reflex was increased [8].

NaHS treatment not only improved the development of neurobehavioral reflexes in the Hcy group but even accelerated the appearance of the righting reflex and acoustic startle reflex in the control group which may be explained by the antioxidant properties of H$_2$S and its contribution for healthy placental vasculature [53, 54]. Therefore, NaHS administration may accelerate the development of reflexes, as shown for the antioxidant agent Mexidol which administration during neonatal period facilitated learning processes of rats [69].

Exploratory behavior is typically assessed in an open field where the inner conflict of the animals to avoid potentially dangerous environments and eagerness to explore it determines their locomotion [70]. In our experimental approach, head rearing, number of crossed squares, rearings, and head dips in the open field test were significantly decreased in the Hcy group, indicating reduced exploratory behavior. Self-grooming behavior reflects the reaction of animals to a stressful environment [71]. Pups from the Hcy group showed an increase of grooming acts in the open field arena, which indicates higher stress susceptibility of animals. Most impressively, NaHS treatment restored all parameters recorded in the open field to the control level. Decreased exploratory behavior and high level of grooming in rats with prenatal hHCY observed in our experiments indicate on the depression and anxiety associated with hHCY conditions. These changes can be explained by decreased dopamine, serotonin, and norepinephrine levels and increased activity of monoamine oxidases in brain tissues [24, 72]. NaHS administration improved grooming and head dips in rats of the hHCY group, indicated its anxiolytic-like effect. Antidepressant and anxiolytic-like effects of H$_2$S were previously shown in forced swimming and tail suspension tests of mice and rats—constituting behavioral models of depression and anxiety [73, 74]. In line with our results, H$_2$S donor increased head dips and lowered the number of grooming of rats in the open field and elevated plus maze [75]. Possible mechanisms of H$_2$S action include the inhibition of the corticotropin-releasing factor secreting from the hypothalamus under stress conditions [76, 77]. Recently, it was shown that H$_2$S inhibits monoamine oxidase activity and restores concentrations of catecholamine and serotonin in the brain of rats with hHCY [24].

Hyperactivation of NMDA receptors with subsequent desensitization impacts on the impairments of brain maturation in prenatal hHCY [7, 8, 78, 79]. In addition, homocysteine increased activity of maxi Ca$^{2+}$-activated K$^+$ channels of rat pituitary tumor cells (GH3) and decreased growth hormone release necessary for growth and development [80]. H$_2$S may prevent excitotoxicity associated with hyperactivation of NMDA receptors [39] as indicated by its inhibitory effects on GluN1/2B receptors, mainly expressed during the neonatal period preventing enhanced neuronal excitability typical for early hippocampal networks [81].

*4.3. H$_2$S Improves Motor Coordination and Muscle Strength of Rats with Prenatal hHCY.* The paw grip endurance (PaGE) test demonstrated that at all tested ages (P4, P14, and P26), the time spent on the grid was lower in hHCY pups indicating diminished muscle strength. Moreover, the decreased latency to fall from the rotating cylinder and shorter rotarod distance indicated impaired fore and hind limb motor coordination and balance which may result from cerebellar dysfunction [82]. Also motor cortex, hippocampus, and basal ganglia play important roles in the performance of this task [83]. These brain areas accumulates homocysteine which induces oxidative stress with subsequent DNA damage and accelerated neuronal apoptosis in fetal brain [7, 84]. It was reported that hHCY conditions in *CBS*$^{+/-}$ mice were detrimental to muscle force generation and responsible for muscle fatigability [85] via oxidative/endoplasmic reticulum (ER) stress [86]. Treatment of hHCY dams with H$_2$S donor restored muscle strength, motor coordination, and balance of pups to control levels which may allude to the importance of endogenous production of H$_2$S in rat skeletal muscle. Beneficial effects of H$_2$S may be explained by the reduction of oxidative and ER stress responses in affected skeletal muscles [38, 86]. In addition, deleterious effects of homocysteine were shown at the level of the neuromuscular junction. Namely, it was shown recently that homocysteine depressed quantal content and largely increases the inhibitory effect of ROS on transmitter release, via NMDA receptors activation [87, 88]. Simultaneously, H$_2$S increased quantal transmitter release in the mammalian neuromuscular junction [14]. Thus, a deficit of $H_2S$ production may be a plausible reason of muscle weakness observed in our study together with oxidative stress induced by hHCY.

## 5. Conclusions

We have shown that homocysteine-evoked oxidative stress during the prenatal period caused delayed brain maturation of the offspring and decreased $H_2S$ levels in brain tissues. Treatment of dams during pregnancy with $H_2S$ reversed the observed developmental impairments, restored muscle strength and coordination, and prevented oxidative stress of the brain tissue. Our data are supported by results obtained in models of acute hHCY in adult animals, where $H_2S$ obliterated homocysteine-induced endoplasmic reticulum stress as well as learning and memory deficits [22], ameliorated cognitive dysfunction, inhibited reactive aldehyde generation, and upregulated glutathione in the hippocampus [23]. Moreover, it was shown that endogenous $H_2S$ is required for healthy placental vasculature to support fetal development and that a decrease in $CSE/H_2S$ activity may contribute to the pathogenesis of preeclampsia [53]. Our findings suggest that $H_2S$ is effective in protection against developmental impairments in prenatal hHCY and has a promising potential role in facilitating a novel strategy to prevent homocysteine/oxidative stress-induced neurotoxicity.

## Acknowledgments

The authors are grateful to Anton Hermann for useful suggestions. This project was supported by the Russian Science Foundation (Grant no. 14-15-00618) and Russian Foundation of Basic Research (Grant no. 18-015-00423).

## References

[1] R. Ansari, A. Mahta, E. Mallack, and J. J. Luo, "Hyperhomocysteinemia and neurologic disorders: a review," *Journal of Clinical Neurology*, vol. 10, no. 4, pp. 281–288, 2014.

[2] D. W. Jacobsen, "Homocysteine and vitamins in cardiovascular disease," *Clinical Chemistry*, vol. 44, no. 8, pp. 1833–1843, 1998.

[3] A. M. Troen, "The central nervous system in animal models of hyperhomocysteinemia," *Progress in Neuro-Psychopharmacology & Biological Psychiatry*, vol. 29, no. 7, pp. 1140–1151, 2005.

[4] A. Agrawal, K. Ilango, P. K. Singh et al., "Age dependent levels of plasma homocysteine and cognitive performance," *Behavioural Brain Research*, vol. 283, pp. 139–144, 2015.

[5] Y. Aubard, N. Darodes, and M. Cantaloube, "Hyperhomocysteinemia and pregnancy — review of our present understanding and therapeutic implications," *European Journal of Obstetrics & Gynecology and Reproductive Biology*, vol. 93, no. 2, pp. 157–165, 2000.

[6] S. T. Koz, N. T. Gouwy, N. Demir, V. S. Nedzvetsky, E. Etem, and G. Baydas, "Effects of maternal hyperhomocysteinemia induced by methionine intake on oxidative stress and apoptosis in pup rat brain," *International Journal of Developmental Neuroscience*, vol. 28, no. 4, pp. 325–329, 2010.

[7] G. Baydas, S. T. Koz, M. Tuzcu, V. S. Nedzvetsky, and E. Etem, "Effects of maternal hyperhomocysteinemia induced by high methionine diet on the learning and memory performance in offspring," *International Journal of Developmental Neuroscience*, vol. 25, no. 3, pp. 133–139, 2007.

[8] S. A. Blaise, E. Nédélec, H. Schroeder et al., "Gestational vitamin B deficiency leads to homocysteine-associated brain apoptosis and alters neurobehavioral development in rats," *The American Journal of Pathology*, vol. 170, no. 2, pp. 667–679, 2007.

[9] A. V. Makhro, A. P. Mashkina, O. A. Solenaya, O. A. Trunova, L. S. Kozina, and A. V. Arutyunian, "Prenatal hyperhomocysteinemia as a model of oxidative stress of the brain skeletal muscle malfunction," *International Journal of Molecular Sciences*, vol. 14, pp. 15074–15091, 2013.

[10] E. Gerasimova, O. Yakovleva, G. Burkhanova, G. Ziyatdinova, N. Khaertdinov, and G. Sitdikova, "Effects of maternal hyperhomocysteinemia on the early physical development and neurobehavioral maturation of rat offspring," *BioNanoScience*, vol. 7, no. 1, pp. 155–158, 2017.

[11] R. S. Beard Jr and S. E. Bearden, "Vascular complications of cystathionine β-synthase deficiency: future directions for homocysteine-to-hydrogen sulfide research," *American Journal of Physiology. Heart and Circulatory Physiology*, vol. 300, no. 1, pp. H13–H26, 2011.

[12] K. Abe and H. Kimura, "The possible role of hydrogen sulfide as an endogenous neuromodulator," *The Journal of Neuroscience*, vol. 16, no. 3, pp. 1066–1071, 1996.

[13] A. V. Yakovlev, E. Kurmashova, A. Zakharov, and G. F. Sitdikova, "Network-driven activity and neuronal excitability in hippocampus of neonatal rats with prenatal hyperhomocysteinemia," *BioNanoScience*, vol. 8, no. 1, pp. 304–309, 2018.

[14] E. Gerasimova, J. Lebedeva, A. Yakovlev, A. Zefirov, R. Giniatullin, and G. Sitdikova, "Mechanisms of hydrogen sulfide (H2S) action on synaptic transmission at the mouse neuromuscular junction," *Neuroscience*, vol. 303, pp. 577–585, 2015.

[15] N. Tyagi, K. S. Moshal, U. Sen et al., "H2S protects against methionine-induced oxidative stress in brain endothelial cells," *Antioxidants & Redox Signaling*, vol. 11, no. 1, pp. 25–33, 2009.

[16] P. K. Kamat, A. Kalani, S. Givvimani, P. B. Sathnur, S. C. Tyagi, and N. Tyagi, "Hydrogen sulfide attenuates neurodegeneration and neurovascular dysfunction induced by intracerebral administered homocysteine in mice," *Neuroscience*, vol. 252, pp. 302–319, 2013.

[17] P. K. Kamat, P. Kyles, A. Kalani, and N. Tyagi, "Hydrogen sulfide ameliorates homocysteine-induced Alzheimer's disease-like pathology, blood–brain barrier disruption, and synaptic disorder," *Molecular Neurobiology*, vol. 53, no. 4, pp. 2451–2467, 2016.

[18] Q. H. Gong, Q. Wang, L. L. Pan, X. H. Liu, H. Xin, and Y. Z. Zhu, "S-Propargyl-cysteine, a novel hydrogen sulfide-modulated agent, attenuates lipopolysaccharide-induced spatial learning and memory impairment: involvement of TNF signaling and NFκB pathway in rats," *Brain, Behavior, and Immunity*, vol. 25, no. 1, pp. 110–119, 2011.

[19] J. He, R. Guo, P. Qiu, X. Su, G. Yan, and J. Feng, "Exogenous hydrogen sulfide eliminates spatial memory retrieval impairment and hippocampal CA1 LTD enhancement caused by acute stress via promoting glutamate uptake," *Neuroscience*, vol. 350, pp. 110–123, 2017.

[20] X. Q. Tang, X. T. Shen, Y. E. Huang et al., "Inhibition of endogenous hydrogen sulfide generation is associated with homocysteine-induced neurotoxicity: role of ERK1/2 activation," *Journal of Molecular Neuroscience*, vol. 45, no. 1, pp. 60–67, 2011.

[21] Z. Cheng, X. Shen, X. Jiang et al., "Hyperhomocysteinemia potentiates diabetes-impaired EDHF-induced vascular relaxation: role of insufficient hydrogen sulfide," *Redox Biology*, vol. 16, pp. 215–225, 2018.

[22] M. H. Li, J. P. Tang, P. Zhang et al., "Disturbance of endogenous hydrogen sulfide generation and endoplasmic reticulum stress in hippocampus are involved in homocysteine-induced defect in learning and memory of rats," *Behavioural Brain Research*, vol. 262, pp. 35–41, 2014.

[23] M. Li, P. Zhang, H. J. Wei et al., "Hydrogen sulfide ameliorates homocysteine-induced cognitive dysfunction by inhibition of reactive aldehydes involving upregulation of ALDH2," *International Journal of Neuropsychopharmacology*, vol. 20, no. 4, pp. 305–315, 2017.

[24] M. Kumar, M. Modi, and R. Sandhir, "Hydrogen sulfide attenuates homocysteine-induced cognitive deficits and neurochemical alterations by improving endogenous hydrogen sulfide levels," *BioFactors*, vol. 43, no. 3, pp. 434–450, 2017.

[25] J. Bełtowski, G. Wójcicka, and A. Wojtak, "Effect of experimental hyperhomocysteinemia on plasma lipid profile, insulin sensitivity and paraoxonase 1 in the rat," *Adipobiology*, vol. 4, pp. 77–84, 2012.

[26] M. G. Alton-Mackey and B. L. Walker, "The physical and neuromotor development of progeny of female rats fed graded levels of pyridoxine during lactation," *The American Journal of Clinical Nutrition*, vol. 31, no. 1, pp. 76–81, 1978.

[27] T. C. B. J. Deiró, R. Manhães-de-Castro, J. E. Cabral-Filho et al., "Sertraline delays the somatic growth and reflex ontogeny in neonate rats," *Physiology & Behavior*, vol. 87, no. 2, pp. 338–344, 2006.

[28] P. Weydt, S. Y. Hong, M. Kliot, and T. Möller, "Assessing disease onset and progression in the SOD1 mouse model of ALS," *Neuroreport*, vol. 14, no. 7, pp. 1051–1054, 2003.

[29] C. Barlow, S. Hirotsune, R. Paylor et al., "Atm-deficient mice: a paradigm of ataxia telangiectasia," *Cell*, vol. 86, no. 1, pp. 159–171, 1996.

[30] T. Karl, R. Pabst, and S. von Hörsten, "Behavioral phenotyping of mice in pharmacological and toxicological research," *Experimental and Toxicologic Pathology*, vol. 55, no. 1, pp. 69–83, 2003.

[31] P. T. Lee, D. Lowinsohn, and R. G. Compton, "Simultaneous detection of homocysteine and cysteine in the presence of ascorbic acid and glutathione using a nanocarbon modified electrode," *Electroanalysis*, vol. 26, no. 7, pp. 1488–1496, 2014.

[32] M. Yusuf, B. T. Kwong Huat, A. Hsu, M. Whiteman, M. Bhatia, and P. K. Moore, "Streptozotocin-induced diabetes in the rat is associated with enhanced tissue hydrogen sulfide biosynthesis," *Biochemical and Biophysical Research Communications*, vol. 333, no. 4, pp. 1146–1152, 2005.

[33] H. Ohkawa, N. Ohishi, and K. Yagi, "Assay for lipid peroxides in animal tissues by thiobarbituric acid reaction," *Analytical Biochemistry*, vol. 95, no. 2, pp. 351–358, 1979.

[34] C. J. Weydert and J. J. Cullen, "Measurement of superoxide dismutase, catalase and glutathione peroxidase in cultured cells and tissue," *Nature Protocols*, vol. 5, no. 1, pp. 51–66, 2010.

[35] M. M. Bradford, "A rapid and sensitive method for the quantitation of microgram quantities of protein utilizing the principle of protein-dye binding," *Analytical Biochemistry*, vol. 72, no. 1-2, pp. 248–254, 1976.

[36] A. V. Pustygina, Y. P. Milyutina, I. V. Zaloznyaya, and A. V. Arutyunyan, "Indices of oxidative stress in the brain of newborn rats subjected to prenatal hyperhomocysteinemia," *Neurochemical Journal*, vol. 9, no. 1, pp. 60–65, 2015.

[37] M. Y. Ali, M. Whiteman, C.-M. Low, and P. K. Moore, "Hydrogen sulphide reduces insulin secretion from HIT-T15 cells by a Katp channel-dependent pathway," *The Journal of Endocrinology*, vol. 195, no. 1, pp. 105–112, 2007.

[38] S. Veeranki and S. Tyagi, "Defective homocysteine metabolism: potential implications for skeletal muscle malfunction," *International Journal of Molecular Sciences*, vol. 14, no. 7, pp. 15074–15091, 2013.

[39] M. Petras, Z. Tatarkova, M. Kovalska et al., "Hyperhomocysteinemia as a risk factor for the neuronal system disorders," *Journal of Physiology and Pharmacology*, vol. 65, no. 1, pp. 15–23, 2014.

[40] W. M. Hague, "Homocysteine and pregnancy," *Best Practice & Research. Clinical Obstetrics & Gynaecology*, vol. 17, no. 3, pp. 459–469, 2003.

[41] A. M. Molloy, J. L. Mills, J. McPartlin, P. N. Kirke, J. M. Scott, and S. Daly, "Maternal and fetal plasma homocysteine concentrations at birth: the influence of folate, vitamin B12, and the 5,10-methylenetetrahydrofolate reductase 677C→T variant," *American Journal of Obstetrics and Gynecology*, vol. 186, no. 3, pp. 499–503, 2002.

[42] E. Tsitsiou, C. P. Sibley, S. W. D'Souza, O. Catanescu, D. W. Jacobsen, and J. D. Glazier, "Homocysteine is transported by the microvillous plasma membrane of human placenta," *Journal of Inherited Metabolic Disease*, vol. 34, no. 1, pp. 57–65, 2011.

[43] Y. G. Acılmış, E. Dikensoy, A. I. Kutlar et al., "Homocysteine, folic acid and vitamin B12 levels in maternal and umbilical cord plasma and homocysteine levels in placenta in pregnant women with pre-eclampsia," *Journal of Obstetrics and Gynaecology Research*, vol. 37, no. 1, pp. 45–50, 2011.

[44] A. Grieve, S. P. Butcher, and R. Griffiths, "Synaptosomal plasma membrane transport of excitatory sulphur amino acid transmitter candidates: kinetic characterisation and analysis of carrier specificity," *Journal of Neuroscience Research*, vol. 32, no. 1, pp. 60–68, 1992.

[45] W. K. C. Lai and M. Y. Kan, "Homocysteine-induced endothelial dysfunction," *Annals of Nutrition & Metabolism*, vol. 67, no. 1, pp. 1–12, 2015.

[46] A. F. Perna, D. Ingrosso, and N. G. De Santo, "Homocysteine and oxidative stress," *Amino Acids*, vol. 45, no. 3-4, pp. 409–417, 2003.

[47] S. Perrone, A. Santacroce, M. Longini, F. Proietti, F. Bazzini, and G. Buonocore, "The free radical diseases of prematurity: from cellular mechanisms to bedside," *Oxidative Medicine and Cellular Longevity*, vol. 2018, Article ID 7483062, 14 pages, 2018.

[48] S. Dayal, K. L. Brown, C. J. Weydert et al., "Deficiency of glutathione peroxidase-1 sensitizes hyperhomocysteinemic mice to endothelial dysfunction," *Arteriosclerosis, Thrombosis, and Vascular Biology*, vol. 22, no. 12, pp. 1996–2002, 2002.

[49] A. L. S. Au, S. W. Seto, S. W. Chan, M. S. Chan, and Y. W. Kwan, "Modulation by homocysteine of the iberiotoxin-sensi-

tive, $Ca^{2+}$-activated $K^+$ channels of porcine coronary artery smooth muscle cells," *European Journal of Pharmacology*, vol. 546, no. 1–3, pp. 109–119, 2006.

[50] C. G. Zou and R. Banerjee, "Homocysteine and redox signaling," *Antioxidants & Redox Signaling*, vol. 7, no. 5-6, pp. 547–559, 2005.

[51] E. Lubos, J. Loscalzo, and D. E. Handy, "Homocysteine and glutathione peroxidase-1," *Antioxidants & Redox Signaling*, vol. 9, no. 11, pp. 1923–1940, 2007.

[52] P. A. Outinen, S. K. Sood, S. I. Pfeifer et al., "Homocysteine-induced endoplasmic reticulum stress and growth arrest leads to specific changes in gene expression in human vascular endothelial cells," *Blood*, vol. 94, no. 3, pp. 959–967, 1999.

[53] K. Wang, S. Ahmad, M. Cai et al., "Dysregulation of hydrogen sulfide producing enzyme cystathionine γ-lyase contributes to maternal hypertension and placental abnormalities in preeclampsia," *Circulation*, vol. 127, no. 25, pp. 2514–2522, 2013.

[54] L. Lu, J. Kingdom, G. J. Burton, and T. Cindrova-Davies, "Placental stem villus arterial remodeling associated with reduced hydrogen sulfide synthesis contributes to human fetal growth restriction," *The American Journal of Pathology*, vol. 187, no. 4, pp. 908–920, 2017.

[55] Z. Z. Xie, Y. Liu, and J. S. Bian, "Hydrogen sulfide and cellular redox homeostasis," *Oxidative Medicine and Cellular Longevity*, vol. 2016, Article ID 6043038, 12 pages, 2016.

[56] M. Whiteman, L. Li, I. Kostetski et al., "Evidence for the formation of a novel nitrosothiol from the gaseous mediators nitric oxide and hydrogen sulphide," *Biochemical and Biophysical Research Communications*, vol. 343, no. 1, pp. 303–310, 2006.

[57] M. Lu, L.-F. Hu, G. Hu, and J.-S. Bian, "Hydrogen sulfide protects astrocytes against $H_2O_2$-induced neural injury via enhancing glutamate uptake," *Free Radical Biology & Medicine*, vol. 45, no. 12, pp. 1705–1713, 2008.

[58] Y. Kimura, Y.-I. Goto, and H. Kimura, "Hydrogen sulfide increases glutathione production and suppresses oxidative stress in mitochondria," *Antioxidants and Redox Signaling*, vol. 12, no. 1, pp. 1–13, 2010.

[59] S. K. Jain, L. Huning, and D. Micinski, "Hydrogen sulfide upregulates glutamate-cysteine ligase catalytic subunit, glutamate-cysteine ligase modifier subunit, and glutathione and inhibits interleukin-1β secretion in monocytes exposed to high glucose levels," *Metabolic Syndrome and Related Disorders*, vol. 12, no. 5, pp. 299–302, 2014.

[60] L. Gan and J. A. Johnson, "Oxidative damage and the Nrf2-ARE pathway in neurodegenerative diseases," *Biochimica et Biophysica Acta*, vol. 1842, no. 8, pp. 1208–1218, 2014.

[61] J. W. Calvert, S. Jha, S. Gundewar et al., "Hydrogen sulfide mediates cardioprotection through Nrf2 signaling," *Circulation Research*, vol. 105, no. 4, pp. 365–374, 2009.

[62] G. Yang, K. Zhao, Y. Ju et al., "Hydrogen sulfide protects against cellular senescence via S-sulfhydration of keap1 and activation of Nrf2," *Antioxidants & Redox Signaling*, vol. 18, no. 15, pp. 1906–1919, 2013.

[63] D. G. Searcy, J. P. Whitehead, and M. J. Maroney, "Interaction of Cu, Zn superoxide dismutase with hydrogen sulfide," *Archives of Biochemistry and Biophysics*, vol. 318, no. 2, pp. 251–263, 1995.

[64] M. Bhatia, J. N. Sidhapuriwala, S. Wei Ng, R. Tamizhselvi, and S. M. Moochhala, "Pro-inflammatory effects of hydrogen sulphide on substance P in caerulein-induced acute pancreatitis," *Journal of Cellular and Molecular Medicine*, vol. 12, no. 2, pp. 580–590, 2008.

[65] R. Tamizhselvi, P. Shrivastava, Y.-H. Koh, H. Zhang, and M. Bhatia, "Preprotachykinin-A gene deletion regulates hydrogen sulfide-induced toll-like receptor 4 signaling pathway in cerulein-treated pancreatic acinar cells," *Pancreas*, vol. 40, no. 3, pp. 444–452, 2011.

[66] S. Kalayarasan, P. N. Prabhu, N. Sriram, R. Manikandan, M. Arumugam, and G. Sudhandiran, "Diallyl sulfide enhances antioxidants and inhibits inflammation through the activation of Nrf2 against gentamicin-induced nephrotoxicity in Wistar rats," *European Journal of Pharmacology*, vol. 606, no. 1–3, pp. 162–171, 2009.

[67] Y.-Y. Liu, B. V. Nagpure, P. T.-H. Wong, and J. S. Bian, "Hydrogen sulfide protects SH-SY5Y neuronal cells against d-galactose induced cell injury by suppression of advanced glycation end products formation and oxidative stress," *Neurochemistry International*, vol. 62, no. 5, pp. 603–609, 2013.

[68] N. Sousa, O. F. X. Almeida, and C. T. Wotjak, "A hitchhiker's guide to behavioral analysis in laboratory rodents," *Genes, Brain and Behavior*, vol. 5, pp. 5–24, 2006.

[69] G. V. Vishnevskaya, E. I. Gern, and T. A. Adzhimolaev, "The effect of the inhibition of free-radical oxidation processes in early postnatal ontogeny on learning in adult rats," *Fiziologicheskiĭ Zhurnal SSSR Imeni I. M. Sechenova*, vol. 76, no. 10, pp. 1393–1396, 1990.

[70] I. Golani, Y. Benjamini, and D. Eilam, "Stopping behavior: constraints on exploration in rats (*Rattus norvegicus*)," *Behavioural Brain Research*, vol. 53, no. 1-2, pp. 21–33, 1993.

[71] W. H. Gispen and R. L. Isaacson, "ACTH-induced excessive grooming in the rat," *Pharmacology & Therapeutics*, vol. 12, no. 1, pp. 209–246, 1981.

[72] P. Bhatia and N. Singh, "Homocysteine excess: delineating the possible mechanism of neurotoxicity and depression," *Fundamental & Clinical Pharmacology*, vol. 29, no. 6, pp. 522–528, 2015.

[73] W.-L. Chen, B. Xie, C. Zhang et al., "Antidepressant-like and anxiolytic-like effects of hydrogen sulfide in behavioral models of depression and anxiety," *Behavioural Pharmacology*, vol. 24, no. 7, pp. 590–597, 2013.

[74] Z. J. Tang, W. Zou, J. Yuan et al., "Antidepressant-like and anxiolytic-like effects of hydrogen sulfide in streptozotocin-induced diabetic rats through inhibition of hippocampal oxidative stress," *Behavioural Pharmacology*, vol. 26, no. 5, pp. 427–435, 2015.

[75] A. F. Donatti, R. N. Soriano, C. R. A. Leite-Panissi, L. G. S. Branco, and A. S. de Souza, "Anxiolytic-like effect of hydrogen sulfide ($H_2S$) in rats exposed and re-exposed to the elevated plus-maze and open field tests," *Neuroscience Letters*, vol. 642, pp. 77–85, 2017.

[76] C. dello Russo, G. Tringali, E. Ragazzoni et al., "Evidence that hydrogen sulphide can modulate hypothalamo-pituitary-adrenal axis function: in vitro and in vivo studies in the rat," *Journal of Neuroendocrinology*, vol. 12, no. 3, pp. 225–233, 2000.

[77] C. Mancuso, P. Navarra, and P. Preziosi, "Roles of nitric oxide, carbon monoxide, and hydrogen sulfide in the regulation of the hypothalamic-pituitary-adrenal axis," *Journal of Neurochemistry*, vol. 113, no. 3, pp. 563–575, 2010.

[78] S. A. Lipton, W. K. Kim, Y. B. Choi et al., "Neurotoxicity associated with dual actions of homocysteine at the N-methyl-D-

aspartate receptor," *Proceedings of the National Academy of Sciences of the United States of America*, vol. 94, no. 11, pp. 5923–5928, 1997.

[79] A. D. Bolton, M. A. Phillips, and M. Constantine-Paton, "Homocysteine reduces NMDAR desensitization and differentially modulates peak amplitude of NMDAR currents, depending on GluN2 subunit composition," *Journal of Neurophysiology*, vol. 110, no. 7, pp. 1567–1582, 2013.

[80] A. S. Gaifullina, A. V. Yakovlev, A. N. Mustafina, T. M. Weiger, A. Hermann, and G. F. Sitdikova, "Homocysteine augments BK channel activity and decreases exocytosis of secretory granules in rat GH3 cells," *FEBS Letters*, vol. 590, no. 19, pp. 3375–3384, 2016.

[81] A. V. Yakovlev, E. D. Kurmasheva, R. Giniatullin, I. Khalilov, and G. F. Sitdikova, "Hydrogen sulfide inhibits giant depolarizing potentials and abolishes epileptiform activity of neonatal rat hippocampal slices," *Neuroscience*, vol. 340, pp. 153–165, 2017.

[82] R. Lalonde, A. N. Bensoula, and M. Filali, "Rotorod sensorimotor learning in cerebellar mutant mice," *Neuroscience Research*, vol. 22, no. 4, pp. 423–426, 1995.

[83] J. Scholz, Y. Niibori, P. W Frankland, and J. P Lerch, "Rotarod training in mice is associated with changes in brain structure observable with multimodal MRI," *NeuroImage*, vol. 107, pp. 182–189, 2015.

[84] P. I. Ho, D. Ashline, S. Dhitavat et al., "Folate deprivation induces neurodegeneration: roles of oxidative stress and increased homocysteine," *Neurobiology of Disease*, vol. 14, no. 1, pp. 32–42, 2003.

[85] N. Tyagi, K. C. Sedoris, M. Steed, A. V. Ovechkin, K. S. Moshal, and S. C. Tyagi, "Mechanisms of homocysteine-induced oxidative stress," *American Journal of Physiology-Heart and Circulatory Physiology*, vol. 289, no. 6, pp. H2649–H2656, 2005.

[86] A. Majumder, M. Singh, J. Behera et al., "Hydrogen sulfide alleviates hyperhomocysteinemia-1 mediated skeletal muscle atrophy via 2 mitigation of oxidative and endoplasmic reticulum stress injury," *American Journal of Physiology-Cell Physiology*, vol. 315, pp. C609–C622, 2018.

[87] E. Bukharaeva, A. Shakirzyanova, V. Khuzakhmetova, G. Sitdikova, and R. Giniatullin, "Homocysteine aggravates ROS-induced depression of transmitter release from motor nerve terminals: potential mechanism of peripheral impairment in motor neuron diseases associated with hyperhomocysteinemia," *Frontiers in Cellular Neuroscience*, vol. 9, no. 391, 2015.

[88] J. S. Wang, D. Bojovic, Y. Chen, and C. A. Lindgren, "Homocysteine sensitizes the mouse neuromuscular junction to oxidative stress by nitric oxide," *Neuroreport*, vol. 29, no. 12, pp. 1030–1035, 2018.

# Neuroprotective Mechanisms of Resveratrol in Alzheimer's Disease: Role of SIRT1

Bruno Alexandre Quadros Gomes,[1] João Paulo Bastos Silva,[1] Camila Fernanda Rodrigues Romeiro,[2] Sávio Monteiro dos Santos,[3] Caroline Azulay Rodrigues,[3] Pricila Rodrigues Gonçalves,[2] Joni Tetsuo Sakai,[3] Paulo Fernando Santos Mendes,[3] Everton Luiz Pompeu Varela,[3] and Marta Chagas Monteiro[1,2,3]

[1]Neuroscience and Cell Biology Graduate Program, Institute of Biological Sciences, Federal University of Pará, Belém, Pará, Brazil
[2]Faculty of Pharmacy, Institute of Health Sciences, Federal University of Pará, Belém, Pará, Brazil
[3]Pharmaceutical Sciences Graduate Program, Institute of Health Sciences, Federal University of Pará, Belém, Pará, Brazil

Correspondence should be addressed to Marta Chagas Monteiro; martachagas2@yahoo.com.br

Academic Editor: Herbenya S. Peixoto

Alzheimer's disease (AD) is a progressive and neurodegenerative disorder of the cortex and hippocampus, which eventually leads to cognitive impairment. Although the etiology of AD remains unclear, the presence of $\beta$-amyloid (A$\beta$) peptides in these learning and memory regions is a hallmark of AD. Therefore, the inhibition of A$\beta$ peptide aggregation has been considered the primary therapeutic strategy for AD treatment. Many studies have shown that resveratrol has antioxidant, anti-inflammatory, and neuroprotective properties and can decrease the toxicity and aggregation of A$\beta$ peptides in the hippocampus of AD patients, promote neurogenesis, and prevent hippocampal damage. In addition, the antioxidant activity of resveratrol plays an important role in neuronal differentiation through the activation of silent information regulator-1 (SIRT1). SIRT1 plays a vital role in the growth and differentiation of neurons and prevents the apoptotic death of these neurons by deacetylating and repressing p53 activity; however, the exact mechanisms remain unclear. Resveratrol also has anti-inflammatory effects as it suppresses M1 microglia activation, which is involved in the initiation of neurodegeneration, and promotes Th2 responses by increasing anti-inflammatory cytokines and SIRT1 expression. This review will focus on the antioxidant and anti-inflammatory neuroprotective effects of resveratrol, specifically on its role in SIRT1 and the association with AD pathophysiology.

## 1. Introduction

Alzheimer's disease (AD) is a neurodegenerative pathology that causes impaired cognitive functioning and memory [1, 2]. Despite the disease being identified over 100 years ago [3], efforts are currently being expended to discover new chemical products (i.e., natural antioxidants) that act at determined points to block the progression of the disease [4, 5]. Resveratrol has been considered as a protector compound for the treatment of neurodegenerative diseases (i.e., AD, Parkinson disease, and amyotrophic lateral sclerosis) that have high levels of oxidative damage due to its antioxidant and anti-inflammatory properties [6]. Moreover, this compound can also modulate different molecular pathways dependent on silent information regulator-1 (SIRT1) in neurodegenerative diseases [6]. However, recent reviews also report other multipathways that are involved in the neuroprotective mechanisms of resveratrol such as inhibition of nuclear factor-κappa B (NF-κB) expression and alteration in the signaling pathways of mitogen-activated protein kinases (P38-MAPK), extracellular signal-regulated kinase 1/2 (ERK1/2) and phosphoinositide 3-kinase (PI3K)/Akt, activation of autophagy, among others [7–10].

Interest in resveratrol has grown recently due to its beneficial effects in several neurological and autoimmune disorders [11, 12]. Resveratrol is a phytoalexin that mainly occurs in grapevine species (Vitis sp.) and other fruits, and attention has been drawn to it due to its versatile biological properties, including its antioxidant, anti-inflammatory, and neuroprotective activities [13–15]. In this sense, resveratrol could indirectly activate SIRT1 expression [16] and lead to neuroprotection in AD cases [17]. SIRT1 regulates the activity of several substrates, including p53 and peroxisome proliferator-activated receptor-gamma coactivator 1α (PGC-1α) [18], which decrease the accumulation of β-amyloid (Aβ) and improve mitochondrial dysfunction [19].

Some studies have shown that resveratrol improves the impaired learning and memory in neurodegenerative disease and protects the memory decline in AD through its antioxidant activity [20]. Resveratrol is also effective at preventing blood-brain barrier (BBB) impairment and inhibiting Aβ1–42 from crossing the BBB and accumulating in the hippocampus [21, 22]. The hippocampus is a critical brain component for cognitive and memory functions, is a region that displays ongoing neurogenesis in adulthood, and is a very sensitive area in AD [23–25]. However, a significant reduction in hippocampal neurodegeneration was observed after intracerebroventricular injection of resveratrol in an animal model, which was associated with a decrease in SIRT1 acetylation [26, 27].

Karuppagounder et al. [28] showed that mice treated with resveratrol for 45 days had reduced Aβ toxicity. This suggests that the onset of neurodegeneration may be delayed by dietary chemopreventive agents (i.e., resveratrol) that protect against Aβ formation and oxidative stress [28]. Wang et al. [29] recently showed that resveratrol protected neurons against Aβ1–42-induced disruption of spatial learning, memory, and synaptic plasticity and rescued the reduction of SIRT1 expression in hippocampal rats. Thus, resveratrol is effective at reducing central nervous system (CNS) damage and decreasing the ischemia and toxicity induced by Aβ peptide, showing its potential therapeutic use in neurodegenerative diseases [30].

One of the major neuroprotective mechanisms of resveratrol is the activation of SIRT1 that is expressed in the adult mammalian brain, predominantly in neurons [31]. Activation of SIRT1 by resveratrol prevents Aβ-induced microglial death and contributes to improved cognitive function [32]. Although the major mechanisms of resveratrol are associated with the overexpression of SIRT1, its subsequent neuroprotective effect remains unknown. However, the overexpression of SIRT1 plays an important role in neuronal protection as it regulates reactive oxygen species (ROS), nitric oxide (NO), proinflammatory cytokine production, and Aβ expression in the brains of AD patients [33–36]. This review discusses the neuroprotective effects of resveratrol that are dependent on its action on SIRT1 and its implications in AD.

## 2. Resveratrol Plant Biosynthesis and Pharmacokinetics

Resveratrol (3,5,4′-trihydroxy-trans-stilbene) is a polyphenol plant secondary metabolite that has a phytoalexin role in high plant species. This metabolite is commonly found in grapevines (Vitis vinifera), grape juice, and wine [37, 38]. Others food sources, including peanuts, pomegranate, spinach, and bananas, also contain high concentrations of resveratrol [39–43]. Table 1 shows the concentration of resveratrol in some food sources.

Resveratrol is synthesized in high plant species using the phenylpropanoid pathway under biotic and abiotic stress conditions (i.e., ultraviolet (UV) light radiation and tissue disruption) and in response to fungal infections (i.e., V. vinifera leaves infected by Plasmopara viticola) [44–46]. The biosynthesis of resveratrol begins with the generation of 4-coumaroyl-CoA units in the phenylpropanoid pathway [47]. At this point, stilbene synthase (STS) and chalcone synthase (CHS) enzymes promote the chain extension of 4-coumaroyl-CoA via the addition of three malonyl-CoA molecules to generate a polyketide compound (Figure 1). Despite both enzymes using the same substrate, STS possesses substantially more amino acids than CHS (the key enzyme in flavonoid biosynthesis), which explains the difference in the end products formed [48, 49].

The polyketide peptide suffers a fold that promotes the generation of aromatic rings in a Claisen-like reaction catalyzed by STS, which produces an unstable intermediate metabolite called stilbene-2-carboxylic acid [50, 51]. The final steps involve the stepwise reactions that promote the decarboxylation, dehydration, and enolization of stilbene-2-carboxylic acid to yield the resveratrol molecule [52]. Resveratrol can undergo other biochemical reactions to produce new stilbenes, including ε-viniferin, t-piceid, t-piceatannol, and t-pterostilbene [53].

Resveratrol is well absorbed but is quickly excreted, mainly by the urinary system [54]. Calliari et al. [55] reported that the pharmacokinetics of resveratrol have been studied in several organs and that its therapeutic effect is mainly dose dependent. After oral consumption, resveratrol is primarily metabolized by phase II enzymes, especially glucuronides and sulfatases, and absorbed in the small gut, predominantly in its glucuronidated form [12, 56]. In addition to the glucuronide metabolite, sulfated products of resveratrol are also commonly found in biological samples [57]; however, only trace amounts of free resveratrol can be detected in plasma [58]. In this regard, Sergides et al. [59] demonstrated higher plasma concentrations of glucuronidated (4083.9 ± 1704.4 ng/ml) and sulfated (1516.0 ± 639.0 ng/ml) resveratrol than its unmetabolized form (71.2 ± 42.4 ng/ml) following the consumption of a single resveratrol (500 mg) tablet in healthy volunteers. Resveratrol is mainly attained by dietary intake; however, there are some concerns regarding its low concentration in food sources and its poor oral bioavailability. This has highlighted the need for strategies that allow biologically active concentrations of resveratrol to reach its target tissues, including the brain [60]. In this regard,

TABLE 1: Resveratrol concentration in food sources.

| Food source | Family | Resveratrol content | Reference |
|---|---|---|---|
| Banana peel (*Musa* sp.) | Musaceae | 38.8 ± 0.1 mg/100 g | [41] |
| Caper bush (*Capparis spinosa*) | Capparidaceae | 235.31 mg/100 g | [42] |
| Whole grapes (*V. vinifera*) | Vitaceae | 8.4 ± 0.2 mg/100 g | [41] |
| White wine (*V. vinifera* cv. Chardonnay) | Vitaceae | 0.04 ± 0.01 mg/l | [43] |
| Red wine (*V. vinifera* cv. Shiraz) | Vitaceae | 0.53 ± 0.06 mg/l | [43] |
| Mulberry wine (*Morus rubra*) | Moraceae | 145.31 ± 8.89 mg/l | [43] |
| Whole Mentha (*Mentha arvensis*) | Lamiaceae | 9.4 ± 0.0 mg/100 g | [41] |
| Boiled peanuts (*Arachis hypogaea*) | Fabaceae | 5.1 ± 2.8 µg/g | [40] |
| Peanut butter (*A. hypogaea*) | Fabaceae | 0.3 ± 0.1 µg/g | [40] |
| Pomegranate pulp (*Punica granatum*) | Punicaceae | 19.9 ± 0.2 mg/100 g | [41] |
| Whole spinach (*Spinacia oleracea*) | Amaranthaceae | 19.3 ± 0.1 mg/100 g | [41] |

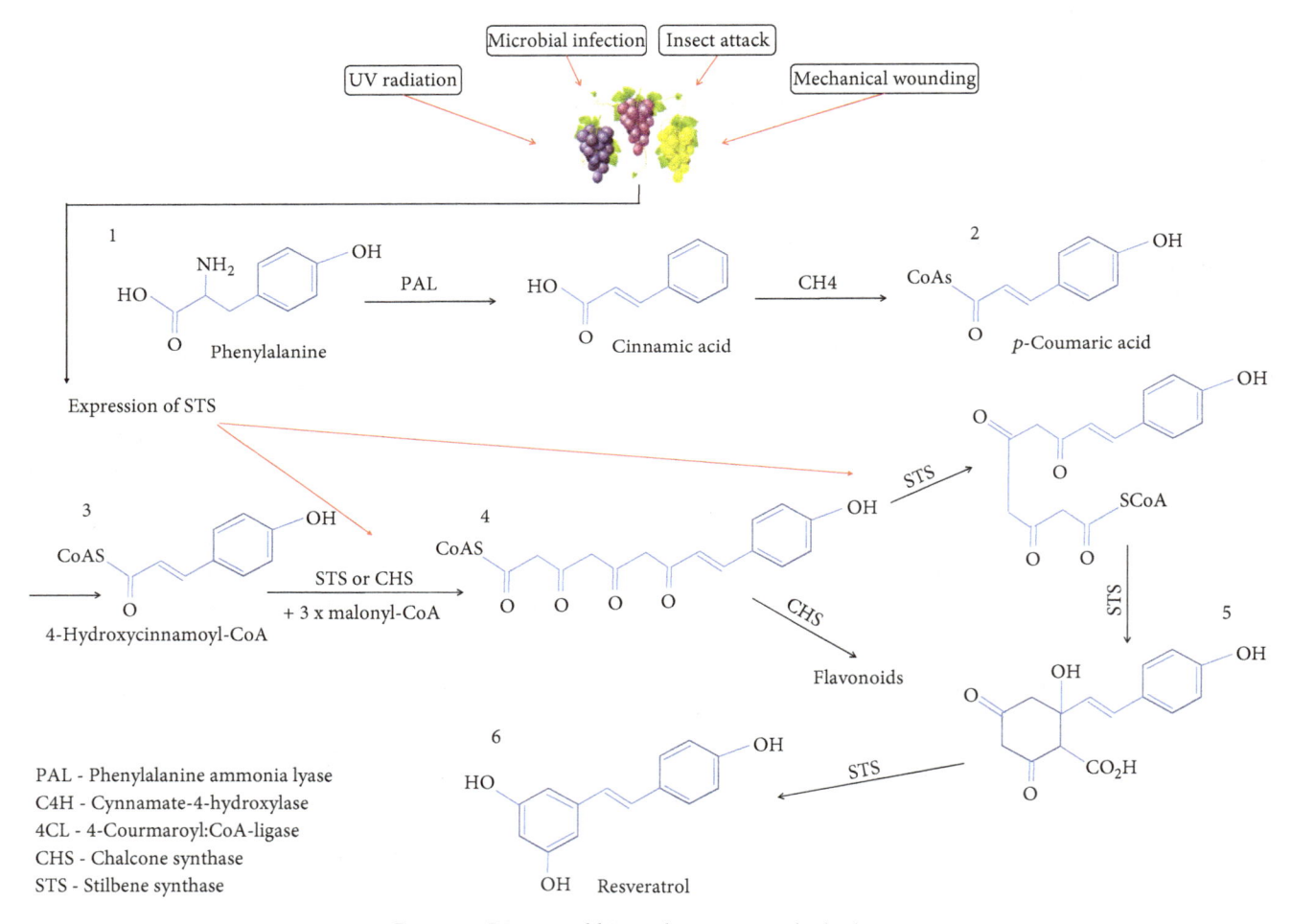

PAL - Phenylalanine ammonia lyase
C4H - Cynnamate-4-hydroxylase
4CL - 4-Courmaroyl:CoA-ligase
CHS - Chalcone synthase
STS - Stilbene synthase

FIGURE 1: Resveratrol biosynthesis route in high plants.

Oliveira et al. [12] reported that the major problem of resveratrol treatment was its low bioavailability, with some human studies reporting that even high-dose resveratrol treatment (500 mg/day) produced low plasma concentrations (10–71.2 ng/ml) of this antioxidant.

The description of resveratrol concentrations in the brain is a challenge that remains to be overcome. Frozza et al. [61] reported that intravenous administration of resveratrol reached satisfactory target brain regions, while oral resveratrol treatment was not well absorbed and resulted in reduced stability, increased photosensitivity, and accelerated metabolism, thus making it difficult to reach the brain. Turner et al. [62] showed that resveratrol and its metabolites crossed the human BBB, and these authors detected resveratrol in both the plasma and cerebrospinal fluid, thus showing its effects on the CNS. Preclinical data suggest that the main metabolite found in the rat brain after resveratrol consumption is resveratrol-3-glucuronic acid, which is also the main metabolite found in plasma [63]. To try to overcome the low oral bioavailability, several researchers focused on the microencapsulation technique or on the creation of prodrugs that, after metabolization, will give rise to resveratrol molecules [12, 64, 65]. Studies with new conjugated particles that improve the pharmacokinetics of resveratrol in the brain are of great importance, as the biologically active concentrations observed in in vitro experiments are much higher than those achieved after oral consumption are. Frozza et al. [61, 66] demonstrated that resveratrol nanoparticles reached the brain at higher concentrations than free resveratrol, resulting in increased bioavailability and possible neuroprotective effects. Resveratrol is considered a low-toxic substance, as humans have used several resveratrol-containing foods for a long time without related toxic effects. Data also confirm the safety of resveratrol on the basis of preclinical tests and clinical trials [67, 68].

Some studies have reported that resveratrol is an activator of SIRT1 [27, 69], although further evidence shows that resveratrol is not a direct activator of SIRT1 [70], and that its role may be related to the activation of substrates of SIRT1 [71]. The overexpression of SIRT1 results in neuroprotection in AD [17]. SIRT1 inhibits NF-$\kappa$B signaling by decreasing A$\beta$-induced toxicity in primary mouse neuronal cultures [32]. SIRT1 may be capable of determining A$\beta$ production by modulating $\beta$-secretase 1 expression through NF-$\kappa$B signaling [32].

## 3. Role of SIRT1 in the Pathophysiology of AD

Oxidative stress and the overproduction of ROS are associated with the pathophysiology of neurodegenerative disorders, including AD, and lead to neural membrane injury and memory impairment [72–75]. Brain tissue is more susceptible to oxidative stress due to its high oxygen consumption rate, low regenerative capability, high polyunsaturated fatty acid content, and low concentration of antioxidants [76, 77]. ROS are major neurotoxic factors released by activated microglia and include superoxide radicals ($O_2^-$), hydroxyl radicals ($^{\cdot}OH$), and hydrogen peroxide ($H_2O_2$). These molecules are highly reactive, and their excessive production can induce lipid peroxidation, (deoxyribonucleic acid) DNA fragmentation, and protein oxidation and result in further cell dysfunction and cell death [78]. Therefore, mitochondria that are damaged during oxidative stress can produce ROS that damage proteins, nucleic acids, and polyunsaturated fatty acid membranes and cause lipid peroxidation, a loss of membrane integrity, and increased calcium ($Ca^{2+}$) permeability. ROS also increase the production of A$\beta$ peptides, which induce oxidative stress both in vitro and in vivo [79]. Thus, a vicious cycle between ROS and A$\beta$ accumulation may accelerate the progression of AD [80]. Studies in vitro and in vivo have shown that ROS increases A$\beta$ production and induces oxidative stress, thus leading to neuronal apoptosis and accelerating the progression of AD [80–82].

AD is a progressive neurodegenerative disorder of the cortex and hippocampus that eventually leads to cognitive impairment. Although the etiology of AD remains unclear, multiple cellular changes have been implicated, including the production and accumulation of A$\beta$ peptides, tau phosphorylation, oxidative stress, mitochondrial dysfunction, synaptic damage, and biometal dyshomeostasis. The neuroinflammatory response via microglial activation and acetylcholine deficits are also considered to play significant roles in the pathophysiology of AD [83, 84]. The main pathogenic event in AD is the cerebral aggregation of A$\beta$ peptides [85]. A$\beta$ is the major constituent of plaques and is generated from amyloid precursor protein (APP) by the action of $\beta$ and $\gamma$-secretases [86]. The accumulation of A$\beta$ could initiate a series of downstream neurotoxic events that result in neuronal dysfunction in AD patients [87, 88]. However, oxidative stress is also an important event in the pathogenesis of AD [89], as the generation and accumulation of ROS and reactive nitrogen species can accelerate fibrillization, increase the toxicity of A$\beta$, and promote neuronal death and neurodegeneration [90–93].

Decreased sirtuin levels, mainly SIRT1 expression levels, were recently correlated with elevated A$\beta$ production and deposition in AD patients [94]. SIRT1 may regulate A$\beta$ metabolism through the modulation of APP processing, and loss of SIRT1 is closely associated with exacerbated A$\beta$ production [95]. However, SIRT1 overexpression decreases A$\beta$ production [95, 96], which may represent an interesting therapeutic approach to block the neurodegeneration and cognitive impairments caused by the disease. SIRT1 is a member of a sirtuin family that utilizes nicotinamide ($NAD^+$) as a substrate to catalyze the deacetylation of various substrates [97]. SIRT1 plays an essential role in regulating cellular homeostasis by influencing neuron survival, insulin sensitivity, glucose metabolism, and mitochondrial biogenesis [98, 99]. In the adult brain, SIRT1 was shown to be essential for synaptic plasticity, cognitive functions [100], and the modulation of learning and memory function [101].

During normal aging, SIRT1 is responsible for the maintenance of neural systems and behavior, including the modulation of synaptic plasticity and memory processes [102]. The absence of SIRT1 expression in hippocampal neurons is correlated with impaired cognitive abilities, including immediate memory, classical conditioning, and spatial learning [100]. SIRT1 can also increase PGC-1$\alpha$ activity, which leads to the inhibition of A$\beta$ production and improved mitochondrial dysfunction [19]. SIRT1 can also deacetylate a large number of other substrates, including p53, NF-$\kappa$B, and Forkhead box O (FOXO), and prevent neuronal apoptosis [103, 104]. Therefore, the pharmacological activation of SIRT1 may represent a promising approach to preventing A$\beta$ deposition and neurodegeneration in AD [105]. Thus inhibiting ROS production may be an important tool for protecting neuronal cells from oxidative damage and a therapeutic

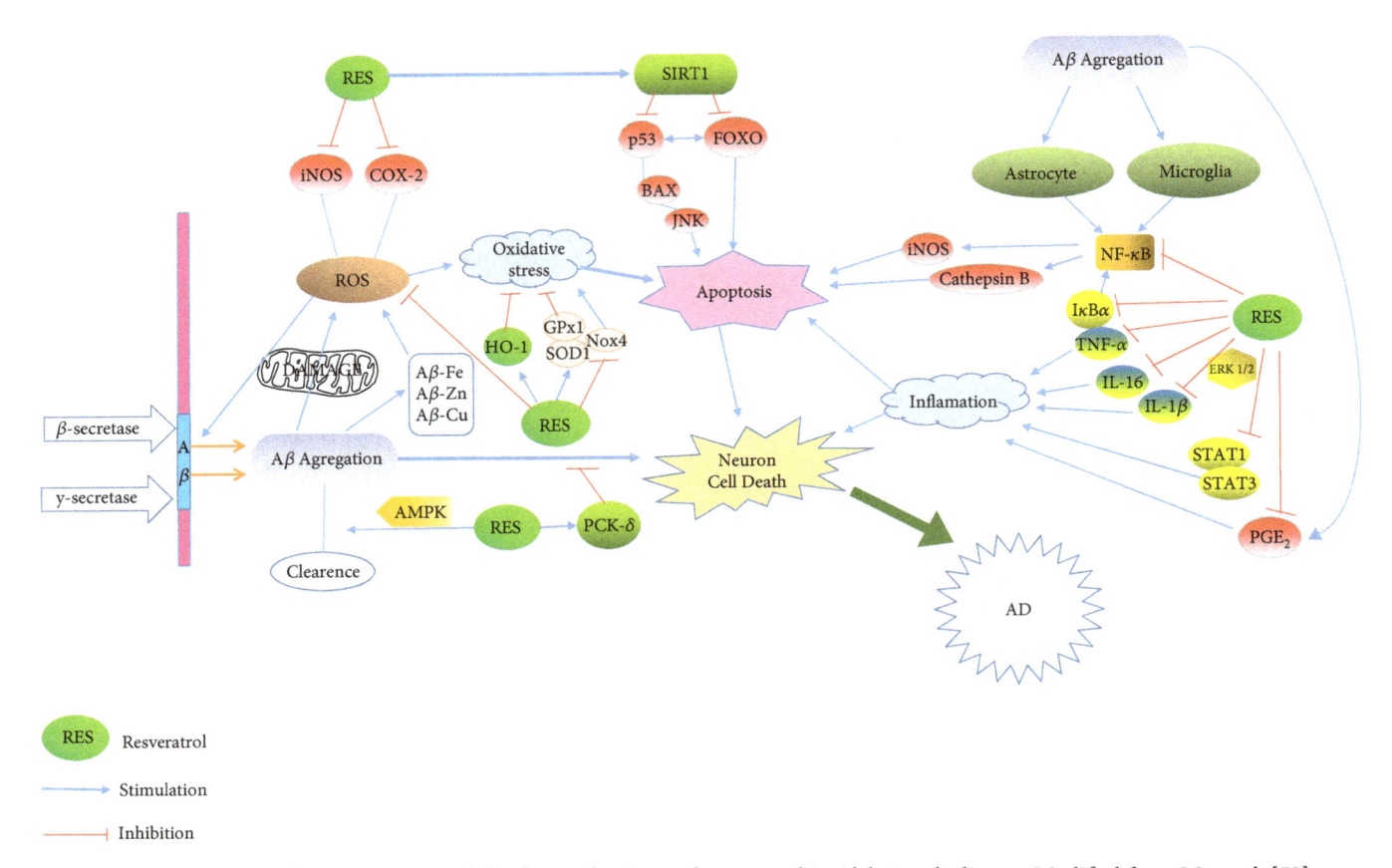

FIGURE 2: Main cellular routes proposed for the mechanisms of resveratrol in Alzheimer's disease. Modified from Ma et al. [72].

strategy in the treatment of neurological disorders [106]. Figure 2 summarizes the pathways by which resveratrol acts on SIRT1 in the pathology of Alzheimer's disease.

*3.1. Antioxidant Mechanisms of Resveratrol in AD: Role of SIRT1.* Oxidative stress induces neuronal damage, modulates intracellular signaling, and leads to neuronal death by apoptosis or necrosis. Therefore, antioxidant products (i.e., resveratrol) are used to protect against neuronal damage in neurodegenerative disorders (i.e., AD) [80]. The antioxidant properties of resveratrol were reported in several studies, which demonstrated that chronic resveratrol treatment reduced the production of malondialdehyde and nitrite and restored glutathione (GSH) levels [107, 108]. Additional antioxidant mechanisms of resveratrol were also described and include SIRT1 activation, A$\beta$ aggregation and toxicity inhibition, metal chelation, and ROS scavenging [106, 108, 109]. These results demonstrate that this compound is an effective therapeutic strategy for AD therapy. Therefore, resveratrol not only plays a role in ROS protection but it can also modulate important glial functions, including glutamate uptake activity, GSH, improved functional recovery, and decreased DNA fragmentation and apoptosis [110–112].

*3.1.1. In Vitro Studies.* Resveratrol can dysregulate the metal ion balance (i.e., copper, zinc, and iron) and play a key role in neurodegeneration, which is related to cellular function changes and neuronal survival dysfunction [27]. These metal ions are able to bind A$\beta$ and neurofibrillary tangles and promote their aggregation [106, 109], enhance the production of ROS, and contribute to AD pathogenesis. Hou et al. [113] demonstrated the interaction between resveratrol and SIRT1 using molecular dynamics simulation. The authors proposed that resveratrol was responsible for enhancing the binding affinity between SIRT1 and the substrate, thus functioning as a binding stabilizer. Nevertheless, Dasgupta and Milbrandt show that resveratrol is a potent activator of AMP-activated protein kinase (AMPK) function, and resveratrol-mediated AMPK activation was independent of SIRT1 [114]. In addition, in cell lines, resveratrol presented a decrease in the acetylation of PGC-1$\alpha$, possibly due to the activation of AMPK [115]. Thus, showing a dose-dependent effect, resveratrol was able to activate AMPK independently of SIRT1 [116]. However, SIRT1 plays a key role in protecting neurons from the oxidative effects of ROS, NO, and A$\beta$ peptides in the brains of AD subjects [117].

*3.1.2. Animal Studies.* One neuroprotective property attributed to resveratrol is the suppression of ROS formation through the inhibition of prooxidative genes (i.e., nicotinamide adenine dinucleotide phosphate oxidase) [118]. Huang et al. [119] showed that the neuroprotective activity of resveratrol included the suppression of inducible nitric oxide synthase (iNOS) production, which is involved in

Aβ-induced lipid peroxidation and heme oxygenase-1 downregulation, thereby protecting the rats from Aβ-induced neurotoxicity [120]. Moreover, resveratrol induced the expression of various antioxidant enzymes, such as superoxide dismutase (SOD), catalase, thioredoxin, and glutathione peroxidase (GPx) [121, 122]. However, Lee et al. [123] showed that resveratrol possesses chelator-metal ion properties to attenuate the metal imbalance and ROS production [124]. Furthermore, the oral administered of resveratrol in mice lowered the Aβ accumulation in the cortex due to the activation of AMPK signaling by enhancing cytosolic $Ca^{2+}$ levels in neuronal cultures [120, 125].

Other studies also showed the neuroprotective action of resveratrol in animal models; for example, Simão et al. [126] evaluated the response to a 7-day resveratrol treatment (30 mg/kg) on postinduced ischemia in rodent models. Cerebral immunohistochemistry showed reduced activation of astrocytes and microglia in the hippocampus and suppression of the inflammatory response mediated by NF-κB, cyclooxygenase 2 (COX-2), and nitric oxide synthetase (NOS) in hippocampal cells, thus suggesting the anti-inflammatory potential of resveratrol in brain damage. Moreover, Wang et al. [127] suggested that resveratrol (200 mg/kg/day for 8 weeks) could act as an AD-adjuvant therapy after human umbilical cord stem cell transplantation. This occurred due to the increased expression of brain-derived neurotrophic factor precursor (BDNF), neuronal growth factor (NGF), and neurotrophin 3 (NT-3), which are associated with neurogenesis, survival, learning, and memory. Thus, resveratrol positively stimulated these cell-protected factors [128]. The overexpression of these neurotrophic factors is related to the ability of resveratrol to increase the activity of SIRT1 [13]. Similarly, resveratrol also induced an increase of SIRT1 in a mice model [129]. Another study also reported the preventive action of resveratrol in decrease the formation of insoluble Aβ plaques in the hippocampus of rats [21], as the etiology of the disease is associated with an imbalance in Aβ homeostasis. Resveratrol effectively reduced the cleavage activation of APP and promoted peptide clearance [10]; therefore, the authors suggested that resveratrol was efficient at reducing the formation of protein aggregates.

*3.1.3. Human Studies.* There are currently studies evaluating the effectiveness of resveratrol in AD; for example, a randomized double-blind placebo-controlled study evaluated the effects of resveratrol in 64 AD patients with a mild form of the disease. A resveratrol dose of 500–1000 mg was administered orally to these patients. However, the results demonstrate that resveratrol and its major metabolites able to cross the BBB and cause weight loss and reactions such as nausea and diarrhea. In addition, brain volume loss was greater in the group receiving resveratrol. Conversely, Imamura et al. [130] demonstrated the antioxidant effect of resveratrol on arterial stiffness in patients with type 2 diabetes mellitus (T2DM). In this randomized double-blind placebo-controlled clinical trial, 50 patients were selected: 25 received resveratrol (100 mg/day) and 25 received a placebo for 12 weeks. Supplementation with resveratrol improved several parameters in the T2DM patients and decreased oxidative stress, which was evaluated through metabolites of reactive oxygen. Mansur et al. [131] also conducted a study to evaluate the effects of resveratrol in humans. Slightly overweight elderly individuals were randomly divided into two groups: group one received 250 mg of resveratrol orally twice daily, while group two received a caloric restriction diet (1000 cal/day). SIRT1 concentrations were determined in both groups at the end of the 30-day treatment period. The serum concentration of SIRT1 was increased in both groups; however, this finding was not correlated with a better profile of metabolic markers for atherosclerotic processes.

*3.2. SIRT1 and Anti-Inflammatory Mechanisms of Resveratrol.* Neuroinflammation is an important contributor to the pathogenesis of AD [132]. Various reports show that inflammatory responses occur in the CNS, including the activation of microglia, astrocytes, lymphocytes, and macrophages that trigger numerous proinflammatory mediators and neurotransmitters [133]. However, the hallmark of brain neuroinflammation is microglia activation, which releases highly proinflammatory cytokines, ROS, and NO and leads to protein oxidation, lipid peroxidation, DNA fragmentation, neuronal inflammation, and cell death [78, 134]. Microglial cells are the resident macrophage-like population within the CNS and are a prime component of the brain immune system. In physiological conditions, microglia actively survey the microenvironment and ensure normal CNS activity by secreting neurotrophic factors (i.e., NGF). Although microglial activation plays an important role in the phagocytosis of dead cells in the CNS, overactivated microglia cause inflammatory responses that lead to neuronal and axonal degeneration and disruption of the immature BBB [135].

Inflammatory mediators such as interleukin-1β (IL-1β), interferon-γ (IFN-γ), tumor necrosis factor-α (TNF-α), and NO are produced by activated microglia and have recently been linked to the pathogenesis of neurological disorders [136]. Therefore, pharmacological interference with the overactivation of microglia may have a therapeutic benefit in the treatment of inflammation-mediated neurological disorders [137]. The activities of resveratrol against neuroinflammation appear to target activated microglia and result in the reduction of proinflammatory factors (i.e., TNF-α, IL-β, prostaglandin E2, cyclooxygenases, and iNOS through the modulation of signal transduction pathways) [138].

Gocmez et al. [139] showed that aging increased the levels of TNF-α and led to chronic neuroinflammation in the hippocampus and impaired spatial learning and memory. However, chronic administration of resveratrol reversed the cognitive deficits and inhibited the production of inflammatory cytokines. In addition, resveratrol also inhibited the activation of signal transducer and activator of transcription (STAT1 and STAT3) and prevented the proinflammatory effect of Aβ and Aβ-triggered microglial activation [140]. However, the role of resveratrol in microglia activation and the molecular mechanisms involved are not fully elucidated. The major pathway seems to involve SIRT1 activation, which promotes Th2 responses by increasing

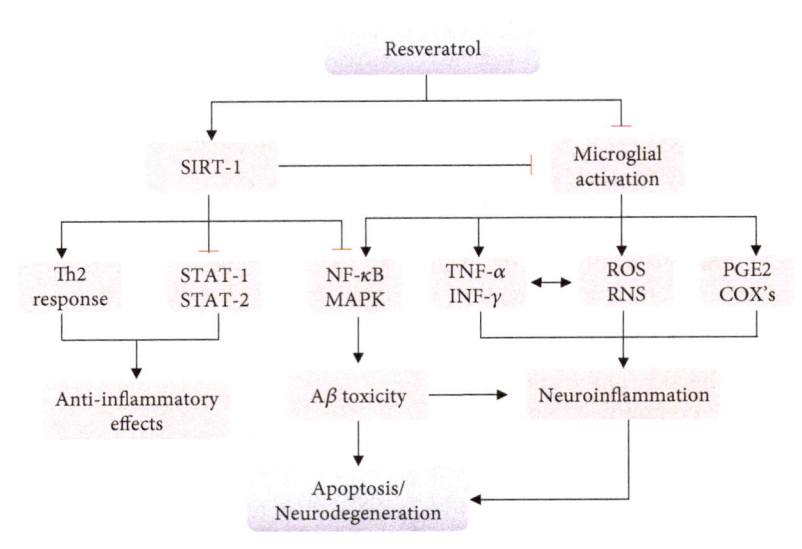

FIGURE 3: Anti-inflammatory effects of resveratrol and the role of SIRT1 in AD.

anti-inflammatory cytokine expression and upregulating PGC-1α (Figure 3) [141, 142].

*3.2.1. In Vitro Studies.* Resveratrol has numerous functions in neuroinflammation, as it induces mitophagy [143, 144]. Wang et al. [80] used a differentiated lineage of cell lymphomas from rat pheochromocytoma as a cellular model of AD treated with Aβ peptide Aβ1–42 (Aβ1–42). Resveratrol decreased the mitophagy-mediated mitochondrial damage and attenuated the oxidative stress caused by Aβ1–42 [141]. Neuroinflammation may also be related to the degradation of the BBB [145]. The BBB is constituted of structural and functional elements such as brain endothelial cells [146, 147]. Thus, Annabi et al. [145] demonstrated that human brain microvascular endothelial cells treated with a carcinogen can signal through NF-κB, allowing release of inflammatory markers such as matrix metalloproteinase 9 (MMP-9) and COX-2. However, resveratrol decreased secretion of MMP-9 and expression of COX-2 [145]. It also activated the expression of SIRT1, which regulated inflammation, inhibited NF-κB signaling, and prevented Aβ-induced degeneration [148].

*3.2.2. Animal Studies.* Several studies suggest that pharmacological activation of SIRT1 may represent a promising approach to prevent amyloid deposition and neurodegeneration in AD [99, 149]. The relationship between SIRT1 and AD is paramount, as a study of the SIRT1 serum concentration in healthy subjects and AD patients showed a reduced serum SIRT1 concentration that correlated with the increasing age of an individual. The decline was much more pronounced in patients with AD [93].

SIRT1 also exhibited therapeutic activity in a transgenic mouse model of AD [150]. Wang et al. [127] assessed an alternative therapy for AD that used mesenchymal stem cells derived from the umbilical cord combined with resveratrol in a mouse model of AD. Resveratrol also favored the formation of neurons and regulated SIRT1 expression in the hippocampus of AD rats [127]. Resveratrol has anti-inflammatory functions and can inhibit Aβ-induced NF-κB signaling in microglia and astrocytes [151]. Another study showed that mice overexpressing SIRT1 exhibited reduced brain inflammation (due to its action in tau phosphorylation) and reduced cognitive defects that were specific to the APP transgenic mouse [149, 150].

*3.2.3. Human Studies.* Some neurodegenerative diseases, such as AD, are associated with oxidative stress and neuroinflammation, and proteins that are closely related to this neurological disorder (i.e., AMPK, SIRT1, and PGC-1α) can be modulated by resveratrol [152]; however, there are few clinical studies on resveratrol in AD patients. Moussa et al. [153] reported that patients treated with resveratrol (1 g/day) for 52 weeks demonstrated reduced MMP-9 levels (an inflammatory marker related to AD) compared to a placebo group. In addition, patients treated with resveratrol had less cerebrospinal fluid decline, which resulted in less Aβ accumulation in the brain. Resveratrol probably strengthened the CNS, hampered the penetration of MMP-9, and reduced the activity of this inflammatory agent [154].

The anti-inflammatory effects of resveratrol are mediated, at least in part, by suppressing the activation of NF-κB, extracellular signal-regulated kinase-1 and kinase-2, and mitogen-activated protein kinase (MAPK) signaling pathways, which are all important upstream modulators of the production of proinflammatory mediators [137]. Resveratrol-mediated overexpression of SIRT1 markedly reduced NF-κB signaling and Aβ-mediated microglial activation and had strong neuroprotective effects [68, 155]. The polymerization of Aβ peptides was markedly inhibited by resveratrol, which stimulated the proteasomal degradation of Aβ peptides [30, 75].

Studies strongly suggest that resveratrol-induced SIRT1 inhibits NF-κB signaling in microglia and astrocytes and protects AD neurons against Aβ-induced toxicity. This

NF-κB signaling controls the expression of iNOS, which mediates apoptosis and neurodegeneration [32]. Resveratrol also effectively suppresses the apoptotic activities of both p53 and FOXO via SIRT1 overexpression and confers neuronal protection in AD [152, 156].

Therefore, the potential anti-inflammatory mechanisms for resveratrol-mediated neuroprotection involve (i) reduction of proinflammatory cytokine expression, (ii) suppression of MAPK signal transduction pathways, and (iii) activation of the SIRT1 pathway, which in turn suppresses the activation of the NF-κB signaling pathway and protects neurons against microglia-dependent Aβ toxicity [134].

In this context, the neuroprotective effects of resveratrol can involve the scavenging of ROS, decreased NO levels, improved antioxidant capacity, NF-κB inhibition, inhibition of inflammatory mediators, promotion of neuronal survival via SIRT1 activation [157, 158], the prevention of DNA lesions, and the prevention of lipid peroxidation in cell membranes [85]. Animal models also indicate that resveratrol improves the spatial memory by decreasing the accumulation of Aβ peptides and lipid peroxidation in the hippocampus, thus protecting against neuronal apoptosis [159].

Therefore, it is also important to emphasize that these neuroprotective effects can also be mediated by other action mechanisms of resveratrol. Another neuroprotective mechanisms of resveratrol include the following: (i) inhibits the tauopathy by interfering with the MID1-PP2A (midline 1-protein phosphatase 2A) complex or by altering or partially inhibiting of the glycogen synthase kinase 3 beta (GSK3β) and p53 interaction [6, 110]; (ii) improves learning and long-term memory formation through the microRNA (microribonucleic acid)-CREB (cAMP response element-binding protein)-BDNF pathway [20]; (iii) protects against Aβ-mediated neuronal impairment (inflammation and oxidative stress) by activation of AMP-activated protein kinase- (AMPK-) dependent signaling and inhibition of NF-κB expression and iNOS levels [160]; (iv) antioxidative activity by reduction in levels of ROS enhances the expression of various antioxidant defensive enzymes (heme oxygenase 1, catalase, glutathione peroxidase, and superoxide dismutase), downregulation of prooxidative stress proteins (i.e., plaque-induced glycogen synthase kinase-3β (GSK-3β), and AMPK [8, 10]; (v) improves cognitive impairment due to inhibition of cholinesterase activity [161]; (vi) inhibits the Aβ plaque synthesis by restoration of normal cellular autophagy via the TyrRS-PARP1 (auto-poly-ADP-ribosylation of poly (ADP-ribose) polymerase 1)-SIRT1 signaling pathway and enhancement of transthyretin (transporter protein) binding to Aβ oligomers [162]; (vii) inhibits mammalian target of rapamycin (mTOR) signaling and induces AMPK, thereby stimulating the clearance of Aβ aggregates [110]; (viii) prevents the neuronal cell death by attenuating apoptosis via Akt/p38 MAPK signaling and inhibits caspase-3 and B cell lymphoma-2 (Bcl-2)/Bcl-2-associated X protein signaling [163, 164]; (ix) increases intracellular calcium levels, promoting the activation of calcium/calmodulin-dependent protein kinase kinase β-CamKKβ-AMPK pathway, which alters mitochondrial function and leads to a decrease in ROS generation [165]; (x) attenuated injury and promoted proliferation of the neural stem cells, at least in part, by upregulating the expression of nuclear factor (erythroid-derived 2)-like 2 (Nrf2), HO-1, and NAD(P)H:quinone oxidoreductase 1 (NQO1) [166]; and (xi) inhibits the neuronal electrical activity by mechanisms associated with large conductance of $Ca^{2+}$ potassium channels and attenuates Aβ-induced early hippocampal neuron excitability impairment [167]. Therefore, resveratrol may be an important tool to protect neuronal cells from oxidative damage and a promising strategy in the treatment of AD.

## 4. Conclusions

Resveratrol is a potential compound for the treatment of AD due to its antioxidant and anti-inflammatory properties. The key neuroprotective mechanism of resveratrol in AD seems to be linked with SIRT1 activation. Although the mechanisms that link resveratrol to the overexpression of SIRT1 and neuroprotection are unknown, this expression may play an important role in neuronal protection from ROS, NF-κB signaling in activated microglia, prevent Aβ toxicity, and contribute to improved learning and memory function. Resveratrol can also effectively suppress the apoptotic activities of both p53 and FOXO via SIRT1 overexpression and confer neuronal protection in AD. Although this review focuses on the importance of SIRT1 activation for the neuroprotective role of resveratrol, it is also important to clarify that these mechanisms are still unclear and fully elucidated. In addition, resveratrol may act on CNS by inhibiting neuroinflammatory and prooxidant mechanisms by multiple action mechanisms that are independent of SIRT-1. These mechanisms are quite complex and involve stimulation or inhibition of multiple signaling pathways or alteration of potassium channels eading to inhibition of neuronal electrical activity. In summary, the major mechanisms that may be associated with the neuroprotective effect of resveratrol, in addition to SIRT1, include stimulation of regulation by microRNA-CREB-BDNF pathway, inhibition of mTOR and AMPK-dependent signaling pathways, inhibition of enzymes (cholinesterase activity), transcription factor (NF-κB) and apoptotic pathways, and stimulation of cellular autophagy and expression of Nrf2, HO-1, NQO1, among others. Therefore, we critically analyze and suggest that SIRT1 is one of the main mechanisms related to the beneficial effects of resveratrol; however, this compound can change multiple pathways simultaneously, and then, there is a need for crosstalk between signaling and regulatory functions to provide improvements in the development and progression of AD. In addition, caution is required in therapies with natural products, since intrinsic aspects of the patient, environmental factors, and characteristics of the compound studied are important for efficacy and therapeutic success.

Despite the neuroprotective potential of resveratrol demonstrated in several in vitro studies, the major limitation currently facing is the lack of information from clinical studies that correlates the SIRT1 activation and the inflammatory and oxidative status reduction associated with improvement in the development and progression of AD. Overall, evidence from clinical trials is weak and largely inconclusive. Most human studies establish a link between consumption of foods rich in resveratrol and reducing the incidence or prevalence of AD, as well as improvement in learning, memory, visual and spatial orientation, and social behavior. However, these observed effects may be the result of complex direct and indirect interactions of the various constituents present in the diet, not only of resveratrol. In addition, other difficulties in clinical trials are the following: (i) the studies are mainly conducted with volunteers, not reflecting the target population, (ii) the participants' age is quite broad between 18 and over 80 years of age, and (iii) sample size is rarely calculated and the slow progression of AD is not investigated because it requires longer clinical time in the trials. Another important issue is the poor bioavailability of resveratrol, which makes it difficult to link with the optimal concentrations achieved in in vitro experiments. Although preclinical studies also indicate that resveratrol is able to cross the blood-brain barrier, low concentrations of this molecule have been detected in the brain, and only higher concentrations of resveratrol and its metabolites have been found in the blood. In addition, it is emphasized that the neuroprotective effects of resveratrol are mainly short term, varying according to dose, dosage form, duration of treatment, pharmacokinetic and pharmacogenetic parameters, food and drug interactions, among others. Thus, we conclude that, to date, evidence based on clinical studies is still insufficient, contradictory, and inconclusive, so we recommend that further clinical trials be conducted to substantiate the neuroprotective effects of resveratrol and its likely mechanisms of action in the body. However, we emphasize that resveratrol is promising in health promotion, not only for its antioxidant activities but also for its anti-inflammatory and neuroprotective properties. Thereby, further studies assessing other routes of administration or pharmaceutical formulations (i.e., nanoencapsulation) are required to improve the tissue-targeting concentration and allow resveratrol to exert its biological activities in AD.

## Authors' Contributions

All authors participated in the design of the study and drafted the manuscript.

## Acknowledgments

The authors were supported by the Brazilian agencies: Conselho Nacional de Desenvolvimento Científico e Tecnológico (CNPq), Coordenação de Aperfeiçoamento de Pessoal de Nível Superior (CAPES), Fundação Amazônia Paraense de Amparo à Pesquisa (FAPESPA), and Federal University of Pará, and MCM thanks the fellowship from CNPq.

## References

[1] B. J. Kelley and R. C. Petersen, "Alzheimer's disease and mild cognitive impairment," *Neurologic Clinics*, vol. 25, no. 3, pp. 577–609, 2007.

[2] H. Jahn, "Memory loss in Alzheimer's disease," *Dialogues in Clinical Neuroscience*, vol. 15, no. 4, pp. 445–454, 2013.

[3] W. Xu, C. Ferrari, and H.-X. Wang, "Epidemiology of Alzheimer's disease," in *Understanding Alzheimer's Disease*, K. Pesek, Ed., InTech, 2013.

[4] Y. Gilgun-Sherki, E. Melamed, and D. Offen, "Antioxidant treatment in Alzheimer's disease: current state," *Journal of Molecular Neuroscience*, vol. 21, no. 1, pp. 1–12, 2003.

[5] Y. Feng and X. Wang, "Antioxidant therapies for Alzheimer's disease," *Oxidative Medicine and Cellular Longevity*, vol. 2012, Article ID 472932, 17 pages, 2012.

[6] E. Tellone, A. Galtieri, A. Russo, B. Giardina, and S. Ficarra, "Resveratrol: a focus on several neurodegenerative diseases," *Oxidative Medicine and Cellular Longevity*, vol. 2015, Article ID 392169, 14 pages, 2015.

[7] R. E. González-Reyes, M. O. Nava-Mesa, K. Vargas-Sánchez, D. Ariza-Salamanca, and L. Mora-Muñoz, "Involvement of astrocytes in Alzheimer's disease from a neuroinflammatory and oxidative stress perspective," *Frontiers in Molecular Neuroscience*, vol. 10, pp. 1–20, 2017.

[8] Y. R. Li, S. Li, and C. C. Lin, "Effect of resveratrol and pterostilbene on aging and longevity," *BioFactors*, vol. 44, no. 1, pp. 69–82, 2018.

[9] P. Sadhukhan, S. Saha, S. Dutta, S. Mahalanobish, and P. C. Sil, "Nutraceuticals: an emerging therapeutic approach against the pathogenesis of Alzheimer's disease," *Pharmacological Research*, vol. 129, pp. 100–114, 2018.

[10] Y. Jia, N. Wang, and X. Liu, "Resveratrol and amyloid-beta: mechanistic insights," *Nutrients*, vol. 9, no. 10, p. 1122, 2017.

[11] R. Mancuso, J. del Valle, L. Modol et al., "Resveratrol improves motoneuron function and extends survival in SOD1$^{G93A}$ ALS mice," *Neurotherapeutics*, vol. 11, no. 2, pp. 419–432, 2014.

[12] A. L. de Brito Oliveira, V. V. S. Monteiro, K. C. Navegantes-Lima et al., "Resveratrol role in autoimmune disease—a mini-review," *Nutrients*, vol. 9, no. 12, article 1306, 2017.

[13] J. Gambini, M. Inglés, G. Olaso et al., "Properties of resveratrol: *in vitro* and *in vivo* studies about metabolism, bioavailability, and biological effects in animal models and humans," *Oxidative Medicine and Cellular Longevity*, vol. 2015, Article ID 837042, 13 pages, 2015.

[14] İ. Gülçin, "Antioxidant properties of resveratrol: a structure–activity insight," *Innovative Food Science & Emerging Technologies*, vol. 11, no. 1, pp. 210–218, 2010.

[15] R. V. Albuquerque, N. S. Malcher, L. L. Amado et al., "In vitro protective effect and antioxidant mechanism of resveratrol induced by dapsone hydroxylamine in human cells," *PLoS One*, vol. 10, no. 8, article e0134768, 2015.

[16] D. Beher, J. Wu, S. Cumine et al., "Resveratrol is not a direct activator of SIRT1 enzyme activity," *Chemical Biology & Drug Design*, vol. 74, no. 6, pp. 619–624, 2009.

[17] D. Kim, M. D. Nguyen, M. M. Dobbin et al., "SIRT1 deacetylase protects against neurodegeneration in models for Alzheimer's disease and amyotrophic lateral sclerosis," *The EMBO Journal*, vol. 26, no. 13, pp. 3169–3179, 2007.

[18] K. Higashida, S. H. Kim, S. R. Jung, M. Asaka, J. O. Holloszy, and D. H. Han, "Effects of resveratrol and SIRT1 on PGC-1α activity and mitochondrial biogenesis: a reevaluation," *PLoS Biology*, vol. 11, no. 7, article e1001603, 2013.

[19] G. Sweeney and J. Song, "The association between PGC-1α and Alzheimer's disease," *Anatomy & Cell Biology*, vol. 49, no. 1, pp. 1–6, 2016.

[20] Y. N. Zhao, W. F. Li, F. Li et al., "Resveratrol improves learning and memory in normally aged mice through microRNA-CREB pathway," *Biochemical and Biophysical Research Communications*, vol. 435, no. 4, pp. 597–602, 2013.

[21] H. F. Zhao, N. Li, Q. Wang, X. J. Cheng, X. M. Li, and T. T. Liu, "Resveratrol decreases the insoluble A$\beta$1-42 level in hippocampus and protects the integrity of the blood-brain barrier in AD rats," *Neuroscience*, vol. 310, pp. 641–649, 2015.

[22] S. H. Omar, "Biophenols pharmacology against the amyloidogenic activity in Alzheimer's disease," *Biomedicine & Pharmacotherapy*, vol. 89, pp. 396–413, 2017.

[23] G. Kempermann, H. Song, and F. H. Gage, "Neurogenesis in the adult hippocampus," *Cold Spring Harbor Perspectives in Biology*, vol. 7, no. 9, pp. 220–226, 2015.

[24] B. Biscaro, O. Lindvall, G. Tesco, C. T. Ekdahl, and R. M. Nitsch, "Inhibition of microglial activation protects hippocampal neurogenesis and improves cognitive deficits in a transgenic mouse model for Alzheimer's disease," *Neurodegenerative Diseases*, vol. 9, no. 4, pp. 187–198, 2012.

[25] J. Thomas, M. L. Garg, and D. W. Smith, "Dietary supplementation with resveratrol and/or docosahexaenoic acid alters hippocampal gene expression in adult C57Bl/6 mice," *The Journal of Nutritional Biochemistry*, vol. 24, no. 10, pp. 1735–1740, 2013.

[26] R. Lalla and G. Donmez, "The role of sirtuins in Alzheimer's disease," *Frontiers in Aging Neuroscience*, vol. 5, p. 16, 2013.

[27] S. D. Rege, T. Geetha, G. D. Griffin, T. L. Broderick, and J. R. Babu, "Neuroprotective effects of resveratrol in Alzheimer disease pathology," *Frontiers in Aging Neuroscience*, vol. 6, pp. 1–27, 2014.

[28] S. S. Karuppagounder, J. T. Pinto, H. Xu, H.-L. Chen, M. F. Beal, and G. E. Gibson, "Dietary supplementation with resveratrol reduces plaque pathology in a transgenic model of Alzheimer's disease," *Neurochemistry International*, vol. 54, no. 2, pp. 111–118, 2009.

[29] R. Wang, Y. Zhang, J. Li, and C. Zhang, "Resveratrol ameliorates spatial learning memory impairment induced by A$\beta_{1-42}$ in rats," *Neuroscience*, vol. 344, pp. 39–47, 2017.

[30] P. Marambaud, H. Zhao, and P. Davies, "Resveratrol promotes clearance of Alzheimer's disease amyloid-$\beta$ peptides," *Journal of Biological Chemistry*, vol. 280, no. 45, pp. 37377–37382, 2005.

[31] N. Guida, G. Laudati, S. Anzilotti et al., "Resveratrol via sirtuin-1 downregulates RE1-silencing transcription factor (REST) expression preventing PCB-95-induced neuronal cell death," *Toxicology and Applied Pharmacology*, vol. 288, no. 3, pp. 387–398, 2015.

[32] J. Chen, Y. Zhou, S. Mueller-Steiner et al., "SIRT1 protects against microglia-dependent amyloid-$\beta$ toxicity through inhibiting NF-$\kappa$B signaling," *Journal of Biological Chemistry*, vol. 280, no. 48, pp. 40364–40374, 2005.

[33] K. C. Morris-Blanco, C. H. Cohan, J. T. Neumann, T. J. Sick, and M. A. Perez-Pinzon, "Protein kinase C epsilon regulates mitochondrial pools of Nampt and NAD following resveratrol and ischemic preconditioning in the rat cortex," *Journal of Cerebral Blood Flow & Metabolism*, vol. 34, no. 6, pp. 1024–1032, 2014.

[34] D. Li, N. Liu, L. Zhao et al., "Protective effect of resveratrol against nigrostriatal pathway injury in striatum via JNK pathway," *Brain Research*, vol. 1654, Part A, pp. 1–8, 2017.

[35] A. Salminen, K. Kaarniranta, and A. Kauppinen, "Crosstalk between oxidative stress and SIRT1: impact on the aging process," *International Journal of Molecular Sciences*, vol. 14, no. 2, pp. 3834–3859, 2013.

[36] D. Albani, L. Polito, S. Batelli et al., "The SIRT1 activator resveratrol protects SK-N-BE cells from oxidative stress and against toxicity caused by $\alpha$-synuclein or amyloid-$\beta$ (1-42) peptide," *Journal of Neurochemistry*, vol. 110, no. 5, pp. 1445–1456, 2009.

[37] L. F. da Silva, C. C. Guerra, D. Klein, and A. M. Bergold, "Solid cation exchange phase to remove interfering anthocyanins in the analysis of other bioactive phenols in red wine," *Food Chemistry*, vol. 227, pp. 158–165, 2017.

[38] J. Popović-Djordjević, B. Pejin, A. Dramićanin et al., "Wine chemical composition and radical scavenging activity of some Cabernet Franc clones," *Current Pharmaceutical Biotechnology*, vol. 18, no. 4, pp. 343–350, 2017.

[39] J. Gabaston, E. Cantos-Villar, B. Biais et al., "Stilbenes from *Vitis vinifera* L. waste: a sustainable tool for controlling *Plasmopara viticola*," *Journal of Agricultural and Food Chemistry*, vol. 65, no. 13, pp. 2711–2718, 2017.

[40] J. Burns, T. Yokota, H. Ashihara, M. E. J. Lean, and A. Crozier, "Plant foods and herbal sources of resveratrol," *Journal of Agricultural and Food Chemistry*, vol. 50, no. 11, pp. 3337–3340, 2002.

[41] J. P. Singh, A. Kaur, K. Shevkani, and N. Singh, "Composition, bioactive compounds and antioxidant activity of common Indian fruits and vegetables," *Journal of Food Science and Technology*, vol. 53, no. 11, pp. 4056–4066, 2016.

[42] N. Tlili, A. Feriani, E. Saadoui, N. Nasri, and A. Khaldi, "*Capparis spinosa* leaves extract: source of bioantioxidants with nephroprotective and hepatoprotective effects," *Biomedicine & Pharmacotherapy*, vol. 87, pp. 171–179, 2017.

[43] A. H. Srikanta, A. Kumar, S. V. Sukhdeo, M. S. Peddha, and V. Govindaswamy, "The antioxidant effect of mulberry and jamun fruit wines by ameliorating oxidative stress in streptozotocin-induced diabetic Wistar rats," *Food & Function*, vol. 7, no. 10, pp. 4422–4431, 2016.

[44] L. Becker, S. Bellow, V. Carré et al., "Correlative analysis of fluorescent phytoalexins by mass spectrometry imaging and fluorescence microscopy in grapevine leaves," *Analytical Chemistry*, vol. 89, no. 13, pp. 7099–7106, 2017.

[45] S. Bruisson, P. Maillot, P. Schellenbaum, B. Walter, K. Gindro, and L. Deglène-Benbrahim, "Arbuscular mycorrhizal symbiosis stimulates key genes of the phenylpropanoid biosynthesis and stilbenoid production in grapevine leaves in response to downy mildew and grey mould infection," *Phytochemistry*, vol. 131, pp. 92–99, 2016.

[46] G. Chitarrini, L. Zulini, D. Masuero, and U. Vrhovsek, "Lipid, phenol and carotenoid changes in "Bianca" grapevine leaves after mechanical wounding: a case study," *Protoplasma*, vol. 254, no. 6, pp. 2095–2106, 2017.

[47] F. Sparvoli, C. Martin, A. Scienza, G. Gavazzi, and C. Tonelli, "Cloning and molecular analysis of structural

genes involved in flavonoid and stilbene biosynthesis in grape (*Vitis vinifera* L.)," *Plant Molecular Biology*, vol. 24, no. 5, pp. 743–755, 1994.

[48] G. Schröder, J. W. S. Brown, and J. Schröder, "Molecular analysis of resveratrol synthase. cDNA, genomic clones and relationship with chalcone synthase," *European Journal of Biochemistry*, vol. 172, no. 1, pp. 161–169, 1988.

[49] J. Schröder and G. Schröder, "Stilbene and chalcone synthases: related enzymes with key functions in plant-specific pathways," *Zeitschrift für Naturforschung C*, vol. 45, no. 1-2, pp. 1–8, 1990.

[50] P. M. Dewick, *Medicinal Natural Products*, John Wiley & Sons, Ltd, Chichester, UK, 2001.

[51] C. Rivière, A. D. Pawlus, and J.-M. Mérillon, "Natural stilbenoids: distribution in the plant kingdom and chemotaxonomic interest in Vitaceae," *Natural Product Reports*, vol. 29, no. 11, pp. 1317–1333, 2012.

[52] N. Rupprich and H. Kindl, "Stilbene synthases and stilbene-carboxylate synthases, I enzymatic synthesis of 3, 5, 4-trihydroxystilbene from p-coumaroyl coenzyme A and malonyl coenzyme A," *Hoppe-Seyler's Zeitschrift für Physiologische Chemie*, vol. 359, no. 2, pp. 165–172, 1978.

[53] E. Hurtado-Gaitán, S. Sellés-Marchart, A. Martínez-Márquez, A. Samper-Herrero, and R. Bru-Martínez, "A focused multiple reaction monitoring (MRM) quantitative method for bioactive grapevine stilbenes by ultra-high-performance liquid chromatography coupled to triple-quadrupole mass spectrometry (UHPLC-QqQ)," *Molecules*, vol. 22, no. 3, p. 418, 2017.

[54] Z. Qiu, J. Yu, Y. Dai et al., "A simple LC-MS/MS method facilitated by salting-out assisted liquid-liquid extraction to simultaneously determine *trans*-resveratrol and its glucuronide and sulfate conjugates in rat plasma and its application to pharmacokinetic assay," *Biomedical Chromatography*, vol. 31, no. 11, 2017.

[55] A. Calliari, N. Bobba, C. Escande, and E. N. Chini, "Resveratrol delays Wallerian degeneration in a NAD$^+$ and DBC1 dependent manner," *Experimental Neurology*, vol. 251, pp. 91–100, 2014.

[56] G. Kuhnle, J. P. E. Spencer, G. Chowrimootoo et al., "Resveratrol is absorbed in the small intestine as resveratrol glucuronide," *Biochemical and Biophysical Research Communications*, vol. 272, no. 1, pp. 212–217, 2000.

[57] A. Courtois, M. Jourdes, A. Dupin et al., "In vitro glucuronidation and sulfation of ε-viniferin, a resveratrol dimer, in humans and rats," *Molecules*, vol. 22, no. 5, p. 733, 2017.

[58] T. Walle, F. Hsieh, M. DeLegge, J. E. Oatis Jr, and U. K. Walle, "High absorption but very low bioavailability of oral resveratrol in humans," *Drug Metabolism and Disposition*, vol. 32, no. 12, pp. 1377–1382, 2004.

[59] C. Sergides, M. Chirilă, L. Silvestro, D. Pitta, and A. Pittas, "Bioavailability and safety study of resveratrol 500 mg tablets in healthy male and female volunteers," *Experimental and Therapeutic Medicine*, vol. 11, no. 1, pp. 164–170, 2016.

[50] C.-H. Cottart, V. Nivet-Antoine, C. Laguillier-Morizot, and J.-L. Beaudeux, "Resveratrol bioavailability and toxicity in humans," *Molecular Nutrition & Food Research*, vol. 54, no. 1, pp. 7–16, 2010.

[61] R. L. Frozza, A. Bernardi, K. Paese et al., "Characterization of *trans*-resveratrol-loaded lipid-core nanocapsules and tissue distribution studies in rats," *Journal of Biomedical Nanotechnology*, vol. 6, no. 6, pp. 694–703, 2010.

[62] R. S. Turner, R. G. Thomas, S. Craft et al., "A randomized, double-blind, placebo-controlled trial of resveratrol for Alzheimer disease," *Neurology*, vol. 85, no. 16, pp. 1383–1391, 2015.

[63] T.-Y. Chen, M. G. Ferruzzi, Q.-L. Wu et al., "Influence of diabetes on plasma pharmacokinetics and brain bioavailability of grape polyphenols and their phase II metabolites in the Zucker diabetic fatty rat," *Molecular Nutrition & Food Research*, vol. 61, no. 10, article 1700111, 2017.

[64] L. Biasutto, A. Mattarei, M. Azzolini et al., "Resveratrol derivatives as a pharmacological tool," *Annals of the New York Academy of Sciences*, vol. 1403, no. 1, pp. 27–37, 2017.

[65] G. Davidov-Pardo and D. J. McClements, "Resveratrol encapsulation: designing delivery systems to overcome solubility, stability and bioavailability issues," *Trends in Food Science & Technology*, vol. 38, no. 2, pp. 88–103, 2014.

[66] R. L. Frozza, A. Bernardi, J. B. Hoppe et al., "Neuroprotective effects of resveratrol against Aβ administration in rats are improved by lipid-core nanocapsules," *Molecular Neurobiology*, vol. 47, no. 3, pp. 1066–1080, 2013.

[67] M. Emília Juan, M. Pilar Vinardell, and J. M. Planas, "The daily oral administration of high doses of *trans*-resveratrol to rats for 28 days is not harmful," *The Journal of Nutrition*, vol. 132, no. 2, pp. 257–260, 2002.

[68] L. Almeida, M. Vaz-da-Silva, A. Falcão et al., "Pharmacokinetic and safety profile of *trans*-resveratrol in a rising multiple-dose study in healthy volunteers," *Molecular Nutrition & Food Research*, vol. 53, Supplement 1, pp. S7–S15, 2009.

[69] K. T. Howitz, K. J. Bitterman, H. Y. Cohen et al., "Small molecule activators of sirtuins extend *Saccharomyces cerevisiae* lifespan," *Nature*, vol. 425, no. 6954, pp. 191–196, 2003.

[70] M. Pacholec, J. E. Bleasdale, B. Chrunyk et al., "SRT 1720, SRT 2183, SRT 1460, and resveratrol are not direct activators of SIRT1," *Journal of Biological Chemistry*, vol. 285, no. 11, pp. 8340–8351, 2010.

[71] M. Kaeberlein, T. McDonagh, B. Heltweg et al., "Substrate-specific activation of sirtuins by resveratrol," *Journal of Biological Chemistry*, vol. 280, no. 17, pp. 17038–17045, 2005.

[72] T. Ma, M.-S. Tan, J.-T. Yu, and L. Tan, "Resveratrol as a therapeutic agent for Alzheimer's disease," *BioMed Research International*, vol. 2014, Article ID 350516, 13 pages, 2014.

[73] C. Fang, L. Gu, D. Smerin, S. Mao, and X. Xiong, "The interrelation between reactive oxygen species and autophagy in neurological disorders," *Oxidative Medicine and Cellular Longevity*, vol. 2017, Article ID 8495160, 16 pages, 2017.

[74] A. Y. Sun, Q. Wang, A. Simonyi, and G. Y. Sun, "Resveratrol as a therapeutic agent for neurodegenerative diseases," *Molecular Neurobiology*, vol. 41, no. 2-3, pp. 375–383, 2010.

[75] L. Kuršvietienė, I. Stanevičienė, A. Mongirdienė, and J. Bernatoniene, "Multiplicity of effects and health benefits of resveratrol," *Medicina*, vol. 52, no. 3, pp. 148–155, 2016.

[76] A. D. Romano, G. Serviddio, A. de Matthaeis, F. Bellanti, and G. Vendemiale, "Oxidative stress and aging," *Journal of Nephrology*, vol. 23, pp. S29–S36, 2010.

[77] F. Li, Q. Gong, H. Dong, and J. Shi, "Resveratrol, a neuroprotective supplement for Alzheimer's disease," *Current Pharmaceutical Design*, vol. 18, no. 1, pp. 27–33, 2012.

[78] B. Liu and J.-S. Hong, "Role of microglia in inflammation-mediated neurodegenerative diseases: mechanisms and strategies for therapeutic intervention," *The Journal of Pharmacology and Experimental Therapeutics*, vol. 304, no. 1, pp. 1–7, 2003.

[79] J. J. Palacino, D. Sagi, M. S. Goldberg et al., "Mitochondrial dysfunction and oxidative damage in parkin-deficient mice," *Journal of Biological Chemistry*, vol. 279, no. 18, pp. 18614–18622, 2004.

[80] H. Wang, T. Jiang, W. Li, N. Gao, and T. Zhang, "Resveratrol attenuates oxidative damage through activating mitophagy in an *in vitro* model of Alzheimer's disease," *Toxicology Letters*, vol. 282, pp. 100–108, 2018.

[81] B. J. Tabner, O. M. A. El-Agnaf, S. Turnbull et al., "Hydrogen peroxide is generated during the very early stages of aggregation of the amyloid peptides implicated in Alzheimer disease and familial British dementia," *Journal of Biological Chemistry*, vol. 280, no. 43, pp. 35789–35792, 2005.

[82] C. Lu, Y. Guo, J. Li et al., "Design, synthesis, and evaluation of resveratrol derivatives as Aß$_{1-42}$ aggregation inhibitors, antioxidants, and neuroprotective agents," *Bioorganic & Medicinal Chemistry Letters*, vol. 22, no. 24, pp. 7683–7687, 2012.

[83] P. H. Reddy, R. Tripathi, Q. Troung et al., "Abnormal mitochondrial dynamics and synaptic degeneration as early events in Alzheimer's disease: implications to mitochondria-targeted antioxidant therapeutics," *Biochimica et Biophysica Acta (BBA) - Molecular Basis of Disease*, vol. 1822, no. 5, pp. 639–649, 2012.

[84] P. H. Reddy, S. Tonk, S. Kumar et al., "A critical evaluation of neuroprotective and neurodegenerative microRNAs in Alzheimer's disease," *Biochemical and Biophysical Research Communications*, vol. 483, no. 4, pp. 1156–1165, 2017.

[85] M. Citron, "Alzheimer's disease: strategies for disease modification," *Nature Reviews Drug Discovery*, vol. 9, no. 5, pp. 387–398, 2010.

[86] F. Wu, M. P. Mattson, and P. J. Yao, "Neuronal activity and the expression of clathrin assembly protein AP180," *Biochemical and Biophysical Research Communications*, vol. 402, no. 2, pp. 297–300, 2010.

[87] M. D. Carter, G. A. Simms, and D. F. Weaver, "The development of new therapeutics for Alzheimer's disease," *Clinical Pharmacology & Therapeutics*, vol. 88, no. 4, pp. 475–486, 2010.

[88] J.-F. Ge, J.-P. Qiao, C.-C. Qi, C.-W. Wang, and J.-N. Zhou, "The binding of resveratrol to monomer and fibril amyloid beta," *Neurochemistry International*, vol. 61, no. 7, pp. 1192–1201, 2012.

[89] A. Nunomura, G. Perry, G. Aliev et al., "Oxidative damage is the earliest event in Alzheimer disease," *Journal of Neuropathology & Experimental Neurology*, vol. 60, no. 8, pp. 759–767, 2001.

[90] C. H.-L. Hung, Y.-S. Ho, and R. C.-C. Chang, "Modulation of mitochondrial calcium as a pharmacological target for Alzheimer's disease," *Ageing Research Reviews*, vol. 9, no. 4, pp. 447–456, 2010.

[91] M. Dumont and M. F. Beal, "Neuroprotective strategies involving ROS in Alzheimer disease," *Free Radical Biology & Medicine*, vol. 51, no. 5, pp. 1014–1026, 2011.

[92] H. F. Stanyon and J. H. Viles, "Human serum albumin can regulate amyloid-β peptide fiber growth in the brain interstitium," *Journal of Biological Chemistry*, vol. 287, no. 33, pp. 28163–28168, 2012.

[93] R. Kumar, P. Chaterjee, P. K. Sharma et al., "Sirtuin1: a promising serum protein marker for early detection of Alzheimer's disease," *PLoS One*, vol. 8, no. 4, article e61560, 2013.

[94] C. Julien, C. Tremblay, V. Émond et al., "Sirtuin 1 reduction parallels the accumulation of tau in Alzheimer disease," *Journal of Neuropathology & Experimental Neurology*, vol. 68, no. 1, pp. 48–58, 2009.

[95] J.-H. Koo, E.-B. Kang, Y.-S. Oh, D.-S. Yang, and J.-Y. Cho, "Treadmill exercise decreases amyloid-β burden possibly via activation of SIRT-1 signaling in a mouse model of Alzheimer's disease," *Experimental Neurology*, vol. 288, pp. 142–152, 2017.

[96] G. Marwarha, S. Raza, C. Meiers, and O. Ghribi, "Leptin attenuates BACE1 expression and amyloid-β genesis via the activation of SIRT1 signaling pathway," *Biochimica et Biophysica Acta (BBA) - Molecular Basis of Disease*, vol. 1842, no. 9, pp. 1587–1595, 2014.

[97] R. Kumar, L. Nigam, A. P. Singh, K. Singh, N. Subbarao, and S. Dey, "Design, synthesis of allosteric peptide activator for human SIRT1 and its biological evaluation in cellular model of Alzheimer's disease," *European Journal of Medicinal Chemistry*, vol. 127, pp. 909–916, 2017.

[98] L. Guarente, "Calorie restriction and sirtuins revisited," *Genes & Development*, vol. 27, no. 19, pp. 2072–2085, 2013.

[99] A. Satoh and S. Imai, "Hypothalamic Sirt1 in aging," *Aging*, vol. 6, no. 1, pp. 1-2, 2014.

[100] S. Michan, Y. Li, M. M.-H. Chou et al., "SIRT1 is essential for normal cognitive function and synaptic plasticity," *Journal of Neuroscience*, vol. 30, no. 29, pp. 9695–9707, 2010.

[101] J. Gao, W.-Y. Wang, Y.-W. Mao et al., "A novel pathway regulates memory and plasticity via SIRT1 and miR-134," *Nature*, vol. 466, no. 7310, pp. 1105–1109, 2010.

[102] A. Z. Herskovits and L. Guarente, "SIRT1 in neurodevelopment and brain senescence," *Neuron*, vol. 81, no. 3, pp. 471–483, 2014.

[103] M. Bernier, R. K. Paul, A. Martin-Montalvo et al., "Negative regulation of STAT3 protein-mediated cellular respiration by SIRT1 protein," *Journal of Biological Chemistry*, vol. 286, no. 22, pp. 19270–19279, 2011.

[104] M. R. Ramis, S. Esteban, A. Miralles, D.-X. Tan, and R. J. Reiter, "Caloric restriction, resveratrol and melatonin: role of SIRT1 and implications for aging and related-diseases," *Mechanisms of Ageing and Development*, vol. 146-148, pp. 28–41, 2015.

[105] J. Wang, H. Fivecoat, L. Ho, Y. Pan, E. Ling, and G. M. Pasinetti, "The role of Sirt 1: at the crossroad between promotion of longevity and protection against Alzheimer's disease neuropathology," *Biochimica et Biophysica Acta (BBA) - Proteins and Proteomics*, vol. 1804, no. 8, pp. 1690–1694, 2010.

[106] S.-Y. Li, X.-B. Wang, and L.-Y. Kong, "Design, synthesis and biological evaluation of imine resveratrol derivatives as multi-targeted agents against Alzheimer's disease," *European Journal of Medicinal Chemistry*, vol. 71, pp. 36–45, 2014.

[107] G. Sadi and D. Konat, "Resveratrol regulates oxidative biomarkers and antioxidant enzymes in the brain of streptozotocin-induced diabetic rats," *Pharmaceutical Biology*, vol. 54, no. 7, pp. 1–8, 2016.

[108] A. Carrizzo, M. Forte, A. Damato et al., "Antioxidant effects of resveratrol in cardiovascular, cerebral and metabolic

[108] diseases," *Food and Chemical Toxicology*, vol. 61, pp. 215–226, 2013.

[109] X. Yang, X. Qiang, Y. Li et al., "Pyridoxine-resveratrol hybrids Mannich base derivatives as novel dual inhibitors of AChE and MAO-B with antioxidant and metal-chelating properties for the treatment of Alzheimer's disease," *Bioorganic Chemistry*, vol. 71, pp. 305–314, 2017.

[110] S. Schweiger, F. Matthes, K. Posey et al., "Resveratrol induces dephosphorylation of tau by interfering with the MID1-PP2A complex," *Scientific Reports*, vol. 7, no. 1, pp. 13753–13713, 2017.

[111] S. D. Rege, T. Geetha, T. L. Broderick, and J. R. Babu, "Resveratrol protects β-amyloid induced oxidative damage and memory associated proteins in H19-7 hippocampal neuronal cells," *Current Alzheimer Research*, vol. 12, no. 2, pp. 147–156, 2015.

[112] C. Lu, Y. Guo, J. Yan et al., "Design, synthesis, and evaluation of multitarget-directed resveratrol derivatives for the treatment of Alzheimer's disease," *Journal of Medicinal Chemistry*, vol. 56, no. 14, pp. 5843–5859, 2013.

[113] X. Hou, D. Rooklin, H. Fang, and Y. Zhang, "Resveratrol serves as a protein-substrate interaction stabilizer in human SIRT1 activation," *Scientific Reports*, vol. 6, no. 1, article 38186, 2016.

[114] B. Dasgupta and J. Milbrandt, "Resveratrol stimulates AMP kinase activity in neurons," *Proceedings of the National Academy of Sciences of the United State of America*, vol. 104, no. 17, pp. 7217–7222, 2007.

[115] C. Cantó, Z. Gerhart-Hines, J. N. Feige et al., "AMPK regulates energy expenditure by modulating $NAD^+$ metabolism and SIRT1 activity," *Nature*, vol. 458, no. 7241, pp. 1056–1060, 2009.

[116] N. L. Price, A. P. Gomes, A. J. Y. Ling et al., "SIRT1 is required for AMPK activation and the beneficial effects of resveratrol on mitochondrial function," *Cell Metabolism*, vol. 15, no. 5, pp. 675–690, 2012.

[117] O. B. Villaflores, Y.-J. Chen, C.-P. Chen, J.-M. Yeh, and T.-Y. Wu, "Curcuminoids and resveratrol as anti-Alzheimer agents," *Taiwanese Journal of Obstetrics and Gynecology*, vol. 51, no. 4, pp. 515–525, 2012.

[118] A. Sahebkar, "Neuroprotective effects of resveratrol: potential mechanisms," *Neurochemistry International*, vol. 57, no. 6, pp. 621-622, 2010.

[119] T.-C. Huang, K.-T. Lu, Y.-Y. P. Wo, Y.-J. Wu, and Y.-L. Yang, "Resveratrol protects rats from Aβ-induced neurotoxicity by the reduction of iNOS expression and lipid peroxidation," *PLoS One*, vol. 6, no. 12, article e29102, 2011.

[120] M. Venigalla, S. Sonego, E. Gyengesi, M. J. Sharman, and G. Münch, "Novel promising therapeutics against chronic neuroinflammation and neurodegeneration in Alzheimer's disease," *Neurochemistry International*, vol. 95, pp. 63–74, 2016.

[121] Y. Liu, X. Chen, and J. Li, "Resveratrol protects against oxidized low-density lipoprotein-induced human umbilical vein endothelial cell apoptosis via inhibition of mitochondrial-derived oxidative stress," *Molecular Medicine Reports*, vol. 15, no. 5, pp. 2457–2464, 2017.

[122] G. Spanier, H. Xu, N. Xia et al., "Resveratrol reduces endothelial oxidative stress by modulating the gene expression of superoxide dismutase 1 (SOD1), glutathione peroxidase 1 (GPx1) and NADPH oxidase subunit (Nox4)," *Journal of Physiology and Pharmacology*, vol. 60, Supplement 4, pp. 111–116, 2009.

[123] J.-G. Lee, J.-M. Yon, C. Lin, A. Y. Jung, K. Y. Jung, and S.-Y. Nam, "Combined treatment with capsaicin and resveratrol enhances neuroprotection against glutamate-induced toxicity in mouse cerebral cortical neurons," *Food and Chemical Toxicology*, vol. 50, no. 11, pp. 3877–3885, 2012.

[124] A. Quincozes-Santos, L. D. Bobermin, A. C. Tramontina et al., "Oxidative stress mediated by NMDA, AMPA/KA channels in acute hippocampal slices: neuroprotective effect of resveratrol," *Toxicology In Vitro*, vol. 28, no. 4, pp. 544–551, 2014.

[125] V. Vingtdeux, L. Giliberto, H. Zhao et al., "AMP-activated protein kinase signaling activation by resveratrol modulates amyloid-β peptide metabolism," *Journal of Biological Chemistry*, vol. 285, no. 12, pp. 9100–9113, 2010.

[126] F. Simão, A. Matté, A. S. Pagnussat, C. A. Netto, and C. G. Salbego, "Resveratrol preconditioning modulates inflammatory response in the rat hippocampus following global cerebral ischemia," *Neurochemistry International*, vol. 61, no. 5, pp. 659–665, 2012.

[127] X. Wang, S. Ma, B. Yang et al., "Resveratrol promotes hUC-MSCs engraftment and neural repair in a mouse model of Alzheimer's disease," *Behavioural Brain Research*, vol. 339, pp. 297–304, 2018.

[128] M. Tajes, J. Gutierrez-Cuesta, J. Folch et al., "Neuroprotective role of intermittent fasting in senescence-accelerated mice P8 (SAMP8)," *Experimental Gerontology*, vol. 45, no. 9, pp. 702–710, 2010.

[129] J. L. Barger, T. Kayo, J. M. Vann et al., "A low dose of dietary resveratrol partially mimics caloric restriction and retards aging parameters in mice," *PLoS One*, vol. 3, no. 6, article e2264, 2008.

[130] H. Imamura, T. Yamaguchi, D. Nagayama, A. Saiki, K. Shirai, and I. Tatsuno, "Resveratrol ameliorates arterial stiffness assessed by cardio-ankle vascular index in patients with type 2 diabetes mellitus," *International Heart Journal*, vol. 58, no. 4, pp. 577–583, 2017.

[131] A. P. Mansur, A. Roggerio, M. F. S. Goes et al., "Serum concentrations and gene expression of sirtuin 1 in healthy and slightly overweight subjects after caloric restriction or resveratrol supplementation: a randomized trial," *International Journal of Cardiology*, vol. 227, pp. 788–794, 2017.

[132] A. J. Nimmo and R. Vink, "Recent patents in CNS drug discovery: the management of inflammation in the central nervous system," *Recent Patents on CNS Drug Discovery*, vol. 4, no. 2, pp. 86–95, 2009.

[133] A. H. Moore and M. K. O'Banion, "Neuroinflammation and anti-inflammatory therapy for Alzheimer's disease," *Advanced Drug Delivery Reviews*, vol. 54, no. 12, pp. 1627–1656, 2002.

[134] F. Zhang, J. Liu, and J.-S. Shi, "Anti-inflammatory activities of resveratrol in the brain: role of resveratrol in microglial activation," *European Journal of Pharmacology*, vol. 636, no. 1–3, pp. 1–7, 2010.

[135] C. Kaur, G. Rathnasamy, and E.-A. Ling, "Roles of activated microglia in hypoxia induced neuroinflammation in the developing brain and the retina," *Journal of Neuroimmune Pharmacology*, vol. 8, no. 1, pp. 66–78, 2013.

[136] D. D. Lofrumento, G. Nicolardi, A. Cianciulli et al., "Neuroprotective effects of resveratrol in an MPTP mouse model

of Parkinson's-like disease: possible role of SOCS-1 in reducing pro-inflammatory responses," *Innate Immunity*, vol. 20, no. 3, pp. 249–260, 2014.

[137] Q. Zhang, L. Yuan, Q. Zhang et al., "Resveratrol attenuates hypoxia-induced neurotoxicity through inhibiting microglial activation," *International Immunopharmacology*, vol. 28, no. 1, pp. 578–587, 2015.

[138] S. Bastianetto, C. Ménard, and R. Quirion, "Neuroprotective action of resveratrol," *Biochimica et Biophysica Acta (BBA) - Molecular Basis of Disease*, vol. 1852, no. 6, pp. 1195–1201, 2015.

[139] S. S. Gocmez, N. Gacar, T. Utkan, G. Gacar, P. J. Scarpace, and N. Tumer, "Protective effects of resveratrol on aging-induced cognitive impairment in rats," *Neurobiology of Learning and Memory*, vol. 131, pp. 131–136, 2016.

[140] H. Capiralla, V. Vingtdeux, H. Zhao et al., "Resveratrol mitigates lipopolysaccharide- and Aβ-mediated microglial inflammation by inhibiting the TLR4/NF-κB/STAT signaling cascade," *Journal of Neurochemistry*, vol. 120, no. 3, pp. 461–472, 2012.

[141] V. K. Nimmagadda, C. T. Bever, N. R. Vattikunta et al., "Overexpression of SIRT1 protein in neurons protects against experimental autoimmune encephalomyelitis through activation of multiple SIRT1 targets," *The Journal of Immunology*, vol. 190, pp. 4595–4607, 2013.

[142] X. Yang, S. Xu, Y. Qian, and Q. Xiao, "Resveratrol regulates microglia M1/M2 polarization via PGC-1α in conditions of neuroinflammatory injury," *Brain, Behavior, and Immunity*, vol. 64, pp. 162–172, 2017.

[143] J. Wu, X. Li, G. Zhu, Y. Zhang, M. He, and J. Zhang, "The role of resveratrol-induced mitophagy/autophagy in peritoneal mesothelial cells inflammatory injury via NLRP3 inflammasome activation triggered by mitochondrial ROS," *Experimental Cell Research*, vol. 341, no. 1, pp. 42–53, 2016.

[144] Y. Zhang, M. Chen, Y. Zhou et al., "Resveratrol improves hepatic steatosis by inducing autophagy through the cAMP signaling pathway," *Molecular Nutrition & Food Research*, vol. 59, no. 8, pp. 1443–1457, 2015.

[145] B. Annabi, S. Lord-Dufour, A. Vézina, and R. Béliveau, "Resveratrol targeting of carcinogen-induced brain endothelial cell inflammation biomarkers MMP-9 and COX-2 is Sirt1-independent," *Drug Target Insights*, vol. 6, article DTI.S9442, 2012.

[146] S. S. Lakka, C. S. Gondi, and J. S. Rao, "Proteases and glioma angiogenesis," *Brain Pathology*, vol. 15, no. 4, pp. 327–341, 2005.

[147] A. Bonoiu, S. D. Mahajan, L. Ye et al., "MMP-9 gene silencing by a quantum dot–siRNA nanoplex delivery to maintain the integrity of the blood brain barrier," *Brain Research*, vol. 1282, pp. 142–155, 2009.

[148] L. Cao, C. Liu, F. Wang, and H. Wang, "SIRT1 negatively regulates amyloid-beta-induced inflammation via the NF-κB pathway," *Brazilian Journal of Medical and Biological Research*, vol. 46, no. 8, pp. 659–669, 2013.

[149] G. M. Pasinetti, J. Wang, P. Marambaud et al., "Neuroprotective and metabolic effects of resveratrol: therapeutic implications for Huntington's disease and other neurodegenerative disorders," *Experimental Neurology*, vol. 232, no. 1, pp. 1–6, 2011.

[150] G. Donmez, D. Wang, D. E. Cohen, and L. Guarente, "SIRT1 suppresses β-amyloid production by activating the α-secretase gene ADAM10," *Cell*, vol. 142, no. 2, pp. 320–332, 2010.

[151] X. Lu, L. Ma, L. Ruan et al., "Resveratrol differentially modulates inflammatory responses of microglia and astrocytes," *Journal of Neuroinflammation*, vol. 7, no. 1, p. 46, 2010.

[152] G. M. Pasinetti, J. Wang, L. Ho, W. Zhao, and L. Dubner, "Roles of resveratrol and other grape-derived polyphenols in Alzheimer's disease prevention and treatment," *Biochimica et Biophysica Acta (BBA) - Molecular Basis of Disease*, vol. 1852, no. 6, pp. 1202–1208, 2015.

[153] C. Moussa, M. Hebron, X. Huang et al., "Resveratrol regulates neuro-inflammation and induces adaptive immunity in Alzheimer's disease," *Journal of Neuroinflammation*, vol. 14, no. 1, pp. 1–10, 2017.

[154] S. Thordardottir, A. Kinhult Ståhlbom, O. Almkvist et al., "The effects of different familial Alzheimer's disease mutations on APP processing in vivo," *Alzheimer's Research & Therapy*, vol. 9, no. 1, article 9, 2017.

[155] F. Yeung, J. E. Hoberg, C. S. Ramsey et al., "Modulation of NF-κB-dependent transcription and cell survival by the SIRT1 deacetylase," *The EMBO Journal*, vol. 23, no. 12, pp. 2369–2380, 2004.

[156] T. S. Anekonda, "Resveratrol—a boon for treating Alzheimer's disease?," *Brain Research Reviews*, vol. 52, no. 2, pp. 316–326, 2006.

[157] J. Moriya, R. Chen, J. Yamakawa, K. Sasaki, Y. Ishigaki, and T. Takahashi, "Resveratrol improves hippocampal atrophy in chronic fatigue mice by enhancing neurogenesis and inhibiting apoptosis of granular cells," *Biological and Pharmaceutical Bulletin*, vol. 34, no. 3, pp. 354–359, 2011.

[158] S. T. Koz, E. O. Etem, G. Baydas et al., "Effects of resveratrol on blood homocysteine level, on homocysteine induced oxidative stress, apoptosis and cognitive dysfunctions in rats," *Brain Research*, vol. 1484, pp. 29–38, 2012.

[159] E.-J. Park and J. M. Pezzuto, "The pharmacology of resveratrol in animals and humans," *Biochimica et Biophysica Acta (BBA) - Molecular Basis of Disease*, vol. 1852, no. 6, pp. 1071–1113, 2015.

[160] M. C. Chiang, C. J. Nicol, and Y. C. Cheng, "Resveratrol activation of AMPK-dependent pathways is neuroprotective in human neural stem cells against amyloid-beta-induced inflammation and oxidative stress," *Neurochemistry International*, vol. 115, pp. 1–10, 2018.

[161] M. H. Jang, X. L. Piao, J. M. Kim, S. W. Kwon, and J. H. Park, "Inhibition of cholinesterase and amyloid-β aggregation by resveratrol oligomers from Vitis amurensis," *Phytotherapy Research*, vol. 22, no. 4, pp. 544–549, 2008.

[162] H. Deng and M. t. Mi, "Resveratrol attenuates $A\beta_{25-35}$ caused neurotoxicity by inducing autophagy through the TyrRS-PARP1-SIRT1 signaling pathway," *Neurochemical Research*, vol. 41, no. 9, pp. 2367–2379, 2016.

[163] W. Hu, E. Yang, J. Ye, W. Han, and Z.-L. Du, "Resveratrol protects neuronal cells from isoflurane-induced inflammation and oxidative stress-associated death by attenuating apoptosis via Akt/p 38 MAPK signaling," *Experimental and Therapeutic Medicine*, vol. 15, pp. 1568–1573, 2018.

[164] T. Huang, D. Gao, X. Jiang, S. Hu, L. Zhang, and Z. Fei, "Resveratrol inhibits oxygen-glucose deprivation-induced MMP-3 expression and cell apoptosis in primary cortical cells via the NF-κB pathway," *Molecular Medicine Reports*, vol. 10, no. 2, pp. 1065–1071, 2014.

[165] S.-J. Park, F. Ahmad, A. Philp et al., "Resveratrol ameliorates aging-related metabolic phenotypes by inhibiting cAMP phosphodiesterases," *Cell*, vol. 148, no. 3, pp. 421–433, 2012.

[166] C. Shen, W. Cheng, P. Yu et al., "Resveratrol pretreatment attenuates injury and promotes proliferation of neural stem cells following oxygen-glucose deprivation/reoxygenation by upregulating the expression of Nrf 2, HO-1 and NQO1 in vitro," *Molecular Medicine Reports*, vol. 14, no. 4, pp. 3646–3654, 2016.

[167] H. Yin, H. Wang, H. Zhang, N. Gao, T. Zhang, and Z. Yang, "Resveratrol attenuates A$\beta$-induced early hippocampal neuron excitability impairment via recovery of function of potassium channels," *Neurotoxicity Research*, vol. 32, no. 3, pp. 311–324, 2017.

# Nox2 Activity is Required in Obesity-Mediated Alteration of Bone Remodeling

Md Mizanur Rahman[1], Amina El Jamali,[2] Ganesh V. Halade,[3] Allal Ouhtit,[1] Haissam Abou-Saleh,[1] and Gianfranco Pintus[4,5,6]

[1]Department of Biological and Environmental Sciences, College of Arts and Sciences, Qatar University, PO Box 2713 Doha, Qatar
[2]Division of Nephrology, Department of Medicine, University of Texas Health Science Center at San Antonio, 7703 Floyd Curl Drive, Texas 78229-3900, USA
[3]Division of Cardiovascular Disease, Department of Medicine, University of Alabama at Birmingham, Birmingham, Alabama 35294, USA
[4]Department of Biomedical Sciences, College of Health Sciences, Qatar University, PO Box 2713 Doha, Qatar
[5]Department of Biomedical Sciences, College of Medicine, University of Sassari, 07100 Sassari, Italy
[6]Biomedical Research Center, Qatar University, Doha, Qatar

Correspondence should be addressed to Md Mizanur Rahman; mrahman@qu.edu.qa

Academic Editor: Carlo Tocchetti

Despite increasing evidence suggesting a role for NADPH oxidases (Nox) in bone pathophysiology, whether Nox enzymes contribute to obesity-mediated bone remodeling remains to be clearly elucidated. Nox2 is one of the predominant Nox enzymes expressed in the bone marrow microenvironment and is a major source of ROS generation during inflammatory processes. It is also well recognized that a high-fat diet (HFD) induces obesity, which negatively impacts bone remodeling. In this work, we investigated the effect of Nox2 loss of function on obesity-mediated alteration of bone remodeling using wild-type (WT) and Nox2-knockout (KO) mice fed with a standard lab chow diet (SD) as a control or a HFD as an obesity model. Bone mineral density (BMD) of mice was assessed at the beginning and after 3 months of feeding with SD or HFD. Our results show that HFD increased bone mineral density to a greater extent in KO mice than in WT mice without affecting the total body weight and fat mass. HFD also significantly increased the number of adipocytes in the bone marrow microenvironment of WT mice as compared to KO mice. The bone levels of proinflammatory cytokines and proosteoclastogenic factors were also significantly elevated in WT-HFD mice as compared to KO-HFD mice. Furthermore, the in vitro differentiation of bone marrow cells into osteoclasts was significantly increased when using bone marrow cells from WT-HFD mice as compared to KO-HFD mice. Our data collectively suggest that Nox2 is implicated in HFD-induced deleterious bone remodeling by enhancing bone marrow adipogenesis and osteoclastogenesis.

## 1. Introduction

Obesity is defined as a body mass index (BMI) greater than or equal to 30. The obesity epidemic correlates with increasing evidence that lipotoxicity and inflammation might be the cause of bone loss [1–3]. Studies in mice and humans indicate that obesity is related to a sustained and elevated inflammatory state of adipose tissue [3, 4]. Several molecular and cellular determinants involved in inflammation- and lipotoxicity-induced bone mass loss have been suggested [1, 5]. Obesity is associated with enhanced tissue inflammation [6]. More recently, it has been shown that a fat increase in the bone marrow (BM) microenvironment [7] may affect bone homeostasis, by inhibiting osteoblast function and increasing osteoclast differentiation/activation [8], through the regulation of inflammatory cytokines [9]. Nevertheless, data concerning the impact of obesity on bone health are still controversial. Indeed, the augmented body mass due to obesity may increase mechanical bone stretching improving both bone mass and mineral density, while increased adiposity in the bone marrow region can lower osteoblast/osteocyte formation, generating low-quality bone and enhanced risk of fracture. However, some studies showed that obesity reduces fracture risk and protects against osteoporosis in adults [10, 11], while others showed that obesity does not protect against fracture in postmenopausal women [12, 13]. Interestingly, one study reported that obesity is a risk factor for fracture in children, while it is protective against fracture in adults [14], suggesting that the effects of obesity on bone parameters may differ with age.

Akin to obesity, a high-fat diet (HFD) can influence the bone microenvironment and remodeling. For instance, decreases in both bone resorption and formation have been reported in HFD-fed mice [15], while increased bone mineral density and decreased osteoclast activity have been shown in rats [16]. We have demonstrated that HFD can promote inflammation of the bone microenvironment and negatively influence bone remodeling toward osteoclastogenesis in mice [17]. The different reported effects in these studies might be due to differences in diet composition and animal models used. Indeed, decreased bone resorption and bone formation have been reported in patients with diabetes [18, 19] suggesting that both diet- and metabolic-related factors may differentially affect the bone microenvironment. It has been shown that knockdown of Nox2 can ameliorate the adipose tissue inflammatory status in HDF mice [20, 21]. In addition, inflammatory cytokines such as tumor necrosis factor alpha (TNF-$\alpha$) can activate Nox2 and increase its activity [22, 23]. Therefore, Nox2 and inflammation are directly interrelated since enhanced inflammation in the bone microenvironment can negatively affect bone remodeling toward bone resorption. Therefore, we may assume that nox2 has a strong role in obesity-mediated bone remodeling.

Oxidative stress is shown to be a key mediator of bone loss [24, 25]. Elevated levels of reactive oxygen species (ROS), specifically superoxide, are consistently associated with increased osteoclastic activity in patients with bone disorders [26, 27] such as bone loss observed in the elderly population or patients with rheumatoid arthritis. The age-associated decrease in sexual hormones in both men and women is also an important risk factor for bone loss that is related to a decrease in sexual hormones. Finally, it is commonly observed that elderly osteoporotic patients are deficient in vitamins with antioxidant properties [27]. Whether the bone loss is due to a ROS overproduction and/or to an antioxidant level reduction, the resulting oxidative stress plays an essential role in osteoporosis. Given the complexity of the various sources of ROS in the bone and the etiology of bone disorders, the identification of the source of ROS for each disease and the elucidation of the mechanism by which ROS are produced and contribute are essential to developing specific and efficient therapy [25, 28].

NADPH oxidases (Nox) are considered key ROS-generating enzymes, whose activity is essential for normal cell function [29]. The Nox family encompass the gp91phox (Nox2), Nox1, Nox3, Nox4, Nox5, Duox-1, and Duox-2. In addition to Nox2, Nox1 and Nox4 are also known to be expressed in the bone microenvironment [25]. However, in this present work, we have mainly studied the role of Nox2 in obesity-mediated bone remodeling using a Nox2-knockout mouse model. In the bone microenvironment, hematopoietic stem cells commit to bone resorbing osteoclastic lineage in response to the receptor activator of NF-$\kappa$B ligand (RANKL) which is produced by osteoblasts, osteocytes, and bone marrow stromal cells [30]. The ligand-mediated activation of the RANK receptor by RANKL triggers a Nox-derived ROS-dependent signaling cascade which is crucial for osteoclast differentiation [31]. In particular, Nox1 and Nox2 are reported to be implicated in osteoclast differentiation in response to RANKL activation, while Nox4 and Nox2 have been reported to contribute to the bone resorption activity of mature osteoclasts [32–35]. By contrast, whether Nox enzymes are involved in osteoblast differentiation and function is still under investigation [36, 37]. In this context, some evidence suggests the implication of Nox4 in osteoblast differentiation and that of Nox2 in osteoblast precursor cell proliferation [25]. However, these findings were not confirmed in vivo in knockout mice.

Studies investigating the role of oxidative stress in bone diseases have been mainly performed by determining the effect of ROS overproduction on bone cell function and bone remodeling. It has been reported that ROS can promote bone loss by lowering osteoblast differentiation and inducing their apoptosis while concomitantly stimulating RANKL-induced osteoclast formation [38]. In addition, an osteoclast-generated superoxide increase is involved in bone matrix degradation [39]. While we showed that the expression level of cathepsin K, which is an indicator of bone resorption activity, is increased by a HFD in a redox-dependent manner [40], the exact role of Nox2 as a major source of ROS remains unknown.

In the present study, we focused upon investigating the role of Nox2 in obesity-mediated inflammation-dependent bone remodeling. Our results show that a HFD augments bone marrow adipogenesis and osteoclastogenesis in a Nox2-dependent manner.

## 2. Materials and Methods

*2.1. Reagents and ELISA Kits.* $\alpha$-Modified minimal essential medium ($\alpha$-MEM), Roswell Park Memorial Institute (RPMI) 1640 medium, phenol red-free $\alpha$-MEM, Hanks' balanced salt solution (HBSS), and fetal bovine serum were purchased from Sigma-Aldrich (St. Louis, MO, USA). Recombinant mouse RANKL and M-CSF were obtained from PeproTech Inc. (Rocky Hill, NJ) and R&D Systems (Minneapolis, MN, USA), respectively. Phorbol 12-myristate 13-acetate (PMA) and superoxide dismutase (SOD) were obtained from Sigma Chemical (St. Louis, MO, USA). The enhancer-containing luminol-based detection system (Diogenes) was obtained from National Diagnostics (Atlanta, GA, USA). Experiments were performed following our previously published procedures as reported in Halade et al. [17].

*2.2. Animals and Diet.* Six-week-old wild-type (WT) and Nox2-knockout (KO) male mice were purchased from Jackson Laboratories (Bar Harbor, Maine 04609, USA) and provided water and standard American Institute of Nutrition (AIN) 93G (a diet recommended by AIN for growth) ad libitum. At 8 weeks, age-matched animals were randomized into two groups, each containing 10 mice. Subsequently, the mice were housed in a standard controlled animal care facility in cages (5 mice/cage) and fed with a high-fat diet (Jackson lab. DIO diet 45% fat; HFD) or a standard diet (SD) ad libitum for 3 months. National Institutes of Health guidelines were strictly followed, and all the studies were approved by the Institutional Laboratory Animal Care and Use Committee of the University of Texas Health Science Center (San Antonio, TX). Body weight was measured weekly. After completion of the 5-month period, animals from both experimental groups were sacrificed under isoflurane anesthesia. Experiments were performed following our previously published procedures as reported in Halade et al. [17].

*2.3. Measurement of Bone Mineral Density (BMD).* Mice were anesthetized with ketamine/xylazine, and a dual-energy X-ray absorptiometry (DXA) scan was performed as described previously [41] before (mice aged 8 weeks) and after the administration of the diet for 3 months (mice aged 20 weeks) using a Lunar PIXImus mouse bone densitometer (General Electric). BMD was analyzed manually with PIXImus software as described previously [41].

*2.4. Preparation and Culture of Primary Bone Marrow (BM) Cells and Osteoclast Differentiation.* At the conclusion of the study, BM cells were isolated from the tibiae and femurs of all groups of mice according to our previously described methods [17, 42]. Briefly, isolated BM cells were cultured in $\alpha$-MEM containing 10% heat-inactivated fetal calf serum (FCS; Invitrogen), 100 U/ml penicillin G, and 100 $\mu$g/ml streptomycin at 37°C for 2 h under 95% air and 5% $HFD_2$. Nonadherent cells were carefully harvested and centrifuged at 2000 rpm for 5 min at room temperature, and viability was determined using the trypan blue exclusion method. They were subsequently cultured in $\alpha$-MEM medium in 24-well ($1 \times 10^6$ cells/well) clear-bottom white culture plates, supplemented with or without 50 ng/ml RANKL and 20 ng/ml M-CSF. A half-volume of the medium, with or without RANKL/M-CSF, was replaced with the fresh medium every 3 days. At day 6, plates were fixed and the number of formed osteoclasts was determined by staining cells with tartrate-resistant acid phosphatase (TRAP) using a TRAP staining kit (Sigma, St. Louis, MO, USA). TRAP-positive multinucleated cells (MNCs) of more than three nuclei were counted as osteoclasts under microscopic observation and expressed as a number of cells per field.

*2.5. Bone Histology.* Left femur specimens from each group were collected and trimmed of excess tissue and were fixed in 10% neutral buffer formalin (NBF) for 48 hours at room temperature (RT). Bone specimens were decalcified in 10% EDTA in water for 2 weeks at RT and were then placed in 70% ETOH, processed and embedded in paraffin, and stained with hematoxylin and eosin (H&E) as previously described [17, 43].

*2.6. Quantitative Real-Time Reverse Transcriptase PCR (RT-PCR).* Experiments were performed following our previously published procedures [17]. Briefly, right whole femurs were crushed under liquid nitrogen conditions using a Kinematica Tissue Pulverizer, and RNA was isolated using the RNeasy Mini Kit following the manufacturer's instructions (Qiagen, Valencia, CA). Total RNA concentration was assessed in NanoDrop™ 1000 spectrophotometer (Thermo Scientific, Wilmington, DE, USA). mRNA expression of genes encoding IL-6, TNF-$\alpha$, RANKL, cathepsin K (ctsk), runt-related transcription factor 2 (RUNX2), peroxisome proliferator-activated receptor (PPAR-$\gamma$), and 18S was measured by real-time RT-PCR carried out using Taq polymerase and SYBR green dye (Applied Biosystems, Foster City, CA) and an ABI Prism 7900HT Sequence Detection System (Applied Biosystems). mRNA Ct values for these genes were normalized to the housekeeping gene 18S.

*2.7. Measurement of Superoxide in Freshly Isolated BM Cells.* As previously reported, the rate of superoxide production by BM cells was determined using a luminol-based chemiluminescent reagent (Diogenes, National Diagnostics, GA) that is specific to superoxide [17, 44]. The cells were washed with PBS and placed in Hanks' balanced salt solution (HBSS). For the assay, a 100 $\mu$l aliquot of the Diogenes reagent was mixed with a maximum of $1 \times 10^5$ cells and incubated at 37°C for 2–4 minutes. Superoxide generation was stimulated with PMA (50 ng/ml) in the presence or absence of SOD (20 $\mu$g/ml). Chemiluminescence was measured every minute for up to 60 minutes using a microplate reader (BioTek Clarity™, BioTek Instruments Inc., VT, USA) and an integration time of 5 seconds.

*2.8. Statistical Analysis.* Data are presented as mean values ± SEM of 3–4 experiments using 4–6 mice/group. The statistical analysis employed for each experiment is reported in the figure legends.

## 3. Results

*3.1. ROS Production in Bone Marrow Cells Isolated from WT and KO Mice.* Superoxide production was examined in freshly isolated bone marrow (BM) cells. A transient PMA-induced production of superoxide was observed in BM cells derived from WT mice. The observed superoxide generation reached its maximum in 5 minutes; then, it sustained up to 30 min and after that decreases to reach the nonstimulated levels in 60 to 90 min. The PMA-elicited superoxide production of BM cells was not significantly affected by the HFD (Figure 1). Abrogation of the chemiluminescence signals by the addition of the superoxide inhibitor superoxide dismutase (SOD) indicated that superoxide anion was specifically detected. As expected, BM cells isolated from KO mice were unable to produce any detectable amount of superoxide. These data indicated that a high-fat diet did not significantly alter superoxide production of BM cells and confirmed the phenotype of Nox2-knockout mice.

*3.2. Effect of Nox2 on HFD Alteration of Bone Mineral Density (BMD).* Three months of feeding of a HFD significantly increased body weight and fat mass of both WT and KO mice. There was no significant difference in body weights (BW) and total fat mass (FM) between SD WT and SD KO

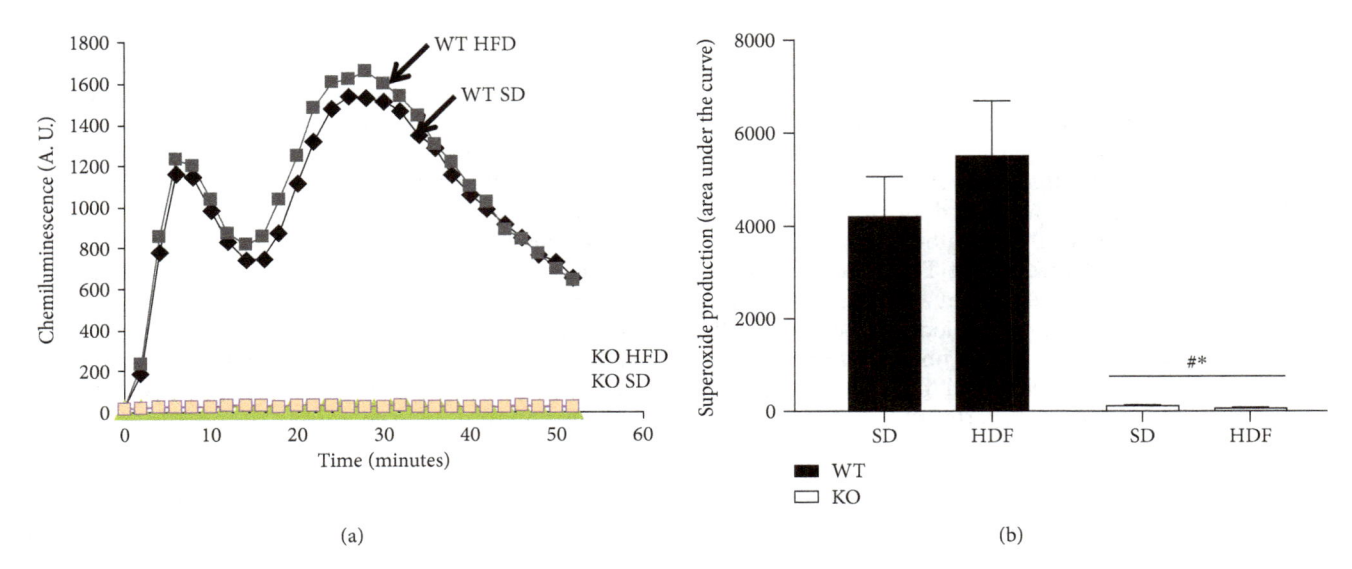

FIGURE 1: The high-fat diet does not alter superoxide production by the bone marrow cells. Superoxide production was measured using bone marrow (BM) cells isolated from wild-type and Nox2-knockout mice fed with either a standard (SD) or a high-fat diet (HFD) for 3 months. BM cells were stimulated with 50 ng/ml PMA, and chemiluminescence was monitored for the indicated time (a). After normalization by subtraction of the zero time value of the chemiluminescence output, the area *under the curve* was calculated as a measure of total superoxide production (b). Data are expressed as mean ± SEM ($n = 4$, 6 mice/group). Two-way ANOVA followed by post hoc Tukey's multiple comparison test was used to assess differences among groups with significance defined as $P < 0.05$. *Significantly different from SD-WT. #Significantly different from HDF-WT.

mice ($26.8 \pm 0.44$ g BW and $3.8 + 0.42$ g total fat mass and $23.7 \pm 1.5$ g BW and $3.3 \pm 0.24$ g FM, respectively) and between HFD WT and HFD KO mice ($41.3 \pm 1.68$ g BW and $8.6 \pm 1.7$ g FM and $36.3 \pm 2.46$ g BW and $9.9 + 2.3$ g FM, respectively). We measured the bone mineral density of all four groups once before starting the experimental diets and again just before sacrifice. Bone mineral density was not significantly reduced in WT versus KO mice fed with a SD with the exception of the distal femoral metaphysis bone "section" (Figure 2(a)). An increase in bone mineral density was observed in both WT and KO animals in all bone sections except distal femoral metaphysis of HDF mice in comparison to that of SD mice (Figure 2(a)). However, the magnitude of this effect was significantly greater in KO mice as compared to WT mice as indicated by the data in Figure 2(b) reporting the percentage change of the difference between HDF and SD. Indeed, this figure clearly depicts that the differences in BMD between HFD- and SD-fed mice (Figure 2(a)) were significantly higher in KO mice as compared to WT mice indicating that HFD-induced BMD alteration depends on Nox2.

*3.3. Nox2 Mediates HFD-Induced Bone Marrow Adiposity.* Alteration of bone mineral density by a HFD is generally paralleled by enhanced bone marrow adiposity. We therefore asked whether Nox2 downregulation could affect bone marrow adiposity. Our results indicate that the HFD increased the number of adipocytes. Interestingly, this effect was dramatically reduced in KO mice (Figure 3(a)). In addition, we found that the expression level of PPAR-$\gamma$, a known marker of adipocyte differentiation and maturation, was increased in the bone of WT HFD mice as compared to WT SD mice (Figure 3(b)), and this effect was abrogated in KO mice. Since the same progenitor cells lead to the formation of osteoblasts and adipocytes, we measured the Runx2 expression level, a known marker of osteoblastogenesis. Interestingly, HFD significantly increased the osteoblastogenic factor Runx2 in both WT and KO mice. These data suggest that Nox2 regulates only bone marrow adiposity and does not alter osteoblastogenesis.

*3.4. Nox2 Mediates HFD-Induced Osteoclastogenesis.* Herein, we wanted to investigate the potential involvement of Nox2 in osteoclastogenesis. To this end, in vitro osteoclastogenesis was induced in BM cells isolated from all the experimental groups by stimulation with RANKL and MCSF. Our data indicated that BM cells from WT HFD mice formed a significantly higher number of TRAP-positive osteoclast-like cells as compared to BM cells isolated from WT SD mice (Figure 4(a)). Interestingly, this HFD-associated effect was significantly reduced when BM cells isolated from KO animals were employed, suggesting the involvement of Nox2 in HFD-induced osteoclastogenesis (Figure 4(a)). These observations are in agreement with the finding that proinflammatory and proosteoclastogenic factors such as TNF-$\alpha$, IL-6, RANKL, and CTSK mRNA levels were significantly increased in the bones of WT HFD mice but not in those of KO HFD mice (Figure 4(b)). These data support the hypothesis that Nox2 is implicated in the increased bone resorption associated with HFD. More importantly, its absence protects the mice against the Nox2-dependent deleterious effect elicited by HFD on the bone. Our data suggest that HFD-induced Nox2-dependent bone deterioration is associated with enhanced osteoclastogenesis.

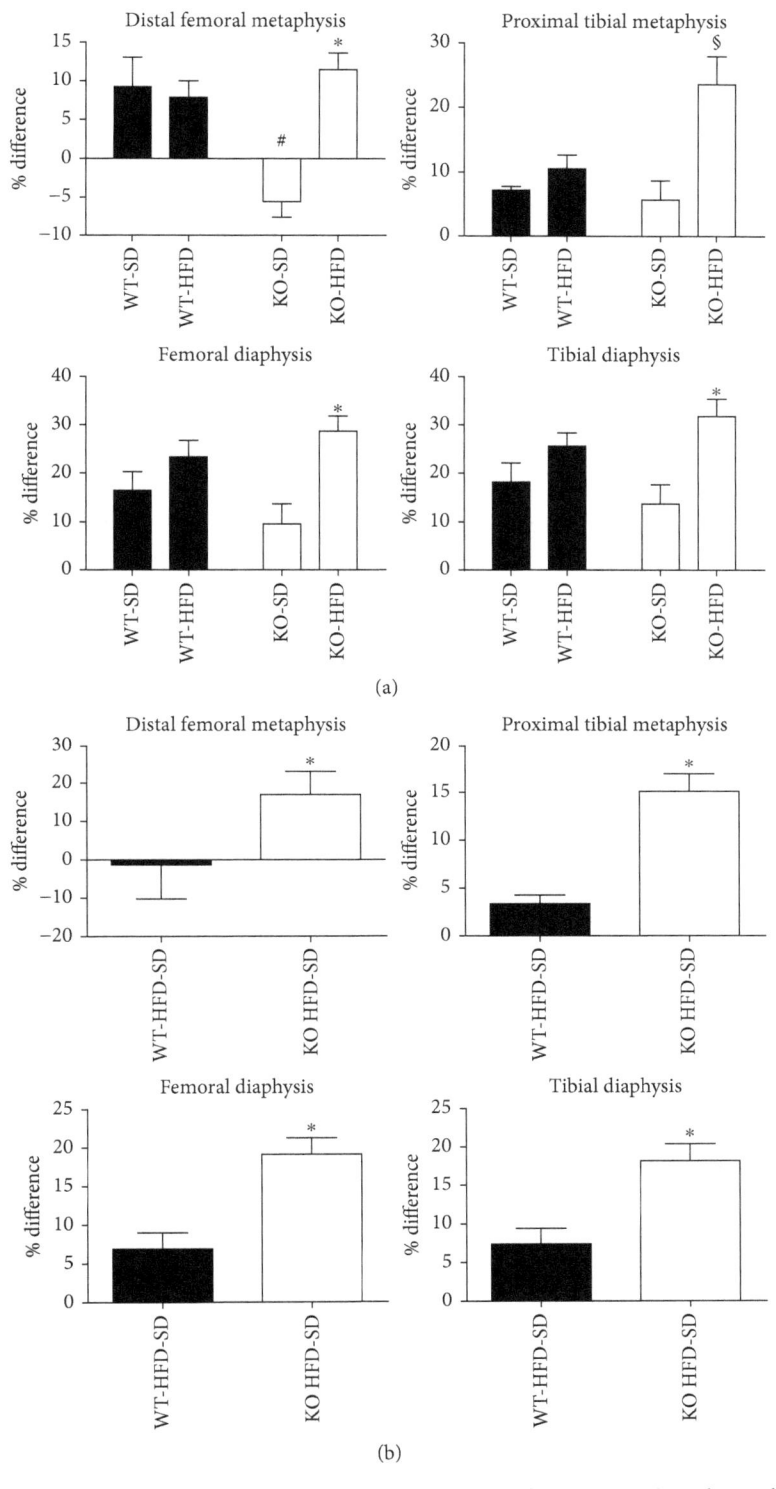

FIGURE 2: Effect of Nox2 on high-fat-diet- (HFD-) induced changes in bone mineral density (BMD). Eight-week-old WT and Nox2-KO mice were fed with either a standard lab chow diet (SD) or HFD for 3 months. Mice were scanned with dual-energy X-ray absorptiometry (DXA) before and after administration of experimental diets. (a) Percent changes in BMD before and after experimental diets in different bone regions. (b) Percentage change of the BMD difference between mice fed with an SD and an HFD in different bone regions. Data are expressed as mean ± SEM ($n = 3$, 4 mice/group). In (a), two-way ANOVA followed by post hoc Tukey's multiple comparison test was used to assess differences among groups with significance defined as $P < 0.05$. #Significantly different from all the group data in the graph. $Significantly different from all the group data in the graph. *Significantly different from SD-KO in all the graphs. In (b), unpaired $t$-test with Welch's correction has been used to assess significance between the groups with significance defined as $P < 0.05$. *Significantly different from each other in all the graph data.

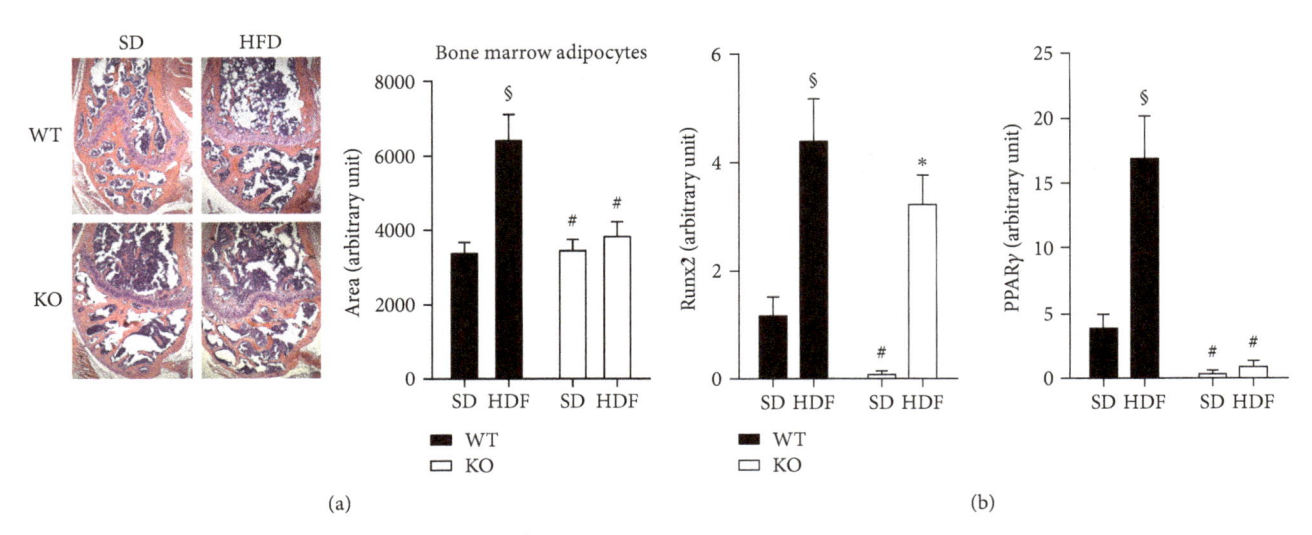

FIGURE 3: Effect of Nox2 on high-fat-diet- (HFD-) induced changes in bone marrow adiposity. Eight-week-old WT and Nox2-KO mice were fed with either an SD or an HFD for 3 months. (a) Left femurs were collected after sacrifice and fixed in 10% neutral-buffered saline (NBF). After processing, sections of bone were embedded in paraffin and stained with hematoxylin and eosin (H&E). Representative microphotographs of stained femur sections from each group are shown. Magnification ×100. The arrow shows adipocytes in the bone marrow cavity. Analysis of white areas (adipocytes) inside the bone marrow has been performed using NIH ImageJ software. (b) Right femurs were crushed under liquid nitrogen, and RNA was isolated and analyzed for mRNA levels for Runx2 and PPAR-$\gamma$ by real-time RT-PCR. mRNA Ct values for these genes were normalized to the housekeeping gene 18S. Data are expressed as mean ± SEM ($n$ = 3, 4 mice/group). Two-way ANOVA followed by post hoc Tukey's multiple comparison test was used to assess differences among groups with significance defined as $P < 0.05$. $^\S$Significantly different from SD-WT. $^\#$Significantly different from HFD-WT. *Significantly different from SD-KO.

## 4. Discussion

Using a corn oil diet in 12-month-old female mice, we previously reported that obesity increased osteoclast formation and lowered osteoblast formation directing the overall bone homeostasis toward bone resorption [17]. In the present work, our results indicate that obesity-mediated bone remodeling in favor of osteoclastogenesis can occur independently from either the aging process or the gender type [17, 45–47]. Our data demonstrated that the magnitude of the increase in BMD was significantly greater in KO mice as compared to WT mice suggesting Nox2 implication in HFD-induced bone alteration.

While the alteration in BMD measured in growing mice cannot be considered bone loss, our data revealed that the HDF-induced bone remodeling occurred in a Nox2-dependent manner. We found that a HFD was able to increase both the number of adipose cells and the expression of PPAR-$\gamma$ in the bone marrow. Interestingly, these HFD-associated effects were significantly reduced in KO mice while body weight and fat mass gain were similar to those observed in wild-type mice. This observation supports the idea that more than body weight or fat mass, the localization of fat cell accumulation is primordial in contributing to tissue lipotoxicity [48–50].

Both adipocytes and osteoblasts are generated from mesenchymal stem cells. Therefore, it has been proposed that a HFD can alter bone marrow cellular composition resulting in an increased adipocyte number and decreased osteoblasts. This imbalanced bone marrow composition would decrease osteogenesis [7], thus reducing BMD and bone mass. Our data indicate that HFD was able to increase Runx2, suggesting that osteoblastogenesis is not reduced, but on the contrary increased by this feeding regime. We think that this could be a primary adaptive mechanism to maintain bone integrity when there is an increase in the total body weight. Indeed, this is in accord with the increase in bone mineral density we observe in HFD mice (Figure 2). More importantly, we observed that only the formation of bone marrow adipocytes was affected in Nox2-knockout mice. Also, the Nox2-dependent alteration of BMD could not be attributed to osteoblastogenesis dysregulation. Since adipocytes secrete proinflammatory cytokines favoring osteoclast formation [40], we examined whether a HFD could affect osteoclastogenesis. Our result highlighted that the expression of proinflammatory and proosteoclastogenic factors such as TNF-$\alpha$, IL-6, RANKL, and CTSK was significantly increased in the bones of WT HFD mice as compared to KO mice.

We also analyzed the formation of osteoclasts in vitro using mouse-derived BM cells. RANKL activation is a pivotal step toward mature osteoclast differentiation [51, 52]. Importantly, NADPH oxidase-mediated superoxide production has been reported to execute a key role in osteoclastogenesis [33], and RANKL is known to stimulate ROS production [32]. Given their macrophage nature, all forms of osteoclasts express high amounts of Nox2 at both the protein and mRNA levels [32–34, 53]. However, osteoclasts also express Nox4 in their mature form and Nox1 in undifferentiated and differentiated preosteoclastic BM cells [35, 54]. Our data indicate that Nox2 is a major producer of ROS by BM cells and that it contributes to RANKL-mediated osteoclast formation in vitro. Interestingly, Nox1 was also shown to be

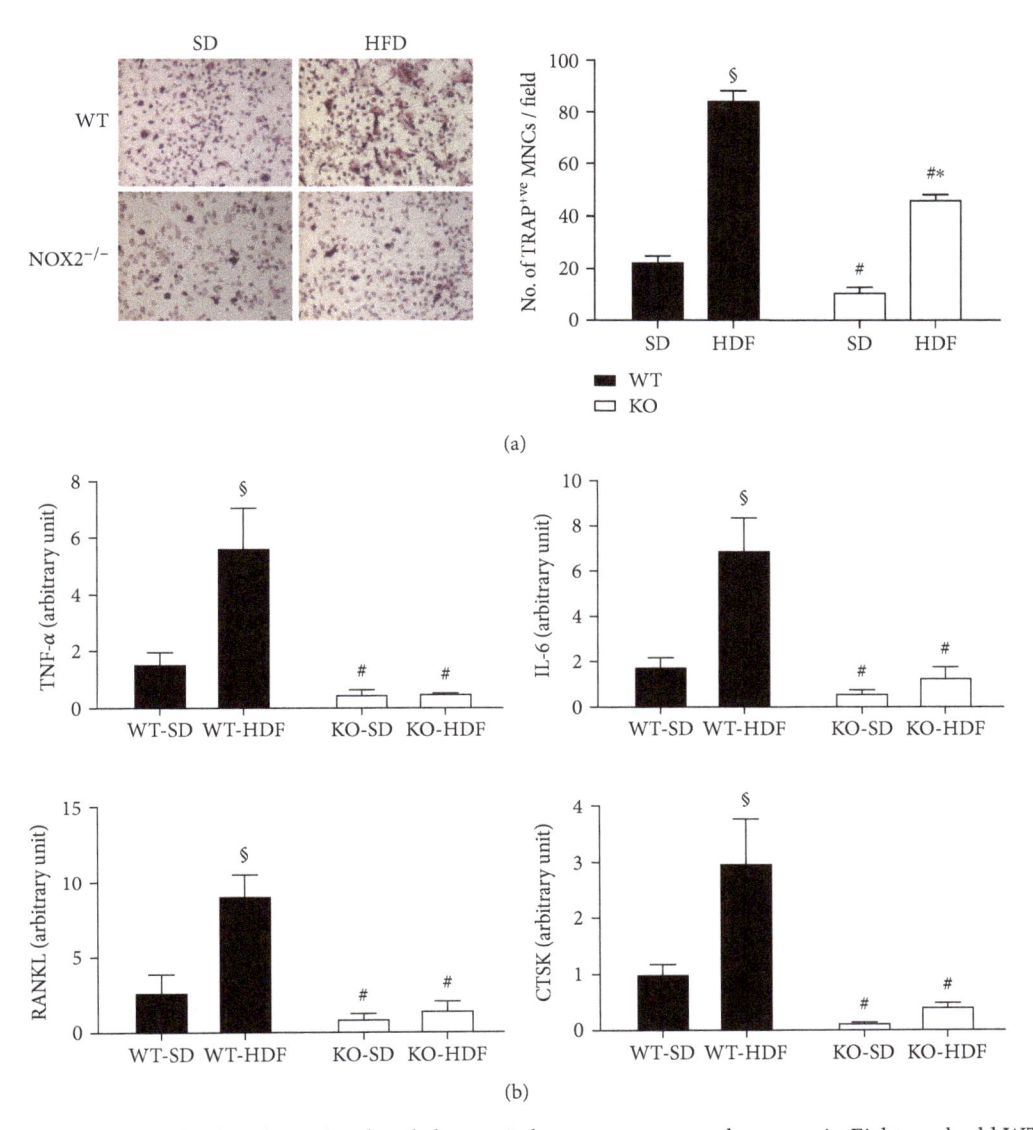

FIGURE 4: Effect of Nox2 on high-fat-diet- (HFD-) induced changes in bone marrow osteoclastogenesis. Eight-week-old WT and Nox2-KO mice were fed with an SD or an HFD for 3 months. (a) Bone marrow cells were isolated from the left femur and tibia of each group. Bone marrow cells ($1 \times 10^6$) were cultured in the presence of RANKL and MCSF for 5 days. Cells were then stained for TRAP and counted under a microscope. TRAP-positive multinucleated cells (MNCs) containing more than 3 nuclei were considered osteoclasts. Bars show the number of TRAP-positive MNCs in each group expressed as a number of cells per field. (b) Right femurs were crushed under liquid nitrogen, and RNA was isolated and analyzed for mRNA levels for TNF-$\alpha$, IL-6, RANKL, and cathepsin K (CTSK) by qRT-PCR. mRNA Ct values for these genes were normalized to the housekeeping gene 18S. Data are expressed as mean ± SEM ($n = 4$, 5 mice/group). Two-way ANOVA followed by post hoc Tukey's multiple comparison test was used to assess differences among groups with significance defined as $P < 0.05$. $^\S$Significantly different from WT-SD. $^\#$Significantly different from WT-HFD. $^*$Significantly different from SD-KO.

essential in osteoclast differentiation stimulated by RANKL [32]. In this study, Nox1 silencing significantly reduced ROS production and differentiation induced by RANKL. Given that ROS are shown to be a key intermediate in osteoclastogenesis [55], this finding is surprising because not only is Nox2 highly expressed in BM cells but also its activation generates a far greater amount of ROS. The explanation for these discrepancies might relate to the possibility that Nox1 siRNA altered the expression of Nox2. In this regard, it has been reported that increased expression of different Nox enzymes may compensate the loss of particular isoforms in osteoclasts [32, 56, 57]. Furthermore, other studies suggest that Nox1 and Nox2 might compensate for one another for RANKL-mediated stimulation of osteoclast formation [33]. Finally, recent works demonstrate that Nox4 expression is increased in Nox2-deficient osteoclasts and contributes to osteoclast formation [35, 54]. In accord with these in vitro studies, the absence of evident bone abnormality in Nox1-, Nox2-, and Nox4-deficient mice suggests that the genome redundancy can guarantee the ROS production that is necessary to maintain bone growth and homeostasis. In our study, we did not observe any significant effect of either the diet or the genotype on Nox1 and Nox4 mRNA expression levels (data not shown), suggesting that the effects observed in KO mice could reasonably be attributed to Nox2 deficiency.

In the present work, we observed that despite any gross alteration of the bone structure, WT KO mice tend to be smaller, and we observed an overall downregulation of genes controlling the bone environment (adiposity, osteoblast, osteoclast, and inflammation). However, our data reveal that under a pathological condition (i.e., HFD), Nox2 deficiency had a beneficial effect on bone. Nox2 deficiency counteracted HFD-induced bone marrow adiposity and osteoclastogenesis. Also, it is possible that the relevance and implication of each Nox enzyme in osteoclast formation and activity might be revealed only under a pathological or stress condition.

To our knowledge, the data reported in this study provide the first evidence that Nox2 mediates HFD-induced bone marrow adiposity and osteoclastogenesis without affecting osteoblastogenesis, therefore affecting bone homeostasis in favor of bone resorption.

## Authors' Contributions

Md Mizanur Rahman and Amina El Jamali contributed equally to this work. Ganesh V. Halade, Allal Ouhtit, Haissam Abou-Saleh, and Gianfranco Pintus were involved in drafting the article or revising it critically for important intellectual content.

## Acknowledgments

The authors thank Yimin Wu, Maria Gamez, and Jacob Crandall for their expert technical assistance. This work was supported by the National Institutes of Health (NIH) grants T32-HL007446, UL1-TR000149 (Clinical and Translational Science Award (CTSA), pilot funding), KL2-TR000118 (CTSA-mentored career development award), K01 DK084297 (career development award), K01-AG034233, and R01 DK033665-24 and Qatar University Grants QUUG-CAS-DBES-15/16-23 and QUCG-CHS-2018\2019. The article processing charge (APC) for the publication of this article was funded by the Qatar National Library.

## References

[1] G. Duque, "Bone and fat connection in aging bone," *Current Opinion in Rheumatology*, vol. 20, no. 4, pp. 429–434, 2008.

[2] M. Zeyda and T. M. Stulnig, "Obesity, inflammation, and insulin resistance–a mini-review," *Gerontology*, vol. 55, no. 4, pp. 379–386, 2009.

[3] R. S. Ahima, "Connecting obesity, aging and diabetes," *Nature Medicine*, vol. 15, no. 9, pp. 996–997, 2009.

[4] D. Wu, Z. Ren, M. Pae et al., "Aging up-regulates expression of inflammatory mediators in mouse adipose tissue," *Journal of Immunology*, vol. 179, no. 7, pp. 4829–4839, 2007.

[5] G. R. Mundy, "Osteoporosis and inflammation," *Nutrition Reviews*, vol. 65, 12 Part 2, pp. S147–S151, 2007.

[6] S. M. Reilly and A. R. Saltiel, "Adapting to obesity with adipose tissue inflammation," *Nature Reviews Endocrinology*, vol. 13, no. 11, pp. 633–643, 2017.

[7] O. Naveiras, V. Nardi, P. L. Wenzel, P. V. Hauschka, F. Fahey, and G. Q. Daley, "Bone-marrow adipocytes as negative regulators of the haematopoietic microenvironment," *Nature*, vol. 460, no. 7252, pp. 259–263, 2009.

[8] S. R. Oh, O. J. Sul, Y. Y. Kim et al., "Saturated fatty acids enhance osteoclast survival," *Journal of Lipid Research*, vol. 51, no. 5, pp. 892–899, 2010.

[9] M. Horowitz, Y. Xi, K. Wilson, and M. Kacena, "Control of osteoclastogenesis and bone resorption by members of the TNF family of receptors and ligands," *Cytokine & Growth Factor Reviews*, vol. 12, no. 1, pp. 9–18, 2001.

[10] I. R. Reid, "Fat and bone," *Archives of Biochemistry and Biophysics*, vol. 503, no. 1, pp. 20–27, 2010.

[11] X. Tang, G. Liu, J. Kang et al., "Obesity and risk of hip fracture in adults: a meta-analysis of prospective cohort studies," *PLoS One*, vol. 8, no. 4, article e55077, 2013.

[12] J. E. Compston, N. B. Watts, R. Chapurlat et al., "Obesity is not protective against fracture in postmenopausal women: GLOW," *The American Journal of Medicine*, vol. 124, no. 11, pp. 1043–1050, 2011.

[13] S. Tanaka, T. Kuroda, M. Saito, and M. Shiraki, "Overweight/obesity and underweight are both risk factors for osteoporotic fractures at different sites in Japanese postmenopausal women," *Osteoporosis International*, vol. 24, no. 1, pp. 69–76, 2013.

[14] P. Dimitri, N. Bishop, J. S. Walsh, and R. Eastell, "Obesity is a risk factor for fracture in children but is protective against fracture in adults: a paradox," *Bone*, vol. 50, no. 2, pp. 457–466, 2012.

[15] J. Wei, M. Ferron, C. J. Clarke et al., "Bone-specific insulin resistance disrupts whole-body glucose homeostasis via decreased osteocalcin activation," *Journal of Clinical Investigation*, vol. 124, no. 4, pp. 1–13, 2014.

[16] C. Lavet, A. Martin, M.-T. Linossier et al., "Fat and sucrose intake induces obesity-related bone metabolism disturbances: kinetic and reversibility studies in growing and adult rats," *Journal of Bone and Mineral Research*, vol. 31, no. 1, pp. 98–115, 2016.

[17] G. V. Halade, A. El Jamali, P. J. Williams, R. J. Fajardo, and G. Fernandes, "Obesity-mediated inflammatory microenvironment stimulates osteoclastogenesis and bone loss in mice," *Experimental Gerontology*, vol. 46, no. 1, pp. 43–52, 2011.

[18] J. N. Farr, M. T. Drake, S. Amin, L. J. Melton III, L. K. McCready, and S. Khosla, "In vivo assessment of bone quality in postmenopausal women with type 2 diabetes," *Journal of Bone and Mineral Research*, vol. 29, no. 4, pp. 787–795, 2014.

[19] P. Gerdhem, A. Isaksson, K. Akesson, and K. J. Obrant, "Increased bone density and decreased bone turnover, but no evident alteration of fracture susceptibility in elderly women with diabetes mellitus," *Osteoporosis International*, vol. 16, no. 12, pp. 1506–1512, 2005.

[20] J. K. Pepping, L. R. Freeman, S. Gupta, J. N. Keller, and A. J. Bruce-Keller, "NOX2 deficiency attenuates markers of adiposopathy and brain injury induced by high-fat diet," *American Journal of Physiology Endocrinology and Metabolism*, vol. 304, no. 4, pp. E392–E404, 2013.

[21] S. X. L. Zhang, A. Khalyfa, Y. Wang et al., "Sleep fragmentation promotes NADPH oxidase 2-mediated adipose tissue inflam-

mation leading to insulin resistance in mice," *International Journal of Obesity*, vol. 38, no. 4, pp. 619–624, 2014.

[22] C. W. Lee, C. C. Lin, I. T. Lee, H. C. Lee, and C. M. Yang, "Activation and induction of cytosolic phospholipase A2 by TNF-α mediated through Nox2, MAPKs, NF-κB, and p300 in human tracheal smooth muscle cells," *Journal of Cellular Physiology*, vol. 226, no. 8, pp. 2103–2114, 2011.

[23] J. L. Wilkinson-Berka, I. Rana, R. Armani, and A. Agrotis, "Reactive oxygen species, Nox and angiotensin II in angiogenesis: implications for retinopathy," *Clinical Science*, vol. 124, no. 10, pp. 597–615, 2013.

[24] C. Wilson, "Bone: oxidative stress and osteoporosis," *Nature Reviews Endocrinology*, vol. 10, p. 3, 2014.

[25] K. Schroder, "NADPH oxidases in bone homeostasis and osteoporosis," *Cellular and Molecular Life Sciences*, vol. 72, no. 1, pp. 25–38, 2015.

[26] M. Almeida, L. Han, M. Martin-Millan et al., "Skeletal involution by age-associated oxidative stress and its acceleration by loss of sex steroids," *Journal of Biological Chemistry*, vol. 282, no. 37, pp. 27285–27297, 2007.

[27] J. M. Lean, J. T. Davies, K. Fuller et al., "A crucial role for thiol antioxidants in estrogen-deficiency bone loss," *Journal of Clinical Investigation*, vol. 112, no. 6, pp. 915–923, 2003.

[28] S. A. Sheweita and K. I. Khoshhal, "Calcium metabolism and oxidative stress in bone fractures: role of antioxidants," *Current Drug Metabolism*, vol. 8, no. 5, pp. 519–525, 2007.

[29] J. D. Lambeth, T. Kawahara, and B. Diebold, "Regulation of Nox and Duox enzymatic activity and expression," *Free Radical Biology & Medicine*, vol. 43, no. 3, pp. 319–331, 2007.

[30] C. A. O'Brien, T. Nakashima, and H. Takayanagi, "Osteocyte control of osteoclastogenesis," *Bone*, vol. 54, no. 2, pp. 258–263, 2013.

[31] G. D. Roodman, "Regulation of osteoclast differentiation," *Annals of the New York Academy of Sciences*, vol. 1068, no. 1, pp. 100–109, 2006.

[32] N. K. Lee, Y. G. Choi, J. Y. Baik et al., "A crucial role for reactive oxygen species in RANKL-induced osteoclast differentiation," *Blood*, vol. 106, no. 3, pp. 852–859, 2005.

[33] H. Sasaki, H. Yamamoto, K. Tominaga et al., "Receptor activator of nuclear factor-kappaB ligand-induced mouse osteoclast differentiation is associated with switching between NADPH oxidase homologues," *Free Radical Biology & Medicine*, vol. 47, no. 2, pp. 189–199, 2009.

[34] H. Sasaki, H. Yamamoto, K. Tominaga et al., "NADPH oxidase-derived reactive oxygen species are essential for differentiation of a mouse macrophage cell line (RAW264.7) into osteoclasts," *The Journal of Medical Investigation*, vol. 56, no. 1,2, pp. 33–41, 2009.

[35] S. Yang, P. Madyastha, S. Bingel, W. Ries, and L. Key, "A new superoxide-generating oxidase in murine osteoclasts," *Journal of Biological Chemistry*, vol. 276, no. 8, pp. 5452–5458, 2001.

[36] C. C. Mandal, S. Ganapathy, Y. Gorin et al., "Reactive oxygen species derived from Nox4 mediate BMP2 gene transcription and osteoblast differentiation," *Biochemical Journal*, vol. 433, no. 2, pp. 393–402, 2011.

[37] Y. Wittrant, Y. Gorin, S. Mohan, B. Wagner, and S. L. Abboud-Werner, "Colony-stimulating factor-1 (CSF-1) directly inhibits receptor activator of nuclear factor-κB ligand (RANKL) expression by osteoblasts," *Endocrinology*, vol. 150, no. 11, pp. 4977–4988, 2009.

[38] F. Wauquier, L. Leotoing, V. Coxam, J. Guicheux, and Y. Wittrant, "Oxidative stress in bone remodelling and disease," *Trends in Molecular Medicine*, vol. 15, no. 10, pp. 468–477, 2009.

[39] L. L. Key Jr, W. C. Wolf, C. M. Gundberg, and W. L. Ries, "Superoxide and bone resorption," *Bone*, vol. 15, no. 4, pp. 431–436, 1994.

[40] G. V. Halade, M. M. Rahman, P. J. Williams, and G. Fernandes, "High fat diet-induced animal model of age-associated obesity and osteoporosis," *The Journal of Nutritional Biochemistry*, vol. 21, no. 12, pp. 1162–1169, 2010.

[41] M. M. Rahman, G. Fernandes, and P. Williams, "Conjugated linoleic acid prevents ovariectomy-induced bone loss in mice by modulating both osteoclastogenesis and osteoblastogenesis," *Lipids*, vol. 49, no. 3, pp. 211–224, 2014.

[42] M. M. Rahman, A. Bhattacharya, J. Banu, J. X. Kang, and G. Fernandes, "Endogenous n-3 fatty acids protect ovariectomy induced bone loss by attenuating osteoclastogenesis," *Journal of Cellular and Molecular Medicine*, vol. 13, no. 8b, pp. 1833–1844, 2009.

[43] M. M. Rahman, J. M. Veigas, P. J. Williams, and G. Fernandes, "DHA is a more potent inhibitor of breast cancer metastasis to bone and related osteolysis than EPA," *Breast Cancer Research and Treatment*, vol. 141, no. 3, pp. 341–352, 2013.

[44] A. El Jamali, A. J. Valente, J. D. Lechleiter et al., "Novel redox-dependent regulation of NOX5 by the tyrosine kinase c-Abl," *Free Radical Biology & Medicine*, vol. 44, no. 5, pp. 868–881, 2008.

[45] K. M. McTigue, R. Hess, and J. Ziouras, "Obesity in older adults: a systematic review of the evidence for diagnosis and treatment," *Obesity*, vol. 14, no. 9, pp. 1485–1497, 2006.

[46] M. Kawai, M. J. Devlin, and C. J. Rosen, "Fat targets for skeletal health," *Nature Reviews Rheumatology*, vol. 5, no. 7, pp. 365–372, 2009.

[47] M. Kawai and C. J. Rosen, "Bone: adiposity and bone accrual-still an established paradigm?," *Nature Reviews Endocrinology*, vol. 6, no. 2, pp. 63–64, 2010.

[48] D. T. Villareal, C. M. Apovian, R. F. Kushner, and S. Klein, "Obesity in older adults: technical review and position statement of the American Society for Nutrition and NAASO, the Obesity Society," *The American Journal of Clinical Nutrition*, vol. 82, no. 5, pp. 923–934, 2005.

[49] D. T. Felson, Y. Zhang, M. T. Hannan, D. P. Kiel, P. Wilson, and J. J. Anderson, "The effect of postmenopausal estrogen therapy on bone density in elderly women," *New England Journal of Medicine*, vol. 329, no. 16, pp. 1141–1146, 1993.

[50] L. J. Zhao, H. Jiang, C. J. Papasian et al., "Correlation of obesity and osteoporosis: effect of fat mass on the determination of osteoporosis," *Journal of Bone and Mineral Research*, vol. 23, no. 1, pp. 17–29, 2008.

[51] X. Feng, "Regulatory roles and molecular signaling of TNF family members in osteoclasts," *Gene*, vol. 350, no. 1, pp. 1–13, 2005.

[52] M. M. Rahman, A. Kukita, T. Kukita, T. Shobuike, T. Nakamura, and O. Kohashi, "Two histone deacetylase inhibitors, trichostatin A and sodium butyrate, suppress differentiation into osteoclasts but not into macrophages," *Blood*, vol. 101, no. 9, pp. 3451–3459, 2003.

[53] S. Yang, W. L. Ries, and L. L. Key Jr, "Nicotinamide adenine dinucleotide phosphate oxidase in the formation of superoxide

in osteoclasts," *Calcified Tissue International*, vol. 63, no. 4, pp. 346–350, 1998.

[54] S. Yang, Y. Zhang, W. Ries, and L. Key, "Expression of Nox4 in osteoclasts," *Journal of Cellular Biochemistry*, vol. 92, no. 2, pp. 238–248, 2004.

[55] I. R. Garrett, B. F. Boyce, R. O. Oreffo, L. Bonewald, J. Poser, and G. R. Mundy, "Oxygen-derived free radicals stimulate osteoclastic bone resorption in rodent bone in vitro and in vivo," *Journal of Clinical Investigation*, vol. 85, no. 3, pp. 632–639, 1990.

[56] G. Gavazzi, B. Banfi, C. Deffert et al., "Decreased blood pressure in NOX1-deficient mice," *FEBS Letters*, vol. 580, no. 2, pp. 497–504, 2006.

[57] K. Matsuno, H. Yamada, K. Iwata et al., "Nox1 is involved in angiotensin II-mediated hypertension: a study in Nox1-deficient mice," *Circulation*, vol. 112, no. 17, pp. 2677–2685, 2005.

# Upregulation of Heme Oxygenase-1 by Hemin Alleviates Sepsis-Induced Muscle Wasting in Mice

Xiongwei Yu,[1,2] Wenjun Han,[1] Changli Wang,[1] Daming Sui,[1,3] Jinjun Bian®,[1] Lulong Bo®,[1] and Xiaoming Deng®[1]

[1]*Faculty of Anesthesiology, Changhai Hospital, Naval Medical University, Shanghai 200433, China*
[2]*Department of Anesthesiology, 285th Hospital of the CPLA, Handan 056001, China*
[3]*Department of Anesthesiology, Chengdu Military General Hospital, Chengdu 610083, China*

Correspondence should be addressed to Lulong Bo; nbastars@126.com and Xiaoming Deng; deng_x@yahoo.com

Guest Editor: Marco Sandri

Hemin, an inducer of heme oxygenase-1 (HO-1), can enhance the activation of HO-1. HO-1 exhibits a variety of activities, such as anti-inflammatory, antioxidative, and antiapoptotic functions. The objective of this study was to investigate the effects of hemin on sepsis-induced skeletal muscle wasting and to explore the mechanisms by which hemin exerts its effects. Cecal ligation and perforation (CLP) was performed to create a sepsis mouse model. Mice were randomly divided into four groups: control, CLP, CLP plus group, and CLP-hemin-ZnPP (a HO-1 inhibitor). The weight of the solei from the mice was measured, and histopathology was examined. Cytokines were measured by enzyme-linked immunosorbent assay (ELISA). Real-time quantitative reverse transcription polymerase chain reaction (qRT-PCR) and Western blotting were used to assess the expression levels of HO-1 and atrogin-1. Furthermore, we investigated the antioxidative effects of HO-1 by detecting malondialdehyde (MDA) levels and superoxide dismutase (SOD) activity. CLP led to dramatic skeletal muscle weakness and atrophy, but pretreatment with hemin protected mice against CLP-mediated muscle atrophy. Hemin also induced high HO-1 expression, which resulted in suppressed proinflammatory cytokine and reactive oxygen species (ROS) production. The expression of MuRF1 and atrogin-1, two ubiquitin ligases of the ubiquitin-proteasome system- (UPS-) mediated proteolysis, was also inhibited by increased HO-1 levels. Hemin-mediated increases in HO-1 expression exert protective effects on sepsis-induced skeletal muscle atrophy at least partly by inhibiting the expression of proinflammatory cytokines, UPS-mediated proteolysis, and ROS activation. Therefore, hemin might be a new treatment target against sepsis-induced skeletal muscle atrophy.

## 1. Introduction

Sepsis is defined as a life-threatening organ dysfunction due to a dysregulated host response to infection [1]. In the United States, nearly 10% of all deaths result from severe sepsis or its related complications every year [2]. Skeletal muscle atrophy and muscle weakness occurring from sepsis have become recognized as important issues in sepsis survivors [3]. A large number of critically ICU patients suffer from severe muscle wasting and impaired muscle function, which can delay respirator weaning and persist long after hospital discharge, thus reducing the patients' quality of life [4, 5].

Muscle atrophy results from an imbalance between muscle proteolysis and protein synthesis. When proteolysis overwhelms protein synthesis, muscle atrophy occurs [6, 7]. Protein degradation within muscle appears to rely on three pathways: ubiquitin-proteasome system- (UPS-) mediated proteolysis, autophagy, and calcium-dependent calpains [8]. However, the pathway that has received the most attention is the UPS-mediated proteolysis, which is believed to play a dominant role in skeletal muscle atrophy [9]. Two ubiquitin ligases, MuRF1 and atrogin-1, are key positive regulators of UPS-mediated proteolysis and are upregulated in all rodent models of skeletal muscle atrophy [10–12]. Additionally, these proteins have been widely used as markers of muscle wasting. Sepsis-induced cytokine secretion can also enhance microvascular permeability, allowing circulating toxins to impair axon activity [13]. The nutrition deficiency in muscle caused by impaired axons may lead to muscle atrophy. As some myofibrillar proteins possess sulfhydryl groups that are sensitive to oxidation, sepsis-induced reactive oxygen species (ROS) appear to contribute to muscle wasting [14]. Thus, inhibiting proinflammatory cytokines and ROS should be an effective method to reverse muscle wasting.

Heme oxygenase-1 (HO-1), also called heat shock protein 32 (Hsp32), is an inducible enzyme that can convert heme into carbon monoxide, biliverdin, and free iron [15, 16]. Recent findings reported that HO-1 and its metabolites exerted anti-inflammatory, antioxidative, and antiapoptotic activities [17, 18]. As metabolites of HO-1, CO and biliverdin were shown to contribute to stimulating the host defense response against sepsis and modulating inflammatory mediators in mice [18]. Previous studies support the beneficial effects of HO-1 and its product in an experimental model of sepsis [19]. We hypothesized that the induction of HO-1 plays a pivotal role in sepsis-induced skeletal muscle wasting.

In our study, we used hemin as an inducer of HO-1 and examined whether hemin exerts a protective effect against septic muscle atrophy in mice. We also investigated the potential mechanism of its protective effect.

## 2. Materials and Methods

*2.1. Sepsis Model.* Cecal ligation and perforation (CLP) was performed on 8-week-old male C57BL/6 mice obtained from the Experimental Animal Center of the Naval Medical University. All animals were fed a standard laboratory diet and water and were acclimatized for at least 1 week before use. All experimental procedures involving animals were approved by the Animal Care and Use Committee of the Second Military Medical University. After we anesthetized the mice with 2%–3% sevoflurane, a midline laparotomy was performed, and the cecum was exposed. The contents of the intestines were extruded to the tip of the cecum, and then the cecum was ligated 1 cm from the tip with a 3-0 silk suture. We performed a double puncture of the cecum wall with a 22-gauge needle. The abdominal wall was closed with a continuous 3-0 silk suture in two layers. Sham-operated mice were subjected to exposure of the peritoneum and cecum but did not undergo ligation or puncture. No antibiotics were used. The mortality of the septic mice is 25% at day 1, while it elevated to 50% at day 7, indicating that the severity is moderate to severe.

*2.2. Chemical Treatment and Experimental Groups.* Hemin and zinc protoporphyrin-IX (ZnPP, a HO-1 inhibitor) (Sigma-Aldrich, St. Louis, MO) were dissolved in dimethyl sulfoxide (DMSO) and then diluted with phosphate-buffered saline (PBS) to a concentration of 10 mg/ml. Mice were randomly divided into four groups: control group (A), CLP group (B), CLP plus hemin group (C), and CLP-hemin-ZnPP group (D). Mice in groups A and B were intraperitoneally (i.p.) injected with 200 μl PBS containing 2% DMSO 1 day before surgery. Mice in group C were injected with the same volume of hemin solution (50 mg/kg, i.p.) 1 day before surgery, and mice in group D were injected with the same volume of a mixture of hemin (50 mg/kg, i.p.) and ZnPP (20 mg/kg, i.p.). We provided the same quantity of food and water to all four groups.

*2.3. Measurements of Muscle Mass and Protein Breakdown Rates.* The skin of the hind limbs was stripped for gross comparisons of muscle mass. The solei of the mice were collected and weighed at baseline and at 1, 3, 5, and 7 days after surgery. We interpreted the weight of soleus as an evaluation of muscle mass loss. Twenty-four hours after CLP or sham surgery, the solei of the mice were dissected and incubated for 2 hours in a shaking water bath at 37°C as previously described [20, 21]. Because tyrosine cannot be synthesized or degraded in muscle, protein breakdown rates were determined by testing the net release of free tyrosine into the incubation medium containing cycloheximide, a compound that prevents the reincorporation of tyrosine into protein.

*2.4. Histopathology Examination.* Tibialis anterior (TA) tissues were fixed with 10% buffered formalin at room temperature, embedded in paraffin, and sliced into 5 μm sections. The sections were then stained with hematoxylin-eosin (H&E) for morphological evaluation under a light microscope (Leica, DM-IL-LED). We observed the transverse sections taken at the midpoints of the legs. The muscle fiber size was recorded by measuring the cross-sectional area (CSA) of the skeletal muscle fiber.

*2.5. Real-Time Quantitative Reverse Transcription Polymerase Chain Reaction (qRT-PCR).* Six mice from each group were randomly killed 1, 4, and 7 days after CLP, after which the TA muscles were harvested. Total RNA was extracted from the TA tissue using TRIzol reagent (Invitrogen) according to the manufacturer's instructions. cDNA was synthesized using a PrimeScript™ 1st strand cDNA synthesis kit (TaKaRa). Quantitative PCR was performed using a SYBR green PCR kit (Invitrogen, Carlsbad, CA). The PCR conditions were as follows: initial heating at 95°C for 30 s, 40 cycles of denaturation at 95°C for 5 s and annealing at 60°C for 30 s, one extension at 95°C for 15 s, and a final ta10extension at 60°C for 30 s. The following primers were used: 5′-TGTGGGTGTATCGGATGGAG-3′ and 5′-GGCAGAGTCTTCCACAGT-3′ for atrogin-1, 5′-TTTGACACCCTCTACGCCAT-3′ and 5′-TTGGCACTTGAGAGAGAGGAAGGT-3′ for MuRF1, and 5′-TTCAGAAGGGTCAGGTCC-3′ and 5′-CAGTGAGGCCCATACCAGAA-3′ for HO-1.

The generation of PCR products was identified by melting-curve analysis. Relative mRNA levels of atrogin-1, MuRF1, and HO-1 were normalized to those of GAPDH.

*2.6. HO-1 Activity Assay.* The activity of HO-1 was identified via bilirubin generation according to the method described by Gong et al. [18]. Briefly, frozen TA samples were homogenized in lysis buffer. A sample fraction was harvested and washed by multiple centrifugations. The pellet was solubilized in 0.1 M $K_2HPO_3$ by sonication and stored at −80°C until extraction with chloroform. The extracted bilirubin

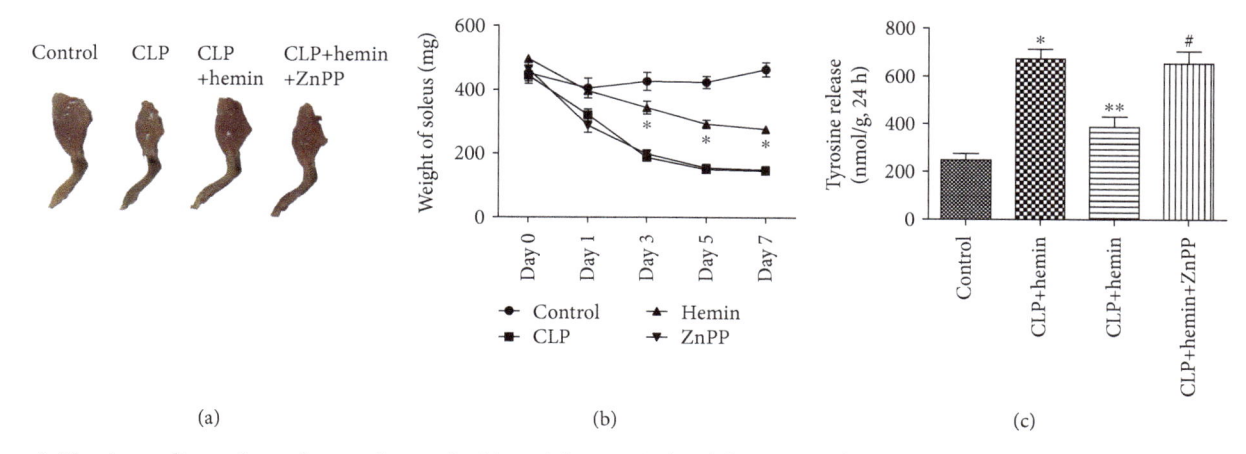

FIGURE 1: Hemin ameliorated muscle mass loss and mitigated the protein breakdown rates. (a) Gross comparisons of muscle mass (8-week-old mice, 3 days after CLP surgery). (b) The weight of the solei from mice in the hemin group was improved, especially on days 3, 5, and 7 after surgery. $^*p < 0.001$ (hemin vs. CLP), $n = 6$. (c) Protein breakdown rates as measured by the tyrosine concentration. $^*p < 0.001$ (control vs. CLP), $^{**}p < 0.001$ (hemin vs. CLP), $^\#p < 0.001$ (ZnPP vs. hemin), $n = 6$.

was measured by the difference in absorbance at 464 nm and 530 nm.

*2.7. Cytokine Analysis.* Blood was collected via cardiac puncture with heparin-treated syringe needles and centrifuged at $1000g$ for 10 min to harvest the serum. Sera were assayed for mouse TNF-$\alpha$ and IL-6 with a murine enzyme-linked immunosorbent assay (ELISA) kit (eBioscience, San Diego, CA) as described by the manufacturer.

*2.8. Western Blotting Analysis.* Briefly, the TA tissues were homogenized, and the proteins were resolved on polyacrylamide SDS gels and electrophoretically transferred to polyvinylidene difluoride membranes. The membranes were blocked with 5% (w/v) fat-free milk in Tris-buffered saline containing 0.05% Tween-20, incubated with Abs (Abcam) against mouse HO-1, atrogin-1, or MuRF1 at 4°C overnight, and finally incubated with secondary Abs. After the immunoreactive protein bands were visualized, the protein levels were normalized to the band density of $\beta$-actin (Abcam), which served as an internal control. The corresponding semiquantitative analysis was performed by measuring the optical density using the ImageJ software, and $\beta$-actin was used as an internal control.

*2.9. Assay of TA Lipid Peroxidation and Antioxidant Enzyme Activities.* Malondialdehyde (MDA), the end product of lipid peroxidation, is a marker of tissue peroxidation. We evaluated the degree of tissue peroxidation by measuring the MDA levels. Superoxide dismutase (SOD) activity was measured to analyze the activity of antioxidant enzymes. At 24 hours after CLP, we harvested the TA muscle and measured the MDA levels and SOD activity according the methods described by Andrianjafiniony et al. [22].

*2.10. Statistical Analysis.* The study results are presented as the mean ± standard deviation (SD). Differences between group means were calculated by one-way analysis of variance (ANOVA) or Student's $t$-test. All data were analyzed using Prism 5.0 (GraphPad Software, USA). $p < 0.05$ was considered statistically significant.

## 3. Results

*3.1. Administration of Hemin Ameliorated the Loss of Muscle Mass.* As shown in Figure 1(a), the size of the hind limbs was significantly atrophic after CLP surgery. However, administration of hemin partly reduced this effect, but treatment with ZnPP, an inhibitor of HO-1, reversed the effects of hemin. The weight of the solei among the groups was not significantly different at baseline. As the time after surgery increased, only slight soleus weight loss was observed in the control group. However, there was a significant deterioration in the weight of the solei from septic mice. Administration of hemin ameliorated this loss in muscle mass, while ZnPP abrogated this protective effect (Figure 1(b)). These results suggested that hemin exerts a protective effect against the loss of muscle mass. Protein breakdown rates in septic mice increased by nearly 2-fold. There was no significant difference in the protein breakdown rates between the control and hemin groups, and the protein breakdown rates in the ZnPP group were similar to those in the CLP group (Figure 1(c)).

*3.2. Hemin Pretreatment Improved Sepsis-Induced Pathological Injury of TA Tissues of Mice.* The TA tissues from the CLP group revealed smaller myofibers and CSA; however, the muscle fibers from the hemin group were in better condition than those from the CLP group (Figure 2(a)). We used the CSA of the muscle fibers to assess the degree of muscle atrophy. The CSA of the fibers from the CLP group was significantly smaller than that of the fibers from the control group. The administration of hemin positively affected the myofiber size, whereas mice in the ZnPP group showed a similar myofiber size to mice in the CLP group (Figure 2(b)).

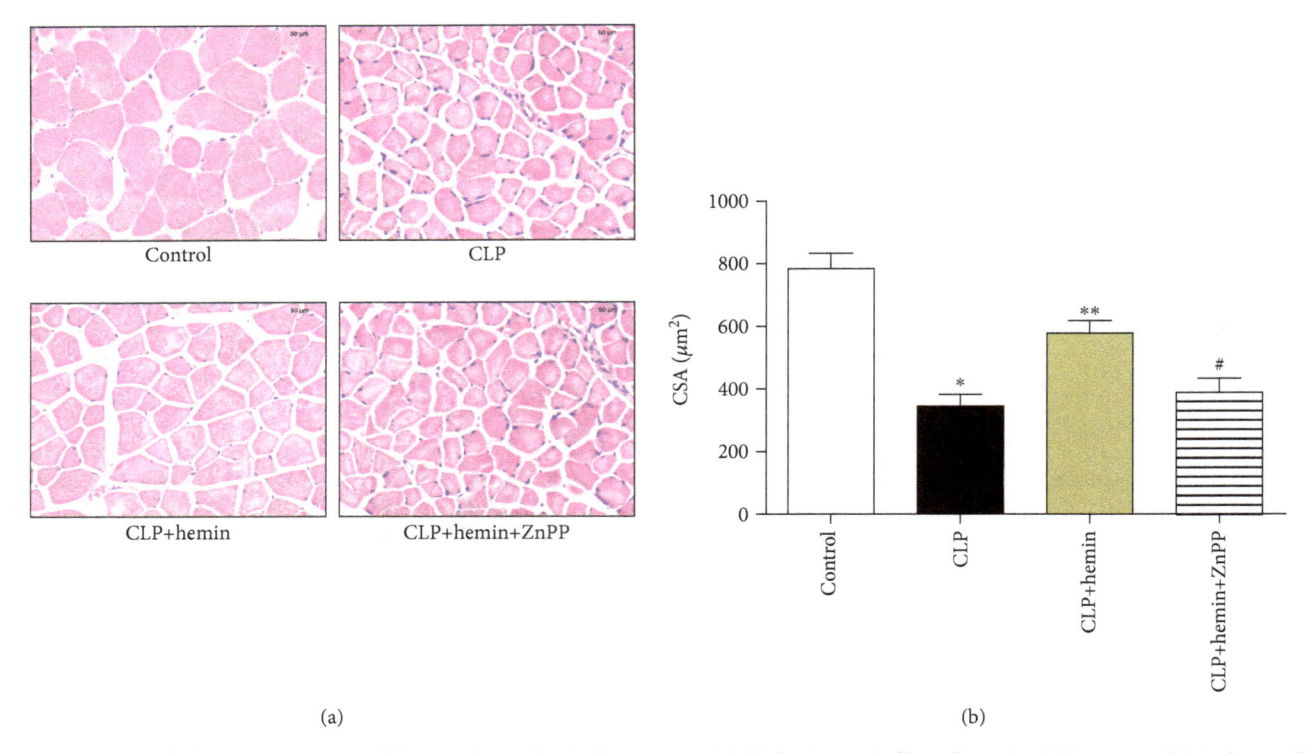

FIGURE 2: Histopathology examination of TA muscles at day 5 after surgery. (a) Skeletal muscle fibers from the CLP group exhibited a smaller cross-sectional area (CSA) than those from the control and hemin groups (H&E staining, 40x magnification, scale bar = 50 μm). (b) Quantification of the CSA of TA muscles from the different groups. *$p < 0.001$ (CLP vs. the control group), **$p < 0.01$ (hemin vs. the CLP group), #$p < 0.05$ (ZnPP vs. the hemin group), $n = 6$.

*3.3. Hemin Upregulated the Expression and Activity of HO-1 in TA.* The qRT-PCR results showed that HO-1 levels and activity were higher in the CLP group than in the sham group ($p < 0.05$), indicating that sepsis could induce HO-1 activation and its activity. We also found that hemin administration significantly increased the HO-1 levels and its activation and that the change in HO-1 expression in the hemin group was higher than that in the control and CLP groups (Figure 3(a)). ZnPP, an inhibitor of HO-1, suppressed HO-1 activation but did not affect hemin-induced upregulation of HO-1 expression (Figure 3(b)). We also found that the protein expression of HO-1 was slightly elevated in the CLP group compared with that in the control group. Pretreatment with hemin significantly enhanced HO-1 protein expression, but ZnPP administration did not affect the increased expression of HO-1 (Figures 3(c) and 3(d)).

*3.4. Hemin Contributed to the Reduction in Proinflammatory Cytokine Levels.* Because changes in proinflammatory cytokine levels play an important role in sepsis and may be involved in muscle atrophy, we examined the production of cytokines. The plasma levels of the proinflammatory cytokines TNF-α and IL-6 were significantly upregulated in mice that underwent the CLP procedure compared with those in mice from the sham group (Figure 4). High HO-1 expression induced by hemin could suppress these increases, and the serum levels of TNF-α and IL-6 did not show a significant difference between the CLP and ZnPP groups.

*3.5. Hemin Enhanced the Antioxidant Defense Response.* Compared with the control group, the CLP group showed a statistically significant increase in the plasma levels of MDA. MDA levels were lower in the hemin-pretreated mice than in mice in the CLP group. MDA levels in mice in the ZnPP group were slightly higher than those in mice in the CLP group, but there was no statistically significant difference (Figure 5(a)). Muscle SOD activity in mice in the hemin group was obviously higher than that in mice in the CLP group, and the protective function of hemin was abrogated by ZnPP (Figure 5(b)).

*3.6. Hemin Inhibited the Expression of Muscle Atrophy Markers.* Atrogin-1 and MuRF1 function as muscle-specific ubiquitin E3 ligases that tag proteins destined for ubiquitin-proteasomal proteolysis; thus, they are two of the most important genes upregulated in the process of muscle atrophy. Figures 6(a) and 6(b) show that the mRNA expression levels of atrogin-1 and MuRF1 were significantly upregulated in the TA tissues from CLP mice, especially at 1 day and 4 days after surgery. The enhanced expression of HO-1 induced by hemin reduced the expression of atrogin-1 and MuRF1 in the hemin group compared with that in the CLP group. The expression of the two ubiquitin E3 ligases was higher in the ZnPP group than in the HO-1 group. As shown in Figures 6(c) and 6(d), the protein expression of atrogin-1 and MuRF1 was elevated in the CLP group compared to the observed levels in the control group. Pretreatment with

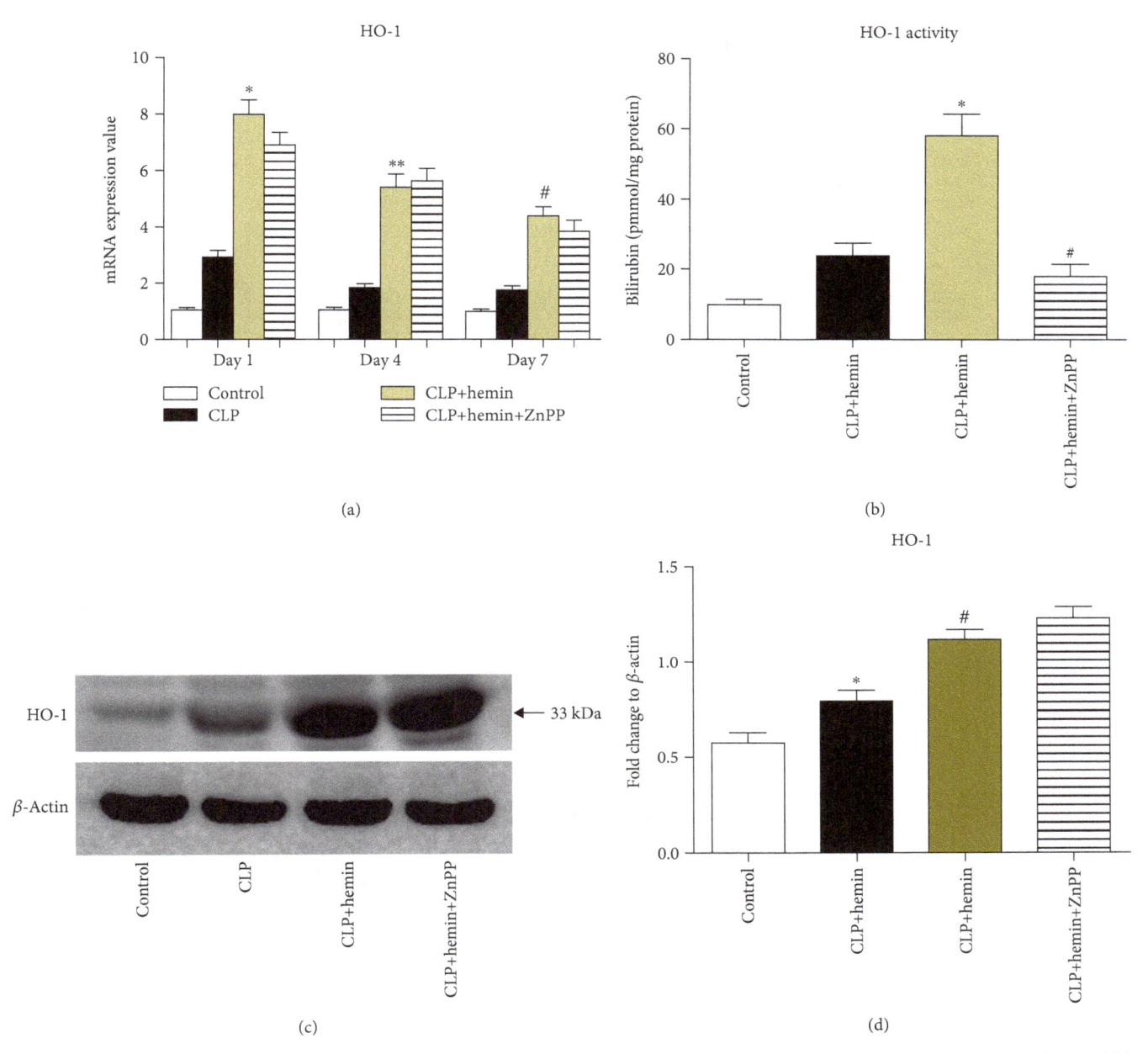

FIGURE 3: Hemin induces high HO-1 expression and increased HO-1 activation. (a) Hemin pretreatment can promote HO-1 mRNA expression at 1, 4, and 7 days after surgery ($^*p < 0.001$ vs. CLP at 1 day, $^{**}p < 0.001$ vs. CLP at 4 days, $^\#p < 0.05$ vs. CLP at 7 days). (b) Hemin enhanced the enzyme activity of HO-1, but ZnPP treatment reversed this effect ($^*p < 0.01$ vs. CLP, $^\#p < 0.01$ vs. CLP + hemin; $n = 6$). (c) Western blot analysis of levels of the HO-1 protein. (d) Results of the corresponding semiquantitative analysis of levels of the HO-1 protein based on the optical density measured using the ImageJ software; the data are presented as means ± SEM and are representative of three separate experiments ($^*p < 0.01$ vs. control, $^\#p < 0.01$ vs. CLP).

hemin dampened the protein expression of these two E3 ligases. However, in the ZnPP group, the protein expression of atrogin-1 and MuRF1 was upregulated.

## 4. Discussion

In our study, we used the CLP mouse model of sepsis to investigate the protective role of hemin in skeletal muscle wasting. Mice in the CLP group showed reduced muscle force and severe pathological atrophy of myosin fibers. Hemin succeeded in promoting the expression of HO-1 in CLP mice. Compared with mice in the CLP group, hemin-pretreated mice showed an apparent improvement in skeletal muscle force and ameliorated muscle atrophy.

Many factors, such as aging, weightlessness, immobilization, corticosteroids, hyperglycemia, and sepsis, may cause loss of muscle mass and muscle atrophy [23]. Previous studies have shown that early initiation of physical rehabilitation and minimization of deep sedation may prevent disuse atrophy and muscle weakness [24]. Muscle atrophy in critically ill

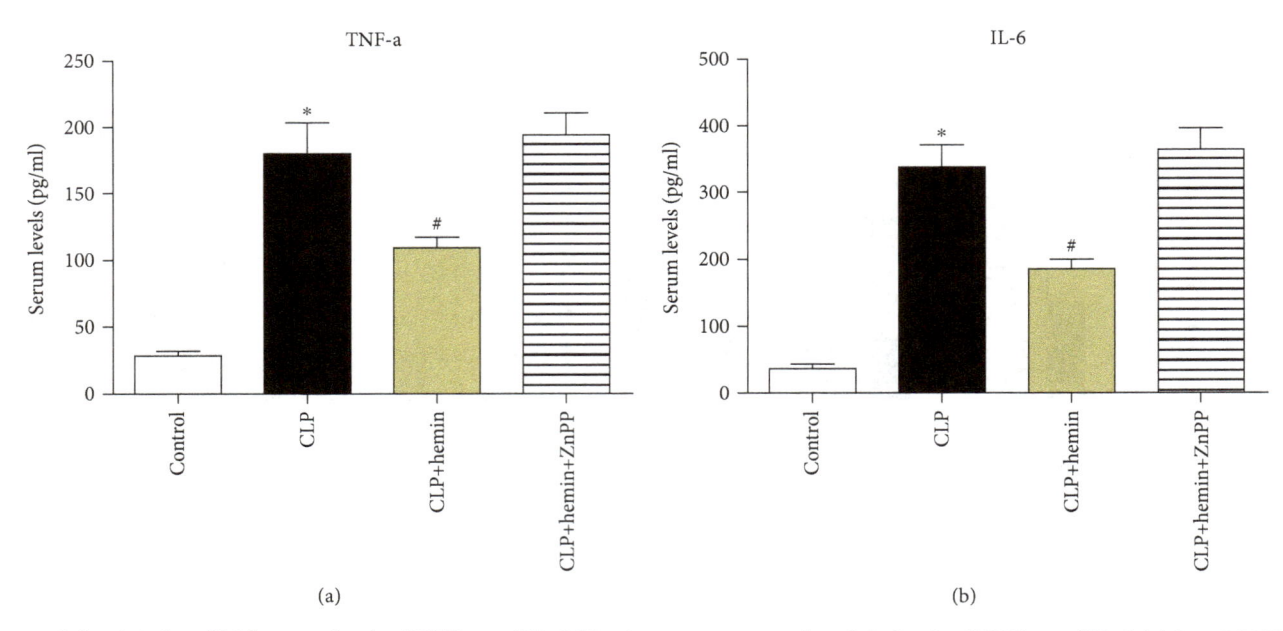

FIGURE 4: Sepsis induced high serum levels of TNF-α and IL-6. Hemin pretreatment reduced the levels of TNF-α and IL-6. (a) *$p < 0.0001$ vs. control, #$p < 0.05$ vs. CLP; $n = 6$. (b) *$p < 0.0001$ vs. control, #$p < 0.01$ vs. CLP; $n = 6$ (1 day after surgery).

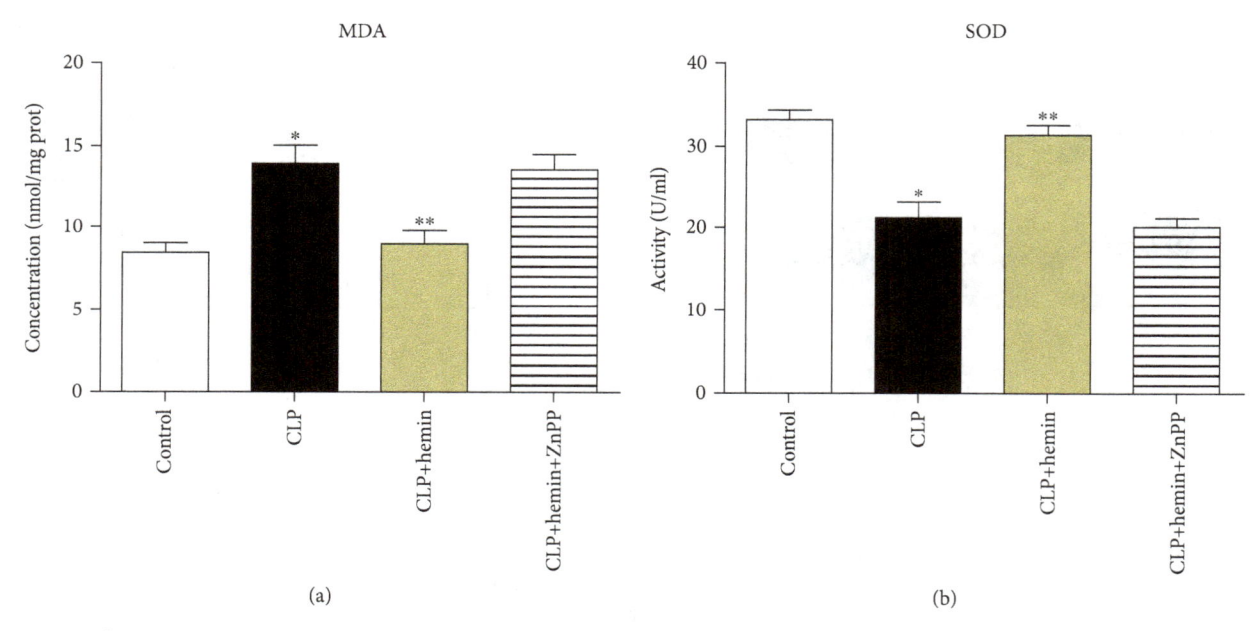

FIGURE 5: Oxidative stress response in TA muscle at 24 hours after CLP. (a) MDA concentration. *$p < 0.05$ vs. the control group; **$p < 0.05$ vs. the CLP group; $n = 6$. (b) SOD activity measurement. *$p < 0.01$ vs. the control group; **$p < 0.01$ vs. the CLP group; $n = 6$.

patients could also be attenuated by intensive insulin therapy [25], suggesting that glycemia control is an effective way to protect against muscle proteolysis.

The mechanisms of muscle breakdown are complicated. The ubiquitin-proteasome system plays a primary role in sepsis-induced muscle atrophy, which suggests that ubiquitin ligases may be involved in the development of muscle atrophy during sepsis [26]. There is persistent loss of myosin heavy chain (MyHC) in the skeletal muscle of severe critically ill patients that manifests shortly after ICU admission [27]. MyHC and many other myofibrillar proteins (such as myosin-binding proteins) were regarded as MuRF1 substrates [28]. The atrogin-1 substrates recognized thus far include MyoD, which is thought to control myoblast identity and differentiation [29]. As a result, atrogin-1 and MuRF1, the two muscle-specific ubiquitin ligases, have become potential therapeutic targets [30]. Sepsis, systemic inflammatory response syndrome, and multiple organ failure seem to play a crucial role in leading to muscle atrophy [31]. Proinflammatory cytokines such as TNF-α and IL-6 influence the blood-nerve barrier and promote endothelial cell leukocyte activation, which damages the axon [32]; this impaired nerve

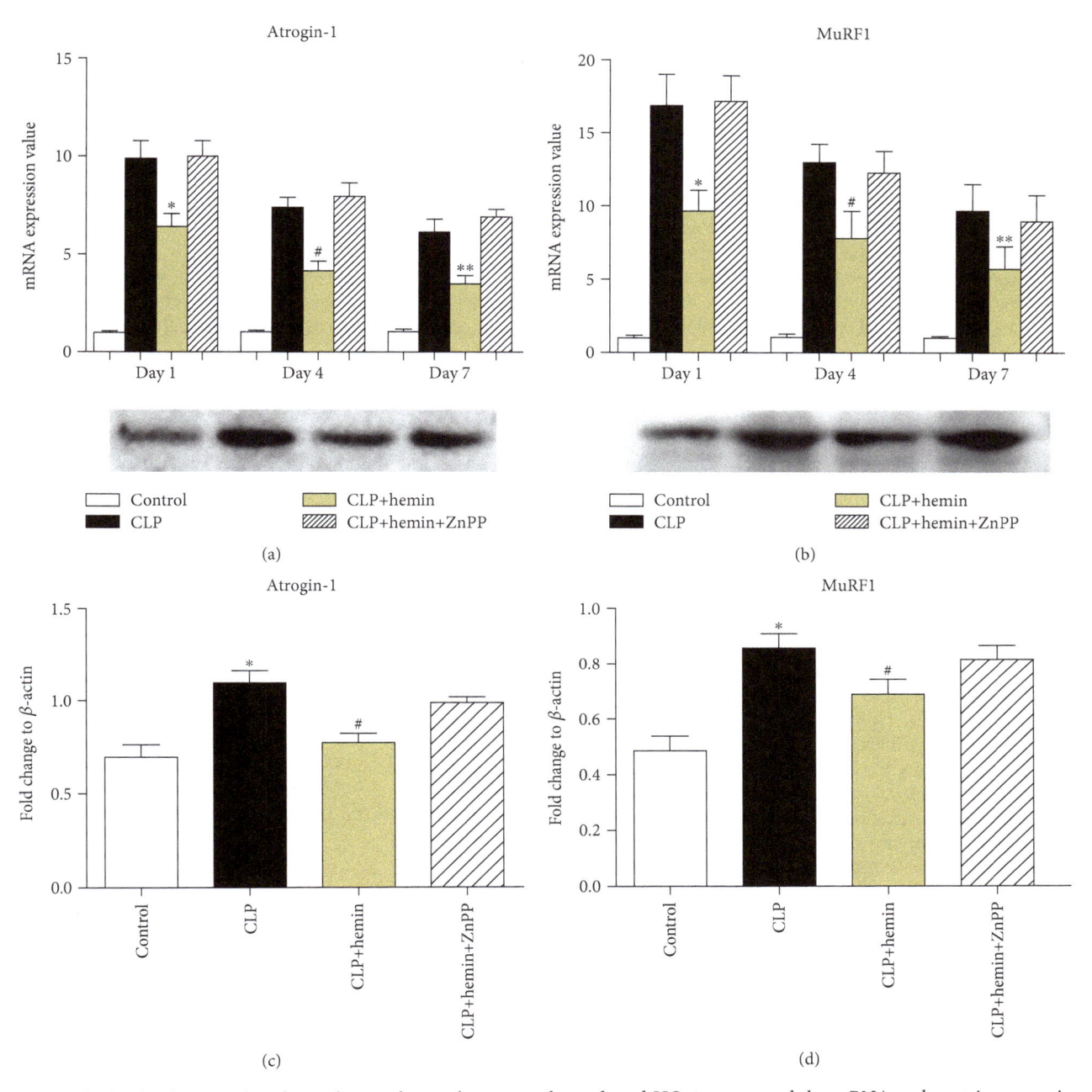

FIGURE 6: The levels of CLP-induced muscle atrophy markers were elevated, and HO-1 suppressed the mRNA and protein expression of atrogin-1 and MuRF1. (a) HO-1 inhibited the mRNA expression of atrogin-1 at 1, 4, and 7 days after surgery ($^*p < 0.01$ vs. CLP at 1 day, $^\#p < 0.05$ vs. CLP at 4 days, $^{**}p < 0.01$ vs. CLP; $n = 6$). (b) HO-1 suppressed the mRNA expression of MuRF1 during sepsis at 1, 4, and 7 days after surgery ($^*p < 0.01$ vs. CLP at 1 day, $^\#p < 0.01$ vs. CLP at 4 days, $^{**}p < 0.01$ vs. CLP at 7 days; $n = 6$). (c, d) Western blot analysis of levels of the atrogin-1 and MuRF1 proteins. Results of the corresponding semiquantitative analysis of levels of the HO-1 protein based on the optical density measured using the ImageJ software; the data are presented as means ± SEM and are representative of three separate experiments ($^*p < 0.01$ vs. control, $^\#p < 0.01$ vs. CLP). The protein expression of atrogin-1 and MuRF1 was upregulated in the CLP group. Hemin suppressed the protein expression of these two ligases; however, ZnPP administration abrogated the protective effects of hemin (1 day after surgery).

function may contribute to muscle atrophy. TNF-$\alpha$, IL-6, and other proinflammatory cytokines can also initiate network cascades that activate the FoxO family, the members of which stimulate the expression of MuRF1 and atrogin-1 [30, 33].

HO-1, the rate-limiting enzyme in heme degradation, was reported to confer cytoprotection against oxidative stress and inflammation in several animal models [34]. The major function of HO-1 is to catabolize the oxidative degradation of heme into biliverdin, CO, and free iron. Biliverdin is an effective antioxidant that eliminates peroxyl radicals, and CO exhibits strongly antiapoptotic and anti-inflammatory activities [35]. Previous in vivo and in vitro studies have demonstrated that HO-1 reduces the production of proinflammatory cytokines, including TNF-$\alpha$ and IL-6, and inhibits the activation of NF-$\kappa$B and MAPK signaling during sepsis [36]. HO-1 may prevent liver fibrosis and alleviate lung pathological injury by suppressing NF-$\kappa$B signaling pathways [37, 38]. Many factors, such as hyperoxia, hypoxia, endotoxins, and nitric oxide, can promote elevated HO-1 expression [39]. In our study, we found that during the early stage of sepsis (24 hours after CLP surgery), there was an obvious increase in HO-1 levels in the CLP group. Hemin, a substrate of HO-1, can enhance the activation of HO-1. In our experiments, we used hemin as an inducer of HO-1 and found that hemin can stimulate HO-1 activation. To identify whether the protective function rooted in HO-1 was mediated by hemin or by other subjects, we included a CLP mouse group treated with ZnPP, an inhibitor of HO-1 that can suppress the enzyme activity of HO-1. We found that mice in the ZnPP group showed no improvements in muscle condition; thus, the protective effects were due to HO-1 but not hemin or other related constituents. Interestingly, we found that ZnPP does not influence the expression of HO-1 but can inhibit its activity, indicating that the protective function of HO-1 may depend on its metabolites.

In our experiments, we found that high levels of HO-1 suppress the gene expression of atrogin-1 and MuRF1, suggesting that HO-1 inhibits muscle fiber atrophy partly by downregulating ubiquitin ligase activation. The CLP model mice showed high levels of TNF-$\alpha$ and IL-6 expression in the skeletal muscle. HO-1 suppressed the levels of these proinflammatory cytokines, indicating that HO-1 may inhibit muscle atrophy partly via its protective role against CIP. Oxidative stress is an important contributor to the etiology of skeletal muscle dysfunction. Appropriate levels of oxidants are required for normal cell adaptation and function, and high levels of ROS negatively affect the structure and function of macromolecules such as proteins, DNA, and lipids [40]. High levels of ROS may also activate proteolytic systems, thus causing enhanced protein breakdown in skeletal muscles, which results in enhanced MuRF1 levels leading to muscle atrophy [41]. Our study shows that HO-1-treated mice exhibited a high level of SOD activity and relatively low MDA levels, indicating that HO-1 exerted a protective role in muscle atrophy partly via its antioxidative function.

Regarding the experimental group design, we considered that after surgery, mice in the CLP group and control group may have different appetites, and food intake can also cause differences in muscle mass. To address this, we offered the same quantity of food to all the groups (the exact food quantity was designated based on the results of our preliminary experiments). There are still some questions our experiments did not elucidate. First, we only explored the protective mechanism of HO-1 in the proteolysis process of muscle atrophy, but we did not investigate whether HO-1 can promote muscle protein synthesis. Second, a recent finding showed that hemin exhibited cytotoxicity in colon-derived epithelial cells in a dose- and time-dependent manner. Hemin can downregulate the expression of the 18 kDa translocator protein, which contributes to cell proliferation [42]. Therefore, we must explore other functions of hemin aside from its protective role against muscle atrophy. There are also some limitations in our experiments. First, the experiments were done in both slow-twitch and fast-twitch muscles. It is then difficult to address whether the effect of hemin is specific or not. Second, although the CLP-induced muscle weakness is evident, we did not perform objective measurement to illustrate the issue.

## 5. Conclusion

In conclusion, our findings showed that hemin may promote the elevated expression of HO-1, which exerts a protective function against sepsis-induced skeletal muscle wasting. HO-1 can reduce muscle proteolysis partly by inhibiting the expression of proinflammatory cytokines and muscle-specific ubiquitin ligases. HO-1 can also exert a protective role by suppressing ROS activation in the skeletal muscle. Therefore, hemin might be a new treatment target against sepsis-induced skeletal muscle atrophy.

## Authors' Contributions

Xiongwei Yu, Wenjun Han, and Changli Wang contributed equally to the article. Xiongwei Yu and Wenjun Han performed the research and analyzed the data. Lulong Bo and Xiaoming Deng designed the research, ensured correct analysis of the data, and wrote the manuscript. Changli Wang, Daming Sui, and Jinjun Bian assisted in the design of the research, oversaw the collection of the data, and contributed to the writing of the manuscript. All authors critically revised the manuscript and gave final approval of the manuscript.

## Acknowledgments

This work was supported by the National Natural Science Foundation of China (nos. 81471845, 81671887, and 81671939), the Shanghai Outstanding Youth Medical Professionals Training Program (no. 2017YQ015), and the Key Developing Discipline Program of Shanghai (no. 2015ZB0102).

## References

[1] C. W. Seymour, V. X. Liu, T. J. Iwashyna et al., "Assessment of clinical criteria for sepsis: for the third international consensus definitions for sepsis and septic shock (sepsis-3)," *The Journal of the American Medical Association*, vol. 315, no. 8, pp. 762–774, 2016.

[2] R. Namas, R. Zamora, R. Namas et al., "Sepsis: something old, something new, and a systems view," *Journal of Critical Care*, vol. 27, no. 3, pp. 314.e1–314.e11, 2012.

[3] J. B. Poulsen, "Impaired physical function, loss of muscle mass and assessment of biomechanical properties in critical ill patients," *Danish Medical Journal*, vol. 59, no. 11, article B4544, 2012.

[4] M. Wilcox and M. Herridge, "Long-term outcomes in patients surviving acute respiratory distress syndrome," *Seminars in Respiratory and Critical Care Medicine*, vol. 31, no. 1, pp. 55–65, 2010.

[5] T. Takei, "Intensive care unit-acquired weakness: development of polyneuropathy and myopathy in critically ill patients," *Brain and Nerve*, vol. 66, no. 2, pp. 161–170, 2014.

[6] P. Bonaldo and M. Sandri, "Cellular and molecular mechanisms of muscle atrophy," *Disease Models & Mechanisms*, vol. 6, no. 1, pp. 25–39, 2012.

[7] J. Batt, C. C. dos Santos, J. I. Cameron, and M. S. Herridge, "Intensive care unit-acquired weakness: clinical phenotypes and molecular mechanisms," *American Journal of Respiratory and Critical Care Medicine*, vol. 187, no. 3, pp. 238–246, 2013.

[8] C. C. Dos Santos and J. Batt, "ICU-acquired weakness: mechanisms of disability," *Current Opinion in Critical Care*, vol. 18, no. 5, pp. 509–517, 2012.

[9] D. Attaix, S. Ventadour, A. Codran, D. Béchet, D. Taillandier, and L. Combaret, "The ubiquitin-proteasome system and skeletal muscle wasting," *Essays in Biochemistry*, vol. 41, pp. 173–186, 2005.

[10] M. Sandri, "Signaling in muscle atrophy and hypertrophy," *Physiology*, vol. 23, no. 3, pp. 160–170, 2008.

[11] S. C. Bodine, E. Latres, S. Baumhueter et al., "Identification of ubiquitin ligases required for skeletal muscle atrophy," *Science*, vol. 294, no. 5547, pp. 1704–1708, 2001.

[12] M. Llano-Diez, A. M. Gustafson, C. Olsson, H. Goransson, and L. Larsson, "Muscle wasting and the temporal gene expression pattern in a novel rat intensive care unit model," *BMC Genomics*, vol. 12, no. 1, p. 602, 2011.

[13] C. F. Bolton, "Neuromuscular manifestations of critical illness," *Muscle & Nerve*, vol. 32, no. 2, pp. 140–163, 2005.

[14] C. Coirault, A. Guellich, T. Barbry, J. L. Samuel, B. Riou, and Y. Lecarpentier, "Oxidative stress of myosin contributes to skeletal muscle dysfunction in rats with chronic heart failure," *American Journal of Physiology. Heart and Circulatory Physiology*, vol. 292, no. 2, pp. H1009–H1017, 2007.

[15] K. Ohta, T. Kikuchi, S. Arai, N. Yoshida, A. Sato, and N. Yoshimura, "Protective role of heme oxygenase-1 against endotoxin-induced uveitis in rats," *Experimental Eye Research*, vol. 77, no. 6, pp. 665–673, 2003.

[16] M. D. Maines, "The heme oxygenase system: past, present, and future," *Antioxidants & Redox Signaling*, vol. 6, no. 5, pp. 797–801, 2004.

[17] D. Morse, L. Lin, A. M. K. Choi, and S. W. Ryter, "Heme oxygenase-1, a critical arbitrator of cell death pathways in lung injury and disease," *Free Radical Biology & Medicine*, vol. 47, no. 1, pp. 1–12, 2009.

[18] X. Gong, L. Zhang, R. Jiang, M. Ye, X. Yin, and J. Wan, "Anti-inflammatory effects of mangiferin on sepsis-induced lung injury in mice via up-regulation of heme oxygenase-1," *The Journal of Nutritional Biochemistry*, vol. 24, no. 6, pp. 1173–1181, 2013.

[19] X. Qian, T. Numata, K. Zhang et al., "Transient receptor potential melastatin 2 protects mice against polymicrobial sepsis by enhancing bacterial clearance," *Anesthesiology*, vol. 121, no. 2, pp. 336–351, 2014.

[20] Q. Chen, N. Li, W. Zhu et al., "Insulin alleviates degradation of skeletal muscle protein by inhibiting the ubiquitin-proteasome system in septic rats," *Journal of Inflammation*, vol. 8, no. 1, p. 13, 2011.

[21] K. Duan, W. Yu, Z. Lin et al., "Insulin ameliorating endotoxaemia-induced muscle wasting is associated with the alteration of hypothalamic neuropeptides and inflammation in rats," *Clinical Endocrinology*, vol. 82, no. 5, pp. 695–703, 2015.

[22] T. Andrianjafiniony, S. Dupré-Aucouturier, D. Letexier, H. Couchoux, and D. Desplanches, "Oxidative stress, apoptosis, and proteolysis in skeletal muscle repair after unloading," *American Journal of Physiology. Cell Physiology*, vol. 299, no. 2, pp. C307–C315, 2010.

[23] P. O. Hasselgren, "β-Hydroxy-β-methylbutyrate (HMB) and prevention of muscle wasting," *Metabolism*, vol. 63, no. 1, pp. 5–8, 2014.

[24] P. A. Mendez-Tellez and D. M. Needham, "Early physical rehabilitation in the ICU and ventilator liberation," *Respiratory Care*, vol. 57, no. 10, pp. 1663–1669, 2012.

[25] G. Biolo, M. de Cicco, S. Lorenzon et al., "Treating hyperglycemia improves skeletal muscle protein metabolism in cancer patients after major surgery," *Critical Care Medicine*, vol. 36, no. 6, pp. 1768–1775, 2008.

[26] C. J. Wray, J. M. V. Mammen, D. D. Hershko, and P. O. Hasselgren, "Sepsis upregulates the gene expression of multiple ubiquitin ligases in skeletal muscle," *The International Journal of Biochemistry & Cell Biology*, vol. 35, no. 5, pp. 698–705, 2003.

[27] T. Wollersheim, J. Woehlecke, M. Krebs et al., "Dynamics of myosin degradation in intensive care unit-acquired weakness during severe critical illness," *Intensive Care Medicine*, vol. 40, no. 4, pp. 528–538, 2014.

[28] S. Cohen, J. J. Brault, S. P. Gygi et al., "During muscle atrophy, thick, but not thin, filament components are degraded by MuRF1-dependent ubiquitylation," *The Journal of Cell Biology*, vol. 185, no. 6, pp. 1083–1095, 2009.

[29] L. A. Tintignac, J. Lagirand, S. Batonnet, V. Sirri, M. P. Leibovitch, and S. A. Leibovitch, "Degradation of MyoD mediated by the SCF (MAFbx) ubiquitin ligase," *The Journal of Biological Chemistry*, vol. 280, no. 4, pp. 2847–2856, 2005.

[30] Z. Puthucheary, S. Harridge, and N. Hart, "Skeletal muscle dysfunction in critical care: wasting, weakness, and rehabilitation strategies," *Critical Care Medicine*, vol. 38, Supplement 10, pp. S676–S682, 2010.

[31] G. Hermans, B. de Jonghe, F. Bruyninckx, and G. Berghe, "Clinical review: critical illness polyneuropathy and myopathy,," *Critical Care*, vol. 12, no. 6, p. 238, 2008.

[32] E. Apostolakis, N. A. Papakonstantinou, N. G. Baikoussis, and G. Papadopoulos, "Intensive care unit-related generalized neuromuscular weakness due to critical illness polyneuropathy/myopathy in critically ill patients," *Journal of Anesthesia*, vol. 29, no. 1, pp. 112–121, 2015.

[33] B. Pajak, S. Orzechowska, B. Pijet et al., "Crossroads of cytokine signaling–the chase to stop muscle cachexia," *Journal of Physiology and Pharmacology*, vol. 59, Supplement 9, pp. 251–264, 2008.

[34] A. Grochot-Przeczek, J. Dulak, and A. Jozkowicz, "Haem oxygenase-1: non-canonical roles in physiology and pathology," *Clinical Science*, vol. 122, no. 3, pp. 93–103, 2012.

[35] S. W. Ryter, D. Morse, and A. M. K. Choi, "Carbon monoxide and bilirubin: potential therapies for pulmonary/vascular

injury and disease," *American Journal of Respiratory Cell and Molecular Biology*, vol. 36, no. 2, pp. 175–182, 2007.

[36] T. S. Lee and L. Y. Chau, "Heme oxygenase-1 mediates the anti-inflammatory effect of interleukin-10 in mice," *Nature Medicine*, vol. 8, no. 3, pp. 240–246, 2002.

[37] H. Yang, L. F. Zhao, Z. F. Zhao, Y. Wang, J. J. Zhao, and L. Zhang, "Heme oxygenase-1 prevents liver fibrosis in rats by regulating the expression of PPARγ and NF-κB," *World Journal of Gastroenterology*, vol. 18, no. 14, pp. 1680–1688, 2012.

[38] K. Kang, C. Nan, D. Fei et al., "Heme oxygenase 1 modulates thrombomodulin and endothelial protein C receptor levels to attenuate septic kidney injury," *Shock*, vol. 40, no. 2, pp. 136–143, 2013.

[39] C. S. T. Origassa and N. O. S. Câmara, "Cytoprotective role of heme oxygenase-1 and heme degradation derived end products in liver injury," *World Journal of Hepatology*, vol. 5, no. 10, pp. 541–549, 2013.

[40] E. Barreiro, "Protein carbonylation and muscle function in COPD and other conditions," *Mass Spectrometry Reviews*, vol. 33, no. 3, pp. 219–236, 2014.

[41] S. Sriram, S. Subramanian, P. K. Juvvuna et al., "Myostatin augments muscle-specific ring finger protein-1 expression through an NF-kB independent mechanism in SMAD3 null muscle," *Molecular Endocrinology*, vol. 28, no. 3, pp. 317–330, 2014.

[42] C. Gemelli, B. M. Dongmo, F. Ferrarini, A. Grande, and L. Corsi, "Cytotoxic effect of hemin in colonic epithelial cell line: involvement of 18 kDa translocator protein (TSPO)," *Life Sciences*, vol. 107, no. 1-2, pp. 14–20, 2014.

# Oxidative Stress, Maternal Diabetes, and Autism Spectrum Disorders

**Barbara Carpita, Dario Muti, and Liliana Dell'Osso**

*Department of Clinical and Experimental Medicine, University of Pisa, Pisa 55100, Italy*

Correspondence should be addressed to Dario Muti; dario.muti1986@gmail.com

Guest Editor: Maria Luca

Autism spectrum disorders (ASD) are a group of early-onset neurodevelopmental conditions characterized by alterations in brain connectivity with cascading effects on neuropsychological functions. To date, in the framework of an increasing interest about environmental conditions which could interact with genetic factors in ASD pathogenesis, many authors have stressed that changes in the intrauterine environment at different stages of pregnancy, such as those linked to maternal metabolic pathologies, may lead to long-term conditions in the newborn. In particular, a growing number of epidemiological studies have highlighted the role of obesity and maternal diabetes as a risk factor for developing both somatic and psychiatric disorders in humans, including ASD. While literature still fails in identifying specific etiopathological mechanisms, a growing body of evidence is available about the presence of a relationship between maternal immune dysregulation, inflammation, oxidative stress, and the development of ASD in the offspring. In this framework, results from high-fat diet animal models about the role played by oxidative stress in shaping offspring neurodevelopment may help in clarifying the pathways through which maternal metabolic conditions are linked with ASD. The aim of this review is to provide an overview of literature about the effects of early life insults linked to oxidative stress which may be involved in ASD etiopathogenesis and how this relationship can be explained in biological terms.

## 1. Introduction

Brain development in a fetus and in the first years of life is pivotal in the shaping of the individual overall neuropsychological performance level [1–3]. Any alteration of the intrauterine environment at different stages of pregnancy, such as maternal metabolic pathologies, may lead to long-term condition in the newborn.

Given the above consideration, it should be noted how the rates of obesity and diabetes have experienced a steep increase in several countries, with a correspondent increasing number of studies in literature devoted to this topic [4]. There is a rich literature regarding the effect of excessive weight before pregnancy, particularly in combination with rapid weight gain during pregnancy: a condition generally described as a risk factor for gestational diabetes (GDM), which is glucose intolerance with the onset or first recognition during pregnancy. Up to 15% of pregnant women worldwide are esteemed to be affected from diabetes, and approximately 87.5% of maternal diabetes are GDM. Only 7.5% are preexisting type 1 diabetes [5]. As such, maternal diabetes (and GDM in particular) is an important risk factor for conditions like, but not limited to, miscarriage, macrosomia [6], and neurodevelopmental impairments in the offspring [7–9]. In particular, several studies suggest that the offspring of GDM mothers presents more often language delay, poor motor development, and impaired recognition memory [10–12]. A more specific group of studies focused on the possible link between autism spectrum disorder (ASD) risk in the offspring of GDM mothers [13–18].

ASD is a neurodevelopmental condition characterized by a deficit in social interaction, communication issues, and repetitive, stereotyped behaviors [19–21]. ASD etiopathogenesis is still unclear [22]. However, good evidence for genetic correlates is available: specific genetic mutations can be, in fact, identified in about 20% of ASD cases, and twin studies estimate a heritability between 64 and 91% [23]. This strong genetic influence but the concomitant lack of full concordance in monozygotic twins points to the relevance of environmental factors in the etiopathogenesis of the disorder [24, 25]. Although studies about ASD are focused mainly on children, the presence of this condition is particularly relevant also in adulthood, due to its quite extensive comorbidity with other psychiatric disorders. The presence of undetected ASD among seeking-treatment inpatients is clearly a subject of clinical relevance, highlighting a need for careful investigation of autistic symptoms both in clinical samples and in the general population. A growing body of studies in fact has shown that also the presence of autistic traits (that is, the presence of autism spectrum symptomatology not necessarily in its full-blown clinical presentation) may not only have an impact on increasing the severity of other mental conditions but also be considered as a risk factor for developing other disorders or toward suicidality [26, 27]. Despite that, autistic symptoms often remain undetected in clinical settings, especially in subjects with little cognitive impairments and moderate symptoms, hided by the manifestations of other comorbid disorders [28–31].

The aim of this paper is to review evidence from literature, regarding whether the effects of early life insults and that linked to oxidative stress in particular might be involved in short- and long-term risks for ASD and how this risk can be explained in biological terms.

## 2. Obesity, GDM, and Neurodevelopment: Toward a Comprehensive Model

A good number of studies on rodent or nonhuman primate models focused on the effects of maternal high-fat diet in the offspring. A regime of high-fat diet might vary depending on the studied specimen (mice, rats, and primates) and also on the specific experimental model. As a result, there is a certain degree of variation among these studies. In order to provide a general picture, a common high-fat chow for murine models ("HF"; Research Diets, D12492) provides 60% kcal from fat [32], while primate high-fat food (Test Diet; 5A1F; Purina Mills) provides 32% of calories from fat, and this lab food might be supplemented with calorically dense treats [33]. As reviewed by Sullivan et al. [34], researches on animal models proved that maternal high-fat diet impacts offspring behavior, socialization, and cognition; moreover, it affects reward pathways. Given this data, several studies focused on the biological pathways on such effects, although with controversial results. Most studies on behavioral effects on the offspring of maternal diet have been conducted in rodents. High-fat diet has been shown influencing offspring behavior by modifying maternal care: in particular, by leading to an increased nursing behavior [32, 35]. It has been suggested [34] that this feature might lead to hyperphagia, hypothalamic reprogramming of energy balance-regulating pathways, and—as a result—increased body weight in the offspring. Also, findings from another rodent model show that consumption of high-energy diet in mothers disrupted hippocampal function and thus impaired learning and memory performance [36]. Moreover, maternal high-fat diet has been associated with heightened anxiety in both nonhuman primates and rodents [33, 37]. According to another study, maternal diet rich in polyunsaturated fatty acid produced a highly aggressive offspring, which displayed hyperlocomotion and decreased immobility in a swim test, due probably to an observed protein kinase C downregulation (in the whole brain except the hypothalamus) [38]. High-fat diet is also correlated with alteration in development of cognitive functions. Male offspring from rats exposed to a diet high in saturated or trans fats, once adults, display a deficit in cognitive spatial functioning [37]. Moreover, offspring from obese mothers show a decrease in hippocampal brain-derived neurotrophic factor production (BDNF) which causes an alteration in neurodevelopment for this area [39]. It is also interesting to note how maternal high-fat diet leads to an alteration in reward-based behavior such as food preference in the offspring, as many studies remark [40–43].

One of the most widely reported data about this topic [34] is the fact that high-fat diet consumption seems to expose the offspring to an increase in inflammatory cytokines, which interact with neural development; despite that, only few studies addressed how effects of inflammatory dysregulation can be modulated by the timing and the duration of exposition to such a diet [37].

Sullivan et al. [44] outline two main pathways by which high-fat diet consumption and obesity influence offspring neurodevelopment and subsequently behavior in animal models. Both immune and (neuro-)endocrine systems should be considered affected by this condition. Inflammation plays a clear role, as obesity is associated with elevated inflammatory cytokine production, due to the increase in adipose tissue, to the point that it has been considered as a state of chronic inflammation [45]. Inflammation exposure during gestation is not only associated with perinatal conditions such as premature birth and low birth weight but also associated with neurodevelopmental disorders like ADHD, ASD, and schizophrenia [46–49]. Inflammatory mediators are able to cross the blood-placenta barrier and interact with fetal neurodevelopment. While evidence from rodent models shows that the proinflammatory cytokine interleukin 6 impacts genes for cortical expression [50], it is known that maternal high-fat diet causes a raise of inflammatory markers, with microglial activation and protoinflammatory cytokines in the offspring hippocampus [44].

However, maternal obesity does expose the fetus not only to proinflammatory cytokines but also to an environment where nutrients and neuroendocrine agents like fatty acids, glucose, triglycerides, and leptin are higher. The most immediate effect of this is fetal hyperglycemia. Since glucose can cross the blood-placenta barrier but maternal insulin cannot, the fetus secretes its own insulin, which is also a growth factor involved in brain development. It has been thus hypothesized that hyperinsulinemia in the prenatal period might lead to an alteration in brain development and regulation [51].

Not only insulin but also leptin shows an increased level in obese and diabetic mothers [52]. Leptin receptors are distributed—in humans—among several brain regions which play a central role in behavioral regulation such as the cortex, hippocampus, amygdala, thalamus, and hypothalamus. While linked to inflammatory response [53], leptin is also active in the hypothalamic pituitary adrenal (HPA) axis, a pivotal structure for stress response [44, 54]. As a result, it is interesting to note that higher levels of leptin were detected in ASD children compared to healthy, age-matched controls (Ashwood et al., 2008; [55]).

Leptin leads to consider the role of the HPA axis and its effect on behavior. The HPA axis is responsible for corticotropin-releasing hormone (CRH), and CRH and arginine vasopressin are synthetized in the paraventricular nucleus of the hypothalamus in response to stress. While maternal diet and extension obesity impact both the HPA axis and extrahypothalamic CRH neurons, the interaction of this pathway with neurodevelopment should be addressed [44]. In rat models, high-fat diet in midpregnancy mothers leads to an offspring with increased postnatal basal corticosterone levels in association with reduction of hippocampal and hypothalamic phospholipid-derived arachidonic acid [56]. The Sasaki et al. [57] rat model showed how high-fat diet consumption during pregnancy and lactation led to an offspring with decreased basal corticosterone levels but heightened response to stress with a slower restoration of baseline corticosterone. This offspring also showed an increase in glucocorticoid receptors in the amygdala, with a concurrent alteration in inflammatory gene expression for the hippocampus and amygdala.

In both human and rats, serotonin plays a crucial role for emotional regulation [44]. Inflammation leads to alteration in serotonin regulation, as the animal model described by Ishikawa et al. [58] suggests by showing how rats treated with cytokine-interferon alpha have decreased serotonergic axon density in the amygdala and the ventral medial prefrontal cortex. Offspring from mothers fed a high-fat diet have been reported displaying alteration in the hippocampus, including not only increased 5-HT1A receptors in the ventral hippocampus but also increased brain-derived neurotrophic factor in the dorsal hippocampus (Peleg-Raibstein et al., 2012). A nonhuman primate model showed that maternal high-fat diet consumption impaired the development of the serotonergic system, leading to a reduction of serotonin synthesis and increased anxiety behaviors in the female offspring [33]. This model also shows increased inflammation levels in the hypothalamus among mothers and offspring [33, 44].

Fewer studies investigated also dopamine pathways. A decrease in mesocorticolimbic dopamine sensitivity, with a concurrent decrease in locomotor activation in response to psychostimulant administration, has been found in rat offspring from mothers who have been fed with a high-fat diet during late gestation and lactation [59]. Another study stressed the involvement of genetic expression in this process, showing in a rat that maternal high-fat diet-induced obesity led to dopamine dysregulation in the offspring by the means of genome-wide methylation and gene regulation for dopamine reuptake trasporter, the $\mu$-opioid receptor and preproenkephalin [60].

## 3. Maternal Metabolic Conditions and ASD

*3.1. Oxidative Stress from GDM.* As Rossignol and Frye [61] have demonstrated in a large systematic review, strong evidence is available in literature on the links between immune dysregulation, inflammation, oxidative stress, and the etiopathogenesis of autism. The above-considered role of oxidative stress in high-fat diet animal models sheds light, through different pathways, on the widely observed correlation in human population between GDM and poor performance of the offspring on standardized IQ tests and motor development assessments (Rizzo et al., 1997; Ratzon et al., 2000), as well as on abnormalities in the limbic system detected in ASD children samples [18].

To date, many epidemiological studies have proven obesity being a risk factor for development of neuropsychiatric disorders in humans; however, no clear causative link has been yet identified [62].

From an epidemiological point of view, the concomitant rise in prevalence of both obesity and neuropsychiatric disorders has been stressed in literature [63, 64], but it is important to remember how the increase of the latter disorder might be explained with the advancement of diagnostic tools and an improved awareness toward this kind of conditions [65].

In a study on 1004 mothers, diabetes and hypertension were more common among mothers of children affected by ASD and among mothers of children affected by developmental delays (DD) without ASD. Diabetes, in particular, was strongly associated with greater deficits in expressive language in children with ASD and to a less extent to impairment in visual reception, adaptive behavior, motor skills, and receptive/expressive language, as measured by the Social Communication Questionnaire, the Mullen Scales of Early Learning, and the Vineland Adaptive Behavior Scales [18].

Another study with longitudinal design, conducted in a population of 308 mothers with singleton pregnancies, reported different results. Among this population, normal weight women (128) have been divided from overweight (58), obese (52), and GDM ones (76). Among the infants, assessed at 6 and 18 months with the Bayley Scales of Infant Development-III (BSID-III), those born from mothers with pregestational obesity had significantly higher scores in both cognitive and language developments at 6 months of age; however, the authors observe that their absolute numbers are quite low thus suggesting a careful generalization [66].

A study on 2734 mother-child pairs, with an average of 6-year follow-up, found that mothers of children with ASD were significantly more often older and affected by GDM than the mothers of children typically developing (TD). Also, mothers of children with ADHD were more likely to be of lower education, to be obese, and to have used alcohol during pregnancy than the mothers of TD children. It should be noted that since intellectual disabilities (ID) but not other DD showed a pattern of risk increasing with obesity and GDM, the author suggests that ASD with ID may be etiologically distinct from ASD without ID [67].

A large study ($N = 165311$) of mother-child pairs, comprising 17,988 diabetic mothers, shows that there is a significant prevalence, even though the odds ratio is rather low, for ID in children of mothers with diabetes. It is noteworthy that this risk factor is independent of other maternal features like tobacco smoke, ethnicity and race, educational level, birth weight, offspring gender, and hypertension. The authors point out that the link between diabetes in mothers and ID should be identified in maternal and fetal inflammation processes, which poses a risk for abnormal fetal brain development [68]. Furthermore, Wang et al. (2015) found that hypertension related more than pregnancy diabetes with ID in a large ($N = 123,922$) sample of mother-child couples. Huang et al. [69] meta-analysis on this topic shows that maternal diabetes (OR 1.15, $p < 0.0001$) and maternal hypertension, preeclampsia, or eclampsia (OR 1.33, $p < 0.0001$) act as a risk factor toward ID, thus strengthening the hypothesis that oxidative stress- and obesity-caused inflammation can disrupt neurodevelopment.

A similar meta-analysis, this time focusing on ASD, shows that maternal diabetes acts as a risk factor (OR 1.48) without significant heterogeneity ($I^2 = 9.1$, $p = 0.35$) (Xu et al., 2015). *In utero* exposure to inflammatory factors and hyperglycemia, exempli gratia, could be linked to an increase of free radical production and an impairment of antioxidant countermeasures, through raising the oxidative stress in the cord blood and the placental tissue [70, 71]. These data are even more interesting in the light of studies that take into account the association between maternal autoimmune diseases and subsequent diagnosis of autism in children. A large body of data has been collected on how a family history of autoimmune disorders has been reported more commonly among ASD children than in healthy controls [72, 73]. The Croen et al. [14] case study on 407 couples of mother-children with ASD and 2095 controls has found that maternal autoimmune conditions were significantly associated with ASD in children and asthma in particular. This is coherent with the reported data about midgestation infections by the influenza virus as a risk factor for autism [74, 75], and since the influenza virus cannot cross the placenta, the focus shifts from viral components to the maternal immune response [76–78].

*3.2. Broadening the Perspective.* As above stated, several contributions point out a relationship among GDM, preeclampsia, autoimmune activation, and, in a broader perspective, oxidative stress and inflammation in mothers of ASD children. However, it is clear from these data that disruption to the normal homeostasis of placental environment may increase the risk but not determine psychiatric disorders; thus, the outcome depends also on the other contextual factors which might be involved in the causative algorithm [79]. Maternal inflammatory states and GDM have the peak of their influence during the central part of the gestation [80], when maternal immune activation is also at full efficiency. According to [79], GDM interaction with maternal immune activation might then disrupt the *in utero* environment thus affecting fetal neurodevelopment, as suggested by the role played by interleukin 6 in several animal models [81–83].

Given the role for inflammatory response, oxidative stress and neurodevelopment disruption, it is also useful to point out that eclampsia and also preeclampsia have been considered a risk factor for ASD in many studies. Gardener et al. [84] extensive meta-analysis shows how preeclampsia has been reported being a significant risk factor for ASD, even if the effect size from the various studies is not always consistent. According to Walker et al. [85], preeclampsia is able to affect fetal neurodevelopment by causing an abnormal trophoblast differentiation during embryogenesis and by limiting fetus intake of nutrient and oxygen. This condition primes the syncytiotrophoblast to release proteins into the maternal bloodstream in an attempt to improve circulation. As a side effect, this release might rise baseline systemic inflammation, insulin resistance, and vascular endothelial changes in the mother. The oxidative stress progressive rise is of particular interest, as it could be integrated with the two-stage model of preeclampsia [85, 86]. In this framework, the first state of preeclampsia, characterized by a poorly perfused placenta, is not sufficient to produce the clinical manifestation of preeclampsia. However, this preliminary condition of poor perfusion and oxidative stress interacts with the maternal immune and vascular systems paving the way for the second stage, where the clinical symptoms of preeclampsia occur [86].

In this framework, a potential integration for etiopathogenetic mechanism may come from contribution that takes into account gene interactions and nutritional factors, which might occur at the time of the inflammation, a field which received rising attention from researchers. It has already been pointed out how deficiency of omega-3 fatty acids (n3FAs) might play a decisive role in the etiology of several neurodevelopmental disorders, namely, ASD and ADHD [87–90]. While fish oil supplementation during pregnancy and in the first 3 months after birth leads, at 4-year follow-up, to higher mental age in probands than in controls [91], lower plasma or rbc levels of n3FAs have been found in ASD patients compared to controls [92–94].

According to Field [95], genetic studies outline several connections between n3FA metabolism and neurodevelopmental disorders. Chromosome site 11q22-23, a location linked to ASD and ADHD, contains the genes for desaturases involved in FA conversion, as well as a dopamine receptor gene linked to bipolar disorder. Sites in the 6p21-23 region containing genes involved with fatty acid metabolism are also associated with ASD, ADHD, schizophrenia, and bipolar disorder. The genes for phospholipase A2s, enzymes responsible for transforming Sn-2 long chain fatty acid into free FA molecules, are near genes involved in ADHD, ASD, and bipolar disorder genes [96]. This led several scholars to question how dietary change, as an environmental genetic interaction, might affect etiopathogenesis of neurodevelopmental disorders ([97, 98]; Stevens et al., 1995).

Field [95] proposed to consider how parental age, a risk factor for ASD (less so for ADHD), might be linked to the aging effect on the maternal metabolism of fatty acids, as somatic cells might have defects in delta 6 desaturase activity causing a poorer conversion of fatty acids in older women. Other interesting evidence arises from studies that take into account pregnancy iron deficiency, a more recent trend than the one on fatty acids. Fe is known for its pivotal role in myelination, synaptogenesis, and nerve cell metabolism as a whole; as such, iron deficiency has already been reported as a risk factor for abnormal development While iron deficiency might be associated with various conditions and given the fact that the placenta has an effective regulatory ability for iron transfer, it should be noted that obesity is a risk factor for iron deficiency [1, 3, 99–102]. Berglund et al. [103] noted how this interaction might act as a confounding factor for the link between oxidative stress, inflammation, and neurodevelopment impairments.

Many genetic pathways have been described for autism ([104]; Bae, Hong, 2018), several of which might be connected to the relationship between maternal metabolism, oxidative stress, and neurodevelopment. Of particular interest is the mTOR pathway. A serine-threonine kinase, mTOR (namely, its two protein complexes mTORC1 and mTORC2) is involved in a complex interaction, whose unifying feature could be identified in the integration of multiple intracellular and extracellular signals to coordinate several responses, including protein synthesis, growth, proliferation, and, in the central nervous system, also synaptic plasticity [21, 105]. In particular, mTORC1 and mTORC2 promote the transcription of genes involved in autophage inhibition, protein translation, carbohydrate metabolism, and lipogenesis. As such, they play a central role in the mechanisms of obesity and autoimmune disorders and also of cancer development and aging [106]. The role of mTor has been studied in a large variety of neuropsychiatric disorders, such as mood disorders, schizophrenia, and drug addiction, and it is involved with an alteration in dopaminergic transmission (Ryskalin et al., 2018). Dysregulation of mTOR is a common feature of several neurodevelopmental disorders, such as tuberous sclerosis, which has been associated with ASD in retrospective, prospective, and also meta-analytic studies [107]. Also, Angelman syndrome [108], Rett syndrome [109], CDKL-5 syndrome [110], and Phelan-McDermid syndrome (associated with SHANK 3 deletion and with a high comorbidity with ASD) [21] have been associated with mTOR pathways. This implication of mTOR in autism led to investigating its role as a potential target to understand and treat ASD [111]. Kalkman and Feuerbach [112] argued that ASD syndromes might be caused by mutations in genes that inhibit mTOR and that mTOR might lead to autophagy inhibition. This evidence may shed more light also on the association between allergies, as well as the wide spectrum of autoimmune disorders, and ASD. An interesting study [113] employed BTBR mice, a strain considered a representative model for ASD behavioral deficits [114], that features upregulation of the mTOR-S6K pathway and synaptic inhibition of the mTOR-ULK1 pathway. This animal model has been treated with acute systemic injection of insulin-like growth factor-II, showing an improvement of many major ASD-like behaviors and specifically of cognitive and social deficits, as well as of repetitive behaviors [113].

## 4. Concluding Remarks

In the framework of a growing interest about possible environmental factors involved in the etiopathogenesis of ASD, in particular during *in utero* life, increasing epidemiological reports highlighted the link between ASD and maternal metabolic conditions. Many studies stressed the crucial role for neurodevelopment of the exposition to oxidative stress, which is linked to inflammation, immune dysregulation, and thus, also to a wide variety of metabolic conditions. Despite that, the mechanisms through which oxidative stress might lead to developing ASD remain unclear and not specific, and maternal metabolic conditions have been associated also with different kinds of both somatic and neuropsychiatric disorders in the offspring. Moreover, most of the authors have focused only on singular mechanisms and/or specific metabolic conditions; thus, there is a lack of studies featuring a concomitant evaluation of a broader range of biochemical pathways.

It might be useful to highlight some of the limitations of the present inquiry. This is not a systematic review on the subject, and as such, it bears possible bias under the methodological and interpretative points of view. Moreover, we consider here only some of the many possible ethiopathogenetic paths of autism. As such, the large number of environmental and genetic variables implied in this process has been restricted. In addition, animal models on high-fat diet provide a good picture of some of the mechanisms that might be acting in ASD development, but due to some practical limitations (high-fat or high-energy diet is defined depending on the species), results from these data should be carefully weighted when comparing them to results from humans. Globally, while human clinical conditions such as diabetes mellitus, preeclampsia, or other metabolic conditions may elicit several neurodevelopmental disruptive pathways, it should be noted how these mechanisms, mostly related with oxidative stress, should be better considered in a wider framework of complex interaction with genetic underpinnings as well as with other environmental and neurobiological factors, not limited to the cases here discussed.

Further studies might allow to clarify this perspective, considering in particular the possible multifactorial etiopathogenesis of ASD, which feature the interaction between genetic and environmental conditions. Increasing the knowledge about this topic is of crucial interest for both clinical and research settings, as it may lead not only to improve therapeutic and prevention strategies but also to shed more light on the relationships between central and peripheral systems and between somatic and neuropsychiatric disorders.

## References

[1] S. E. Cusick and M. K. Georgieff, "Nutrient supplementation and neurodevelopment: timing is the key," *Archives of Pediatrics & Adolescent Medicine*, vol. 166, no. 5, pp. 481-482, 2012.

[2] M. Hadders-Algra, "Prenatal and early postnatal supplementation with long-chain polyunsaturated fatty acids: neurodevelopmental considerations," *The American Journal*

[3] P. Tomalski and M. H. Johnson, "The effects of early adversity on the adult and developing brain," *Current Opinion in Psychiatry*, vol. 23, no. 3, pp. 233–238, 2010.

[4] M. M. Finucane, G. A. Stevens, M. J. Cowan et al., "National, regional, and global trends in body-mass index since 1980: systematic analysis of health examination surveys and epidemiological studies with 960 country-years and 9·1 million participants," *The Lancet*, vol. 377, no. 9765, pp. 557–567, 2011.

[5] G. Xu, J. Jing, K. Bowers, B. Liu, and W. Bao, "Maternal diabetes and the risk of autism spectrum disorders in the offspring: a systematic review and meta-analysis," *Journal of Autism and Developmental Disorders*, vol. 44, no. 4, pp. 766–775, 2014.

[6] E. Herrera and H. Ortega-Senovilla, "Disturbances in lipid metabolism in diabetic pregnancy – are these the cause of the problem?," *Best Practice & Research Clinical Endocrinology & Metabolism*, vol. 24, no. 4, pp. 515–525, 2010.

[7] J. L. Anderson, D. K. Waller, M. A. Canfield, G. M. Shaw, M. L. Watkins, and M. M. Werler, "Maternal obesity, gestational diabetes, and central nervous system birth defects," *Epidemiology*, vol. 16, no. 1, pp. 87–92, 2005.

[8] M. K. Georgieff, "The effect of maternal diabetes during pregnancy on the neurodevelopment of offspring," *Minnesota Medicine*, vol. 89, no. 3, pp. 44–47, 2006.

[9] A. Ornoy, N. Ratzon, C. Greenbaum, A. Wolf, and M. Dulitzky, "School-age children born to diabetic mothers and to mothers with gestational diabetes exhibit a high rate of inattention and fine and gross motor impairment," *Journal of Pediatric Endocrinology and Metabolism*, vol. 14, Supplement, 2001.

[10] T. D. Clausen, E. L. Mortensen, L. Schmidt et al., "Cognitive function in adult offspring of women with gestational diabetes–the role of glucose and other factors," *PLoS One*, vol. 8, no. 6, article e67107, 2013.

[11] T. DeBoer, S. Wewerka, P. J. Bauer, M. K. Georgieff, and C. A. Nelson, "Explicit memory performance in infants of diabetic mothers at 1 year of age," *Developmental Medicine and Child Neurology*, vol. 47, no. 8, pp. 525–531, 2005.

[12] A. Fraser, S. M. Nelson, C. Macdonald-Wallis, and D. A. Lawlor, "Associations of existing diabetes, gestational diabetes, and glycosuria with offspring IQ and educational attainment: the Avon Longitudinal Study of Parents and Children," *Experimental Diabetes Research*, vol. 2012, Article ID 963735, 7 pages, 2012.

[13] S. Buchmayer, S. Johansson, A. Johansson, C. M. Hultman, P. Sparén, and S. Cnattingius, "Can association between preterm birth and autism be explained by maternal or neonatal morbidity?," *Pediatrics*, vol. 124, no. 5, pp. e817–e825, 2009.

[14] L. A. Croen, J. K. Grether, C. K. Yoshida, R. Odouli, and J. Van de Water, "Maternal autoimmune diseases, asthma and allergies, and childhood autism spectrum disorders: a case-control study," *Archives of Pediatrics & Adolescent Medicine*, vol. 159, no. 2, pp. 151–157, 2005.

[15] L. Dodds, D. B. Fell, S. Shea, B. A. Armson, A. C. Allen, and S. Bryson, "The role of prenatal, obstetric and neonatal factors in the development of autism," *Journal of Autism and Developmental Disorders*, vol. 41, no. 7, pp. 891–902, 2011.

[16] M. A. A. Elhameed, A. E. O. A. Elbaky, and E. A. Kamel, "A controlled study of the risk factors and clinical picture of children with autism in an Egyptian sample," *Egypt Journal of Neurology, Neurosurgery and Psychiatry*, vol. 48, pp. 271–276, 2011.

[17] C. M. Hultman, P. Sparén, and S. Cnattingius, "Perinatal risk factors for infantile autism," *Epidemiology*, vol. 13, no. 4, pp. 417–423, 2002.

[18] P. Krakowiak, C. K. Walker, A. A. Bremer et al., "Maternal metabolic conditions and risk for autism and other neurodevelopmental disorders," *Pediatrics*, vol. 129, no. 5, pp. e1121–e1128, 2012.

[19] APA, *Diagnostic and Statistical Manual of Mental Disorders*, American Psychiatric Publishing, Arlington, TX, USA, 2013.

[20] S. E. Levy, D. S. Mandell, and R. T. Schultz, "Autism," *The Lancet*, vol. 374, no. 9701, pp. 1627–1638, 2009.

[21] K. D. Winden, D. Ebrahimi-Fakhari, and M. Sahin, "Abnormal mTOR activation in autism," *Annual Review of Neuroscience*, vol. 41, no. 1, pp. 1–23, 2018.

[22] L. Billeci, S. Calderoni, E. Conti et al., "The broad autism (endo)phenotype: neurostructural and neurofunctional correlates in parents of individuals with autism spectrum disorders," *Frontiers in Neuroscience*, vol. 10, p. 346, 2016.

[23] B. Tick, P. Bolton, F. Happé, M. Rutter, and F. Rijsdijk, "Heritability of autism spectrum disorders: a meta-analysis of twin studies," *Journal of Child Psychology and Psychiatry*, vol. 57, no. 5, pp. 585–595, 2016.

[24] H. Lehn, E. M. Derks, J. J. Hudziak, P. Heutink, T. C. E. M. van Beijsterveldt, and D. I. Boomsma, "Attention problems and attention-deficit/hyperactivity disorder in discordant and concordant monozygotic twins: evidence of environmental mediators," *Journal of the American Academy of Child & Adolescent Psychiatry*, vol. 46, no. 1, pp. 83–91, 2007.

[25] C. J. Newschaffer, L. A. Croen, J. Daniels et al., "The epidemiology of autism spectrum disorders," *Annual Review of Public Health*, vol. 28, no. 1, pp. 235–258, 2007.

[26] K. Kato, K. Mikami, F. Akama et al., "Clinical features of suicide attempts in adults with autism spectrum disorders," *General Hospital Psychiatry*, vol. 35, no. 1, pp. 50–53, 2013.

[27] K. Takara and T. Kondo, "Autism spectrum disorder among first-visit depressed adult patients: diagnostic clues from backgrounds and past history," *General Hospital Psychiatry*, vol. 36, no. 6, pp. 737–742, 2014.

[28] L. Dell'Osso, R. Dalle Luche, C. Cerliani, C. A. Bertelloni, C. Gesi, and C. Carmassi, "Unexpected subthreshold autism spectrum in a 25-year-old male stalker hospitalized for delusional disorder: a case report," *Comprehensive Psychiatry*, vol. 61, pp. 10–14, 2015.

[29] L. Karjalainen, C. Gillberg, M. Råstam, and E. Wentz, "Eating disorders and eating pathology in young adult and adult patients with ESSENCE," *Comprehensive Psychiatry*, vol. 66, pp. 79–86, 2016.

[30] H. Mito, N. Matsuura, K. Mukai et al., "The impacts of elevated autism spectrum disorder traits on clinical and psychosocial features and long-term treatment outcome in adult patients with obsessive-compulsive disorder," *Comprehensive Psychiatry*, vol. 55, no. 7, pp. 1526–1533, 2014.

[31] N. Yirmiya and M. Shaked, "Psychiatric disorders in parents of children with autism: a meta-analysis," *Journal of Child Psychology and Psychiatry*, vol. 46, no. 1, pp. 69–83, 2005.

[32] R. H. Purcell, B. Sun, L. L. Pass, M. L. Power, T. H. Moran, and K. L. K. Tamashiro, "Maternal stress and high-fat diet effect on maternal behavior, milk composition, and pup ingestive behavior," *Physiology & Behavior*, vol. 104, no. 3, pp. 474–479, 2011.

[33] E. L. Sullivan, B. Grayson, D. Takahashi et al., "Chronic consumption of a high-fat diet during pregnancy causes perturbations in the serotonergic system and increased anxiety-like behavior in nonhuman primate offspring," *The Journal of Neuroscience*, vol. 30, no. 10, pp. 3826–3830, 2010.

[34] E. L. Sullivan, E. K. Nousen, and K. A. Chamlou, "Maternal high fat diet consumption during the perinatal period programs offspring behavior," *Physiology & Behavior*, vol. 123, pp. 236–242, 2014.

[35] M. Bertino, "Effects of high fat, protein supplemented diets on maternal behavior in rats," *Physiology & Behavior*, vol. 29, no. 6, pp. 999–1005, 1982.

[36] S. E. Kanoski, Y. Zhang, W. Zheng, and T. L. Davidson, "The effects of a high-energy diet on hippocampal function and blood-brain barrier integrity in the rat," *Journal of Alzheimer's Disease*, vol. 21, no. 1, pp. 207–219, 2010.

[37] S. D. Bilbo and V. Tsang, "Enduring consequences of maternal obesity for brain inflammation and behavior of offspring," *The FASEB Journal*, vol. 24, no. 6, pp. 2104–2115, 2010.

[38] M. Raygada, E. Cho, and L. Hilakivi-Clarke, "High maternal intake of polyunsaturated fatty acids during pregnancy in mice alters offsprings' aggressive behavior, immobility in the swim test, locomotor activity and brain protein kinase C activity," *The Journal of Nutrition*, vol. 128, no. 12, pp. 2505–2511, 1998.

[39] Y. Tozuka, M. Kumon, E. Wada, M. Onodera, H. Mochizuki, and K. Wada, "Maternal obesity impairs hippocampal BDNF production and spatial learning performance in young mouse offspring," *Neurochemistry International*, vol. 57, no. 3, pp. 235–247, 2010.

[40] S. A. Bayol, S. J. Farrington, and N. C. Stickland, "A maternal 'junk food' diet in pregnancy and lactation promotes an exacerbated taste for 'junk food' and a greater propensity for obesity in rat offspring," *British Journal of Nutrition*, vol. 98, no. 4, pp. 843–851, 2007.

[41] Y. Nakashima, "Fish-oil high-fat diet intake of dams after day 5 of pregnancy and during lactation guards against excessive fat consumption of their weaning pups," *Journal of Nutritional Science and Vitaminology*, vol. 54, no. 1, pp. 46–53, 2008.

[42] Z. Y. Ong and B. S. Muhlhausler, "Maternal "junk-food" feeding of rat dams alters food choices and development of the mesolimbic reward pathway in the offspring," *The FASEB Journal*, vol. 25, no. 7, pp. 2167–2179, 2011.

[43] C. D. Walker, L. Naef, E. d'Asti et al., "Perinatal maternal fat intake affects metabolism and hippocampal function in the offspring," *Annals of the New York Academy of Sciences*, vol. 1144, no. 1, pp. 189–202, 2008.

[44] E. L. Sullivan, K. M. Riper, R. Lockard, and J. C. Valleau, "Maternal high-fat diet programming of the neuroendocrine system and behavior," *Hormones and Behavior*, vol. 76, pp. 153–161, 2015.

[45] M. Visser, L. M. Bouter, G. McQuillan, M. H. Wener, and T. B. Harris, "Elevated C-reactive protein levels in overweight and obese adults," *JAMA*, vol. 282, no. 22, pp. 2131–2135, 1999.

[46] A. Angelidou, S. Asadi, K.-D. Alysandratos, A. Karagkouni, S. Kourembanas, and T. C. Theoharides, "Perinatal stress, brain inflammation and risk of autism-review and proposal," *BMC Pediatrics*, vol. 12, no. 1, 2012.

[47] E. R. Blackmore, J. A. Moynihan, D. R. Rubinow, E. K. Pressman, M. Gilchrist, and T. G. O'Connor, "Psychiatric symptoms and proinflammatory cytokines in pregnancy," *Psychosomatic Medicine*, vol. 73, no. 8, pp. 656–663, 2011.

[48] S. L. Buka, M. T. Tsuang, E. F. Torrey, M. A. Klebanoff, R. L. Wagner, and R. H. Yolken, "Maternal cytokine levels during pregnancy and adult psychosis," *Brain, Behavior, and Immunity*, vol. 15, no. 4, pp. 411–420, 2001.

[49] R. Donev and J. Thome, "Inflammation: good or bad for ADHD?," *ADHD Attention Deficit and Hyperactivity Disorders*, vol. 2, no. 4, pp. 257–266, 2010.

[50] S. E. P. Smith, J. Li, K. Garbett, K. Mirnics, and P. H. Patterson, "Maternal immune activation alters fetal brain development through interleukin-6," *The Journal of Neuroscience*, vol. 27, no. 40, pp. 10695–10702, 2007.

[51] R. B. Simerly, "Hypothalamic substrates of metabolic imprinting," *Physiology & Behavior*, vol. 94, no. 1, pp. 79–89, 2008.

[52] J. Lepercq, S. Hauguel-De Mouzon, J. Timsit, and P. M. Catalano, "Fetal macrosomia and maternal weight gain during pregnancy," *Diabetes & Metabolism*, vol. 28, no. 4, pp. 323–328, 2002.

[53] M. Lappas, M. Permezel, and G. E. Rice, "Leptin and adiponectin stimulate the release of proinflammatory cytokines and prostaglandins from human placenta and maternal adipose tissue via nuclear factor-$\kappa$B, peroxisomal proliferator-activated receptor-$\gamma$ and extracellularly regulated kinase 1/2," *Endocrinology*, vol. 146, no. 8, pp. 3334–3342, 2005.

[54] L. Arborelius, M. J. Owens, P. M. Plotsky, and C. B. Nemeroff, "The role of corticotropin-releasing factor in depression and anxiety disorders," *Journal of Endocrinology*, vol. 160, no. 1, pp. 1–12, 1999.

[55] P. Blardi, A. de Lalla, L. Ceccatelli, G. Vanessa, A. Auteri, and J. Hayek, "Variations of plasma leptin and adiponectin levels in autistic patients," *Neuroscience Letters*, vol. 479, no. 1, pp. 54–57, 2010.

[56] E. D'Asti, H. Long, J. Tremblay-Mercier et al., "Maternal dietary fat determines metabolic profile and the magnitude of endocannabinoid inhibition of the stress response in neonatal rat offspring," *Endocrinology*, vol. 151, no. 4, pp. 1685–1694, 2010.

[57] A. Sasaki, W. C. De Vega, P. Pan, and P. O. McGowan, "Perinatal high fat diet alters glucocorticoid signaling and anxiety behavior in adulthood," *Neuroscience*, vol. 240, pp. 1–12, 2013.

[58] J. Ishikawa, A. Ishikawa, and S. Nakamura, "Interferon-$\alpha$ reduces the density of monoaminergic axons in the rat brain," *NeuroReport*, vol. 18, no. 2, pp. 137–140, 2007.

[59] L. Naef, L. Srivastava, A. Gratton, H. Hendrickson, S. M. Owens, and C. D. Walker, "Maternal high fat diet during the perinatal period alters mesocorticolimbic dopamine in the adult rat offspring: reduction in the behavioral responses to repeated amphetamine administration," *Psychopharmacology*, vol. 197, no. 1, pp. 83–94, 2008.

[60] Z. Vucetic, J. Kimmel, K. Totoki, E. Hollenbeck, and T. M. Reyes, "Maternal high-fat diet alters methylation and gene

expression of dopamine and opioid-related genes," *Endocrinology*, vol. 151, no. 10, pp. 4756-4764, 2010.

[61] D. A. Rossignol and R. E. Frye, "A review of research trends in physiological abnormalities in autism spectrum disorders: immune dysregulation, inflammation, oxidative stress, mitochondrial dysfunction and environmental toxicant exposures," *Molecular Psychiatry*, vol. 17, no. 4, pp. 389–401, 2012.

[62] H. M. Rivera, K. J. Christiansen, and E. L. Sullivan, "The role of maternal obesity in the risk of neuropsychiatric disorders," *Frontiers in Neuroscience*, vol. 9, p. 194, 2015.

[63] C. A. Boyle, S. Boulet, L. A. Schieve et al., "Trends in the prevalence of developmental disabilities in US children, 1997–2008," *Pediatrics*, vol. 127, no. 6, pp. 1034-1042, 2011.

[64] M. Olfson, C. Blanco, S. Wang, G. Laje, and C. U. Correll, "National trends in the mental health care of children, adolescents, and adults by office-based physicians," *JAMA Psychiatry*, vol. 71, no. 1, pp. 81–90, 2014.

[65] D. L. Christensen, D. A. Bilder, W. Zahorodny et al., "Prevalence and characteristics of autism spectrum disorder among 4-year-old children in the autism and developmental disabilities monitoring network," *Journal of Developmental & Behavioral Pediatrics*, vol. 37, no. 1, pp. 1–8, 2016.

[66] F. J. Torres-Espinola, S. K. Berglund, L. M. García-Valdés et al., "Maternal obesity, overweight and gestational diabetes affect the offspring neurodevelopment at 6 and 18 months of age – a follow up from the PREOBE cohort," *PLoS One*, vol. 10, no. 7, article e0133010, 2015.

[67] M. Li, M. D. Fallin, A. Riley et al., "The association of maternal obesity and diabetes with autism and other developmental disabilities," *Pediatrics*, vol. 137, no. 2, article e20152206, 2016.

[68] J. R. Mann, C. Pan, G. A. Rao, S. McDermott, and J. W. Hardin, "Children born to diabetic mothers may be more likely to have intellectual disability," *Maternal and Child Health Journal*, vol. 17, no. 5, pp. 928–932, 2013.

[69] J. Huang, T. Zhu, Y. Qu, and D. Mu, "Prenatal, perinatal and neonatal risk factors for intellectual disability: a systemic review and meta-analysis," *PLoS One*, vol. 11, no. 4, article e0153655, 2016.

[70] A. Biri, A. Onan, E. Devrim, F. Babacan, M. Kavutcu, and I. Durak, "Oxidant status in maternal and cord plasma and placental tissue in gestational diabetes," *Placenta*, vol. 27, no. 2-3, pp. 327–332, 2006.

[71] X. Chen and T. O. Scholl, "Oxidative stress: changes in pregnancy and with gestational diabetes mellitus," *Current Diabetes Reports*, vol. 5, no. 4, pp. 282–288, 2005.

[72] A. M. Comi, A. W. Zimmerman, V. H. Frye, P. A. Law, and J. N. Peeden, "Familial clustering of autoimmune disorders and evaluation of medical risk factors in autism," *Journal of Child Neurology*, vol. 14, no. 6, pp. 388–394, 1999.

[73] T. L. Sweeten, S. L. Bowyer, D. J. Posey, G. M. Halberstadt, and C. J. McDougle, "Increased prevalence of familial autoimmunity in probands with pervasive developmental disorders," *Pediatrics*, vol. 112, no. 5, pp. e420–e420, 2003.

[74] H. Ó. Atladóttir, P. Thorsen, L. Østergaard et al., "Maternal infection requiring hospitalization during pregnancy and autism spectrum disorders," *Journal of Autism and Developmental Disorders*, vol. 40, no. 12, pp. 1423–1430, 2010.

[75] A. S. Brown, M. D. Begg, S. Gravenstein et al., "Serologic evidence of prenatal influenza in the etiology of schizophrenia," *Archives of General Psychiatry*, vol. 61, no. 8, pp. 774–780, 2004.

[76] K. A. Garbett, E. Y. Hsiao, S. Kalman, P. H. Patterson, and K. Mirnics, "Effects of maternal immune activation on gene expression patterns in the fetal brain," *Translational Psychiatry*, vol. 2, no. 4, p. e98, 2012.

[77] L. Shi, S. H. Fatemi, R. W. Sidwell, and P. H. Patterson, "Maternal influenza infection causes marked behavioral and pharmacological changes in the offspring," *The Journal of Neuroscience*, vol. 23, no. 1, pp. 297–302, 2003.

[78] L. Shi, N. Tu, and P. H. Patterson, "Maternal influenza infection is likely to alter fetal brain development indirectly: the virus is not detected in the fetus," *International Journal of Developmental Neuroscience*, vol. 23, no. 2-3, pp. 299–305, 2005.

[79] K. M. Money, T. L. Barke, A. Serezani et al., "Gestational diabetes exacerbates maternal immune activation effects in the developing brain," *Molecular Psychiatry*, 2017.

[80] A. H. Xiang, X. Wang, M. P. Martinez et al., "Association of maternal diabetes with autism in offspring," *JAMA*, vol. 313, no. 14, pp. 1425–1434, 2015.

[81] N. Goeden, J. Velasquez, K. A. Arnold et al., "Maternal inflammation disrupts fetal neurodevelopment via increased placental output of serotonin to the fetal brain," *The Journal of Neuroscience*, vol. 36, no. 22, pp. 6041–6049, 2016.

[82] E. Y. Hsiao and P. H. Patterson, "Activation of the maternal immune system induces endocrine changes in the placenta via IL-6," *Brain, Behavior, and Immunity*, vol. 25, no. 4, pp. 604–615, 2011.

[83] H. P. Li, X. Chen, and M. Q. Li, "Gestational diabetes induces chronic hypoxia stress and excessive inflammatory response in murine placenta," *International Journal of Clinical and Experimental Pathology*, vol. 6, no. 4, pp. 650–659, 2013.

[84] H. Gardener, D. Spiegelman, and S. L. Buka, "Prenatal risk factors for autism: comprehensive meta-analysis," *The British Journal of Psychiatry*, vol. 195, no. 01, pp. 7–14, 2009.

[85] C. K. Walker, P. Krakowiak, A. Baker, R. L. Hansen, S. Ozonoff, and I. Hertz-Picciotto, "Preeclampsia, placental insufficiency, and autism spectrum disorder or developmental delay," *JAMA Pediatrics*, vol. 169, no. 2, pp. 154–162, 2015.

[86] J. M. Roberts and C. A. Hubel, "The two stage model of preeclampsia: variations on the theme," *Placenta*, vol. 30, pp. 32–37, 2009.

[87] S. S. Field, "Omega-3 fatty acids, prematurity, and autism," *Pediatrics*, vol. 122, no. 6, pp. 1416-1417, 2008.

[88] A. J. Richardson, "Omega-3 fatty acids in ADHD and related neurodevelopmental disorders," *International Review of Psychiatry*, vol. 18, no. 2, pp. 155–172, 2006.

[89] L. F. Saugstad, "Infantile autism: a chronic psychosis since infancy due to synaptic pruning of the supplementary motor area," *Nutrition and Health*, vol. 20, no. 3-4, pp. 171–182, 2011.

[90] L. J. Stevens, S. S. Zentall, M. L. Abate, T. Kuczek, and J. R. Burgess, "Omega-3 fatty acids in boys with behavior, learning, and health problems," *Physiology & Behavior*, vol. 59, no. 4-5, pp. 915–920, 1996.

[91] I. B. Helland, L. Smith, K. Saarem, O. D. Saugstad, and C. A. Drevon, "Maternal supplementation with very-long-chain n-

[91] ...3 fatty acids during pregnancy and lactation augments children's IQ at 4 years of age," *Pediatrics*, vol. 111, no. 1, pp. e39–e44, 2003.

[92] J. G. Bell, E. E. MacKinlay, J. R. Dick, D. J. MacDonald, R. M. Boyle, and A. C. A. Glen, "Essential fatty acids and phospholipase A₂ in autistic spectrum disorders," *Prostaglandins, Leukotrienes and Essential Fatty Acids*, vol. 71, no. 4, pp. 201–204, 2004.

[93] N. A. Meguid, H. M. Atta, A. S. Gouda, and R. O. Khalil, "Role of polyunsaturated fatty acids in the management of Egyptian children with autism," *Clinical Biochemistry*, vol. 41, no. 13, pp. 1044–1048, 2008.

[94] M. M. Wiest, J. B. German, D. J. Harvey, S. M. Watkins, and I. Hertz-Picciotto, "Plasma fatty acid profiles in autism: a case-control study," *Prostaglandins, Leukotrienes and Essential Fatty Acids*, vol. 80, no. 4, pp. 221–227, 2009.

[95] S. S. Field, "Interaction of genes and nutritional factors in the etiology of autism and attention deficit/hyperactivity disorders: a case control study," *Medical Hypotheses*, vol. 82, no. 6, pp. 654–661, 2014.

[96] D. F. Horrobin and C. N. Bennett, "New gene targets related to schizophrenia and other psychiatric disorders: enzymes, binding proteins and transport proteins involved in phospholipid and fatty acid metabolism," *Prostaglandins, Leukotrienes and Essential Fatty Acids (PLEFA)*, vol. 60, no. 3, pp. 141–167, 1999.

[97] I. Hertz-Picciotto, P. G. Green, L. Delwiche, R. Hansen, C. Walker, and I. N. Pessah, "Blood mercury concentrations in CHARGE Study children with and without autism," *Environmental Health Perspectives*, vol. 118, no. 1, pp. 161–166, 2009.

[98] D. V. Keen, F. D. Reid, and D. Arnone, "Autism, ethnicity and maternal immigration," *The British Journal of Psychiatry*, vol. 196, no. 04, pp. 274–281, 2010.

[99] T. Anjos, S. Altmäe, P. Emmett et al., "Nutrition and neurodevelopment in children: focus on NUTRIMENTHE project," *European Journal of Nutrition*, vol. 52, no. 8, pp. 1825–1842, 2013.

[100] J. L. Beard, "Why iron deficiency is important in infant development," *The Journal of Nutrition*, vol. 138, no. 12, pp. 2534–2536, 2008.

[101] M. K. Georgieff, "Nutrition and the developing brain: nutrient priorities and measurement," *The American Journal of Clinical Nutrition*, vol. 85, no. 2, pp. 614S–620S, 2007.

[102] M. K. Georgieff, "The role of iron in neurodevelopment: fetal iron deficiency and the developing hippocampus," *Biochemical Society Transactions*, vol. 36, no. 6, pp. 1267–1271, 2008.

[103] S. K. Berglund, F. J. Torres-Espínola, L. García-Valdés et al., "The impacts of maternal iron deficiency and being overweight during pregnancy on neurodevelopment of the offspring," *British Journal of Nutrition*, vol. 118, no. 7, pp. 533–540, 2017.

[104] J. A. S. Vorstman, J. R. Parr, D. Moreno-de-Luca, R. J. L. Anney, J. I. Nurnberger Jr, and J. F. Hallmayer, "Autism genetics: opportunities and challenges for clinical translation," *Nature Reviews Genetics*, vol. 18, no. 6, pp. 362–376, 2017.

[105] R. A. Saxton and D. M. Sabatini, "mTOR signaling in growth, metabolism, and disease," *Cell*, vol. 168, no. 6, pp. 960–976, 2017.

[106] A. Perl, "mTOR activation is a biomarker and a central pathway to autoimmune disorders, cancer, obesity, and aging," *Annals of the New York Academy of Sciences*, vol. 1346, no. 1, pp. 33–44, 2015.

[107] S. S. Jeste, K. J. Varcin, G. S. Hellemann et al., "Symptom profiles of autism spectrum disorder in tuberous sclerosis complex," *Neurology*, vol. 87, no. 8, pp. 766–772, 2016.

[108] J. Sun, Y. Liu, J. Tran, P. O'Neal, M. Baudry, and X. Bi, "mTORC1-S6K1 inhibition or mTORC2 activation improves hippocampal synaptic plasticity and learning in Angelman syndrome mice," *Cellular and Molecular Life Sciences*, vol. 73, no. 22, pp. 4303–4314, 2016.

[109] A. E. Pohodich and H. Y. Zoghbi, "Rett syndrome: disruption of epigenetic control of postnatal neurological functions," *Human Molecular Genetics*, vol. 24, no. R1, pp. R10–R16, 2015.

[110] E. Schroeder, L. Yuan, E. Seong et al., "Neuron-type specific loss of CDKL5 leads to alterations in mTOR signaling and synaptic markers," *Molecular Neurobiology*, pp. 1–12, 2018.

[111] A. Sato, "mTOR, a potential target to treat autism spectrum disorder," *CNS & Neurological Disorders - Drug Targets*, vol. 15, no. 5, pp. 533–543, 2016.

[112] H. Kalkman and D. Feuerbach, "Microglia M2A polarization as potential link between food allergy and autism spectrum disorders," *Pharmaceuticals*, vol. 10, no. 4, p. 95, 2017.

[113] A. B. Steinmetz, S. A. Stern, A. S. Kohtz, G. Descalzi, and C. M. Alberini, "Insulin-like growth factor II targets the mTOR pathway to reverse autism-like phenotypes in mice," *The Journal of Neuroscience*, vol. 38, no. 4, pp. 1015–1029, 2018.

[114] F. I. Roullet and J. N. Crawley, "Mouse models of autism: testing hypotheses about molecular mechanisms," *Current Topics in Behavioral Neurosciences*, vol. 7, pp. 187–212, 2011.

# Mitoproteomics: Tackling Mitochondrial Dysfunction in Human Disease

María Gómez-Serrano[1,2] Emilio Camafeita,[2,3] Marta Loureiro,[1] and Belén Peral[4]

[1]*Laboratory of Cardiovascular Proteomics, Centro Nacional de Investigaciones Cardiovasculares (CNIC), 28029 Madrid, Spain*
[2]*Centro de Investigación Biomédica en Red sobre Enfermedades Cardiovasculares (CIBERCV), Spain*
[3]*Proteomics Unit, Centro Nacional de Investigaciones Cardiovasculares (CNIC), 28029 Madrid, Spain*
[4]*Instituto de Investigaciones Biomédicas Alberto Sols (IIBM), Consejo Superior de Investigaciones Científicas & Universidad Autónoma de Madrid (CSIC-UAM), 28029 Madrid, Spain*

Correspondence should be addressed to María Gómez-Serrano; maria.gomez@cnic.es and Belén Peral; bperal@iib.uam.es

Academic Editor: Mario Zoratti

Mitochondria are highly dynamic and regulated organelles that historically have been defined based on their crucial role in cell metabolism. However, they are implicated in a variety of other important functions, making mitochondrial dysfunction an important axis in several pathological contexts. Despite that conventional biochemical and molecular biology approaches have provided significant insight into mitochondrial functionality, innovative techniques that provide a global view of the mitochondrion are still necessary. Proteomics fulfils this need by enabling accurate, systems-wide quantitative analysis of protein abundance. More importantly, redox proteomics approaches offer unique opportunities to tackle oxidative stress, a phenomenon that is intimately linked to aging, cardiovascular disease, and cancer. In addition, cutting-edge proteomics approaches reveal how proteins exert their functions in complex interaction networks where even subtle alterations stemming from early pathological states can be monitored. Here, we describe the proteomics approaches that will help to deepen the role of mitochondria in health and disease by assessing not only changes to mitochondrial protein composition but also alterations to their redox state and how protein interaction networks regulate mitochondrial function and dynamics. This review is aimed at showing the reader how the application of proteomics approaches during the last 20 years has revealed crucial mitochondrial roles in the context of aging, neurodegenerative disorders, metabolic disease, and cancer.

## 1. Introduction

Mitochondria represent the metabolic dynamism of the cell, and they are present in the cytoplasm of all eukaryotic cells relying on aerobic metabolism [1]. These intracellular organelles are involved in a number of vital processes including the generation of adenosine triphosphate (ATP) by oxidative phosphorylation (OXPHOS), the determination of the redox state through the generation and detoxification of reactive nitrogen and oxygen species (RNOS), the regulation of calcium homeostasis, and cell death [2, 3]. Mitochondrial activity comprises various processes, including the activity of the mitochondrial electron transport chain (ETC) complexes, substrate oxidation through the tricarboxylic acid (TCA) cycle and ATP synthesis, $\beta$-oxidation of fatty acids, ketogenesis, and biosynthesis of pyrimidines, heme groups, and urea [4, 5]. Additionally, mitochondria sense oxygen and nutrient levels, providing the energy required for other cellular processes such as muscle contraction or heat production, and maintain ionic gradients and membrane potentials in excitable cells, allowing the secretion of hormones and neurotransmitters [6].

Due to the huge number of cellular processes in which this organelle is implicated, even in the closed context of cellular respiration, the term mitochondrial function can refer to a variety of features. In fact, the definition of mitochondrial capacity has been broadened during the last decade to include additional features such as mitochondrial dynamics, which includes fusion and fission events, turnover (biogenesis and mitophagy), and plasticity [7]. The consideration of these mitochondrial capabilities has advanced our understanding of how the cell may respond to metabolic disturbances such as oxidative stress, inflammation, and gluco- and lipotoxicity [8], among others. Moreover, mitochondrial dynamics is influenced by energy expenditure and nutrient supply, and changes in mitochondrial architecture have emerged as a new mechanism of adaptation to metabolic demand [9]. Therefore, the mitochondrion is both origin and target of several metabolic signals which orchestrate cellular function and homeostasis [4].

Mitochondrial function can be tackled through *in vivo*, *ex vivo*, and *in vitro* methods, which have been extensively reviewed [7, 10–13]. In general, measuring aspects of mitochondrial function *in vivo* is preferable to *in vitro* assessments, although this approach might be less specific and even become unattainable depending on the type of sample under study. Despite that these *in vivo*, *ex vivo*, and *in vitro* approaches provide invaluable data on specific mitochondrial functions (e.g., submaximal and maximal ADP-stimulated OXPHOS capacity, TCA cycle flux, rates of fatty acid uptake, and RNOS production), they have limitations. Thus, the invasiveness of the method, the indirect measurement of the parameter, the impossibility to discriminate between mitochondrial and cytosolic shared functions, the expensiveness of the hardware/methodology, and/or the lack of data on mitochondrial density usually hamper the application of these methodologies in the study of mitochondrial function [7]. In addition, while the aforementioned approaches address mitochondrial functions individually, the central role of mitochondria in cellular metabolism and maintenance demands the application of additional innovative techniques that have the capacity to provide a global view of mitochondria in a single experiment. Although in an indirect manner, proteomics offers this functional snapshot thanks to the robustness, velocity, sensitiveness, and precision that mass spectrometry (MS) has acquired during the last decades. MS is recognized as the method of choice for both protein identification and quantification, and it has been the driving force for proteomics research development [14, 15].

Here, we describe innovative approaches to the quantitation of the mitochondrial proteome, with an emphasis on the oxidative modifications involved. In addition, we include cutting-edge experiments aimed at unveiling the mitochondrial interactome, since to deepen the role of mitochondria in health and disease it is essential to monitor not only the alterations to mitochondrial protein composition, but also how these proteins organize in interaction networks that regulate mitochondrial function and dynamics. The scope of the present review is to show the reader how the application of proteomics methodologies during the last 20 years has discovered key mitochondrial roles in the context of aging, neurodegenerative disorders, metabolic disease, and cancer.

## 2. The Mitochondrial Proteome

As a reminiscence of its prokaryotic past, the mitochondrion hosts its own genome. In mammalian cells, the mitochondrial DNA (mtDNA) only encodes 13 proteins belonging to the mitochondrial respiratory chain, 2 ribosomal RNAs, and 22 transfer RNAs, determining its intramitochondrial translation code, which is markedly different from the extramitochondrial one and is tightly regulated to maintain mitochondrial function [16]. The remaining mitochondrial proteins, i.e., the components of the TCA cycle, $\beta$-oxidation and protein transport, the other respiratory chain subunits, and the apoptotic factors, are all nuclear-encoded, which makes the definition of the mitochondrial proteome extremely challenging.

Since most mitochondrial proteins have to be imported into the organelle, there are several mechanisms whereby they can be transported [17, 18]. The import of cytosolically synthesized mitochondrial proteins usually requires cytosolic chaperones, such as heat shock proteins 70 and 90. These chaperones guide pre-proteins to mitochondrial outer membrane receptors (primarily TOM70 and TOM20) of the TOM (translocase of the outer membrane) complex [19]. The successive import into the diverse suborganelle compartments (that is, outer membrane (OM), intermembrane space (IMS), inner membrane (IM), or mitochondrial matrix) depends on several targeting signals recognized by various mitochondrial protein import complexes (reviewed in [17]). Of note, the characterization of key components and pathways involved in protein targeting has been extensively described over the last three decades, including oxidative folding machinery in the IMS, which contributes to the redox-dependent control of proteostasis [20]. Moreover, protein sorting could be differentially regulated among different organisms and physiological or pathological conditions [21], currently constituting an area of intensive research [20, 22, 23].

*2.1. Key Aspects of the Mitochondrial Proteome.* To date, the wealth of data that have largely contributed to define the mitochondrial proteome comes from proteomics studies performed in a number of eukaryotic organisms. In the human context, the majority of the studies have been carried out in connection with disease research [24]. Hence, in the last 20 years, the study of the mitochondrial proteome has elucidated several properties about this extraordinary organelle, some of which are outlined below.

*2.1.1. Evolutionary Origin and Mitochondrial Proteome Composition.* Evidence suggests that current mitochondrion derives from an $\alpha$-proteobacterial endosymbiont which transferred its DNA to the host nucleus [25]. It is estimated that only 15–20% of the current human mitochondrial proteome derives from the original endosymbiont DNA [26], although mammalian mitochondrial proteins are more conserved (nearly 75%) compared to the rest of cellular components (around 48% of evolutionary conservation) [27, 28]. In fact, phylogenetic profiling has been very useful in elucidating the function of several mitochondrial proteins [28, 29].

At present, the number of mammalian mitochondrial proteins ranges from 1100 to 1900 depending on the database and resource used. In the last decade, despite of the broad dynamic range (up to six orders of magnitude) that mitochondrial proteins exhibit, which hinders the detection of less abundant proteins, MS analyses have significantly contributed to the description of the mitochondrial proteome [30, 31]. These results have underscored remarkable quantitative differences across different tissues.

*2.1.2. Tissue Heterogeneity.* Although the mitochondrial function is essential for almost all cell types (with the exception of red blood cells, which lack mitochondria), there are tissue-specific differences. Large-scale studies have provided valuable information about tissue or cell specificity. Thus, heart mitochondria are not only over 5-fold more abundant than adipose tissue mitochondria but also qualitatively different. In their seminal proteomics comparison across tissues, Mootha et al. found that mitochondria from different tissues shared approximately 75% of their components [32], and these data have been repeatedly confirmed with the subsequent extensive analyses of the organelle. As later revealed by Pagliarini and colleagues, mitochondria from developmentally related organs tend to share more proteins [28]. Out of the ~1100 mitochondrial proteins they initially described, nearly half of them appear to be core components virtually found in all tissues; however, some of these pathways showed striking patterns of tissue diversity. According to their results, the most representative mitochondrial proteome could be found in skeletal muscle followed by heart tissue, since both proteomes include the least proportion of less abundant mitochondrial proteins [28, 33]. Interestingly, respiratory complexes I (CI), II (CII), III (CIII), and V (CV) are found in high abundance in all the tissues surveyed in contrast to complex IV (CIV), which seems to have a reasonable number of subunits expressed in a tissue-dependent manner. In fact, protein expression strongly differs in complex IV and large ribosomal subunits of mitochondria between heart and adipose tissue, most probably responding to the differences in the energetic demand between these tissues [28, 33].

Apart from the differences in ETC complexes, some mitochondrial proteins and isoenzymes are preferentially, or even uniquely, expressed in certain cell types. For example, uncoupling protein-1 (UCP1) is a distinctive marker of mitochondria from brown adipocytes [34]. This protein supports the nonshivering thermogenesis by the uncoupling of mitochondrial respiration which characterizes brown adipose tissue.

Remarkably, the characterization of tissue-specific mitochondrial proteomes has enabled the description of the biosynthetic capacities of different organs and the construction of flux balance analysis models [35–37]. These innovative studies represent the use of the mitochondrial proteome as a framework for systems physiology analyses [38–40].

*2.1.3. Dual Localization of Mitochondrial Proteins.* Since different compartments can exhibit similar functions, it is not surprising that proteins can be dually localized in the cell. Generally, dual localization refers to those protein products from a single gene locus that are present in multiple subcellular locations [41]. In the case of mitochondria, it is noteworthy that they actively interact with other subcellular compartments like the endoplasmic reticulum, the nucleus, or the peroxisomes, which favours the dual localization of proteins. Thus, it has been estimated that around 15–30% of mitochondrial proteins are dual-localized [30, 42]. This dual localization can be achieved through several mechanisms, including alternative splicing of the same gene product or alternative start sites from the same gene locus, which produce distinct proteins. Additionally, some proteins may intermittently interact in the surface of mitochondria (e.g., hexokinase binds to the channel protein VDAC1 in order to release intramitochondrial ATP for glucose phosphorylation [43]), while others may translocate from the mitochondria to the cytosol in certain conditions (a typical example is the release of cytochrome c under proapoptotic conditions [44]) or vice versa (e.g., BID protein transfers from the cytosol into the mitochondrion after cell death stimulus [45]). In addition, certain proteins can localize to two different compartments under similar physiological conditions (e.g., the enzyme complex pyruvate dehydrogenase, which participates in the TCA cycle in the mitochondria, whereas in the nucleus it produces histone acetylation as part of the epigenetic regulation [46]).

In this complex scenario, the reports on the mitochondrial localization of proteins not usually associated with mitochondria or the localization of well-known mitochondrial proteins outside the organelle must be carefully considered. The validation of these results could be quite challenging and should involve at least two independent methods. In this regard, quantitative proteomics analysis of the total proteome and the corresponding mitochondrial subproteome could reveal even subtle quantitative differences originated by dual localization of mitochondrial proteins [24]. Additionally, other resources such as The Human Protein Atlas (http://www.proteinatlas.org) could provide orthogonal evidence to proteomic results based on subcellular localization data obtained through antibody-based techniques [47].

*2.1.4. The Mitochondrial Subproteome.* As expected from cellular complexity, not only organellar localization but also the precise compartmentalization within the proper organelle is important for protein function, adding an additional level of complexity to the study of the mitochondrial proteome. For instance, the IMS is key to ATP production by storing the electromotive force generated by the ETC, which in turn localizes to the OM (and proteins localized to membranes have very different properties from those found in the mitochondrial matrix or the cytosol). Nevertheless, up to date only a few proteomic studies have analysed the human mitochondrial subproteome. In this regard, Rhee and colleagues have greatly contributed to the description of the mitochondrial matrix [38] and IMS proteomes [48] by applying MS-compatible tagging techniques as APEX [49]. Among other advantages, APEX strategy selectively tags proteins that localize to any specific region in living cells, circumventing the requirement for organellar purification when applied to membrane-bound compartments. This *in vivo* system is based on ascorbate peroxidase (APX) activity, which produces biotin-phenol radicals that can conjugate to the side chain of amino acids in proteins only in its immediate proximity. Thus, once biotin-labelled proteins are obtained, they can be pulled down and subsequently analysed by MS [49]. Thanks to this approach, Rhee and coworkers described 31 [38] and 9 proteins [48] that had not been previously ascribed to the mitochondrion.

*2.2. Mitochondrial Databases.* Due to the critical roles that mitochondria play in both cell life and cell death, lots of efforts have been dedicated to obtaining a complete map of mitochondrial proteins from different tissues and organisms. To date, the application of MS methods has been the most successful strategy to define the mitochondrial proteome. Pioneering studies combining two-dimensional electrophoresis (2-DE) with MS methods date back to 1998 and 2001, with the analysis of mitochondrial extracts from human placenta [50] and a human neuroblastoma cell line [51], respectively. In 1999, MITOP database was released [52]. It was the first database for both nuclear- and mitochondrial-encoded proteins, comprising genetic and functional information on five species (including mouse and human), with annotated data derived from a variety of online resources and literature. Three years later, Taylor et al. published the first comprehensive human mitochondrial proteome, extending previous information about mitochondrial proteins identified by MS [53]. This study constituted the first step for the achievement of the MitoProteome Project (http://www.mitoproteome.org/) [54]. In its initial release, the Human Mitochondrial Protein Database contained 847 human mitochondrial protein sequences derived from public sequence databases and MS analyses [54]. This set was substantially revised in 2015, and, in the current version (as of February 2016), it covers information for 1705 genes and 3625 proteins. Additional reference databases have been released that list mitochondrial proteins identified via multiple approaches, including MS analysis, literature curations, and bioinformatics evaluations [55]. Some examples are MitoP2 [56] or MitoMiner [57]. The latter contains green fluorescence protein (GFP) localization images, is updated regularly, and provides useful lists of functionally related genes in up to 12 different organisms [58]. Currently, MitoMiner describes 1837 human proteins, providing experimental or bioinformatic prediction for mitochondrial localization. In 2008, Pagliarini et al. reported the largest compendium of mammalian mitochondrial proteins to date [28], the MitoCarta database (http://www.broadinstitute.org/pubs/MitoCarta), which encompassed 1098 genes coding for mouse mitochondrial proteins unveiled by experimental MS-based identification, bioinformatics analyses, and literature curation. This inventory was updated in 2016 after the use of improved transcript models, MS search algorithms, new database versions, and homology detection methods [27]. The current version (MitoCarta 2.0) consists of 1158 human and 1158 mouse genes encoding mitochondrial proteins whose localization and distribution across 14 tissues are provided [27].

Other interesting sources for mitochondrial annotation currently available are the UniProt database (http://www.uniprot.org) [59] and the Gene Ontology (GO) functional repository (http://www.geneontology.org) [60].

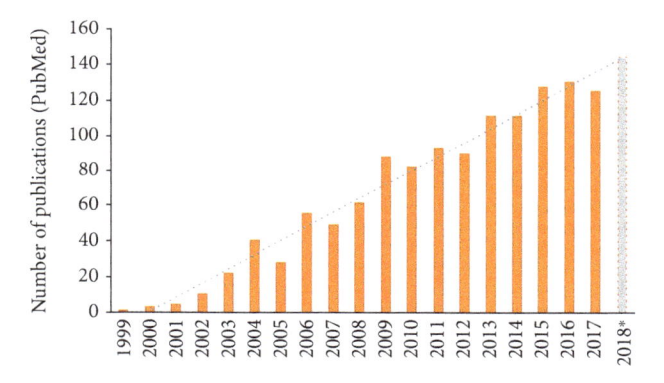

FIGURE 1: Timeline for "mitoproteomics" publications. Bars show the number of publications per year corresponding to the keywords "mitochondria" and "proteomics" in PubMed database (https://www.ncbi.nlm.nih.gov/pubmed). Note that the first proteomic study of human mitochondria [50] is not indexed. Results for 2018 are not included since they should be constantly updated during the writing process. According to the linear progression (dashed line), in 2018 we will probably reach a total of 144 publications related to mitochondrial proteomics (*grey bar).

Recently, large-scale studies like the Human Proteome Map (HPM) initiative (http://humanproteomemap.org/) [61], the Human Protein Atlas (http://www.proteinatlas.org) [47], and COPaKB (the Cardiac Organellar Protein Atlas Knowledgebase) (http://www.heartproteome.org/) [62] have also provided highly specific new data on the subcellular localization of mitochondrial proteins.

## 3. Addressing Mitochondrial Dysfunction through High-Throughput Proteomics Techniques

It is agreed that the first attempt to address the mitochondrial proteome was carried out by Rabilloud and colleagues 20 years ago [50]. Since then, the interest on mitochondrial proteomics has gradually increased, as reflected by a simple query on the PubMed database (https://www.ncbi.nlm.nih.gov/pubmed) (Figure 1). The publication of landmark proteomics studies describing the mitochondria of human [53, 63], rat [64, 65], mouse [28, 66], and yeast [67, 68] species, as well as those related to different tissues [32] and cells [69], have critically contributed to the definition of the mitochondrial proteome to date.

*3.1. Isolation of Mitochondria.* The starting point in the study of the mitochondrial proteome is mitochondrion isolation. Mitochondrion-enriched fractions have been used as starting material until differential centrifugation [70] became the method of choice. Three key steps can be distinguished in mitochondrial enrichment: (i) permeabilization and disruption of plasma membrane and release of mitochondria without compromising their integrity, (ii) differential centrifugation to remove cellular debris and larger organelles, and (iii) high-speed centrifugation to pellet the mitochondria [71]. The first step is usually performed through mechanical methods (e.g., mortar and pestle) under well-controlled buffer compositions. Some of the disadvantages of mechanical methods are excessive heat generation and stress on mitochondria, as well as increased intersample variability when executed manually [71]. In contrast, differential centrifugation should not compromise the quality or the yield of the preparation. Furthermore, additional gradient centrifugation protocols can be applied after differential centrifugation to obtain mitochondria with higher purity [72].

There are currently a few well-established protocols for isolating mitochondria [71, 73–75], although certain sample types might require specific procedures, as is the case of brain [76], adrenal glands [77], and adipose tissue [78]. Of note, the alternative use of magnetic devices for mitochondrial enrichment has gained popularity in recent years [78, 79].

The most common limitation of isolation strategies is the presence of carried-over nonmitochondrial proteins in the mitochondrial extracts, a phenomenon that is expected by virtue of the physical connection between the mitochondrion and other subcellular organelles [80]. However, high coverage of the total cellular proteome can be achieved thanks to recent improvements in protein identification methods based on liquid chromatography coupled with mass spectrometry (LC-MS) [24]. When combined with bioinformatics sorting of mitochondrial proteins, these data on protein identification reduce the necessity of purified mitochondria, which points out the combination of high-throughput MS analyses and database knowledge as a successful strategy, as previously described [78].

*3.2. Sample Preparation for Mitochondrial Proteomics.* As the identification of mitochondrial proteins is technically challenging due to their number and dynamic range, sample fractionation is desirable to reduce the complexity before MS analysis. While the first mitochondrial proteomics studies resorted to protein separation by 2-DE and peptide mass fingerprinting MS analysis to identify only a few dozen proteins [50, 51], later developments in LC-MS have facilitated the large-scale identification of mitochondrial proteins. In 2003, Taylor et al. identified 615 mitochondrial proteins from human heart by LC-MS (instead of 2-DE) analysis of 11 sucrose fractions [53], which represented a breakthrough in the identification of mitochondrial proteins. Later that year, gel filtration chromatography was used to separate mitochondrial proteins from brain, heart, kidney, and liver of C57BL6/J mice [32]. From each tissue, the authors obtained 15–20 fractions that were subjected to LC-MS analysis, which allowed the identification of 133 proteins not previously associated with this organelle. However, Mootha et al. did not overlook that their analysis was biased toward abundant molecules and that improvements in sample preparation, protein separation, and sensitiveness of MS were necessary. Many researchers have resorted to protein separation by sodium dodecyl sulfate (SDS) polyacrylamide gel electrophoresis (PAGE) followed by gel slicing and analysis of each band by high-performance LC-MS [28, 64, 66, 81]. Other less frequent fractionation methods have been applied prior to LC-MS, like the aforementioned sucrose gradients [53] and gel filtration chromatography [32].

*3.3. Identification of Mitochondrial Proteins.* Most proteomics experiments involve the enzymatic digestion of a complex protein extract prior to peptide separation, ionization, and analysis by LC-MS. For the last 15 years, LC-MS in data-dependent acquisition (DDA) mode has been the standard for proteomics identification [82, 83]. DDA is a milestone for high-throughput proteomics due to its efficiency at identifying peptides and proteins. In this method, peptide peaks (precursor ions) that overcome a threshold intensity in a survey MS scan are selected for fragmentation in the $MS^2$ mode. This fragmentation generates an experimental fragment ion pattern characteristic of the peptide amino acid sequence that can be matched against theoretical patterns derived from protein databases for peptide identification [84]. In spite of the power of the DDA method, random peptide sampling for fragmentation entails a bias towards those with the strongest signal, which hampers the reproducible quantitation of especially low-abundance peptides.

Conversely, targeted acquisition methods, most notably selected reaction monitoring (SRM) [85], focus the mass spectrometer capabilities on the fragmentation of a limited number of peptides representing proteins of interest with very high sensitivity and quantitative accuracy. However, owing to the limited fraction of the proteome amenable to SRM analysis, this method is not suitable for most proteomics experiments, where detailed protein composition is required. So far, only a few studies have resorted to targeted methods for the study of human mitochondrial protein profiles [86, 87]. Other authors have applied targeted methods to the study of such posttranslational modifications (PTMs) as acetylation [88, 89] and phosphorylation [90, 91] in the mitochondrial proteome.

In recent years, data-independent acquisition (DIA), which combines the advantages of DDA and SRM, has raised increasing interest. The DIA method performs the fragmentation of all the peptides within a defined mass-to-charge ($m/z$) window; the window is then progressively shifted until the full $m/z$ range is covered. This enables accurate peptide quantitation without being limited to profiling a predefined set of peptides of interest, but at the cost of losing the link between precursor and fragment ions thereof. An implementation of the DIA method, SWATH, relies on spectral libraries that collect $m/z$ values, relative intensities, retention times, etc., to trace the so-obtained compound fragmentation maps for specific peptides. Since these libraries must be carefully built up based on additional DDA experiments [92], only a few studies have relied on SWATH to address mitochondrial disturbances in disease models [93–95]. A recent development based on DIA, DiS (data-independent scanning), resorts to narrower isolation windows to facilitate peptide identification based on the conventional database search methods developed for DDA [96]. DiS has been successfully used to generate a highly detailed structural map of OXPHOS mitochondrial supercomplexes in several models, including the characterization of protein species and PTMs thereof that regulate complex and supercomplex assembly [96]. DIA methods have enormous potential for mitoproteomics, where highly curated databases (e.g., MitoCarta 2.0) will help improve mapping and quantification of mitochondrial proteins.

*3.4. Quantification of Mitochondrial Proteins.* Depending on the separation strategy selected, protein quantification can be performed through gel-based or MS-based (*gel-free*) approaches. In gel-based approaches, the quantification of differentially abundant proteins is conducted by comparison of gel images, and MS methods are set aside for protein identification. Paramount among gel-based strategies is difference gel electrophoresis (DIGE), an improvement on 2-DE in which the proteins from two samples are labelled with different fluorescent dyes and separated in the same gel, allowing a more accurate relative quantification as compared to 2-DE [97].

In contrast, *gel-free* strategies rely on mass spectrometric data for both protein identification and protein quantitation. At this point, it should be noted that *gel-free* approaches have higher yield and well-established advantages in quantification over gel-based techniques. The quantification method in MS-based approaches includes label-free and stable isotope labelling (SIL) methods. Most often, label-free quantification relies on the counting of $MS^2$ spectra from the peptides of a given protein [98]. The label-free approach has been used in numerous large-scale studies, as it provides a simple way to carry out relative comparisons across different samples with improved dynamic range of quantification [99]. Conversely, in SIL strategies, the relative quantification is based on the addition of chemically equivalent differential mass tags, which allow the comparison of up to ten samples in a single experiment. Despite that some limitations have been pointed out for SIL-based quantification, namely, limited linear dynamic range, complex sample preparation, and high cost of the reagents [100], this approach has great advantages in quantification accuracy and reproducibility. Some examples of SIL techniques successfully employed in comparative mitoproteomics are (i) metabolic labelling with stable isotope labelling by amino acids in cell culture (SILAC) [101, 102], (ii) enzymatic labelling with $^{18}O$ during proteolytic digestion [103, 104], and (iii) chemical labelling employing isotope-coded affinity tags (ICAT) [105, 106], isobaric tags for relative and absolute quantitation (iTRAQ) [78, 107–109], or tandem mass tags (TMT) [110, 111].

*3.5. Mitoproteomics in Human Disease.* Early mitoproteomics studies relied mainly on gel-based techniques for protein separation, such as 2-DE, in combination with a variety of MS methods. After 2-DE separation, the protein spots of interest must be sliced off from the gel and digested for later off-line MS analysis. In these pioneering studies, the underrepresentation of extreme pI, low abundant, or very hydrophobic proteins, which has been pointed out as an important drawback inherent to 2-DE, has hindered the identification of a higher number of mitochondrial proteins [50, 65, 112].

During the first decade after mitoproteomics implementation, alternative separation methods based on SDS-PAGE coupled with LC-MS were developed that allowed significant advances in the investigation of the mitochondrial proteome (reviewed in [100]). In the last decade, there has been a dramatic increase in the number of publications related to mitoproteomics (e.g., until 2008, the total of publications according to PubMed indexing was 273, a number that was overcome in just the last two years) (Figure 1).

Most mitoproteomics studies have been performed on animal models, as these offer more possibilities for mitochondrial isolation and purification. Results on these models give insight into basic functions of proteins and reveal the mechanisms of adaptation under loss of function of the respective proteins (e.g., knockout models), altered nutritional status (e.g., obesity), pharmacological treatment, or physiopathological stressors, shedding light on mitochondrial responses. In the case of human specimens, there are significant limitations regarding sample availability, some of which are related to obvious difficulties in collecting the material, whereas others derive from medical and ethical issues. Despite this bottleneck, during the last years the increased robustness of MS approaches and the application of multiplexing methods (such as iTRAQ or TMT), which substantially reduce the amount of sample needed, have counteracted the limitations inherent to working with human samples, bringing mitoproteomics closer to translational research [78].

Furthermore, the combination of mitoproteomics with other high-resolution techniques has provided new insights into mitochondrial disease. For instance, Walheim and coworkers have recently described the combination of quantitative mitochondrial respirometry (through Seahorse technology) and proteomics (LC-MS-based total protein approach) to understand how protein changes relate to mitochondrial energy capacity during diet-induced obesity [113]. This approach, termed "respiromics," has shown not only that CI concentration controls mitochondrial lipid oxidation in the liver, but also that obesity affects the functional capacity of this complex by regulating the nuclear- but not mitochondrial-encoded subunit concentrations.

In the following sections, the reader will find a selection of recent illustrative mitoproteomics studies in different pathological contexts, emphasizing the global perspective obtained on mitochondrial research thanks to these cutting-edge techniques.

*3.5.1. Aging.* Mitochondrial dysfunction is considered one of the hallmarks of aging [114], where several aspects of mitochondrial physiology are involved, such as mitochondrial biogenesis and turnover, calcium dynamics, energy sensing, RNOS production, and apoptosis [115]. In addition, aging is recognized as an independent risk factor for different pathologies such as cardiovascular disease (CVD), type 2 diabetes mellitus (T2DM), neurodegenerative disorders, or cancer [116, 117]. Several mitoproteomics studies have addressed the effect of aging in the context of these pathologies: see [78, 118] for metabolic disease, [119–121] for cardiac disturbances, and [122, 123] for neurodegenerative disorders. Downregulation of complex subunits and mitochondrial respiratory capacity is a common feature of normal aging mitochondria, although particular changes have been highlighted depending on the model, the age differences, and the particular tissue under study. For instance, in C57BL/6 mice, CI subunits have been described both to decrease [124] and to increase [123] with aging, and Stauch et al. have reported different patterns of metabolic and ETC modifications according to the age of mice in both synapses [125] and whole-brain mitochondria [126]. In the context of obesity, a pioneering study on human adipocyte mitochondria showed a general decrease in metabolic enzymes and ETC subunits in older compared to younger individuals, despite that all individuals were middle-aged (from 32 to 52 years) [78].

Of note, mitochondrial dysfunction has been addressed by proteomics in mouse models of premature aging (also known as Hutchinson-Gilford progeria syndrome (HGPS)) [127]. Although the genetic components of this disease are well-known (i.e., a dominant mutation in the LMNA gene leading to the expression of progerin), the mechanisms underlying cellular damage, senescence, and accelerated aging in HGPS are only partly understood. Resorting to SILAC and LC-MS analysis, the authors observed a pronounced downregulation of several components of the mitochondrial ATPase complex together with an upregulation of glycolytic enzymes. Interestingly, the analysis of different tissues revealed that mitochondrial dysfunction was time- and dose-dependent in the HPGS mouse models [127]. Altogether, these results underline the tight regulation undergone by mitochondria with aging, pointing out the importance of the surrounding context for this organelle.

*3.5.2. Neurodegenerative Diseases.* As previously introduced, mitoproteomics studies in the context of the nervous system have been closely related to aging research. This fact most probably responds to the observation of a decline in brain energy metabolism with aging [128]. This metabolic change is considered part of the physiological aging contributing to age-related neurodegeneration (e.g., Alzheimer's disease (AD) or Parkinson's disease (PD)), where different evidences support a key role of mitochondria in its progression [123, 129, 130]. Several proteomics approaches, including redox proteomics (see *Redox Mitoproteomics* Section in this review), have pointed out specific mitochondrial proteins as key regulators, providing new insights into their role in aberrant mitochondrial function and structure in AD and PD, among others (for an excellent review, see [131]). Interestingly, alterations to TCA enzymes and OXPHOS subunits are an early feature of mitochondrial dysfunction that can be observed prior to the deposition of significant amyloid plaques and neurofibrillary tangles in AD [132]. Causal links for mitochondrial abnormalities and oxidative damage in synaptic degeneration and amyloid-$\beta$ deposition have been proposed [133, 134]. Moreover, substantial differences between synaptic and nonsynaptic mitochondria have been previously reported [135, 136]. In the mitoproteomics field, Graham et al. have recently evidenced distinct molecular fingerprints between mitochondrial subpopulations that are likely to influence the synaptic morphology *in vivo*, highlighting mitochondrial CI as an upstream regulator in neurodegenerative pathologies [137]. Thus, mitoproteomics research is regarded as a promising tool for the investigation of preclinical and clinical biomarkers of neurodegeneration and its potential treatments [131].

*3.5.3. CVD.* Heart tissue is rich in mitochondria and highly dependent on the catabolism of branched-chain amino acids and fatty acids within this organelle. Mitochondrial function has been closely related to cardiomyopathies [138] and ischemic heart disease [139]. Although the precise underlying mechanisms are not fully understood, it is clear that mitochondrial metabolism is heavily involved in both cardiac injury [140] and cardiac protection [141].

Heart ischemia causes tissue injury through an acute phase followed by secondary damage. Notably, ischemic preconditioning (brief episodes of ischemia preceding prolonged ischemia) provides robust protection against cell death resulting from ischemia and reperfusion (I/R) injury in the heart and is currently explored as a new target for cardiac treatment [142]. In fact, a combination of mitoproteomics with respiratory assays and electron microscopy (EM) described a "mitochondrial preconditioning" independent of cytosol that confers protection against I/R-induced respiratory failure and oxidative damage [143].

The cardiac mitochondrial proteome has been widely explored in models of I/R [144–147]. Most of these studies have also relied on redox approaches (see *Redox Mitoproteomics* Section in this review), since mitochondrial RNOS generation has been described as a double-edged sword contributing to myocardial infarction and stroke as well as ischemic preconditioning [148].

*3.5.4. Metabolic Disease: Obesity and Diabetes.* In essence, obesity consists in an excessive fat accumulation due to an imbalance between energy expenditure and uptake. Sedentary lifestyle and overnutrition have greatly contributed to the present epidemic of obesity, which is also aggravated by the aging of population [149]. Mitochondrial dysfunction has long been associated to the development of insulin resistance (IR) and T2DM, since changes in physical activity and overnutrition have a direct impact upon mitochondrial function [4]. During the last decades, most mitoproteomics studies addressing IR and T2DM pathologies were based on skeletal muscle samples, therefore overlooking the relevance of mitochondria in adipose tissue and the systemic implications of its impairment (for an excellent review of mitoproteomics in the context of metabolic disease, see [118]). Most of these studies relied on animal or cellular models, which imposed additional limitations to the extent of their conclusions [150]. The key role for white adipose tissue in regulating energy expenditure and IR was clearly recognized during the first decade of the 2000s [5, 151], but by that time the only comparative study between white and brown adipocyte mitochondria had been carried out in mice [102]. Therefore, the study of human adipocyte mitochondria was of outmost importance.

In 2015, Lindinger et al. published the first mitoproteomics study on this area [152]. Despite the limitations of their analysis (e.g., only 62 putative mitochondrial proteins were identified), the authors validated four proteins (citrate synthase, mitofilin, HADHA, and LETM1) that were inversely associated with body mass index (BMI) in white adipocyte mitochondria from 76 patients. Their results confirmed previous studies regarding whole adipose tissue biopsies

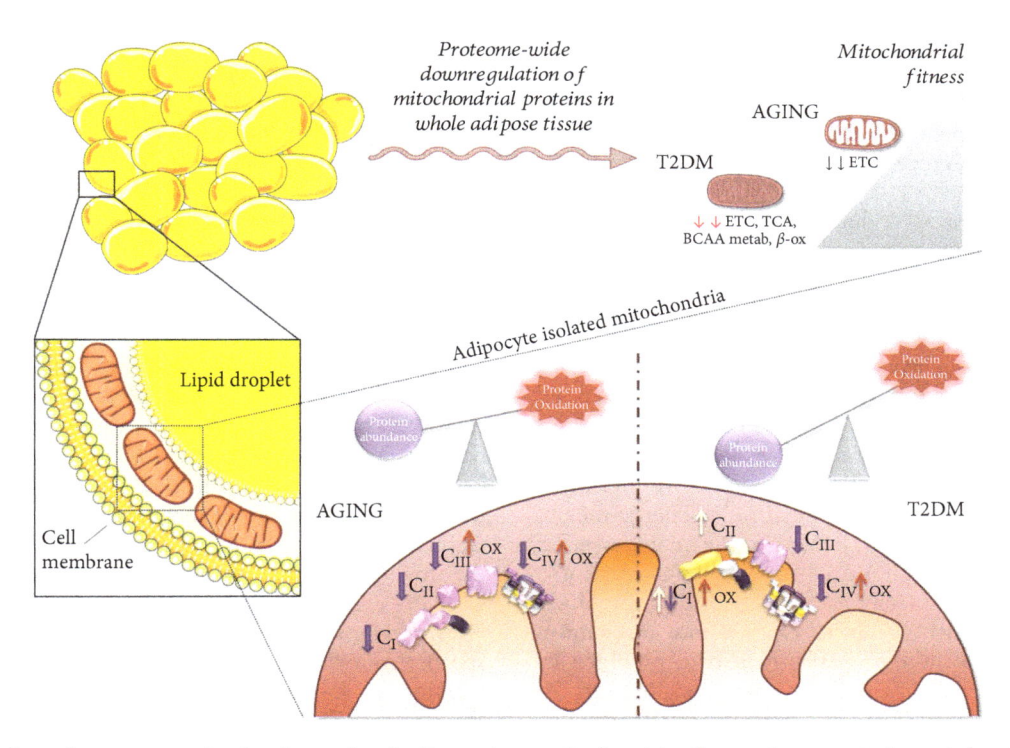

FIGURE 2: High-throughput proteomics for the study of adipose tissue mitochondria. Gomez-Serrano et al. carried out a deep study of mitochondrial alterations in human fat. In a first study based on whole adipose tissue samples, they described proteome-wide downregulation of the identified mitochondrial proteins in both aging and T2DM processes thanks to the application of systems biology analyses (*upper panel*). Alterations in relation to aging were mostly circumscribed to the electron transport chain (ETC), whereas disturbances in diabetic patients included also other metabolic pathways such as tricarboxylic acid (TCA) cycle, branched-chain amino acid (BCAA), metabolism, or $\beta$-oxidation, most likely resembling a decrease in mitochondrial mass with severe consequences on mitochondrial fitness [155]. In a second study, the authors proceeded to the isolation of adipocyte mitochondria in order to further analyse this organelle (*lower panel*). Resorting to a redox approach, they described thiol oxidative modifications as well as protein abundance changes [78]. The authors found that thiol protein oxidation was inversely correlated to protein levels in adipocyte mitochondria and that this relationship was more dramatic in T2DM compared to the aging process. Additionally, OXPHOS mitochondrial- *vs.* nuclear-encoded protein modules were altered in T2DM (note that complex modules are represented in different colours to indicate upregulation (yellow) and downregulation (purple), respectively). Thus, their results underscored defects in respiratory capacity and protein import in aging and T2DM. Of note, CIV emerged as a common target of oxidative damage connecting aging and T2DM development. Graphical elements from this figure were taken and adapted from Servier Medical Art Powerpoint image bank (http://smart.servier.com/). Servier Medical Art by Servier is licensed under a Creative Commons Attribution 3.0 Unported License.

[153, 154]. A few years later, Peral and colleagues published two complementary works that constitute the most comprehensive depiction of the human adipocyte mitochondrial proteome to date. In their first study, they described a global reduction of mitochondrial proteins (~150 proteins) in whole visceral adipose tissue biopsies from T2DM individuals by means of a high-throughput multiplexed proteomics approach based on iTRAQ technology [155]. In their second study, the authors deepened the proteome changes in adipocyte mitochondria with an improved redox approach [78] (see *Redox Mitoproteomics* Section for further details). Their results revealed impaired assembly of mitochondrial complexes together with defective protein import as a new potential cause of mitochondrial dysfunction in T2DM (Figure 2). In fact, the authors underlined the alteration of mitochondrial import pathways at different levels, like the TOM complex or the mitochondrial intermembrane space assembly (MIA) pathway [78], extending their previous results regarding the alteration of heat shock protein 70 in diabetic patients [154]. Additionally, the multiplexing capacity of their proteomics approach allowed the authors to address also the differences associated to aging in the same experiment, pointing out CIV as a common target for mitochondrial remodelling in both the aging and T2DM processes [78].

*3.5.5. Liver Disease.* Also, in close relation to aging or metabolic pathologies, we could find excellent works performed in the context of liver disease. Nonalcoholic fatty liver disease (NAFLD) is a chronic liver condition strongly associated with obesity, IR, and T2DM [156]. It can gradually progress to nonalcoholic steatohepatitis (NASH) in 12–40% of cases, cirrhosis (15% of NASH patients), and hepatic cancer (10%) [157], and oxidative damage on mitochondria has been tightly related to this fatal progression [158]. Thus, NAFLD and NASH attract a great interest among the scientific proteomics community (reviewed elsewhere [159]). These studies have revealed the importance of mitochondrial proteins in liver disease together with enzymes involved in methionine metabolism [160, 161], whose alteration may lead to an increase in both protein and lipid oxidative damage influencing the development of NAFLD and NASH. Furthermore, the beneficial action of metformin treatment in NAFLD has been also revealed in a mitoproteomic study of a mouse model [162].

Interestingly, ethanol-induced stress on liver has been also addressed through these approaches [163, 164], revealing an important modification of the mitochondrial proteome under these xenobiotic stimuli.

*3.5.6. Cancer.* Since the mitochondrion is the powerhouse of the cell, this organelle has long been related to carcinogenesis, cancer progression, and metastasis, although its specific role is still disputed [165]. It has been suggested that mitochondrial pathways such as RNOS signalling or $Ca^{2+}$ homeostasis could play a critical role in the alteration of energy metabolism in cancer cells [166, 167]. Besides, the Warburg effect [168], a shift from mitochondrial respiration towards glycolysis in cancer cell metabolism, constitutes a canonical observation for almost all cancer types whose mechanism remains obscure. Of note, the wealth of data provided by cancer mitoproteomics has provided new insights into the Warburg effect (outstandingly reviewed in [169]), although proteomic studies in cancer have been mostly used for tumor subtyping [170].

The heterogeneity of tumor tissue and cell composition hampers the analysis of the corresponding mitochondria, hindering the interpretation of the results. Notwithstanding the above, altered structure and protein levels have been reported for mitochondria from cancer cells, and mitochondria have also been found affected by antioncogenic treatments (reviewed in [170]). Interestingly, a recent study of Casal's group has revealed the mechanisms leading to OXPHOS downregulation in colorectal cancer [171]. By using SILAC technology, subcellular fractionation, LC-MS analysis, and bioinformatics, the authors described spatial proteome alterations at the subcellular level that help elucidate the molecular mechanisms underlying colorectal cancer metastasis [171]. Apart from mitochondrial dysfunction, their data suggest that proteins belonging to the TCA cycle and OXPHOS might have secondary localizations such as the nucleus and the cytoskeleton, confirming previous results in MCF7 breast cancer cells [172].

*3.5.7. Other Pathologies.* Because of the great interest in mitochondrial proteomics shown in the last decade, many examples of models or diseases addressed through these approaches can be found. Thus, proteomics studies have also revealed the importance of mitochondrial proteins in hypertension [173], osteoarthritis [174], and male infertility [175, 176], among others.

## 4. A Further Step: Redox Mitoproteomics

Since mitochondria contain major producers of RNOS, including components of the respiratory chain and other redox enzymes [177], together with powerful antioxidative defence systems [178], measuring redox damage in this organelle is central to deciphering disease pathogenesis. Thus, significantly increased mitochondrial RNOS levels have been associated with CVD [179–181], cancer [182, 183], neurodegenerative disease [184, 185], and aging [55].

Among many other chemically feasible oxidative PTMs that may occur in proteins, redox reactions involving Cys thiol (-S-H) groups are frequent, mostly reversible, enzymatically controlled, and selective *in vivo* [186, 187] and are considered the main way whereby oxidants integrate into cellular signal transduction pathways [188]. Common reversible Cys oxoforms include disulphide bonds (-S-S-), S-glutathionylation (-S-SG-), S-nitrosation (-S-NO), and S-sulfenylation (-S-OH), whereas sulfinic ($S-O_2H$) and sulfonic acid ($S-O_3H$) constitute irreversible Cys oxoforms (except for eukaryotic 2-Cys peroxiredoxins, where sulfinic acid may be reduced by sulfiredoxin [189]). Despite that the mechanisms that regulate these processes in the mitochondrion remain obscure, the respiratory chain should play an essential role, as it harbours the main producers of mitochondrial RNOS, CI [190], CII [191], and CIII [192]. Still, the development of redox mitoproteomics, i.e., the large-scale characterization of the specific mitochondrial Cys residues that sense the oxidative milieu, has been obstructed by the lability and low abundance of Cys oxoforms, which for decades could only be tackled by site-directed mutagenesis [193]. Proteomics approaches to protein Cys thiol oxidation provide a variety of methods for the enrichment, detection, and quantitation of such modifications, which can be classified into gel-based and gel-free methods.

*4.1. Gel-Based Redox Mitoproteomics.* Reversible S-nitrosation (-S-NO) affects a number of mitochondrial proteins and is thought to be a major way in which nitric oxide (NO) metabolism modulates mitochondrial physiology [194]. Galkin et al. used fluorescein-N-ethylmaleimide (NEM) labelling and Blue Native (BN) separation to identify by MS the specific residue of mitochondrially encoded ND3 subunit that becomes accessible to chemical modification only in the deactive form of mitochondrial CI from bovine heart mitochondria [195]. This enzyme had been previously suggested to undergo modification by S-nitrosation under pathological conditions during hypoxia or when the $NO:O_2$ ratio increases [196]. Under these conditions, RNOS and NO generation by the respiratory chain complexes and NO synthases, respectively, could cause the accumulation of the endogenous nitrosating agent peroxynitrite and result in permanent deactivation of the enzyme [196]. Shortly afterwards, S-nitrosation of the Cys switch of CI active/deactive transition was associated with marked cardioprotective effects in mice [197]. In 2007, Hurd et al. adapted DIGE [97] for the detection of proteins differentially oxidized by mitochondrial RNOS in mitochondria isolated from rat heart [198]. Based on the oxidation of a small subset of mitochondrial thiol proteins in the absence of bulk thiol changes, the authors suggested a possible link between mitochondrial RNOS production and the modulation of mitochondrial fatty acid and carbohydrate metabolism. Chouchani et al. further developed the DIGE method for the selective analysis of S-nitrosated proteins in mitochondria from rat heart and liver, where S-nitrosated proteins could be discriminated from proteins oxidized due to NO metabolism [199]. The authors identified 13 S-nitrosated mitochondrial proteins plus four that were oxidized, probably due to evanescent S-nitrosation relaxing to a reversible thiol modification. Interestingly, S-nitrosation was found to selectively and reversibly inhibit the activity of enzymes central to mitochondrial metabolism, namely, aconitase, mitochondrial aldehyde dehydrogenase, and α-ketoglutarate dehydrogenase. The contribution of NO-mediated S-nitrosation to the nerve injury-evoked pathology was shown in mouse spinal cord by Scheving et al., who used DIGE to reveal abundance changes of S-nitrosated proteins after sciatic nerve injury upon pretreatment with a NO synthase inhibitor [200]. A DIGE-based study also enabled quantitation of S-nitrosation in mitochondrial CI in normoxic and ischemic intact mouse hearts [201], where S-nitrosation had been associated with marked cardioprotective effects [197].

It has been suggested that changes to the sulfenome, which encompasses Cys sulfenic (-S-OH) groups derived from the initial oxidation of Cys thiol (-S-H) groups, play a central role in redox signalling. Thus, identifying the sulfenome under oxidative stress has been pointed out as a way to detect potential redox sensors. To date, the sulfenome has only been addressed in *Arabidopsis thaliana* cell suspensions exposed to $H_2O_2$ stress [202–204]. Using the dimedone derivative 4-(pent-4-yn-1-yl)cyclohexane-1,3-dione as a probe, Akter et al. identified by MS more than 200 sulfenylated proteins after the oxidative treatment of *Arabidopsis thaliana* cells, 14 of which located to the mitochondrion and were involved in the TCA cycle and the ETC [203].

*4.2. MS-Based Redox Mitoproteomics.* Early MS-based redox proteomics approaches resorted to the so-called biotin switch assay, whereby labile S-nitrosated Cys are converted to stable biotinylated Cys and the corresponding proteins purified by affinity chromatography and detected by immunoblotting or MS [205–208]. This assay allowed Hao et al. to pinpoint by LC-MS 68 S-nitrosation sites from 56 rat cerebellar proteins after treatment with S-nitrosoglutathione [205]. Some of these S-nitrosated Cys were involved in important mitochondrial processes like the formation of the permeability transition pore, the inactivation of sarcomeric creatine kinase, and mitochondrial relocation and neuroprotective activity of protein/nucleic acid deglycase DJ-1 [205]. The biotin switch assay has also been adapted for the analysis of additional Cys oxoforms (e.g., sulfenic acid [209–211]). Moreover, MS-based identification of S-glutathionylated proteins isolated from mouse liver helped to reveal a potential mechanism of resistance to acetaminophen-induced hepatotoxicity [212]. A combination of chemical enrichment methodologies with MS allowed Gould et al. to map 2596 sites of Cys modification including S-nitrosylation, S-glutathionylation, S-acylation, and S-sulfenylation, in wild-type mouse livers under normal physiological conditions [213]. Structural analysis localized these modifications in unique, evolutionary conserved protein segments outside commonly annotated functional regions.

Quantitative LC-MS approaches to Cys oxidation often start by stabilizing the redox status of the biological sample, usually by blocking free thiols with highly thiol-reactive reagents (e.g., iodoacetamide (IAM)). Thereafter, reversibly oxidized Cys residues are treated with a strong reducing agent like dithiothreitol (DTT) and the so-obtained free thiols alkylated with a second reagent to enable specific detection of Cys oxoforms. Differential Cys tagging with SIL reagents produces either chemically indiscernible light and heavy or isobaric peptides, which coelute in the chromatographic separation and exhibit identical ionization properties. Quantitation is then achieved by measuring the intensity of either neighbouring heavy and light precursor mass signals in the MS spectrum or reporter ions in the $MS^2$ spectrum [214–218]. Weerapana et al. used click chemistry conjugation and affinity enrichment to characterize Cys functionality in mouse and human proteomes based on the levels of Cys oxoforms, some of which pertained to mitochondrial proteins [219]. In 2008, Leichert et al. proposed the OxICAT method [220], based on the 2-plex ICAT (isotope-coded affinity tag) technology, to address the thiol redoxome (Figure 3, *left*). A comparative study of redoxome analysis based on DIGE [198] and OxICAT identified 63 proteins in mouse heart that were prone to thiol oxidation upon $H_2O_2$ treatment, many of which were mitochondrial TCA cycle members [221]. Results showed superior sensitivity of the OxICAT approach, which has been progressively used for investigating the thiol redoxome in a variety of nonmitochondrial biological samples, including *Saccharomyces cerevisiae* [222], *Schizosaccharomyces pombe* [223, 224], *Escherichia coli* [225], *Caenorhabditis elegans* [226], *Phaeodactylum tricornutum* [227], *Drosophila melanogaster* [228], and *Mus musculus* heart [180]. Of note, Go et al. investigated the effect of acute cadmium exposure on the redox proteome and metabolome of mitochondria isolated from mouse liver [229]. OxICAT unveiled significantly increased oxidation relative to controls in 1247 Cys residues from 547 proteins that were mapped to mitochondrial pathways like OXPHOS and pyruvate metabolism. A recent work by Topf et al. used OxICAT to quantify oxidation in more than 4300 Cys residues pertaining to ~2200 proteins from *S. cerevisiae* upon exogenous and intracellular mitochondria-derived oxidative stress [230]. Site-specifically mapping these redox-active Cys to the cellular translation machinery helped the authors identify redox switches for global translation modulation by mitochondrially produced RNOS. SIL labelling of peptides derived from S-glutathionylated mouse liver proteins based on 6-plex TMT has evidenced that this Cys modification has a fundamental role in energy metabolism [231]. Chouchani et al. also resorted to TMT-based, 6-plex peptide tagging to label reversibly oxidized thiols in mouse brown adipose tissue, where mitochondrial RNOS induction was identified as a mechanism that supports UCP1-dependent thermogenesis and whole-body energy expenditure [232]. Recently, alterations to S-glutathionylated mitochondrial protein levels were linked to fatiguing contractions in mice [233]. For that, the authors resorted to selective reduction with a glutaredoxin enzyme cocktail and resin-assisted enrichment followed by 10-plex TMT labelling of total

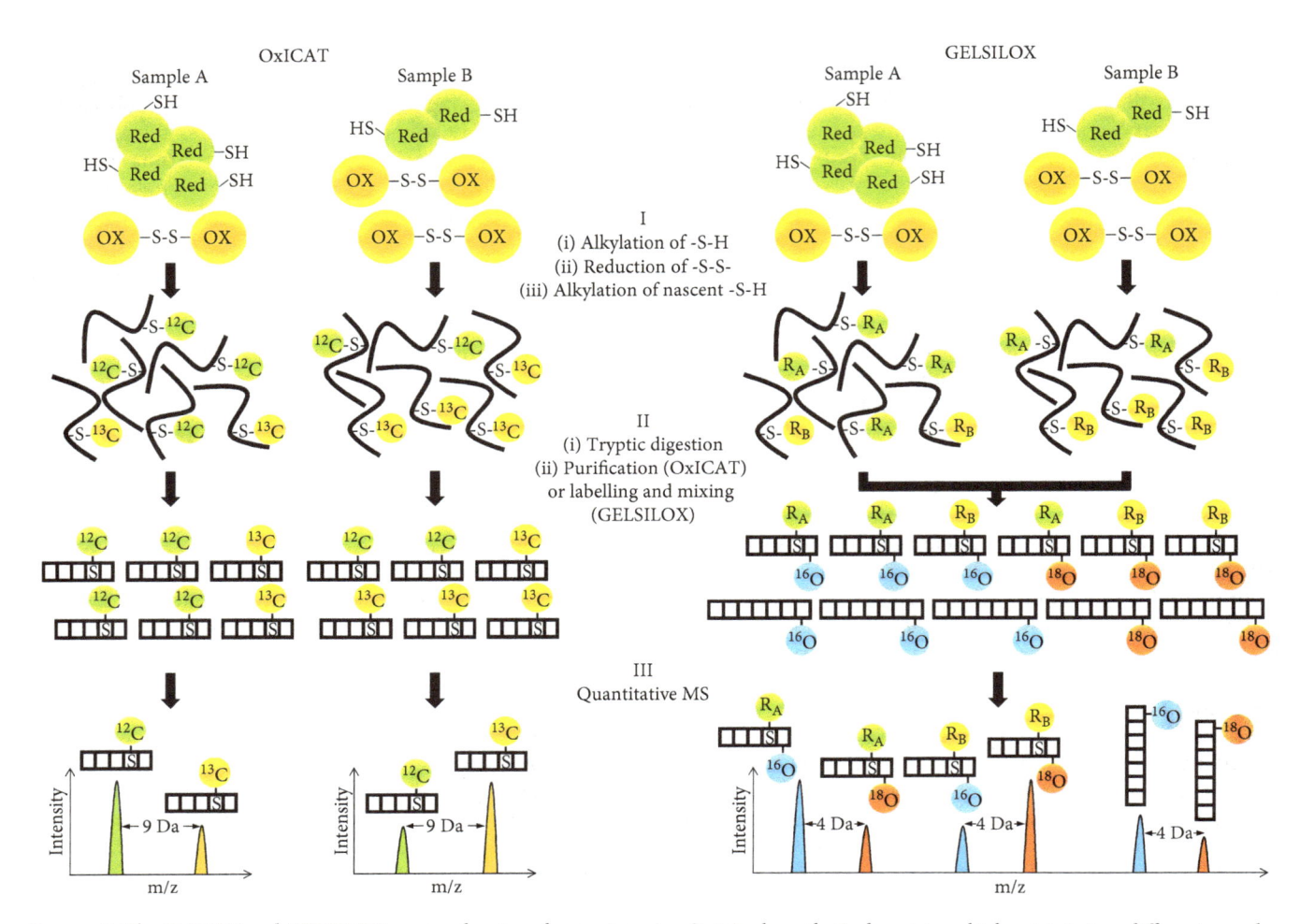

FIGURE 3: The OxICAT and GELSILOX approaches to redox proteomics. *Step I*: a hypothetical protein, which exists in two different samples A and B in both the reduced (green) and oxidized (disulfide-linked, yellow) form, is incubated with an alkylating agent ($^{12}$C-IAM for OxICAT, *left*, and, e.g., IAM, for GELSILOX, *right*) to block free sulfhydryl groups. Then, a reducing agent is used to cleave the disulfide bonds and the nascent thiol groups blocked using a second alkylating agent ($^{13}$C-IAM for OxICAT and, e.g., NEM, for GELSILOX). *Step II*: the proteins are digested with trypsin and the so-obtained peptides either affinity-purified (OxICAT) or tagged with $^{18}$O labelling prior to mixing (GELSILOX). *Step III*: two independent quantitative MS assays reveal the extent of thiol modification in samples A and B (OxICAT), whereas one single MS assay yields not only the relative amount of the reduced and oxidized forms but also the relative quantitation of the corresponding protein between the two samples based on the MS signals from noncysteine peptides.

thiol and S-glutathionylated peptides to unveil about 2200 S-glutathionylation sites.

These quantitative, MS-based methods have enabled not only the identification of redox-regulated thiol proteins, most of them of mitochondrial origin, but also to pinpoint the precise reactive Cys and to quantify their oxidation level. Nevertheless, their dependence on trapping and enrichment of Cys-containing peptides prevents parallel assessment of alterations in protein abundance. Given that redox signalling or oxidative stress could trigger changes in protein abundance owing to protein degradation or protein translocation, their assessment is essential to providing an integrated view of the underlying processes.

*4.3. The GELSILOX Approach.* To circumvent the above-described deficiencies, in 2012, Martínez-Acedo et al. proposed GELSILOX (gel-based stable isotope labelling of oxidized Cys, Figure 3, *right*) [234]. By combining differential alkylation of reduced and oxidized Cys with SIL of both Cys-containing and non-Cys peptides, GELSILOX provides simultaneous quantification of reversible Cys oxoforms and protein abundance. For that, the method relies on a statistical framework newly developed for SIL-based quantitative proteomics, the weighted spectrum, peptide, and protein (WSPP) model [235]. GELSILOX was initially applied to characterize the main redox targets of $H_2O_2$ in human endothelial cells as well as to study alterations to the thiol redoxome in mitochondria isolated from rat heart cardiomyocytes after the I/R insult [234]. The method was later applied by Ruiz-Meana et al. to show that decreased subsarcolemmal mitochondrial respiration in heart rat after I/R is paralleled by increased Cys oxidation, with ischemic preconditioning preventing both phenomena [143]. In their investigation of the role of altered mitochondrial function in aging, Fernandez-Sanz et al. resorted to GELSILOX to demonstrate increased oxidation of different subunits of mitochondrial

ATP synthase in hearts from old mice upon I/R [120]. Moreover, GELSILOX allowed Guaras et al. to hypothesize that specific Cys oxidation is linked to the differential stability and degradation of mitochondrial CI during reoxygenation. For that, the authors assessed the oxidation state of more than 1000 Cys residues from 784 mitochondrial proteins of mouse fibroblasts [96]. Of note, a combination of redox approaches allowed Gomez-Serrano et al. to point out defects in protein translocation to the mitochondrion as responsible for mitochondrial dysfunction in T2DM and aging. For that, the authors relied on a GELSILOX-based method to quantify 244 Cys oxidation sites within a set of 116 *bona fide* mitochondrial proteins that evidenced for the first time a link between increased thiol oxidation and decreased protein abundance under both conditions (Figure 2).

Despite tremendous advances of recent years, several challenges remain to be addressed by redox mitoproteomics. First, some of the aforementioned methodologies have been demonstrated appropriate for *in vitro* studies, and therefore their suitability to tackle the mitochondrial redoxosome *in vivo* is in question. For this reason, and also because redox proteomics studies based on human mitochondria are still scarce, the experimental evidence for physiological readouts of mitochondrial Cys thiol oxidation is yet limited. In this regard, care must be taken if reagents with widespread use like DTT and dimedone are incorporated to *in vivo* assays, as the former can stimulate RNOS production by mitochondrial enzymes [236] and the latter can react rapidly with sulfenyl amides [237]. Another hurdle to overcome when investigating the mitochondrial redoxome is the isolation of the organelle, which has traditionally been carried out by ultracentrifugation [238] and has the drawback of carrying over organelles with similar size profiles like lysosomes. In this regard, the effective enrichment of mitochondria is often a lengthy procedure, which is detrimental to the preservation of oxidative PTMs. Furthermore, as stated above, prevalent methods like OxICAT fail to provide parallel assessment of protein abundance changes in the same assay that unveils the alterations of Cys oxoforms, which hampers data interpretation.

Finally, notwithstanding the importance of oxidative modifications, many other PTMs take place in the mitochondrion that are responsive to the stressful and changing environmental conditions in which this organelle exerts its functions, e.g., phosphorylation, acetylation, ubiquitination, and succinylation. The functional significance of these PTMs and their interplay in the mitochondrion are still under investigation.

## 5. The Mitochondrial Interactome

Another step forward on defining the mitochondrial proteome was to understand the roles and consequences of protein-protein interactions (PPIs), as proteins do not usually function in isolation but interact with other molecules. Understanding PPIs is fundamental for the development of systems biology and novel therapeutics. A number of public databases provide comprehensive lists of physical PPIs in human as well as in other species [239]. The STRING (Search Tool for the Retrieval of Interacting Genes/Proteins) database (http://string-db.org/) contains both known and predicted protein-protein interactions [240]. IntAct (http://www.ebi.ac.uk/intact) and MINT databases provide molecular interaction data and curation tools for annotation [241]. Both IntAct and MINT are active contributors to the IMEx consortium (http://www.imexconsortium.org), which provides a nonredundant set of physical molecular interaction data from a broad taxonomic range of organisms. The Database of Interacting Proteins (DIP, http://dip.doe-mbi.ucla.edu/dip/) is an integration of experimentally determined PPIs on different organisms and contains an interesting visualization tool which provides a graphical illustration of these interactions [242]. In The Human Protein Reference Database (HPRD) the information of PPI is manually curated from the published literature and it also integrates the information of domain architecture, PTMs, interaction networks, and disease association for each protein in the human proteome [243].

The information about the physical interactions contained in these and other nonmentioned databases was derived from several methodologies applied to dissect PPIs. Methods based on immunoprecipitation or two-hybrid system in combination with Western blotting have been widely used to identify interaction partners of specific proteins. However, these approaches are limited to a reduced number of proteins.

The recognition that proteomes are highly interconnected networks, or interactomes, has changed biology's view of cause and effect, simultaneously highlighting the importance of systems biology and necessitating its use in deciphering complex biological phenomena for the development of novel therapeutics [64, 244–246]. This concept of connectivity modulating systems explains how functionally dissimilar systems, such as tissues, can have largely similar compositions [61, 247, 248].

In this regard, the combination of affinity purification and MS (AP-MS) has emerged as a powerful approach to delineate biological processes. The method consists in tagging a bait protein with an affinity tag such as His-tag, flag-tag, or tandem affinity purification (TAP) tag for expression *in vivo*. The group of partners interacting with the bait protein can be purified from the cell lysate by affinity or immunoaffinity techniques and analysed by MS for identification of the interacting components [249]. This technique has been applied to study interactions in a high-throughput manner [246, 250, 251], but also subsets of specific networks have been analysed, such as the interaction partners of COX4 for the study of CIV assembly [252], the characterization of ATPase protein components [253], and the interaction partners of the inner mitochondrial membrane protein LETM1 [254]. In the neurodegeneration research field this technique has unveiled parkin-associated proteins that are implicated in the mechanism of mitophagy of defective mitochondria [255], and a map of mitochondrial proteins involved in neurodegenerative diseases was obtained in 2017 [256]. Other recent works have described the bcl-2 interactome in human lung adenocarcinoma cells [257], as well as the interaction network of a set of 50 mitochondrial proteins that lacked significant functional annotation [258] to gain insight into their role in health and disease.

However, the largest set of AP-MS experiments to date is Bio-Plex [259, 260], consisting of more than 5891 individual experiments with the human ORFeome (the collection of open reading frames (ORFs)). The data can be explored through the web application Bio-Plex (http://bioplex.hms.harvard.edu/index.php) [261]. This platform allows the exploration of individual networks, highly interconnected network communities, subthreshold interactions, and specific experimental information like peptide spectral match counts for individual AP-MS experiments [260].

In spite of the power of AP-MS for studying interactions, it still suffers from some limitations. The global study of PPIs in a given system requires tagging every ORF of interest to provide a measurable readout or to enable purification and identification of the protein complex. In addition, protein tagging can be time-consuming and can disrupt interactions or alter the localization of the protein complex [262]. Moreover, these approaches generally trigger perturbations to the system, such as genetic modification of the native cells or disruption of the original cellular context, which increases the rate of both false-positive and false-negative results [263]. The selection of few cell types under artificial expression conditions also poses limitations to the interactome characterization [246, 259]. For these reasons, current networks remain of limited use in interpreting biological phenomena, as the observed interactions may not reflect the interactome landscapes within other systems or in disease. Alternative approaches are necessary for mapping native, endogenous interactomes under multiple conditions [264], as well as to survey the global organization of protein complexes within mitochondria. For the last years, protein complexes have gathered attention, as proteins arrange to form macromolecular assemblies that facilitate their function and regulation. Although macromolecular complexes are essential players in numerous biological processes, a limited number of methods suitable for defining the composition and functional dynamics of such assemblies have been described. Thus, the comprehensive mapping of these complex networks of stable and transient associations remains a key goal.

To study macromolecular complexes, it is necessary to apply mild, nondestructive techniques during sample processing to obtain the native complexes. Density gradient centrifugation, size exclusion chromatography (SEC), and native electrophoresis have been applied to define large multiprotein complexes, e.g., the respiratory chain supercomplexes in mitochondria [265, 266] and pyruvate dehydrogenase and other complexes from chloroplasts [267]. Density gradient centrifugation requires large protein amounts and suffers from low resolution. SEC provides a robust workflow for the separation of cytoplasmic complexes; however, it is not compatible with membrane complexes as they are extremely sensitive to the separation conditions [264, 268, 269], and size exclusion columns lead to sample dilution and cannot be used for complexes larger than 5 MDa [270]. Notwithstanding these drawbacks, interesting results have been achieved (e.g., the combination of quantitative proteomics and SEC to map 291 coeluting complexes, where the use of triplex labelling enabled monitoring of interactome rearrangements [262]). Due to these reasons, BN electrophoresis has been the preferred method for fractionating mitochondrial membrane protein complexes. Below is a description of the methods available for the study of the mitochondrial interactome.

*5.1. BN-PAGE.* BN-PAGE was developed by Schägger and von Jagow [271] and was firstly used for the isolation and characterization of the respiratory complexes from bovine mitochondria [272, 273]. BN electrophoresis is a special case of native electrophoresis for high-resolution separation of enzymatically active protein complexes from tissue homogenates and cell fractions, between 10 and 10,000 kDa [274]. Higher resolution can be attained by decreasing the pore size in the polyacrylamide gradient gel. The principle of BN-PAGE is simple, yet powerful: the protein complexes interact with Coomassie dye through hydrophobic interactions and, since the dye is negatively charged, become negatively charged as well. With this negative charge, the protein complexes migrate, at neutral pH, towards the anode, the same way that SDS forces proteins to run towards the anode in SDS-PAGE [249]. The difference between SDS-based SDS-PAGE and Coomassie-based BN-PAGE is that SDS is a strong denaturing agent, while Coomassie dye does not affect the integrity of the native protein complexes. Therefore, due to the external charge induced by Coomassie dye, BN-PAGE separates native protein complexes according to their molecular mass in the 100–1500 (or even higher) kDa mass range. BN-PAGE provides a vast amount of information: complex size, subunit composition and stoichiometry, assembly of protein complexes into supercomplexes, and relative abundance and stability of subcomplexes within protein complexes. In addition, the relative abundance of these protein complexes can also be determined [249].

BN electrophoresis in combination with MS analysis [275] emerged as one of the methods of choice for the analysis of the "complexome" (protein complex proteome) when in 2009 Wessels and coworkers reported the first "complexome profiling" [265]. The authors analysed a mitochondrial enriched fraction from a human HEK cell line and identified 48 of the 71 canonical subunits of the respiratory chain complexes (CI–CIV). Manually cut slices from BN gels were subjected to label-free, semiquantitative LC-MS, and protein abundance profiles were determined across the gel. Protein correlation profiling (PCP) was performed to identify potentially interacting proteins. Later that year, another work confirmed the potential of this technique studying how the composition and formation of respiratory chain complexes change under anaerobic conditions in mitochondrial protein extracts from S. cerevisiae [276].

*5.2. PCP-MS-Based Approaches.* PCP is based on the generation of individual protein elution or migration profiles, where comparison with other profiles by computational clustering and other approaches allows the identification of putative interacting proteins on the basis of similarities in their elution profiles [277, 278]. Since its discovery, lots of works performed on mitochondria have been published [264, 270, 279–282], and the methodology has been gradually improved: first, regarding the quality of native gel separation and gel slicing [281], and second, based on the evolution of mass spectrometers, which are faster and more sensitive [283].

Combined PCP-MS approaches have been used to characterize mitochondrial complexomes, leading to the discovery of TMEM126B as an assembly factor of respiratory chain CI [270], the analysis of human mitochondrial ribosomal complexes [282], the identification of novel assemblies of voltage-dependent anion channels/porins and TOM proteins [281], the analysis of changes occurring in cytoplasmic and mitochondrial interactomes in response to apoptosis initiation by caspase activity [264], the description of the assembly mechanism for CI [280], and the superassembly of CIII and CIV [279]. Although the aforementioned works performed the correlation profiles based on protein migration in a BN gel, elution profiles derived from other techniques, such as biochemical fractionation or SEC, are also feasible [262, 284, 285].

A potential advantage of the PCP approach is that hundreds to thousands of protein complexes can be simultaneously and rapidly analysed. A disadvantage of the PCP approach is that currently only soluble complexes with interactions that are not markedly weakened by the buffers used can be analysed [278].

*5.3. Alternative Techniques: An Insight into Cross-Linking.* Finally, another outstanding MS approach that can provide new evidences on the systems-wide organization of proteins in intact mitochondria is cross-linking mass spectrometry (XL-MS). In XL-MS, native protein contacts are captured using a cross-linker, which is typically a small organic molecule composed of a spacer arm and two functional groups that are reactive toward specific residue side chains. After proteolytic digestion of the cross-linked sample, residue-to-residue cross-links can be localized by MS-based peptide sequencing. A cross-link can only occur if the cross-linker can bridge the distance between the residues. Detected cross-links, therefore, reveal maximum residue-to-residue distance constraints within and in between proteins, providing insights into protein conformations, protein complex architectures, and protein interaction networks [286].

Although the identification of cross-linked peptides is not trivial even for purified protein complexes, which may be available in large quantity [263], recent advances in cross-linking methods have enabled high-throughput identification of protein interactions in complex mixtures and living cells [287], allowing, e.g., the determination of the molecular structure of the mammalian mitochondrial ribosome [288]. XL-MS provides data about the location of the interaction site in the biomolecule under study, an information that cannot be derived from the other methods for studying PPIs. Many works have contributed to improving (i) methods to enrich cross-links [289–291], (ii) cross-linking chemistries [263, 292–297], (iii) detection and identification strategies [292, 296, 298–305], and (iv) tools to perform structural analysis based on cross-linking sites [306–310]. The most widely used cross-linkers are active esters, such as disuccinimidyl suberate and bis(sulfosuccinimidyl) suberate, which induce nucleophilic attacks on primary amines and thus rely on the coupling of Lys residues [311].

Recent works have performed high-throughput investigation of protein structures and interactions at the mitochondrial proteome-wide level using XL-MS approaches [286, 287]. By combining cross-link identifications from 11 mitochondrial preparations from 4 different tissues, Schweppe et al. reported 1920 unique residue-to-residue connections and found intercomplex cross-linked peptides, supporting the existence of the respirosome in an excellent agreement with cryo-EM models [287]. Very recently, Liu et al. analysed intact mouse heart mitochondria and identified 3322 unique residue-to-residue contacts, the largest survey of mitochondrial protein interactions reported so far. In addition, the authors revealed the mitochondrial localization of four proteins not yet included in the MitoCarta database [286].

The major limitation of XL-MS from the structural point of view is that it cannot directly determine the relative stoichiometry of subunits in a complex, although this information can be derived from complementary quantitative MS methods. In addition, the field remains heterogeneous, with various experimental protocols and software applications, and this variety may seem intimidating to newcomers. Nevertheless, XL-MS studies of partially purified complexes will certainly provide new insights about protein interaction networks and their changes on perturbation, e.g., as a result of mutations connected to diseases. If this technique becomes optimized for complex samples, it will solidify the relevance of cross-linking-based methods not only in structural biology but also in systems biology, thus advancing the convergence of structural and cell biology [311].

## 6. Outlook

These are exciting times for exploring the human mitochondrial proteome. Since this highly active organelle contains major producers of RNOS, encompassing redox enzymes and OXPHOS proteins, measuring the redox damage with innovative redox proteomics-based approaches will be crucial to puzzle out human pathology in the near future. Mitochondria and energy metabolism are key components in life and death regulation. It is not surprising that dysfunctional mitochondria lie beneath the physiological process of aging and the unhealthy aging. Despite that research efforts have resulted in significant advances in the knowledge of the mitochondrial proteome, the identification of low-abundance proteins and the detection of very subtle expression changes under human physiological and pathological conditions is still challenging. In the coming years, improved MS instrumentation together with the combination of advanced labelling and fractionation methods will likely render a deeper view of the mitochondrial proteome that will shed light on the human diseases connected to mitochondria. So far, it is noteworthy that the application of systems biology analyses to big data sets accomplished through proteomics has greatly contributed to the depiction of the mitochondrial network. This achievement is particularly interesting, as it enables to assess whether and how protein abundance changes occur in a coordinated manner, a biological event that is well-known [312]. In the near future, interactomics will provide an upper level of information about complex regulation both inside and outside the mitochondrion, revealing the role of intricate protein networks in the whole cell metabolism. Thus, proteomics implementation for the study of the mitochondrial interactome constitutes a step forward that must be taken in the near future, especially in the context of human disease, where such kind of studies is scarce.

Furthermore, to understand the regulation of these proteins and networks, the high-throughput analysis of PTMs in health and disease is also of outmost importance. The unprecedented capacity of current MS-based approaches provides a vast amount of biologically relevant information, producing a molecular picture of the complex interplay between proteins and regulators. Of note, many PTMs have been described as molecular switches directly involved in protein activation and/or inhibition. Oxidative modifications constitute an exciting area of research due to the implications of RNOS in cell signalling and oxidative damage. Given that redox mitoproteomics has just started to be explored, the application of these approaches in different pathological models will probably elucidate novel mechanisms of redox regulation in human disease. Other PTM analyses have been performed in the mitochondrion, notably phosphorylation, acetylation, succinylation, and ubiquitination (reviewed in [313]); nevertheless, high-throughput studies have been hindered due to technical issues (e.g., usually only one, at most two, type of modification can be addressed in a single experiment). Therefore, the development of advanced algorithms for PTM discovery and their application in mitoproteomics will probably constitute a new milestone in mitochondrial research.

## Abbreviations

| | |
|---|---|
| 2-DE: | Two-dimensional gel electrophoresis |
| AD: | Alzheimer's disease |
| AP-MS: | Affinity purification and tandem mass spectrometry |
| ATP: | Adenosine triphosphate |
| BMI: | Body mass index |
| BN: | Blue native |
| CI: | Complex I, NADH dehydrogenase |
| CII: | Complex II, succinate dehydrogenase |
| CIII: | Complex III, cytochrome c-oxidoreductase |
| CIV: | Complex IV, cytochrome c oxidase |
| CV: | Complex V or ATP synthase |
| CVD: | Cardiovascular disease |
| DIA: | Data-independent acquisition |
| DIGE: | Difference gel electrophoresis |
| DiS: | Data-independent scanning |
| DDA: | Data-dependent acquisition |
| DTT: | Dithiothreitol |
| EM: | Electron microscopy |
| ETC: | Electron transport chain |
| GELSILOX: | Gel-based stable isotope labelling of oxidized Cys |
| GFP: | Green fluorescence protein |
| GO: | Gene ontology |
| HPLC: | High-performance liquid chromatography |
| I/R: | Ischemia/reperfusion |
| IAA: | Iodoacetic acid |
| IAM: | Iodoacetamide |
| ICAT: | Isotope-coded affinity tag |
| IM: | Inner membrane |
| IMS: | Intermembrane space |
| IR: | Insulin resistance |
| iTRAQ: | Isobaric tags for relative and absolute quantitation |
| LC: | Liquid chromatography |
| LC-MS: | Liquid chromatography coupled with mass spectrometry |
| MIA: | Mitochondrial intermembrane space assembly |
| $MS^2$: | Fragmentation spectrum |
| MS: | Mass spectrometry |
| mtDNA: | Mitochondrial DNA |
| *m/z*: | Mass-to-charge |
| NEM: | N-Ethylmaleimide |
| NO: | Nitric oxide |
| OM: | Outer membrane |
| ORF: | Open reading frame |
| OXPHOS: | Oxidative phosphorylation |
| PAGE: | Polyacrylamide gel electrophoresis |
| PD: | Parkinson disease |
| PCP: | Protein correlation profiling |
| PPIs: | Protein-protein interactions |
| PTM: | Posttranslational modification |
| RNOS: | Reactive nitrogen and oxygen species |
| SDS: | Sodium dodecyl sulfate |
| SEC: | Size-exclusion chromatography |
| SIL: | Stable isotope labelling |
| SILAC: | Stable isotope labelling by amino acids in cell culture |
| SRM: | Selected reaction monitoring |
| T2DM: | Type 2 diabetes mellitus |
| TAP: | Tandem affinity purification |
| TCA: | Tricarboxylic acid |
| TCEP: | Tris(2-carboxyethyl)phosphine |
| TMT: | Tandem mass tags |
| TOM: | Translocase of the outer membrane |
| UCP1: | Uncoupling protein-1 |
| XL-MS: | Cross-linking mass spectrometry. |

## Acknowledgments

The proteomic data supporting this systematic review are from previously reported studies and datasets, which have been cited throughout the text. This work was supported by grant SAF2012-33014 from the Ministry of Economy, Industry and Competitiveness (MEIC), Spain, and partially financed with FEDER funds (to B.P.). CNIC is supported by the Ministerio de Ciencia, Innovación y Universidades, and the Pro CNIC Foundation and is a Severo Ochoa Center of Excellence (SEV-2015-0505).

# References

[1] B. Alberts, A. Johnson, J. Lewis, P. Walter, M. Raff, and K. Roberts, *Molecular Biology of the Cell*, International Student Edition, Routledge, 4th edition, 2002.

[2] Z. Cheng and M. Ristow, "Mitochondria and metabolic homeostasis," *Antioxidants & Redox Signaling*, vol. 19, no. 3, pp. 240–242, 2013.

[3] A. A. Starkov, "The role of mitochondria in reactive oxygen species metabolism and signaling," *Annals of the New York Academy of Sciences*, vol. 1147, no. 1, pp. 37–52, 2008.

[4] A. W. Gao, C. Canto, and R. H. Houtkooper, "Mitochondrial response to nutrient availability and its role in metabolic disease," *EMBO Molecular Medicine*, vol. 6, no. 5, pp. 580–589, 2014.

[5] C. M. Kusminski and P. E. Scherer, "Mitochondrial dysfunction in white adipose tissue," *Trends in Endocrinology and Metabolism*, vol. 23, no. 9, pp. 435–443, 2012.

[6] M. R. Duchen, "Mitochondria in health and disease: perspectives on a new mitochondrial biology," *Molecular Aspects of Medicine*, vol. 25, no. 4, pp. 365–451, 2004.

[7] C. Koliaki and M. Roden, "Alterations of mitochondrial function and insulin sensitivity in human obesity and diabetes mellitus," *Annual Review of Nutrition*, vol. 36, no. 1, pp. 337–367, 2016.

[8] C. A. Galloway and Y. Yoon, "Mitochondrial morphology in metabolic diseases," *Antioxidants & Redox Signaling*, vol. 19, no. 4, pp. 415–430, 2013.

[9] M. Liesa and O. S. Shirihai, "Mitochondrial dynamics in the regulation of nutrient utilization and energy expenditure," *Cell Metabolism*, vol. 17, no. 4, pp. 491–506, 2013.

[10] B. Kalyanaraman, V. Darley-Usmar, K. J. A. Davies et al., "Measuring reactive oxygen and nitrogen species with fluorescent probes: challenges and limitations," *Free Radical Biology & Medicine*, vol. 52, no. 1, pp. 1–6, 2012.

[11] G. J. Kemp and K. M. Brindle, "What do magnetic resonance-based measurements of Pi→ATP flux tell us about skeletal muscle metabolism?," *Diabetes*, vol. 61, no. 8, pp. 1927–1934, 2012.

[12] C. G. R. Perry, D. A. Kane, I. R. Lanza, and P. D. Neufer, "Methods for assessing mitochondrial function in diabetes," *Diabetes*, vol. 62, no. 4, pp. 1041–1053, 2013.

[13] J. Szendroedi, E. Phielix, and M. Roden, "The role of mitochondria in insulin resistance and type 2 diabetes mellitus," *Nature Reviews Endocrinology*, vol. 8, no. 2, pp. 92–103, 2011.

[14] N. Gregersen, J. Hansen, and J. Palmfeldt, "Mitochondrial proteomics—a tool for the study of metabolic disorders," *Journal of Inherited Metabolic Disease*, vol. 35, no. 4, pp. 715–726, 2012.

[15] M. Nikolov, C. Schmidt, and H. Urlaub, "Quantitative mass spectrometry-based proteomics: an overview," *Methods in Molecular Biology*, vol. 893, pp. 85–100, 2012.

[16] C. M. Gustafsson, M. Falkenberg, and N. G. Larsson, "Maintenance and expression of mammalian mitochondrial DNA," *Annual Review of Biochemistry*, vol. 85, no. 1, pp. 133–160, 2016.

[17] A. Chacinska, C. M. Koehler, D. Milenkovic, T. Lithgow, and N. Pfanner, "Importing mitochondrial proteins: machineries and mechanisms," *Cell*, vol. 138, no. 4, pp. 628–644, 2009.

[18] D. Stojanovski, P. Bragoszewski, and A. Chacinska, "The MIA pathway: a tight bond between protein transport and oxidative folding in mitochondria," *Biochimica et Biophysica Acta (BBA) - Molecular Cell Research*, vol. 1823, no. 7, pp. 1142–1150, 2012.

[19] W. Neupert and J. M. Herrmann, "Translocation of proteins into mitochondria," *Annual Review of Biochemistry*, vol. 76, no. 1, pp. 723–749, 2007.

[20] L. MacPherson and K. Tokatlidis, "Protein trafficking in the mitochondrial intermembrane space: mechanisms and links to human disease," *Biochemical Journal*, vol. 474, no. 15, pp. 2533–2545, 2017.

[21] A. M. Sokol, M. E. Sztolsztener, M. Wasilewski, E. Heinz, and A. Chacinska, "Mitochondrial protein translocases for survival and wellbeing," *FEBS Letters*, vol. 588, no. 15, pp. 2484–2495, 2014.

[22] A. B. Harbauer, R. P. Zahedi, A. Sickmann, N. Pfanner, and C. Meisinger, "The protein import machinery of mitochondria-a regulatory hub in metabolism, stress, and disease," *Cell Metabolism*, vol. 19, no. 3, pp. 357–372, 2014.

[23] M. Opalinska and C. Meisinger, "Metabolic control via the mitochondrial protein import machinery," *Current Opinion in Cell Biology*, vol. 33, pp. 42–48, 2015.

[24] J. Palmfeldt and P. Bross, "Proteomics of human mitochondria," *Mitochondrion*, vol. 33, pp. 2–14, 2017.

[25] S. G. E. Andersson, A. Zomorodipour, J. O. Andersson et al., "The genome sequence of *Rickettsia prowazekii* and the origin of mitochondria," *Nature*, vol. 396, no. 6707, pp. 133–140, 1998.

[26] M. A. Huynen, M. de Hollander, and R. Szklarczyk, "Mitochondrial proteome evolution and genetic disease," *Biochimica et Biophysica Acta (BBA) - Molecular Basis of Disease*, vol. 1792, no. 12, pp. 1122–1129, 2009.

[27] S. E. Calvo, K. R. Clauser, and V. K. Mootha, "MitoCarta2.0: an updated inventory of mammalian mitochondrial proteins," *Nucleic Acids Research*, vol. 44, D1, pp. D1251–D1257, 2016.

[28] D. J. Pagliarini, S. E. Calvo, B. Chang et al., "A mitochondrial protein compendium elucidates complex I disease biology," *Cell*, vol. 134, no. 1, pp. 112–123, 2008.

[29] T. Gabaldon, D. Rainey, and M. A. Huynen, "Tracing the evolution of a large protein complex in the eukaryotes, NADH:ubiquinone oxidoreductase (complex I)," *Journal of Molecular Biology*, vol. 348, no. 4, pp. 857–870, 2005.

[30] L. J. Foster, C. L. de Hoog, Y. Zhang et al., "A mammalian organelle map by protein correlation profiling," *Cell*, vol. 125, no. 1, pp. 187–199, 2006.

[31] T. Kislinger, B. Cox, A. Kannan et al., "Global survey of organ and organelle protein expression in mouse: combined proteomic and transcriptomic profiling," *Cell*, vol. 125, no. 1, pp. 173–186, 2006.

[32] V. K. Mootha, J. Bunkenborg, J. V. Olsen et al., "Integrated analysis of protein composition, tissue diversity, and gene regulation in mouse mitochondria," *Cell*, vol. 115, no. 5, pp. 629–640, 2003.

[33] S. E. Calvo and V. K. Mootha, "The mitochondrial proteome and human disease," *Annual Review of Genomics and Human Genetics*, vol. 11, no. 1, pp. 25–44, 2010.

[34] B. Cannon and J. Nedergaard, "Brown adipose tissue: function and physiological significance," *Physiological Reviews*, vol. 84, no. 1, pp. 277–359, 2004.

[35] D. T. Johnson, R. A. Harris, P. V. Blair, and R. S. Balaban, "Functional consequences of mitochondrial proteome heterogeneity," *American Journal of Physiology-Cell Physiology*, vol. 292, no. 2, pp. C698–C707, 2007.

[36] D. T. Johnson, R. A. Harris, S. French et al., "Tissue heterogeneity of the mammalian mitochondrial proteome," *American Journal of Physiology-Cell Physiology*, vol. 292, no. 2, pp. C689–C697, 2007.

[37] T. D. Vo, H. J. Greenberg, and B. O. Palsson, "Reconstruction and functional characterization of the human mitochondrial metabolic network based on proteomic and biochemical data," *Journal of Biological Chemistry*, vol. 279, no. 38, pp. 39532–39540, 2004.

[38] H. W. Rhee, P. Zou, N. D. Udeshi et al., "Proteomic mapping of mitochondria in living cells via spatially restricted enzymatic tagging," *Science*, vol. 339, no. 6125, pp. 1328–1331, 2013.

[39] I. Thiele, N. D. Price, T. D. Vo, and B. O. Palsson, "Candidate metabolic network states in human mitochondria. Impact of diabetes, ischemia, and diet," *Journal of Biological Chemistry*, vol. 280, no. 12, pp. 11683–11695, 2005.

[40] E. G. Williams, Y. Wu, P. Jha et al., "Systems proteomics of liver mitochondria function," *Science*, vol. 352, no. 6291, article aad0189, 2016.

[41] O. Yogev and O. Pines, "Dual targeting of mitochondrial proteins: mechanism, regulation and function," *Biochimica et Biophysica Acta (BBA) - Biomembranes*, vol. 1808, no. 3, pp. 1012–1020, 2011.

[42] R. Ben-Menachem, M. Tal, T. Shadur, and O. Pines, "A third of the yeast mitochondrial proteome is dual localized: a question of evolution," *Proteomics*, vol. 11, no. 23, pp. 4468–4476, 2011.

[43] D. J. Roberts and S. Miyamoto, "Hexokinase II integrates energy metabolism and cellular protection: Akting on mitochondria and TORCing to autophagy," *Cell Death & Differentiation*, vol. 22, no. 2, pp. 248–257, 2015.

[44] X. Liu, C. N. Kim, J. Yang, R. Jemmerson, and X. Wang, "Induction of apoptotic program in cell-free extracts: requirement for dATP and cytochrome c," *Cell*, vol. 86, no. 1, pp. 147–157, 1996.

[45] X. Luo, I. Budihardjo, H. Zou, C. Slaughter, and X. Wang, "Bid, a Bcl2 interacting protein, mediates cytochrome c release from mitochondria in response to activation of cell surface death receptors," *Cell*, vol. 94, no. 4, pp. 481–490, 1998.

[46] G. Sutendra, A. Kinnaird, P. Dromparis et al., "A nuclear pyruvate dehydrogenase complex is important for the generation of acetyl-CoA and histone acetylation," *Cell*, vol. 158, no. 1, pp. 84–97, 2014.

[47] M. Uhlen, L. Fagerberg, B. M. Hallstrom et al., "Proteomics. Tissue-based map of the human proteome," *Science*, vol. 347, no. 6220, article 1260419, 2015.

[48] V. Hung, P. Zou, H. W. Rhee et al., "Proteomic mapping of the human mitochondrial intermembrane space in live cells via ratiometric APEX tagging," *Molecular Cell*, vol. 55, no. 2, pp. 332–341, 2014.

[49] J. D. Martell, T. J. Deerinck, Y. Sancak et al., "Engineered ascorbate peroxidase as a genetically encoded reporter for electron microscopy," *Nature Biotechnology*, vol. 30, no. 11, pp. 1143–1148, 2012.

[50] T. Rabilloud, S. Kieffer, V. Procaccio et al., "Two-dimensional electrophoresis of human placental mitochondria and protein identification by mass spectrometry: toward a human mitochondrial proteome," *Electrophoresis*, vol. 19, no. 6, pp. 1006–1014, 1998.

[51] N. K. Scheffler, S. W. Miller, A. K. Carroll et al., "Two-dimensional electrophoresis and mass spectrometric identification of mitochondrial proteins from an SH-SY5Y neuroblastoma cell line," *Mitochondrion*, vol. 1, no. 2, pp. 161–179, 2001.

[52] C. Scharfe, P. Zaccaria, K. Hoertnagel et al., "MITOP: database for mitochondria-related proteins, genes and diseases," *Nucleic Acids Research*, vol. 27, no. 1, pp. 153–155, 1999.

[53] S. W. Taylor, E. Fahy, B. Zhang et al., "Characterization of the human heart mitochondrial proteome," *Nature Biotechnology*, vol. 21, no. 3, pp. 281–286, 2003.

[54] D. Cotter, P. Guda, E. Fahy, and S. Subramaniam, "MitoProteome: mitochondrial protein sequence database and annotation system," *Nucleic Acids Research*, vol. 32, no. 90001, pp. 463D–4467, 2004.

[55] H. Cui, Y. Kong, and H. Zhang, "Oxidative stress, mitochondrial dysfunction, and aging," *Journal of Signal Transduction*, vol. 2012, Article ID 646354, 13 pages, 2012.

[56] M. Elstner, C. Andreoli, U. Ahting et al., "MitoP2: an integrative tool for the analysis of the mitochondrial proteome," *Molecular Biotechnology*, vol. 40, no. 3, pp. 306–315, 2008.

[57] A. C. Smith and A. J. Robinson, "MitoMiner, an integrated database for the storage and analysis of mitochondrial proteomics data," *Molecular & Cellular Proteomics*, vol. 8, no. 6, pp. 1324–1337, 2009.

[58] A. C. Smith and A. J. Robinson, "MitoMiner v3.1, an update on the mitochondrial proteomics database," *Nucleic Acids Research*, vol. 44, D1, pp. D1258–D1261, 2016.

[59] The UniProt Consortium, "UniProt: a hub for protein information," *Nucleic Acids Research*, vol. 43, pp. D204–D212, 2015.

[60] M. Ashburner, C. A. Ball, J. A. Blake et al., "Gene ontology: tool for the unification of biology," *Nature Genetics*, vol. 25, no. 1, pp. 25–29, 2000.

[61] M.-S. Kim, S. M. Pinto, D. Getnet et al., "A draft map of the human proteome," *Nature*, vol. 509, no. 7502, pp. 575–581, 2014.

[62] N. C. Zong, H. Li, H. Li et al., "Integration of cardiac proteome biology and medicine by a specialized knowledgebase," *Circulation Research*, vol. 113, no. 9, pp. 1043–1053, 2013.

[63] N. Lefort, Z. Yi, B. Bowen et al., "Proteome profile of functional mitochondria from human skeletal muscle using one-dimensional gel electrophoresis and HPLC-ESI-MS/MS," *Journal of Proteomics*, vol. 72, no. 6, pp. 1046–1060, 2009.

[64] F. Forner, L. J. Foster, S. Campanaro, G. Valle, and M. Mann, "Quantitative proteomic comparison of rat mitochondria from muscle, heart, and liver," *Molecular & Cellular Proteomics*, vol. 5, no. 4, pp. 608–619, 2006.

[65] M. Fountoulakis, P. Berndt, H. Langen, and L. Suter, "The rat liver mitochondrial proteins," *Electrophoresis*, vol. 23, no. 2, pp. 311–328, 2002.

[66] J. Zhang, X. Li, M. Mueller et al., "Systematic characterization of the murine mitochondrial proteome using functionally validated cardiac mitochondria," *Proteomics*, vol. 8, no. 8, pp. 1564–1575, 2008.

[67] H. Prokisch, C. Scharfe, D. G. Camp et al., "Integrative analysis of the mitochondrial proteome in yeast," *PLoS Biology*, vol. 2, no. 6, article e160, 2004.

[68] A. Sickmann, J. Reinders, Y. Wagner et al., "The proteome of *Saccharomyces cerevisiae* mitochondria," *Proceedings of the National Academy of Sciences of the United States of America*, vol. 100, no. 23, pp. 13207–13212, 2003.

[69] T. Geiger, A. Wehner, C. Schaab, J. Cox, and M. Mann, "Comparative proteomic analysis of eleven common cell lines reveals ubiquitous but varying expression of most proteins," *Molecular & Cellular Proteomics*, vol. 11, no. 3, article M111.014050, 2012.

[70] G. H. Hogeboom, W. C. Schneider, and G. E. Pallade, "Cytochemical studies of mammalian tissues; isolation of intact mitochondria from rat liver; some biochemical properties of mitochondria and submicroscopic particulate material," *Journal of Biological Chemistry*, vol. 172, no. 2, pp. 619–635, 1948.

[71] J. Wettmarshausen and F. Perocchi, "Isolation of functional mitochondria from cultured cells and mouse tissues," *Methods in Molecular Biology*, vol. 1567, pp. 15–32, 2017.

[72] J. M. Graham, "Purification of a crude mitochondrial fraction by density-gradient centrifugation," *Current Protocols in Cell Biology*, vol. 4, pp. 3.4.1–3.4.22, 2001.

[73] C. Frezza, S. Cipolat, and L. Scorrano, "Organelle isolation: functional mitochondria from mouse liver, muscle and cultured fibroblasts," *Nature Protocols*, vol. 2, no. 2, pp. 287–295, 2007.

[74] F. Pallotti and G. Lenaz, "Isolation and subfractionation of mitochondria from animal cells and tissue culture lines," *Methods in Cell Biology*, vol. 80, pp. 3–44, 2007.

[75] S. Schulz, J. Lichtmannegger, S. Schmitt et al., "A protocol for the parallel isolation of intact mitochondria from rat liver, kidney, heart, and brain," *Methods in Molecular Biology*, vol. 1295, pp. 75–86, 2015.

[76] C. Chinopoulos, S. F. Zhang, B. Thomas, V. Ten, and A. A. Starkov, "Isolation and functional assessment of mitochondria from small amounts of mouse brain tissue," *Methods in Molecular Biology*, vol. 793, pp. 311–324, 2011.

[77] P. Solinas, H. Fujioka, B. Tandler, and C. L. Hoppel, "Isolation of rat adrenocortical mitochondria," *Biochemical and Biophysical Research Communications*, vol. 427, no. 1, pp. 96–99, 2012.

[78] M. Gomez-Serrano, E. Camafeita, J. A. Lopez et al., "Differential proteomic and oxidative profiles unveil dysfunctional protein import to adipocyte mitochondria in obesity-associated aging and diabetes," *Redox Biology*, vol. 11, pp. 415–428, 2017.

[79] B. Tang, L. Zhao, R. Liang, Y. Zhang, and L. Wang, "Magnetic nanoparticles: an improved method for mitochondrial isolation," *Molecular Medicine Reports*, vol. 5, no. 5, pp. 1271–1276, 2012.

[80] C. Giorgi, D. De Stefani, A. Bononi, R. Rizzuto, and P. Pinton, "Structural and functional link between the mitochondrial network and the endoplasmic reticulum," *The International Journal of Biochemistry & Cell Biology*, vol. 41, no. 10, pp. 1817–1827, 2009.

[81] J. Adachi, C. Kumar, Y. Zhang, and M. Mann, "In-depth analysis of the adipocyte proteome by mass spectrometry and bioinformatics," *Molecular & Cellular Proteomics*, vol. 6, no. 7, pp. 1257–1273, 2007.

[82] A. Hu, W. S. Noble, and A. Wolf-Yadlin, "Technical advances in proteomics: new developments in data-independent acquisition," *F1000Res*, vol. 5, 2016.

[83] T. C. Walther and M. Mann, "Mass spectrometry–based proteomics in cell biology," *Journal of Cell Biology*, vol. 190, no. 4, pp. 491–500, 2010.

[84] M. P. Washburn, D. Wolters, and J. R. Yates, "Large-scale analysis of the yeast proteome by multidimensional protein identification technology," *Nature Biotechnology*, vol. 19, no. 3, pp. 242–247, 2001.

[85] V. Lange, P. Picotti, B. Domon, and R. Aebersold, "Selected reaction monitoring for quantitative proteomics: a tutorial," *Molecular Systems Biology*, vol. 4, p. 222, 2008.

[86] P. Fernández-Guerra, R. I. D. Birkler, B. Merinero et al., "Selected reaction monitoring as an effective method for reliable quantification of disease-associated proteins in maple syrup urine disease," *Molecular Genetics & Genomic Medicine*, vol. 2, no. 5, pp. 383–392, 2014.

[87] E. Rosello-Lleti, E. Tarazon, M. G. Barderas et al., "Heart mitochondrial proteome study elucidates changes in cardiac energy metabolism and antioxidant PRDX3 in human dilated cardiomyopathy," *PLoS One*, vol. 9, no. 11, article e112971, 2014.

[88] F. Hosp, I. Lassowskat, V. Santoro et al., "Lysine acetylation in mitochondria: from inventory to function," *Mitochondrion*, vol. 33, pp. 58–71, 2017.

[89] M. J. Rardin, J. C. Newman, J. M. Held et al., "Label-free quantitative proteomics of the lysine acetylome in mitochondria identifies substrates of SIRT3 in metabolic pathways," *Proceedings of the National Academy of Sciences of the United States of America*, vol. 110, no. 16, pp. 6601–6606, 2013.

[90] R. Kruse and K. Hojlund, "Mitochondrial phosphoproteomics of mammalian tissues," *Mitochondrion*, vol. 33, pp. 45–57, 2017.

[91] M. P. Y. Lam, S. B. Scruggs, T. Y. Kim et al., "An MRM-based workflow for quantifying cardiac mitochondrial protein phosphorylation in murine and human tissue," *Journal of Proteomics*, vol. 75, no. 15, pp. 4602–4609, 2012.

[92] O. T. Schubert, L. C. Gillet, B. C. Collins et al., "Building high-quality assay libraries for targeted analysis of SWATH MS data," *Nature Protocols*, vol. 10, no. 3, pp. 426–441, 2015.

[93] L. Grois, J. Hupf, J. Reinders et al., "Combined inhibition of the renin-angiotensin system and neprilysin positively influences complex mitochondrial adaptations in progressive experimental heart failure," *PLoS One*, vol. 12, no. 1, article e0169743, 2017.

[94] L. M. Villeneuve, P. R. Purnell, M. D. Boska, and H. S. Fox, "Early expression of Parkinson's disease-related mitochondrial abnormalities in PINK1 knockout rats," *Molecular Neurobiology*, vol. 53, no. 1, pp. 171–186, 2016.

[95] L. M. Villeneuve, K. L. Stauch, and H. S. Fox, "Proteomic analysis of the mitochondria from embryonic and postnatal rat brains reveals response to developmental changes in energy demands," *Journal of Proteomics*, vol. 109, pp. 228–239, 2014.

[96] A. Guaras, E. Perales-Clemente, E. Calvo et al., "The CoQH2/CoQ ratio serves as a sensor of respiratory chain efficiency," *Cell Reports*, vol. 15, no. 1, pp. 197–209, 2016.

[97] M. Unlu, M. E. Morgan, and J. S. Minden, "Difference gel electrophoresis: a single gel method for detecting changes in

protein extracts," *Electrophoresis*, vol. 18, no. 11, pp. 2071–2077, 1997.

[98] M. Bantscheff, S. Lemeer, M. M. Savitski, and B. Kuster, "Quantitative mass spectrometry in proteomics: critical review update from 2007 to the present," *Analytical and Bioanalytical Chemistry*, vol. 404, no. 4, pp. 939–965, 2012.

[99] F. Calderon-Celis, J. R. Encinar, and A. Sanz-Medel, "Standardization approaches in absolute quantitative proteomics with mass spectrometry," *Mass Spectrometry Reviews*, vol. 37, no. 6, pp. 715–737, 2018.

[100] X. Chen, J. Li, J. Hou, Z. Xie, and F. Yang, "Mammalian mitochondrial proteomics: insights into mitochondrial functions and mitochondria-related diseases," *Expert Review of Proteomics*, vol. 7, no. 3, pp. 333–345, 2010.

[101] D. F. Bogenhagen, A. G. Ostermeyer-Fay, J. D. Haley, and M. Garcia-Diaz, "Kinetics and mechanism of mammalian mitochondrial ribosome assembly," *Cell Reports*, vol. 22, no. 7, pp. 1935–1944, 2018.

[102] F. Forner, C. Kumar, C. A. Luber, T. Fromme, M. Klingenspor, and M. Mann, "Proteome differences between brown and white fat mitochondria reveal specialized metabolic functions," *Cell Metabolism*, vol. 10, no. 4, pp. 324–335, 2009.

[103] M. Bienengraeber, M. Pellitteri-Hahn, N. Hirata, T. M. Baye, Z. J. Bosnjak, and M. Olivier, "Quantitative characterization of changes in the cardiac mitochondrial proteome during anesthetic preconditioning and ischemia," *Physiological Genomics*, vol. 45, no. 5, pp. 163–170, 2013.

[104] J. R. Smith, I. R. Matus, D. A. Beard, and A. S. Greene, "Differential expression of cardiac mitochondrial proteins," *Proteomics*, vol. 8, no. 3, pp. 446–462, 2008.

[105] Y. Liu, J. He, S. Ji et al., "Comparative studies of early liver dysfunction in senescence-accelerated mouse using mitochondrial proteomics approaches," *Molecular & Cellular Proteomics*, vol. 7, no. 9, pp. 1737–1747, 2008.

[106] M. A. Lovell, S. Xiong, W. R. Markesbery, and B. C. Lynn, "Quantitative proteomic analysis of mitochondria from primary neuron cultures treated with amyloid beta peptide," *Neurochemical Research*, vol. 30, no. 1, pp. 113–122, 2005.

[107] B. W. Newton, S. M. Cologna, C. Moya, D. H. Russell, W. K. Russell, and A. Jayaraman, "Proteomic analysis of 3T3-L1 adipocyte mitochondria during differentiation and enlargement," *Journal of Proteome Research*, vol. 10, no. 10, pp. 4692–4702, 2011.

[108] Z. Shi, W. Long, C. Zhao, X. Guo, R. Shen, and H. Ding, "Comparative proteomics analysis suggests that placental mitochondria are involved in the development of preeclampsia," *PLoS One*, vol. 8, no. 5, article e64351, 2013.

[109] Z. Xu, X. Jin, W. Cai et al., "Proteomics analysis reveals abnormal electron transport and excessive oxidative stress cause mitochondrial dysfunction in placental tissues of early-onset preeclampsia," *Proteomics Clinical applications*, vol. 12, no. 5, article e1700165, 2018.

[110] Y. Cheng, T. Hou, J. Ping, G. Chen, and J. Chen, "Quantitative succinylome analysis in the liver of non-alcoholic fatty liver disease rat model," *Proteome Science*, vol. 14, no. 1, p. 3, 2016.

[111] W. Gao, J. Xu, F. Wang et al., "Mitochondrial proteomics approach reveals voltage-dependent anion channel 1 (VDAC1) as a potential biomarker of gastric cancer," *Cellular Physiology and Biochemistry*, vol. 37, no. 6, pp. 2339–2354, 2015.

[112] M. Fountoulakis and E. J. Schlaeger, "The mitochondrial proteins of the neuroblastoma cell line IMR-32," *Electrophoresis*, vol. 24, no. 12, pp. 260–275, 2003.

[113] E. Walheim, J. R. Wisniewski, and M. Jastroch, "Respiromics - an integrative analysis linking mitochondrial bioenergetics to molecular signatures," *Molecular Metabolism*, vol. 9, pp. 4–14, 2018.

[114] C. Lopez-Otin, M. A. Blasco, L. Partridge, M. Serrano, and G. Kroemer, "The hallmarks of aging," *Cell*, vol. 153, no. 6, pp. 1194–1217, 2013.

[115] M. Gonzalez-Freire, R. de Cabo, M. Bernier et al., "Reconsidering the role of mitochondria in aging," *The Journals of Gerontology Series A, Biological Sciences and Medical Sciences*, vol. 70, no. 11, pp. 1334–1342, 2015.

[116] Q. Bao, J. Pan, H. Qi et al., "Aging and age-related diseases – from endocrine therapy to target therapy," *Molecular and Cellular Endocrinology*, vol. 394, no. 1-2, pp. 115–118, 2014.

[117] T. Niccoli and L. Partridge, "Ageing as a risk factor for disease," *Current Biology*, vol. 22, no. 17, pp. R741–R752, 2012.

[118] J. R. Peinado, A. Diaz-Ruiz, G. Fruhbeck, and M. M. Malagon, "Mitochondria in metabolic disease: getting clues from proteomic studies," *Proteomics*, vol. 14, no. 4-5, pp. 452–466, 2014.

[119] B. Chakravarti, M. Oseguera, N. Dalal et al., "Proteomic profiling of aging in the mouse heart: altered expression of mitochondrial proteins," *Archives of Biochemistry and Biophysics*, vol. 474, no. 1, pp. 22–31, 2008.

[120] C. Fernandez-Sanz, M. Ruiz-Meana, J. Castellano et al., "Altered FoF1 ATP synthase and susceptibility to mitochondrial permeability transition pore during ischaemia and reperfusion in aging cardiomyocytes," *Thrombosis and Haemostasis*, vol. 113, no. 03, pp. 441–451, 2015.

[121] C. Fernandez-Sanz, M. Ruiz-Meana, E. Miro-Casas et al., "Defective sarcoplasmic reticulum-mitochondria calcium exchange in aged mouse myocardium," *Cell Death & Disease*, vol. 5, no. 12, article e1573, 2014.

[122] T. Ingram and L. Chakrabarti, "Proteomic profiling of mitochondria: what does it tell us about the ageing brain?," *Aging*, vol. 8, no. 12, pp. 3161–3179, 2016.

[123] A. Pollard, F. Shephard, J. Freed, S. Liddell, and L. Chakrabarti, "Mitochondrial proteomic profiling reveals increased carbonic anhydrase II in aging and neurodegeneration," *Aging*, vol. 8, no. 10, pp. 2425–2436, 2016.

[124] L. Mao, C. Zabel, M. A. Wacker et al., "Estimation of the mtDNA mutation rate in aging mice by proteome analysis and mathematical modeling," *Experimental Gerontology*, vol. 41, no. 1, pp. 11–24, 2006.

[125] K. L. Stauch, P. R. Purnell, and H. S. Fox, "Aging synaptic mitochondria exhibit dynamic proteomic changes while maintaining bioenergetic function," *Aging*, vol. 6, no. 4, pp. 320–334, 2014.

[126] K. L. Stauch, P. R. Purnell, L. M. Villeneuve, and H. S. Fox, "Proteomic analysis and functional characterization of mouse brain mitochondria during aging reveal alterations in energy metabolism," *Proteomics*, vol. 15, no. 9, pp. 1574–1586, 2015.

[127] J. Rivera-Torres, R. Acin-Perez, P. Cabezas-Sanchez et al., "Identification of mitochondrial dysfunction in Hutchinson–Gilford progeria syndrome through use of stable isotope labeling with amino acids in cell culture," *Journal of Proteomics*, vol. 91, pp. 466–477, 2013.

[128] A. Boveris and A. Navarro, "Brain mitochondrial dysfunction in aging," *IUBMB Life*, vol. 60, no. 5, pp. 308–314, 2008.

[129] H. E. Moon and S. H. Paek, "Mitochondrial dysfunction in Parkinson's disease," *Experimental Neurobiology*, vol. 24, no. 2, pp. 103–116, 2015.

[130] I. G. Onyango, J. Dennis, and S. M. Khan, "Mitochondrial dysfunction in Alzheimer's disease and the rationale for bioenergetics based therapies," *Aging and Disease*, vol. 7, no. 2, pp. 201–214, 2016.

[131] D. A. Butterfield, E. M. Palmieri, and A. Castegna, "Clinical implications from proteomic studies in neurodegenerative diseases: lessons from mitochondrial proteins," *Expert Review of Proteomics*, vol. 13, no. 3, pp. 259–274, 2016.

[132] J. L. Chou, D. V. Shenoy, N. Thomas et al., "Early dysregulation of the mitochondrial proteome in a mouse model of Alzheimer's disease," *Journal of Proteomics*, vol. 74, no. 4, pp. 466–479, 2011.

[133] A. Nunomura, G. Perry, G. Aliev et al., "Oxidative damage is the earliest event in Alzheimer disease," *Journal of Neuropathology and Experimental Neurology*, vol. 60, no. 8, pp. 759–767, 2001.

[134] R. H. Swerdlow, J. M. Burns, and S. M. Khan, "The Alzheimer's disease mitochondrial cascade hypothesis," *Journal of Alzheimer's Disease*, vol. 20, Supplement 2, pp. S265–S279, 2010.

[135] C. Bertoni-Freddari, P. Fattoretti, T. Casoli, C. Spagna, W. Meier-Ruge, and J. Ulrich, "Morphological plasticity of synaptic mitochondria during aging," *Brain Research*, vol. 628, no. 1-2, pp. 193–200, 1993.

[136] M. R. Brown, P. G. Sullivan, and J. W. Geddes, "Synaptic mitochondria are more susceptible to $Ca^{2+}$ overload than nonsynaptic mitochondria," *Journal of Biological Chemistry*, vol. 281, no. 17, pp. 11658–11668, 2006.

[137] L. C. Graham, S. L. Eaton, P. J. Brunton et al., "Proteomic profiling of neuronal mitochondria reveals modulators of synaptic architecture," *Molecular Neurodegeneration*, vol. 12, no. 1, p. 77, 2017.

[138] C. Brunel-Guitton, A. Levtova, and F. Sasarman, "Mitochondrial diseases and cardiomyopathies," *The Canadian Journal of Cardiology*, vol. 31, no. 11, pp. 1360–1376, 2015.

[139] M. D. Hirschey, R. DeBerardinis, A. M. E. Diehl et al., "Dysregulated metabolism contributes to oncogenesis," *Seminars in Cancer Biology*, vol. 35, pp. S129–S150, 2015.

[140] A. P. Wojtovich, S. M. Nadtochiy, P. S. Brookes, and K. Nehrke, "Ischemic preconditioning: the role of mitochondria and aging," *Experimental Gerontology*, vol. 47, no. 1, pp. 1–7, 2012.

[141] O. Ertracht, A. Malka, S. Atar, and O. Binah, "The mitochondria as a target for cardioprotection in acute myocardial ischemia," *Pharmacology & Therapeutics*, vol. 142, no. 1, pp. 33–40, 2014.

[142] M. G. Perrelli, P. Pagliaro, and C. Penna, "Ischemia/reperfusion injury and cardioprotective mechanisms: role of mitochondria and reactive oxygen species," *World Journal of Cardiology*, vol. 3, no. 6, pp. 186–200, 2011.

[143] M. Ruiz-Meana, E. Nunez, E. Miro-Casas et al., "Ischemic preconditioning protects cardiomyocyte mitochondria through mechanisms independent of cytosol," *Journal of Molecular and Cellular Cardiology*, vol. 68, pp. 79–88, 2014.

[144] J. A. Boylston, J. Sun, Y. Chen, M. Gucek, M. N. Sack, and E. Murphy, "Characterization of the cardiac succinylome and its role in ischemia-reperfusion injury," *Journal of Molecular and Cellular Cardiology*, vol. 88, pp. 73–81, 2015.

[145] Q. Chen, M. Younus, J. Thompson, Y. Hu, J. M. Hollander, and E. J. Lesnefsky, "Intermediary metabolism and fatty acid oxidation: novel targets of electron transport chain-driven injury during ischemia and reperfusion," *American Journal of Physiology-Heart and Circulatory Physiology*, vol. 314, no. 4, pp. H787–H795, 2018.

[146] E. T. Chouchani, A. M. James, C. Methner et al., "Identification and quantification of protein S-nitrosation by nitrite in the mouse heart during ischemia," *Journal of Biological Chemistry*, vol. 292, no. 35, pp. 14486–14495, 2017.

[147] L. Lichardusova, Z. Tatarkova, A. Calkovska et al., "Proteomic analysis of mitochondrial proteins in the guinea pig heart following long-term normobaric hyperoxia," *Molecular and Cellular Biochemistry*, vol. 434, no. 1-2, pp. 61–73, 2017.

[148] T. Kalogeris, Y. Bao, and R. J. Korthuis, "Mitochondrial reactive oxygen species: a double edged sword in ischemia/reperfusion vs preconditioning," *Redox Biology*, vol. 2, pp. 702–714, 2014.

[149] W. P. T. James, "The epidemiology of obesity: the size of the problem," *Journal of Internal Medicine*, vol. 263, no. 4, pp. 336–352, 2008.

[150] X. Chen, S. Wei, and F. Yang, "Mitochondria in the pathogenesis of diabetes: a proteomic view," *Protein & Cell*, vol. 3, no. 9, pp. 648–660, 2012.

[151] K. Sun, C. M. Kusminski, and P. E. Scherer, "Adipose tissue remodeling and obesity," *The Journal of Clinical Investigation*, vol. 121, no. 6, pp. 2094–2101, 2011.

[152] P. W. Lindinger, M. Christe, A. N. Eberle et al., "Important mitochondrial proteins in human omental adipose tissue show reduced expression in obesity," *Journal of Proteomics*, vol. 124, pp. 79–87, 2015.

[153] S. J. Kim, S. Chae, H. Kim et al., "A protein profile of visceral adipose tissues linked to early pathogenesis of type 2 diabetes mellitus," *Molecular & Cellular Proteomics*, vol. 13, no. 3, pp. 811–822, 2014.

[154] R. Perez-Perez, E. Garcia-Santos, F. J. Ortega-Delgado et al., "Attenuated metabolism is a hallmark of obesity as revealed by comparative proteomic analysis of human omental adipose tissue," *Journal of Proteomics*, vol. 75, no. 3, pp. 783–795, 2012.

[155] M. Gomez-Serrano, E. Camafeita, E. Garcia-Santos et al., "Proteome-wide alterations on adipose tissue from obese patients as age-, diabetes- and gender-specific hallmarks," *Scientific Reports*, vol. 6, no. 1, article 25756, 2016.

[156] R. J. Perry, V. T. Samuel, K. F. Petersen, and G. I. Shulman, "The role of hepatic lipids in hepatic insulin resistance and type 2 diabetes," *Nature*, vol. 510, no. 7503, pp. 84–91, 2014.

[157] P. Golabi, C. T. Locklear, P. Austin et al., "Effectiveness of exercise in hepatic fat mobilization in non-alcoholic fatty liver disease: systematic review," *World Journal of Gastroenterology*, vol. 22, no. 27, pp. 6318–6327, 2016.

[158] G. Paradies, V. Paradies, F. M. Ruggiero, and G. Petrosillo, "Oxidative stress, cardiolipin and mitochondrial dysfunction in nonalcoholic fatty liver disease," *World Journal of Gastroenterology*, vol. 20, no. 39, pp. 14205–14218, 2014.

[159] N. Nuno-Lambarri, V. J. Barbero-Becerra, M. Uribe, and N. C. Chavez-Tapia, "Mitochondrial molecular pathophysiology of nonalcoholic fatty liver disease: a proteomics

[160] A. Thomas, M. S. Klein, A. P. Stevens et al., "Changes in the hepatic mitochondrial and membrane proteome in mice fed a non-alcoholic steatohepatitis inducing diet," *Journal of Proteomics*, vol. 80, pp. 107–122, 2013.

[161] A. Valle, V. Catalan, A. Rodriguez et al., "Identification of liver proteins altered by type 2 diabetes mellitus in obese subjects," *Liver International*, vol. 32, no. 6, pp. 951–961, 2012.

[162] A. Stachowicz, M. Suski, R. Olszanecki, J. Madej, K. Okon, and R. Korbut, "Proteomic analysis of liver mitochondria of apolipoprotein E knockout mice treated with metformin," *Journal of Proteomics*, vol. 77, pp. 167–175, 2012.

[163] L. Shi, Y. Wang, S. Tu et al., "The responses of mitochondrial proteome in rat liver to the consumption of moderate ethanol: the possible roles of aldo-keto reductases," *Journal of Proteome Research*, vol. 7, no. 8, pp. 3137–3145, 2008.

[164] A. Venkatraman, A. Landar, A. J. Davis et al., "Modification of the mitochondrial proteome in response to the stress of ethanol-dependent hepatotoxicity," *Journal of Biological Chemistry*, vol. 279, no. 21, pp. 22092–22101, 2004.

[165] M. Potter, E. Newport, and K. J. Morten, "The Warburg effect: 80 years on," *Biochemical Society Transactions*, vol. 44, no. 5, pp. 1499–1505, 2016.

[166] S. W. Kang, S. Lee, and E. K. Lee, "ROS and energy metabolism in cancer cells: alliance for fast growth," *Archives of Pharmacal Research*, vol. 38, no. 3, pp. 338–345, 2015.

[167] R. Uzhachenko, A. Shanker, W. G. Yarbrough, and A. V. Ivanova, "Mitochondria, calcium, and tumor suppressor Fus1: at the crossroad of cancer, inflammation, and autoimmunity," *Oncotarget*, vol. 6, no. 25, pp. 20754–20772, 2015.

[168] O. Warburg, "On the origin of cancer cells," *Science*, vol. 123, no. 3191, pp. 309–314, 1956.

[169] W. Zhou, L. A. Liotta, and E. F. Petricoin, "The Warburg effect and mass spectrometry-based proteomic analysis," *Cancer Genomics Proteomics*, vol. 14, no. 4, pp. 211–218, 2017.

[170] Y. Jiang and X. Wang, "Comparative mitochondrial proteomics: perspective in human diseases," *Journal of Hematology & Oncology*, vol. 5, no. 1, p. 11, 2012.

[171] M. Mendes, A. Pelaez-Garcia, M. Lopez-Lucendo et al., "Mapping the spatial proteome of metastatic cells in colorectal cancer," *Proteomics*, vol. 17, no. 19, 2017.

[172] A. T. Qattan, M. Radulovic, M. Crawford, and J. Godovac-Zimmermann, "Spatial distribution of cellular function: the partitioning of proteins between mitochondria and the nucleus in MCF7 breast cancer cells," *Journal of Proteome Research*, vol. 11, no. 12, pp. 6080–6101, 2012.

[173] A. Lopez-Campistrous, L. Hao, W. Xiang et al., "Mitochondrial dysfunction in the hypertensive rat brain: respiratory complexes exhibit assembly defects in hypertension," *Hypertension*, vol. 51, no. 2, pp. 412–419, 2008.

[174] C. Ruiz-Romero, V. Calamia, J. Mateos et al., "Mitochondrial dysregulation of osteoarthritic human articular chondrocytes analyzed by proteomics: a decrease in mitochondrial superoxide dismutase points to a redox imbalance," *Molecular & Cellular Proteomics*, vol. 8, no. 1, pp. 172–189, 2009.

[175] T. An, Y. F. Wang, J. X. Liu et al., "Comparative analysis of proteomes between diabetic and normal human sperm: insights into the effects of diabetes on male reproduction based on the regulation of mitochondria-related proteins," *Molecular Reproduction and Development*, vol. 85, no. 1, pp. 7–16, 2018.

[176] L. Samanta, A. Agarwal, N. Swain et al., "Proteomic signatures of sperm mitochondria in varicocele: clinical use as biomarkers of varicocele associated infertility," *The Journal of Urology*, vol. 200, no. 2, pp. 414–422, 2018.

[177] M. P. Murphy, "How mitochondria produce reactive oxygen species," *Biochemical Journal*, vol. 417, no. 1, pp. 1–13, 2009.

[178] M. P. Murphy, "Mitochondrial thiols in antioxidant protection and redox signaling: distinct roles for glutathionylation and other thiol modifications," *Antioxidants & Redox Signaling*, vol. 16, no. 6, pp. 476–495, 2012.

[179] A. P. Halestrap, S. J. Clarke, and I. Khaliulin, "The role of mitochondria in protection of the heart by preconditioning," *Biochimica et Biophysica Acta (BBA) - Bioenergetics*, vol. 1767, no. 8, pp. 1007–1031, 2007.

[180] V. Kumar, T. Kleffmann, M. B. Hampton, M. B. Cannell, and C. C. Winterbourn, "Redox proteomics of thiol proteins in mouse heart during ischemia/reperfusion using ICAT reagents and mass spectrometry," *Free Radical Biology & Medicine*, vol. 58, pp. 109–117, 2013.

[181] D. M. Yellon and D. J. Hausenloy, "Myocardial reperfusion injury," *The New England Journal of Medicine*, vol. 357, no. 11, pp. 1121–1135, 2007.

[182] D. C. Wallace, "Mitochondria and cancer," *Nature Reviews Cancer*, vol. 12, no. 10, pp. 685–698, 2012.

[183] F. Weinberg and N. S. Chandel, "Reactive oxygen species-dependent signaling regulates cancer," *Cellular and Molecular Life Sciences*, vol. 66, no. 23, pp. 3663–3673, 2009.

[184] D. A. Butterfield, A. Gnjec, H. F. Poon et al., "Redox proteomics identification of oxidatively modified brain proteins in inherited Alzheimer's disease: an initial assessment," *Journal of Alzheimer's Disease*, vol. 10, no. 4, pp. 391–397, 2006.

[185] M. T. Lin and M. F. Beal, "Mitochondrial dysfunction and oxidative stress in neurodegenerative diseases," *Nature*, vol. 443, no. 7113, pp. 787–795, 2006.

[186] J. M. Held and B. W. Gibson, "Regulatory control or oxidative damage? Proteomic approaches to interrogate the role of cysteine oxidation status in biological processes," *Molecular & Cellular Proteomics*, vol. 11, no. 4, article R111.013037, 2012.

[187] M. Lindahl, A. Mata-Cabana, and T. Kieselbach, "The disulfide proteome and other reactive cysteine proteomes: analysis and functional significance," *Antioxidants & Redox Signaling*, vol. 14, no. 12, pp. 2581–2642, 2011.

[188] C. C. Winterbourn and M. B. Hampton, "Thiol chemistry and specificity in redox signaling," *Free Radical Biology & Medicine*, vol. 45, no. 5, pp. 549–561, 2008.

[189] S. G. Rhee, W. Jeong, T. S. Chang, and H. A. Woo, "Sulfiredoxin, the cysteine sulfinic acid reductase specific to 2-Cys peroxiredoxin: its discovery, mechanism of action, and biological significance," *Kidney International*, vol. 72, pp. S3–S8, 2007.

[190] F. L. Muller, Y. Liu, M. A. Abdul-Ghani et al., "High rates of superoxide production in skeletal-muscle mitochondria respiring on both complex I- and complex II-linked substrates," *Biochemical Journal*, vol. 409, no. 2, pp. 491–499, 2008.

[191] S. Drose, "Differential effects of complex II on mitochondrial ROS production and their relation to cardioprotective pre-

and postconditioning," *Biochimica et Biophysica Acta (BBA) - Bioenergetics*, vol. 1827, no. 5, pp. 578–587, 2013.

[192] P. Lanciano, B. Khalfaoui-Hassani, N. Selamoglu, A. Ghelli, M. Rugolo, and F. Daldal, "Molecular mechanisms of superoxide production by complex III: a bacterial versus human mitochondrial comparative case study," *Biochimica et Biophysica Acta (BBA) - Bioenergetics*, vol. 1827, no. 11-12, pp. 1332–1339, 2013.

[193] Y. M. W. Janssen-Heininger, B. T. Mossman, N. H. Heintz et al., "Redox-based regulation of signal transduction: principles, pitfalls, and promises," *Free Radical Biology & Medicine*, vol. 45, no. 1, pp. 1–17, 2008.

[194] S. Moncada and J. D. Erusalimsky, "Does nitric oxide modulate mitochondrial energy generation and apoptosis?," *Nature Reviews Molecular Cell Biology*, vol. 3, no. 3, pp. 214–220, 2002.

[195] A. Galkin, B. Meyer, I. Wittig et al., "Identification of the mitochondrial ND3 subunit as a structural component involved in the active/deactive enzyme transition of respiratory complex I," *Journal of Biological Chemistry*, vol. 283, no. 30, pp. 20907–20913, 2008.

[196] A. Galkin and S. Moncada, "S-nitrosation of mitochondrial complex I depends on its structural conformation," *Journal of Biological Chemistry*, vol. 282, no. 52, pp. 37448–37453, 2007.

[197] T. A. Prime, F. H. Blaikie, C. Evans et al., "A mitochondria-targeted S-nitrosothiol modulates respiration, nitrosates thiols, and protects against ischemia-reperfusion injury," *Proceedings of the National Academy of Sciences of the United States of America*, vol. 106, no. 26, pp. 10764–10769, 2009.

[198] T. R. Hurd, T. A. Prime, M. E. Harbour, K. S. Lilley, and M. P. Murphy, "Detection of reactive oxygen species-sensitive thiol proteins by redox difference gel electrophoresis: implications for mitochondrial redox signaling," *The Journal of Biological Chemistry*, vol. 282, no. 30, pp. 22040–22051, 2007.

[199] E. T. Chouchani, T. R. Hurd, S. M. Nadtochiy et al., "Identification of S-nitrosated mitochondrial proteins by S-nitrosothiol difference in gel electrophoresis (SNO-DIGE): implications for the regulation of mitochondrial function by reversible S-nitrosation," *Biochemical Journal*, vol. 430, no. 1, pp. 49–59, 2010.

[200] R. Scheving, I. Wittig, H. Heide et al., "Protein S-nitrosylation and denitrosylation in the mouse spinal cord upon injury of the sciatic nerve," *Journal of Proteomics*, vol. 75, no. 13, pp. 3987–4004, 2012.

[201] E. T. Chouchani, C. Methner, S. M. Nadtochiy et al., "Cardioprotection by S-nitrosation of a cysteine switch on mitochondrial complex I," *Nature Medicine*, vol. 19, no. 6, pp. 753–759, 2013.

[202] S. Akter, S. Carpentier, F. Van Breusegem, and J. Messens, "Identification of dimedone-trapped sulfenylated proteins in plants under stress," *Biochemistry and Biophysics Reports*, vol. 9, pp. 106–113, 2017.

[203] S. Akter, J. Huang, N. Bodra et al., "DYn-2 based identification of *Arabidopsis* Sulfenomes," *Molecular & Cellular Proteomics*, vol. 14, no. 5, pp. 1183–1200, 2015.

[204] C. Waszczak, S. Akter, D. Eeckhout et al., "Sulfenome mining in Arabidopsis thaliana," *Proceedings of the National Academy of Sciences of the United States of America*, vol. 111, no. 31, pp. 11545–11550, 2014.

[205] G. Hao, B. Derakhshan, L. Shi, F. Campagne, and S. S. Gross, "SNOSID, a proteomic method for identification of cysteine S-nitrosylation sites in complex protein mixtures," *Proceedings of the National Academy of Sciences of the United States of America*, vol. 103, no. 4, pp. 1012–1017, 2006.

[206] S. R. Jaffrey, H. Erdjument-Bromage, C. D. Ferris, P. Tempst, and S. H. Snyder, "Protein S-nitrosylation: a physiological signal for neuronal nitric oxide," *Nature Cell Biology*, vol. 3, no. 2, pp. 193–197, 2001.

[207] S. R. Jaffrey and S. H. Snyder, "The biotin switch method for the detection of S-nitrosylated proteins," *Science's STKE*, vol. 2001, article pl1, 2001.

[208] C. I. Murray, H. Uhrigshardt, R. N. O'Meally, R. N. Cole, and J. E. Van Eyk, "Identification and quantification of S-nitrosylation by cysteine reactive tandem mass tag switch assay," *Molecular & Cellular Proteomics*, vol. 11, no. 2, article M111 013441, 2012.

[209] R. L. Charles, E. Schröder, G. May et al., "Protein sulfenation as a redox sensor: proteomics studies using a novel biotinylated dimedone analogue," *Molecular & Cellular Proteomics*, vol. 6, no. 9, pp. 1473–1484, 2007.

[210] V. Gupta and K. S. Carroll, "Sulfenic acid chemistry, detection and cellular lifetime," *Biochimica et Biophysica Acta (BBA) - General Subjects*, vol. 1840, no. 2, pp. 847–875, 2014.

[211] L. B. Poole, C. Klomsiri, S. A. Knaggs et al., "Fluorescent and affinity-based tools to detect cysteine sulfenic acid formation in proteins," *Bioconjugate Chemistry*, vol. 18, no. 6, pp. 2004–2017, 2007.

[212] D. J. McGarry, P. Chakravarty, C. R. Wolf, and C. J. Henderson, "Altered protein S-glutathionylation identifies a potential mechanism of resistance to acetaminophen-induced hepatotoxicity," *The Journal of Pharmacology and Experimental Therapeutics*, vol. 355, no. 2, pp. 137–144, 2015.

[213] N. S. Gould, P. Evans, P. Martinez-Acedo et al., "Site-specific proteomic mapping identifies selectively modified regulatory cysteine residues in functionally distinct protein networks," *Chemistry & Biology*, vol. 22, no. 7, pp. 965–975, 2015.

[214] S. P. Gygi, B. Rist, S. A. Gerber, F. Turecek, M. H. Gelb, and R. Aebersold, "Quantitative analysis of complex protein mixtures using isotope-coded affinity tags," *Nature Biotechnology*, vol. 17, no. 10, pp. 994–999, 1999.

[215] S. E. Ong, B. Blagoev, I. Kratchmarova et al., "Stable isotope labeling by amino acids in cell culture, SILAC, as a simple and accurate approach to expression proteomics," *Molecular & Cellular Proteomics*, vol. 1, no. 5, pp. 376–386, 2002.

[216] P. L. Ross, Y. N. Huang, J. N. Marchese et al., "Multiplexed protein quantitation in *Saccharomyces cerevisiae* using amine-reactive isobaric tagging reagents," *Molecular & Cellular Proteomics*, vol. 3, no. 12, pp. 1154–1169, 2004.

[217] I. I. Stewart, T. Thomson, and D. Figeys, "18O labeling: a tool for proteomics," *Rapid Communications in Mass Spectrometry*, vol. 15, no. 24, pp. 2456–2465, 2001.

[218] A. Thompson, J. Schafer, K. Kuhn et al., "Tandem mass tags: a novel quantification strategy for comparative analysis of complex protein mixtures by MS/MS," *Analytical Chemistry*, vol. 75, no. 8, pp. 1895–1904, 2003.

[219] E. Weerapana, C. Wang, G. M. Simon et al., "Quantitative reactivity profiling predicts functional cysteines in proteomes," *Nature*, vol. 468, no. 7325, pp. 790–795, 2010.

[220] L. I. Leichert, F. Gehrke, H. V. Gudiseva et al., "Quantifying changes in the thiol redox proteome upon oxidative stress

*in vivo*," *Proceedings of the National Academy of Sciences of the United States of America*, vol. 105, no. 24, pp. 8197–8202, 2008.

[221] C. Fu, J. Hu, T. Liu, T. Ago, J. Sadoshima, and H. Li, "Quantitative analysis of redox-sensitive proteome with DIGE and ICAT," *Journal of Proteome Research*, vol. 7, no. 9, pp. 3789–3802, 2008.

[222] N. Brandes, D. Reichmann, H. Tienson, L. I. Leichert, and U. Jakob, "Using quantitative redox proteomics to dissect the yeast redoxome," *The Journal of Biological Chemistry*, vol. 286, no. 48, pp. 41893–41903, 2011.

[223] S. Garcia-Santamarina, S. Boronat, A. Domenech, J. Ayte, H. Molina, and E. Hidalgo, "Monitoring in vivo reversible cysteine oxidation in proteins using ICAT and mass spectrometry," *Nature Protocols*, vol. 9, no. 5, pp. 1131–1145, 2014.

[224] S. Garcia-Santamarina, S. Boronat, G. Espadas, J. Ayte, H. Molina, and E. Hidalgo, "The oxidized thiol proteome in fission yeast—optimization of an ICAT-based method to identify $H_2O_2$-oxidized proteins," *Journal of Proteomics*, vol. 74, no. 11, pp. 2476–2486, 2011.

[225] C. Lindemann and L. I. Leichert, "Quantitative redox proteomics: the NOxICAT method," *Methods in Molecular Biology*, vol. 893, pp. 387–403, 2012.

[226] D. Knoefler, M. Thamsen, M. Koniczek, N. J. Niemuth, A. K. Diederich, and U. Jakob, "Quantitative in vivo redox sensors uncover oxidative stress as an early event in life," *Molecular Cell*, vol. 47, no. 5, pp. 767–776, 2012.

[227] S. Rosenwasser, S. Graff van Creveld, D. Schatz et al., "Mapping the diatom redox-sensitive proteome provides insight into response to nitrogen stress in the marine environment," *Proceedings of the National Academy of Sciences of the United States of America*, vol. 111, no. 7, pp. 2740–2745, 2014.

[228] K. E. Menger, A. M. James, H. M. Cocheme et al., "Fasting, but not aging, dramatically alters the redox status of cysteine residues on proteins in *Drosophila melanogaster*," *Cell Reports*, vol. 13, no. 6, p. 1285, 2015.

[229] Y. M. Go, J. R. Roede, M. Orr, Y. Liang, and D. P. Jones, "Integrated redox proteomics and metabolomics of mitochondria to identify mechanisms of Cd toxicity," *Toxicological Sciences*, vol. 139, no. 1, pp. 59–73, 2014.

[230] U. Topf, I. Suppanz, L. Samluk et al., "Quantitative proteomics identifies redox switches for global translation modulation by mitochondrially produced reactive oxygen species," *Nature Communications*, vol. 9, no. 1, p. 324, 2018.

[231] D. J. McGarry, W. Chen, P. Chakravarty, D. L. Lamont, C. R. Wolf, and C. J. Henderson, "Proteome-wide identification and quantification of S-glutathionylation targets in mouse liver," *Biochemical Journal*, vol. 469, no. 1, pp. 25–32, 2015.

[232] E. T. Chouchani, L. Kazak, M. P. Jedrychowski et al., "Mitochondrial ROS regulate thermogenic energy expenditure and sulfenylation of UCP1," *Nature*, vol. 532, no. 7597, pp. 112–116, 2016.

[233] P. A. Kramer, J. Duan, M. J. Gaffrey et al., "Fatiguing contractions increase protein S-glutathionylation occupancy in mouse skeletal muscle," *Redox Biology*, vol. 17, pp. 367–376, 2018.

[234] P. Martínez-Acedo, E. Núñez, F. J. S. Gómez et al., "A novel strategy for global analysis of the dynamic thiol redox proteome," *Molecular & Cellular Proteomics*, vol. 11, no. 9, pp. 800–813, 2012.

[235] P. Navarro, M. Trevisan-Herraz, E. Bonzon-Kulichenko et al., "General statistical framework for quantitative proteomics by stable isotope labeling," *Journal of Proteome Research*, vol. 13, no. 3, pp. 1234–1247, 2014.

[236] R. J. Mailloux, D. Gardiner, and M. O'Brien, "2-Oxoglutarate dehydrogenase is a more significant source of $O_2^{\cdot-}/H_2O_2$ than pyruvate dehydrogenase in cardiac and liver tissue," *Free Radical Biology & Medicine*, vol. 97, pp. 501–512, 2016.

[237] H. J. Forman, M. J. Davies, A. C. Kramer et al., "Protein cysteine oxidation in redox signaling: caveats on sulfenic acid detection and quantification," *Archives of Biochemistry and Biophysics*, vol. 617, pp. 26–37, 2017.

[238] C. Welter, E. Meese, and N. Blin, "Rapid step-gradient purification of mitochondrial DNA," *Molecular Biology Reports*, vol. 13, no. 2, pp. 117–120, 1988.

[239] M. Caldera, P. Buphamalai, F. Müller, and J. Menche, "Interactome-based approaches to human disease," *Current Opinion in Systems Biology*, vol. 3, pp. 88–94, 2017.

[240] D. Szklarczyk, J. H. Morris, H. Cook et al., "The STRING database in 2017: quality-controlled protein-protein association networks, made broadly accessible," *Nucleic Acids Research*, vol. 45, D1, pp. D362–D368, 2017.

[241] S. Orchard, M. Ammari, B. Aranda et al., "The MIntAct project—IntAct as a common curation platform for 11 molecular interaction databases," *Nucleic Acids Research*, vol. 42, D1, pp. D358–D363, 2014.

[242] L. Salwinski, C. S. Miller, A. J. Smith, F. K. Pettit, J. U. Bowie, and D. Eisenberg, "The database of interacting proteins: 2004 update," *Nucleic Acids Research*, vol. 32, no. 90001, pp. 449D–4451D, 2004.

[243] S. K. Miryala, A. Anbarasu, and S. Ramaiah, "Discerning molecular interactions: a comprehensive review on biomolecular interaction databases and network analysis tools," *Gene*, vol. 642, pp. 84–94, 2018.

[244] A.-L. Barabási, N. Gulbahce, and J. Loscalzo, "Network medicine: a network-based approach to human disease," *Nature Reviews Genetics*, vol. 12, no. 1, pp. 56–68, 2011.

[245] A.-L. Barabási and Z. N. Oltvai, "Network biology: understanding the cell's functional organization," *Nature Reviews Genetics*, vol. 5, no. 2, pp. 101–113, 2004.

[246] R. M. Ewing, P. Chu, F. Elisma et al., "Large-scale mapping of human protein-protein interactions by mass spectrometry," *Molecular Systems Biology*, vol. 3, p. 89, 2007.

[247] T. Geiger, A. Velic, B. Macek et al., "Initial quantitative proteomic map of 28 mouse tissues using the SILAC mouse," *Molecular & Cellular Proteomics*, vol. 12, no. 6, pp. 1709–1722, 2013.

[248] M. Wilhelm, J. Schlegl, H. Hahne et al., "Mass-spectrometry-based draft of the human proteome," *Nature*, vol. 509, no. 7502, pp. 582–587, 2014.

[249] A. G. Ngounou Wetie, I. Sokolowska, A. G. Woods, U. Roy, J. A. Loo, and C. C. Darie, "Investigation of stable and transient protein-protein interactions: past, present and future," *Proteomics*, vol. 13, no. 3-4, pp. 538–557, 2013.

[250] M. O. Collins and J. S. Choudhary, "Mapping multiprotein complexes by affinity purification and mass spectrometry," *Current Opinion in Biotechnology*, vol. 19, no. 4, pp. 324–330, 2008.

[251] M. Selbach and M. Mann, "Protein interaction screening by quantitative immunoprecipitation combined with

[251] knockdown (QUICK)," *Nature Methods*, vol. 3, no. 12, pp. 981–983, 2006.

[252] L. Böttinger, B. Guiard, S. Oeljeklaus et al., "A complex of Cox4 and mitochondrial Hsp70 plays an important role in the assembly of the cytochrome c oxidase," *Molecular Biology of the Cell*, vol. 24, no. 17, pp. 2609–2619, 2013.

[253] M. J. Runswick, J. V. Bason, M. G. Montgomery, G. C. Robinson, I. M. Fearnley, and J. E. Walker, "The affinity purification and characterization of ATP synthase complexes from mitochondria," *Open Biology*, vol. 3, no. 2, 2013.

[254] S. Austin, K. L. Bennett, and K. Nowikovsky, "Interactome of LETM1 using miniaturised affinity purification mass spectrometry," *Biochimica et Biophysica Acta (BBA) - Bioenergetics*, vol. 1837, article e107, 2014.

[255] Y. Sun, A. A. Vashisht, J. Tchieu, J. A. Wohlschlegel, and L. Dreier, "Voltage-dependent anion channels (VDACs) recruit parkin to defective mitochondria to promote mitochondrial autophagy," *Journal of Biological Chemistry*, vol. 287, no. 48, pp. 40652–40660, 2012.

[256] R. H. Malty, H. Aoki, A. Kumar et al., "A map of human mitochondrial protein interactions linked to neurodegeneration reveals new mechanisms of redox homeostasis and NF-κB signaling," *Cell Systems*, vol. 5, no. 6, pp. 564–577.e12, 2017.

[257] D. Trisciuoglio, M. Desideri, V. Farini et al., "Affinity purification-mass spectrometry analysis of bcl-2 interactome identified SLIRP as a novel interacting protein," *Cell Death & Disease*, vol. 7, no. 2, article e2090, 2016.

[258] B. J. Floyd, E. M. Wilkerson, M. T. Veling et al., "Mitochondrial protein interaction mapping identifies regulators of respiratory chain function," *Molecular Cell*, vol. 63, no. 4, pp. 621–632, 2016.

[259] E. L. Huttlin, R. J. Bruckner, J. A. Paulo et al., "Architecture of the human interactome defines protein communities and disease networks," *Nature*, vol. 545, no. 7655, pp. 505–509, 2017.

[260] E. L. Huttlin, L. Ting, R. J. Bruckner et al., "The BioPlex network: a systematic exploration of the human interactome," *Cell*, vol. 162, no. 2, pp. 425–440, 2015.

[261] D. K. Schweppe, E. L. Huttlin, J. W. Harper, and S. P. Gygi, "BioPlex display: an interactive suite for large-scale AP-MS protein–protein interaction data," *Journal of Proteome Research*, vol. 17, no. 1, pp. 722–726, 2018.

[262] A. R. Kristensen, J. Gsponer, and L. J. Foster, "A high-throughput approach for measuring temporal changes in the interactome," *Nature Methods*, vol. 9, no. 9, pp. 907–909, 2012.

[263] X. Tang and J. E. Bruce, "A new cross-linking strategy: protein interaction reporter (PIR) technology for protein–protein interaction studies," *Molecular BioSystems*, vol. 6, no. 6, pp. 939–947, 2010.

[264] N. E. Scott, L. D. Rogers, A. Prudova et al., "Interactome disassembly during apoptosis occurs independent of caspase cleavage," *Molecular Systems Biology*, vol. 13, no. 1, p. 906, 2017.

[265] H. J. C. T. Wessels, R. O. Vogel, L. van den Heuvel et al., "LC-MS/MS as an alternative for SDS-PAGE in blue native analysis of protein complexes," *Proteomics*, vol. 9, no. 17, pp. 4221–4228, 2009.

[266] I. Wittig, R. Carrozzo, F. M. Santorelli, and H. Schägger, "Supercomplexes and subcomplexes of mitochondrial oxidative phosphorylation," *Biochimica et Biophysica Acta (BBA) - Bioenergetics*, vol. 1757, no. 9-10, pp. 1066–1072, 2006.

[267] P. D. B. Olinares, L. Ponnala, and K. J. van Wijk, "Megadalton complexes in the chloroplast stroma of Arabidopsis thaliana characterized by size exclusion chromatography, mass spectrometry, and hierarchical clustering," *Molecular & Cellular Proteomics*, vol. 9, no. 7, pp. 1594–1615, 2010.

[268] M. Babu, J. Vlasblom, S. Pu et al., "Interaction landscape of membrane-protein complexes in Saccharomyces cerevisiae," *Nature*, vol. 489, no. 7417, pp. 585–589, 2012.

[269] D. Drew, S. Newstead, Y. Sonoda, H. Kim, G. von Heijne, and S. Iwata, "GFP-based optimization scheme for the overexpression and purification of eukaryotic membrane proteins in Saccharomyces cerevisiae," *Nature Protocols*, vol. 3, no. 5, pp. 784–798, 2008.

[270] H. Heide, L. Bleier, M. Steger et al., "Complexome profiling identifies TMEM126B as a component of the mitochondrial complex I assembly complex," *Cell Metabolism*, vol. 16, no. 4, pp. 538–549, 2012.

[271] H. Schägger and G. von Jagow, "Blue native electrophoresis for isolation of membrane protein complexes in enzymatically active form," *Analytical Biochemistry*, vol. 199, no. 2, pp. 223–231, 1991.

[272] H. Schägger, "[12] Native electrophoresis for isolation of mitochondrial oxidative phosphorylation protein complexes," *Methods in Enzymology*, vol. 260, pp. 190–202, 1995.

[273] H. Schagger, W. A. Cramer, and G. Vonjagow, "Analysis of molecular masses and oligomeric states of protein complexes by blue native electrophoresis and isolation of membrane protein complexes by two-dimensional native electrophoresis," *Analytical Biochemistry*, vol. 217, no. 2, pp. 220–230, 1994.

[274] V. Reisinger and L. A. Eichacker, "Analysis of membrane protein complexes by blue native PAGE," *Proteomics*, vol. 6, no. S2, pp. 6–15, 2006.

[275] I. Wittig, H.-P. Braun, and H. Schagger, "Blue native PAGE," *Nature Protocols*, vol. 1, no. 1, pp. 418–428, 2006.

[276] A. O. Helbig, M. J. L. de Groot, R. A. van Gestel et al., "A three-way proteomics strategy allows differential analysis of yeast mitochondrial membrane protein complexes under anaerobic and aerobic conditions," *Proteomics*, vol. 9, no. 20, pp. 4787–4798, 2009.

[277] J. S. Andersen, C. J. Wilkinson, T. Mayor, P. Mortensen, E. A. Nigg, and M. Mann, "Proteomic characterization of the human centrosome by protein correlation profiling," *Nature*, vol. 426, no. 6966, pp. 570–574, 2003.

[278] M. Larance and A. I. Lamond, "Multidimensional proteomics for cell biology," *Nature Reviews Molecular Cell Biology*, vol. 16, no. 5, pp. 269–280, 2015.

[279] S. Cogliati, E. Calvo, M. Loureiro et al., "Mechanism of superassembly of respiratory complexes III and IV," *Nature*, vol. 539, no. 7630, pp. 579–582, 2016.

[280] S. Guerrero-Castillo, F. Baertling, D. Kownatzki et al., "The assembly pathway of mitochondrial respiratory chain complex I," *Cell Metabolism*, vol. 25, no. 1, pp. 128–139, 2017.

[281] C. S. Müller, W. Bildl, A. Haupt et al., "Cryo-slicing blue native-mass spectrometry (csBN-MS), a novel technology for high resolution complexome profiling," *Molecular & Cellular Proteomics*, vol. 15, no. 2, pp. 669–681, 2016.

[282] H. J. C. T. Wessels, R. O. Vogel, R. N. Lightowlers et al., "Analysis of 953 human proteins from a mitochondrial HEK293 fraction by complexome profiling," *PLoS One*, vol. 8, no. 7, article e68340, 2013.

[283] N. Iwamoto and T. Shimada, "Recent advances in mass spectrometry-based approaches for proteomics and biologics: great contribution for developing therapeutic antibodies," *Pharmacology & Therapeutics*, vol. 185, pp. 147–154, 2018.

[284] P. C. Havugimana, G. T. Hart, T. Nepusz et al., "A census of human soluble protein complexes," *Cell*, vol. 150, no. 5, pp. 1068–1081, 2012.

[285] K. J. Kirkwood, Y. Ahmad, M. Larance, and A. I. Lamond, "Characterization of native protein complexes and protein isoform variation using size-fractionation-based quantitative proteomics," *Molecular & Cellular Proteomics*, vol. 12, no. 12, pp. 3851–3873, 2013.

[286] F. Liu, P. Lössl, B. M. Rabbitts, R. S. Balaban, and A. J. R. Heck, "The interactome of intact mitochondria by cross-linking mass spectrometry provides evidence for coexisting respiratory supercomplexes," *Molecular & Cellular Proteomics*, vol. 17, no. 2, pp. 216–232, 2018.

[287] D. K. Schweppe, J. D. Chavez, C. F. Lee et al., "Mitochondrial protein interactome elucidated by chemical cross-linking mass spectrometry," *Proceedings of the National Academy of Sciences of the United States of America*, vol. 114, no. 7, pp. 1732–1737, 2017.

[288] B. J. Greber, D. Boehringer, A. Leitner et al., "Architecture of the large subunit of the mammalian mitochondrial ribosome," *Nature*, vol. 505, no. 7484, pp. 515–519, 2014.

[289] R. Fritzsche, C. H. Ihling, M. Götze, and A. Sinz, "Optimizing the enrichment of cross-linked products for mass spectrometric protein analysis," *Rapid Communications in Mass Spectrometry*, vol. 26, no. 6, pp. 653–658, 2012.

[290] A. Leitner, R. Reischl, T. Walzthoeni et al., "Expanding the chemical cross-linking toolbox by the use of multiple proteases and enrichment by size exclusion chromatography," *Molecular & Cellular Proteomics*, vol. 11, no. 3, article M111.014126, 2012.

[291] M. Trester-Zedlitz, K. Kamada, S. K. Burley, D. Fenyö, B. T. Chait, and T. W. Muir, "A modular cross-linking approach for exploring protein interactions," *Journal of the American Chemical Society*, vol. 125, no. 9, pp. 2416–2425, 2003.

[292] A. Kao, C. L. Chiu, D. Vellucci et al., "Development of a novel cross-linking strategy for fast and accurate identification of cross-linked peptides of protein complexes," *Molecular & Cellular Proteomics*, vol. 10, no. 1, article M110.002212, 2011.

[293] K. Kölbel, C. H. Ihling, and A. Sinz, "Analysis of peptide secondary structures by photoactivatable amino acid analogues," *Angewandte Chemie International Edition*, vol. 51, no. 50, pp. 12602–12605, 2012.

[294] A. Leitner, L. A. Joachimiak, P. Unverdorben et al., "Chemical cross-linking/mass spectrometry targeting acidic residues in proteins and protein complexes," *Proceedings of the National Academy of Sciences*, vol. 111, no. 26, pp. 9455–9460, 2014.

[295] M. Q. Müller, F. Dreiocker, C. H. Ihling, M. Schäfer, and A. Sinz, "Fragmentation behavior of a thiourea-based reagent for protein structure analysis by collision-induced dissociative chemical cross-linking," *Journal of Mass Spectrometry*, vol. 45, no. 8, pp. 880–891, 2010.

[296] O. Rinner, J. Seebacher, T. Walzthoeni et al., "Identification of cross-linked peptides from large sequence databases," *Nature Methods*, vol. 5, no. 4, pp. 315–318, 2008.

[297] C. Yu, W. Kandur, A. Kao, S. Rychnovsky, and L. Huang, "Developing new isotope-coded mass spectrometry-cleavable cross-linkers for elucidating protein structures," *Analytical Chemistry*, vol. 86, no. 4, pp. 2099–2106, 2014.

[298] F. Chu, P. R. Baker, A. L. Burlingame, and R. J. Chalkley, "Finding chimeras: a bioinformatics strategy for identification of cross-linked peptides," *Molecular & Cellular Proteomics*, vol. 9, no. 1, pp. 25–31, 2010.

[299] M. Götze, J. Pettelkau, S. Schaks et al., "StavroX—a software for analyzing crosslinked products in protein interaction studies," *Journal of the American Society for Mass Spectrometry*, vol. 23, no. 1, pp. 76–87, 2012.

[300] M. R. Hoopmann, C. R. Weisbrod, and J. E. Bruce, "Improved strategies for rapid identification of chemically cross-linked peptides using protein interaction reporter technology," *Journal of Proteome Research*, vol. 9, no. 12, pp. 6323–6333, 2010.

[301] M. R. Hoopmann, A. Zelter, R. S. Johnson et al., "Kojak: efficient analysis of chemically cross-linked protein complexes," *Journal of Proteome Research*, vol. 14, no. 5, pp. 2190–2198, 2015.

[302] F. Liu, D. T. S. Rijkers, H. Post, and A. J. R. Heck, "Proteome-wide profiling of protein assemblies by cross-linking mass spectrometry," *Nature Methods*, vol. 12, no. 12, pp. 1179–1184, 2015.

[303] C. R. Weisbrod, J. D. Chavez, J. K. Eng, L. Yang, C. Zheng, and J. E. Bruce, "*In vivo* protein interaction network identified with a novel real-time cross-linked peptide identification strategy," *Journal of Proteome Research*, vol. 12, no. 4, pp. 1569–1579, 2013.

[304] H. Xu, P.-H. Hsu, L. Zhang, M.-D. Tsai, and M. A. Freitas, "Database search algorithm for identification of intact cross-links in proteins and peptides using tandem mass spectrometry," *Journal of Proteome Research*, vol. 9, no. 7, pp. 3384–3393, 2010.

[305] B. Yang, Y. J. Wu, M. Zhu et al., "Identification of cross-linked peptides from complex samples," *Nature Methods*, vol. 9, no. 9, pp. 904–906, 2012.

[306] C. W. Combe, L. Fischer, and J. Rappsilber, "xiNET: cross-link network maps with residue resolution," *Molecular & Cellular Proteomics*, vol. 14, no. 4, pp. 1137–1147, 2015.

[307] M. Grimm, T. Zimniak, A. Kahraman, and F. Herzog, "xVis: a web server for the schematic visualization and interpretation of crosslink-derived spatial restraints," *Nucleic Acids Research*, vol. 43, W1, pp. W362–W369, 2015.

[308] J. Kosinski, A. von Appen, A. Ori, K. Karius, C. W. Müller, and M. Beck, "Xlink analyzer: software for analysis and visualization of cross-linking data in the context of three-dimensional structures," *Journal of Structural Biology*, vol. 189, no. 3, pp. 177–183, 2015.

[309] D. K. Schweppe, J. D. Chavez, and J. E. Bruce, "XLmap: an R package to visualize and score protein structure models based on sites of protein cross-linking," *Bioinformatics*, vol. 32, no. 2, pp. 306–308, 2016.

[310] C. Zheng, C. R. Weisbrod, J. D. Chavez et al., "XLink-DB: database and software tools for storing and visualizing protein interaction topology data," *Journal of Proteome Research*, vol. 12, no. 4, pp. 1989–1995, 2013.

[311] A. Leitner, M. Faini, F. Stengel, and R. Aebersold, "Crosslinking and mass spectrometry: an integrated technology to understand the structure and function of molecular machines," *Trends in Biochemical Sciences*, vol. 41, no. 1, pp. 20–32, 2016.

[312] E. Sprinzak, S. J. Cokus, T. O. Yeates, D. Eisenberg, and M. Pellegrini, "Detecting coordinated regulation of multiprotein complexes using logic analysis of gene expression," *BMC Systems Biology*, vol. 3, no. 1, p. 115, 2009.

[313] A. Hofer and T. Wenz, "Post-translational modification of mitochondria as a novel mode of regulation," *Experimental Gerontology*, vol. 56, pp. 202–220, 2014.

# Resveratrol Decreases Oxidative Stress by Restoring Mitophagy and Improves the Pathophysiology of Dystrophin-Deficient *mdx* Mice

Rio Sebori, Atsushi Kuno, Ryusuke Hosoda, Takashi Hayashi, and Yoshiyuki Horio

*Department of Pharmacology, Sapporo Medical University School of Medicine, S1, W 17, Chu-ouku, Sapporo 060-8556, Japan*

Correspondence should be addressed to Yoshiyuki Horio; horio@sapmed.ac.jp

Academic Editor: Reiko Matsui

We previously showed that treatment with resveratrol (3,5,4′-trihydroxy-*trans*-stilbene), an activator of the $NAD^+$-dependent deacetylase SIRT1 at 4 g/kg food for 32 weeks, significantly decreased the muscular reactive oxygen species (ROS) levels and ameliorated the pathology of *mdx* mice, an animal model of *Duchenne* muscular dystrophy (DMD). Here, we treated *mdx* mice with various doses of resveratrol (0.04, 0.4, and 4 g/kg food) for 56 weeks and examined the effects on serum creatine kinase levels and physical activities. Because resveratrol promotes autophagy, we also investigated whether autophagy including mitochondrial autophagy (mitophagy) is involved in resveratrol's effects. Autophagy/mitophagy-related genes and autophagic flux were downregulated in the muscle of *mdx* mice, and these phenomena were reversed by resveratrol with significant ROS reduction. Resveratrol at 4 g/kg food reduced the number of immature myofibers containing central nuclei and fine fibers < 400 $\mu m^2$ and increased that of thicker myofibers in the quadriceps, suggesting that resveratrol decreased myofiber wasting and promoted muscular maturation. Accordingly, resveratrol at 0.4 g/kg food reduced the creatine kinase levels to one-third of those in untreated *mdx* mice and significantly increased the animals' physical activities. In C2C12 myoblast cells, resveratrol promoted mitophagy and eliminated mitochondria containing high superoxide levels. The clearance of damaged mitochondria and ROS reduction by resveratrol was completely suppressed by an autophagy inhibitor (chloroquine) and by knocking down *Atg5* or *Pink1*, essential genes for autophagy and mitophagy, respectively. Thus, resveratrol is a potential therapeutic agent for DMD, and the clearance of damaged mitochondria probably contributes to its action.

## 1. Introduction

Duchenne muscular dystrophy (DMD) is a severe type of muscular dystrophy, in which mutations of the dystrophin gene lead to progressive muscle wasting and degeneration [1]. Few treatments exist for DMD except for glucocorticoids, which increase muscle strength and functional measures in the short term, although their ability to extend walking ability for more than two years is unclear [2]. Furthermore, the long-term glucocorticoid use may cause prediabetes and osteoporosis, and glucocorticoids are not thought to improve myogenesis or fibrosis [2, 3].

SIRT1, an $NAD^+$-dependent protein deacetylase, regulates transcription machineries and plays pivotal roles in controlling metabolism, inflammation, differentiation, and DNA repair [4]. SIRT1 promotes cell survival by reducing oxidative stress and by increasing mitochondrial biogenesis by deacetylating and activating Forkhead Box O transcription factors (FOXOs) and peroxisome proliferator-activated receptor gamma coactivator 1-alpha (PGC-1$\alpha$) [4]. Resveratrol (3,5,4′-trihydroxy-*trans*-stilbene), a natural polyphenol found in grapes and red wine, is an activator of SIRT1 [5]. We found that resveratrol at a dose of 4 g/kg food ameliorates the skeletal muscle and cardiac pathologies of dystrophin-deficient *mdx* mice [6, 7]. Beneficial effects of resveratrol in *mdx* mice have also been reported by other groups [8–10], and SIRT1 overexpression in *mdx* mice was shown to reduce muscle damage and improve function [11].

In the heart, resveratrol induces reactive oxygen species- (ROS-) detoxifying enzyme superoxide dismutase 2 (SOD2)

by activating nuclear SIRT1, thereby decreasing oxidative damage [12]. Resveratrol also inhibits myocardial hypertrophy and fibrosis by promoting SIRT1's deacetylation of coactivator p300, which then undergoes ubiquitin-dependent degradation [7].

Surprisingly, SOD2 levels in the skeletal muscle of *mdx* mice were not significantly elevated by resveratrol, possibly because SIRT1 was not concentrated in the nuclei of myofibers [6]. Although resveratrol suppressed the upregulation of NADPH oxidase subunits, resveratrol and SIRT1 may use another mechanism to reduce ROS levels in the muscle of *mdx* mice.

Autophagy is a process that digests unnecessary or dysfunctional components in cells. SIRT1 promotes autophagy by deacetylating and activating autophagic components such as Atg5, Atg7, and LC3 [13, 14], and resveratrol induces autophagy by activating cytoplasmic SIRT1 [15, 16]. Damaged or dysfunctional mitochondria, the major source of ROS in most cells [17], are eliminated by an autophagic process called mitophagy [18–20]. The loss of membrane potential in damaged mitochondria causes PTEN-induced putative kinase 1 (Pink1) to accumulate on their outer membrane, where Pink1 recruits, phosphorylates, and activates parkin, a ubiquitin ligase. The activated parkin recruits p62, an autophagy adaptor protein, to the damaged mitochondria, leading to encapsulation of the damaged mitochondria by LC3 in autophagosomes; the mitochondria are then degraded in lysosomes [18–20]. Autophagy insufficiency induced by the knockout of autophagy/mitophagy-related genes such as *Atg3*, *Atg5*, *Atg7*, *LC3B*, and *Pink1* causes a significant increase in cellular ROS [19], suggesting that ROS are liberated from damaged mitochondria that escape mitophagy. In addition, the muscle-specific knockout of *Atg5* or *Atg7* results in muscle atrophy, dysfunction, and myopathy [21, 22]. Notably, mitochondria in the muscle of *Atg7*-null mice are morphologically and functionally abnormal [23]. Loss-of-function mutants of *Pink1* or *parkin*, mitophagy-related genes, show mitochondrial dysfunction and flight muscle degeneration in *Drosophila* [24, 25]. Thus, mitophagy may have a role in the pathology of muscular dystrophies.

Here, we examined autophagy/mitophagy in *mdx* mice and in C2C12 myoblast cells. Because resveratrol can act as a mitochondrial depolarizing agent [26], and because a low dose of resveratrol (2.5 mg/kg/day) improves insulin resistance in mice [27], we investigated the effects of lower doses of resveratrol, i.e., 0.04 and 0.4 g/kg food, as well as 4 g/kg food, on *mdx* mice. Resveratrol increased the expression of autophagy/mitophagy-related genes and autophagic flux and reduced ROS levels in the muscle of *mdx* mice. Furthermore, resveratrol improved the muscular pathology and physical strength of the *mdx* mice. We further showed that mitophagy was indispensable for the ROS reduction caused by resveratrol in C2C12 myoblast cells.

## 2. Materials and Methods

*2.1. Reagents and Antibodies.* Resveratrol (185-01721) and Hoechst 33342 (346-07951) were from Wako Pure Chemicals (Osaka, Japan). Food grade resveratrol for mouse treatment was from ChromaDex (ASB-00018089-101, Irvine, CA). FITC-conjugated wheat germ agglutinin (WGA) lectin (W834), dihydroethidium (DHE) (D1168), MitoSOX Red (M36008), MitoTracker Red (MTR, M7512), and Lipofectamine RNAiMAX Transfection Reagent (13778-150) were from Thermo Fisher Scientific (Rockford, IL). The RNeasy Fibrous Tissue Mini Kit (74704) was from Qiagen (Valencia, CA). The GoScript Reverse Transcription System (A6010), GoTaq qPCR Master Mix (A600A), and ViaFect Transfection Reagent (E4982) were from Promega (Madison, WI). Antimycin A (A8674) and chloroquine (CQ) (C6628) were from Sigma-Aldrich (St. Louis, MO). Plasmid EGFP-LC3 Expression Vector was from Addgene (#11546). siRNAs against mouse Atg5 (SASI_Mm01_00089196), mouse Pink1 (SASI_Mm02_00331134), and MISSION siRNA Universal Negative Control (SIC-001) were from Sigma Genosys Japan (Ishikari, Japan). Antibodies used were as follows: anti-LC3AB (#12741), anti-phoso-Ser65-4EBP1 (#9451), anti-total 4EBP1 (#9452), and anti-ubiquitin (#3936) from Cell Signaling Technology (Beverly, MA), anti-p62 (GP62-C) from Progen (Heidelberg, Germany), and anti-GAPDH (G8795) and anti-α-tubulin (T5168) from Sigma-Aldrich.

*2.2. Animals and Experimental Design.* All *in vivo* experiments were conducted in strict accordance with the Guide for the Care and Use of Laboratory Animals (Institute of Laboratory Animal Resources, 1996) and approved by the Animal Care and Use Committee of Sapporo Medical University. Male C57BL/10ScSn-Dmdmdx/J mice (*mdx* mice) and age-matched C57BL10 mice were purchased from Oriental Yeast Co. Ltd. (Tokyo, Japan). C57BL10 mice served as WT. Muscle tissue samples were prepared at 22 weeks of age. In a series of experiments, the effects of resveratrol were analyzed in 24 *mdx* mice. *mdx* mice were orally given 0, 0.04, 0.4, or 4 g resveratrol/kg food ad libitum from 9 weeks to 65 weeks of age (6 mice for each dose). At 65 weeks of age, the mice were sacrificed and their quadriceps were frozen in liquid nitrogen-cooled isopentane (Nacalai Tesque, Kyoto, Japan) and stored at −80°C until use.

*2.3. Gene Expression Assay.* Total RNA was prepared from quadriceps muscles using the RNeasy Fibrous Tissue Mini Kit. Complementary DNA generated with the GoScript Reverse Transcription System was analyzed by the StepOne Real-Time PCR System (Applied Biosystems, Foster City, CA) using the GoTaq qPCR Master Mix. Each sample was run in duplicate, and the mean value was used to calculate the mRNA level of the gene of interest. All data were normalized to 18s ribosomal RNA using the standard curve method. The primer sequences are listed in Supplemental Table 1.

*2.4. Western Blotting.* Frozen quadriceps muscles were powdered by mortar and pestle, lysed in ice-cold CelLytic M Tissue Lysis Reagent (C3228, Sigma-Aldrich) with a 1% protease inhibitor cocktail (25955-11, Nacalai Tesque, Kyoto, Japan) and 1% phosphatase inhibitor cocktail (07574-61, Nacalai Tesque), and centrifuged at 10,000*g* for 10 min at 4°C. C2C12 samples were homogenized in ice-cold CelLytic M Cell Lysis Reagent (C3228, Sigma-Aldrich) with the above-described protease inhibitor and phosphatase inhibitor cocktails. The protein concentration of the supernatant was measured using the Protein Quantification Kit-Rapid (PQ01, Dojindo, Kumamoto, Japan). Supernatant fractions of equal protein concentration were analyzed by Western blotting as described previously [6].

*2.5. Histological Analyses.* Frozen muscles were embedded in optimal cutting temperature compound (Tissue-Tek, Torrance, CA), and blocks were cross-sectioned mid-belly at $5\,\mu m$ by cryostat at $-20°C$. To monitor tissue ROS levels, sections of quadriceps muscles were incubated with $5\,\mu M$ DHE (Thermo Fisher Scientific) for 30 min at $37°C$ and washed twice with PBS. The digital images were captured using an inverted confocal laser scanning microscope (LSM510META; Zeiss, Germany) at $512 \times 512$ pixels, with a $63\times$ oil immersion objective lens. The DHE fluorescence intensity was quantified by the ImageJ software (National Institutes of Health, Bethesda, MD). The fluorescence intensity was measured from 6 randomly selected images of each muscle, and the average of 4 mice in each group was determined.

To analyze cross-sectional areas, sections of quadriceps muscles were labeled with FITC-conjugated WGA. Nuclei were stained with Hoechst 33342. The digital images were captured by an LSM510META inverted confocal laser canning microscope, and the cross-sectional areas and central nuclei were quantified by the ImageJ software. The cross-sectional areas of approximately 300 randomly selected myofibers per muscle were measured. The percentage of fibers with centrally located nuclei was analyzed in 480–500 myofibers per muscle in each group.

*2.6. Cell Culture.* C2C12 myoblast cells were cultured in Dulbecco's modified Eagle's medium (Wako Pure Chemical) supplemented with a 1% antibiotic-antimycotic mixed stock solution (Nacalai Tesque) and 10% fetal bovine serum (MP Biomedicals, Solon, OH).

*2.7. Transfection of siRNA.* Lipofectamine RNAiMAX Transfection Reagent was used to transfect siRNAs (30 nM) targeting *Atg5* and *Pink1*, according to the manufacturer's instructions. Cells were analyzed 48 h after transfection.

*2.8. Analysis of Mitophagy in C2C12 Cells.* C2C12 cells were transfected with EGFP-LC3 using ViaFect Transfection Reagent (Promega) according to the manufacturer's instructions and then were stained with 200 nM MTR 42 h after transfection. The cells were incubated with vehicle or $30\,\mu M$ resveratrol for 6 h. In some samples, $50\,\mu M$ CQ was added before the resveratrol treatment. After fixation, the colocalization of EGFP-LC3 with mitochondria was analyzed by confocal laser microscopy. After mitochondria take-up MTR in a membrane potential-dependent manner, the MTR fluorescence is retained, even if the mitochondrial potential is lost during fixation. The number of EGFP-LC3 dots colocalized with MTR was counted in at least 30 randomly selected cells in each group, and 3 independent experiments were performed.

*2.9. Detection of Mitochondria and Mitochondrial ROS Levels.* The mitochondrial superoxide levels in C2C12 cells were detected by MitoSOX Red staining according to the manufacturer's protocol, and the fluorescence was analyzed by the ImageJ software. Twenty-four images were selected randomly and analyzed in each group. Four independent experiments were carried out.

*2.10. Measurement of Serum CK-MM Isoenzyme Levels.* Blood was collected from the tail vein of mice at 23 and 65 weeks of age. The samples were incubated at room temperature for 20 min to allow clotting and then centrifuged at $1000g$ for 20 min. The serum was collected and stored at $-80°C$ until use. The serum level of the muscular isoform of creatine kinase (CK-MM) was measured in duplicate, using the CK-MM ELISA Kit (MBS705327, MyBioSource, San Diego, CA) according to the manufacturer's instructions.

*2.11. Four-Limb Hanging Test.* The four-limb hanging test was performed using mice at 37, 38, and 39 weeks of age. Mice were placed on a net, and then the net was inverted by hand. The hanging time was measured in 5 consecutive trials separated by 1 min intervals. Three independent experiments were performed, and the results are shown as the mean of 3 trials per group.

*2.12. Rotarod Test.* To assess motor coordination, *mdx* mice were tested on the rotarod (Ugo Basile, Mount Laurel, NJ) at 40 weeks of age. For training, mice were placed on the rotarod at 10 rpm for 5 min, on 3 consecutive days before the beginning of the experiment. The rotarod was accelerated from 10 to 50 rpm in 2 min, and the time at which the mouse fell off was recorded. Each mouse underwent 5 consecutive trials separated by 5 min intervals, and the results are shown as the mean values of 5 trials per group.

*2.13. Data Analysis.* Data are presented as means ± SEM. Statistical significance was determined using an unpaired Student's two-tailed $t$-test for 2 datasets. Differences between multiple groups were assessed by one-way analysis of variance (ANOVA) followed by the Tukey post hoc test. For all tests, $P < 0.05$ was considered statistically significant. All analyses were performed with the SigmaStat software (Systat Software Inc., San Jose, CA).

## 3. Results

*3.1. Impaired Autophagy/Mitophagy and Increased ROS Levels in the Muscle of mdx Mice.* Impaired autophagic flux has been reported in the muscle of *mdx* mice [28]. We examined the mRNA levels of mitophagy- and autophagy-related genes in the quadriceps of 22-week-old *mdx* mice and compared them with those in age-matched wild type (WT) mice (Figure 1(a)). The mRNA expression levels of mitophagy-related genes, including *Pink1*, *parkin*, *Bnip3*, and *Fundc1*, were significantly reduced in the muscle of *mdx* mice. The expression levels of *Becn1*, *Atg5*, *Map1lc3b*, and *p62*, which are necessary for autophagy as well as mitophagy, were also lower in *mdx* than in WT mice (Figure 1(a)). The expression level of *transcription factor EB (Tfeb)*, a positive regulator of

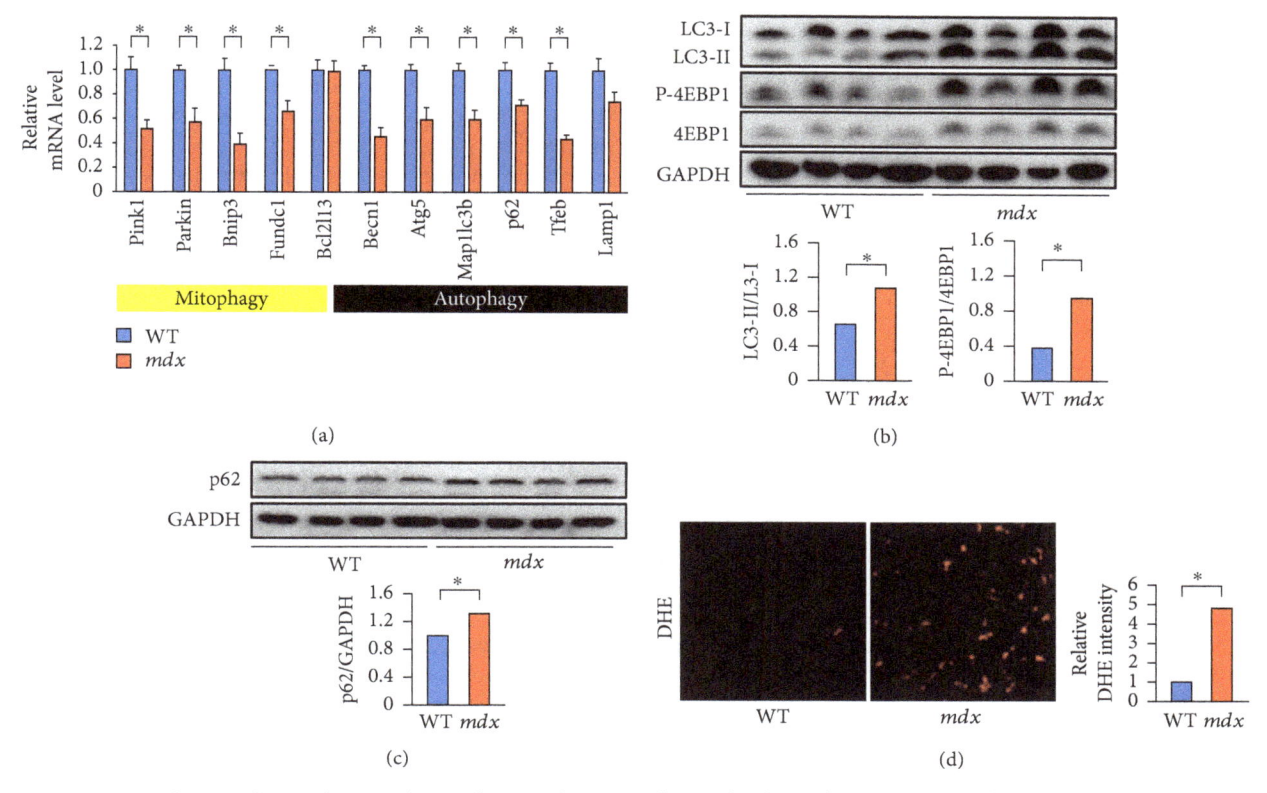

FIGURE 1: Downregulation of mitophagy and autophagy and increased ROS levels in the muscle of *mdx* mice. (a) Expression levels of mitophagy- and autophagy-related genes in the quadriceps muscle of WT and *mdx* mice at 22 weeks of age. Data were normalized to 18s ribosomal RNA. $n = 4$. (b) Representative Western blots for LC3, phosho-Ser65-4EBP1 (P-4EBP1), and total 4EBP1 in the quadriceps muscle (upper). Summary data of the LC3-II/LC3-I ratio and P-4EBP1 levels normalized to the total 4EBP1 level in the muscle (lower). $n = 4$. (c) Representative Western blots for p62 in the quadriceps muscle (upper) and summary data (lower). $n = 4$. (d) Representative dihydroethidium (DHE) staining in the quadriceps (left) and summary data of DHE fluorescence (right). Six randomly selected images were captured in each muscle section, and 4 mice were analyzed in each group. Scale bar: 50 μm. *$P < 0.05$. WT: wild type.

autophagy-related genes [18], in *mdx* mice was downregulated to less than half the level in control mice (Figure 1(a)).

To analyze the autophagic flux in the muscle of *mdx* mice, the protein levels of LC3-I, LC3-II, and p62 were monitored. LC3-I is processed to LC3-II during autophagosome formation, and then LC3-II is degraded after the autophagosome fuses with a lysosome. A low ratio of LC3-II to LC3-I levels (LC3-II/LC3-I) indicates an insufficiency in autophagosome formation, whereas a high LC3-II/LC3-I indicates insufficient autophagosome degradation or enhanced autophagosome formation. p62 is necessary for autophagy, and inhibiting autophagy increases the p62 protein level. As shown in Figure 1(b), the LC3-II/LC3-I was significantly higher in *mdx* than in WT mice. Although the mRNA levels of p62 were decreased in *mdx* mice (Figure 1(a)), the p62 protein levels were higher in *mdx* than in WT mice (Figure 1(c)). These results indicated that autophagy/mitophagy was suppressed in the muscle of *mdx* mice.

mTORC1, a major negative regulator of autophagy, is activated in *mdx* mice [28]. Since mTORC1 phosphorylates eukaryotic initiation factor 4E-binding protein 1 (4EBP1), we examined the phosphorylation levels of 4EBP1 (P-4EBP1). We found that the P-4EBP1 levels were significantly increased, suggesting that mTORC1 is activated, in the muscle of *mdx* mice (Figure 1(b)).

Suppressing mitophagy increases ROS levels [19]. To monitor ROS levels, sections of skeletal muscle were stained with dihydroethidium (DHE). Cellular superoxide converts DHE to ethidium bromide, which stains nuclear DNA with red fluorescence. The DHE fluorescence levels were much higher (4.8-fold) in *mdx* than in WT mice (Figure 1(d)). Together, these findings indicated that defects in autophagy/mitophagy could be involved in the increased ROS levels in the muscle of *mdx* mice.

3.2. Restoration of Autophagy/Mitophagy by Resveratrol in the Muscle of mdx Mice. *mdx* mice were treated with 0.04, 0.4, or 4 g resveratrol/kg food, and the effect of resveratrol on autophagy/mitophagy was examined. The administration of resveratrol to *mdx* mice was started at 9 weeks of age, and the muscle tissues were examined at 65 weeks of age. During the experiment, two untreated *mdx* mice and 1, 1, and 2 mice receiving resveratrol at 0.04, 0.4, and 4 g/kg food, respectively, died of muscle tumors or unknown causes. No difference in the mean body weight was found among the untreated and resveratrol-treated mice at 34, 41, or 65 weeks

TABLE 1: Effect of resveratrol on body weight in *mdx* mice.

| Age (weeks) | Dose of resveratrol (g/kg food) | | | |
|---|---|---|---|---|
| | 0 | 0.04 | 0.4 | 4 |
| | Body weight (g) | | | |
| 34 | 34 ± 1 | 33 ± 0 | 32 ± 1 | 34 ± 1 |
| 41 | 35 ± 1 | 34 ± 0 | 33 ± 2 | 34 ± 1 |
| 65 | 33 ± 2 | 31 ± 0 | 30 ± 1 | 31 ± 1 |

No significant difference was observed in the body weight of *mdx* mice treated with 0, 0.04, 0.4, or 4 g resveratrol/kg food at any age.

of age (Table 1). The expression levels of mitophagy- and autophagy-related genes were significantly increased in the quadriceps of resveratrol-treated *mdx* mice compared with those of control *mdx* mice (Figure 2(a)). Approximately 2-fold increases in the *Pink1*, *parkin*, and *p62* mRNA levels were found in mice treated with all three doses of resveratrol in food. The *Bnip3*, *Fundc1*, *Atg5*, *Becn1*, *Map1lc3b*, *Tfeb*, and *Lamp1* levels were significantly upregulated by resveratrol at 0.04 and 4 g/kg food. The Bcl2l13 level was slightly but significantly elevated in the muscle of *mdx* mice treated with 4 g/kg food. Thus, resveratrol significantly increased the expression levels of mitophagy- and autophagy-related genes. In addition, resveratrol administered at 0.4 g/kg food significantly increased the SIRT1 mRNA levels by 2- to 3-fold in the quadriceps, diaphragm, and tibialis anterior (Supplemental Figure 1(a)).

An impairment in autophagic flux increases the ubiquitinated protein levels in tissues [19]. Western blot analysis showed that the administration of resveratrol dose-dependently decreased the levels of ubiquitinated proteins (Figure 2(b)), suggesting that resveratrol promoted the removal of ubiquitinated proteins from the muscle by inducing autophagic flux. Actually, we observed that resveratrol administration increased the LC3-II/LC3-I at 0.4 and 4 g/kg food (Figure 2(c)), but it did not significantly increase the p62 protein level (Figure 2(d)). These observations indicated that resveratrol increased the autophagic flux in *mdx* mice. The phosphorylation levels of 4EBP1 were not reduced by resveratrol (Supplemental Figure 1(b)), indicating that resveratrol did not inhibit the mTORC1 activity. DHE staining of muscular sections showed that resveratrol significantly decreased the ROS levels in the muscle of *mdx* mice. The ROS levels in the quadriceps of *mdx* mice treated with 0.04, 0.4, and 4 g/kg were 29, 26, and 11% of those in untreated *mdx* mice, respectively (Figure 2(e)).

*3.3. Improvements in Skeletal Muscle Damage and Function in mdx Mice by Resveratrol.* The autophagy/mitophagy restoration and ROS reduction by resveratrol may affect muscle degeneration and regeneration. Since regenerating myofibers contain central nuclei, sections of the quadriceps from *mdx* mice at 65 weeks of age were treated with Hoechst 33342 and FITC-conjugated WGA to stain nuclei and plasma membranes, respectively, and examined by confocal microscopy (Figures 3(a) and 3(b)). Resveratrol treatment at all three doses significantly decreased the number of myofibers with central nuclei, showing that resveratrol reduced the number of newly generated myofibers. Analysis of the cross-sectional areas of myofibers revealed that resveratrol dose-dependently decreased the number of fine fibers (under 400 μm² in cross-sectional area) compared with the number in untreated *mdx* mice (Figure 3(c)). In contrast, the number of wider myofibers (1000 to 1199 μm² in cross-sectional area) was significantly increased by resveratrol administered at 4 g/kg food (Figure 3(c)). These results suggested that resveratrol decelerated the turnover rates and promoted the maturation of myofibers in *mdx* mice.

Whether resveratrol affected number of satellite cells, mRNA levels of Pax 7, a marker of satellite cells, were measured in the quadriceps and soleus of *mdx* mice. However, Pax7 mRNA levels were not affected by resveratrol administration (Supplemental Figure 2). Recently, AMPK activation in satellite cells has been shown to inhibit apoptosis and promote muscle repair [29]. To detect AMPK activation, phosphorylation levels of AMPK in the quadriceps were examined, but we could not detect significant increase of activated AMPK levels by resveratrol (Supplemental Figure 3).

The serum levels of CK-MM reflect skeletal muscle cell damage. Thus, to examine whether muscle injuries were attenuated by resveratrol, the serum CK-MM levels were examined. Resveratrol administered at 0.4 g/kg food to *mdx* mice significantly decreased the CK-MM levels to about one-third of those in untreated *mdx* mice at 23 and 65 weeks of age (Figures 3(d) and 3(e)). At 65 weeks of age, the administration of resveratrol at 4 g/kg food also significantly decreased the CK-MM levels, which were less than half those observed in untreated *mdx* mice (Figure 3(e)).

To investigate whether resveratrol improves skeletal muscle motor function, *mdx* mice were examined by the inverted hang test and the rotarod test, which reflect fatigue resistance and muscular coordination. At 37 weeks of age, the average hanging time in untreated *mdx* mice was 69 sec, and resveratrol treatment at 0.4 and 4 g/kg food significantly extended the time to 121 sec and 114 sec, respectively. At 40 weeks of age, an approximate 2-fold extension of riding time on the rotating rod was observed in *mdx* mice treated with resveratrol as low as 0.04 g/kg food (Figure 3(g)). Resveratrol at 0.4 g/kg and 4 g/kg food extended the riding time on the rotarod to durations similar to those seen with 0.04 g/kg food (Figure 3(g)). Therefore, treatment with resveratrol, especially at a dose of 0.4 g/kg food, decreased the muscle injury and improved physical activities of *mdx* mice.

*3.4. Mitophagy Induction by Resveratrol in C2C12 Cells.* Resveratrol induces autophagy [15, 16]. To examine whether resveratrol also promotes mitophagy, we examined the effect of resveratrol on mitophagy using C2C12 myoblast cells. Autophagosome formation can be monitored by the appearance of EGFP-LC3 dots in cells [30]. EGFP-LC3 was expressed in C2C12 cells, and the number of EGFP-LC3 dots was counted. Treating the cells with resveratrol significantly increased the number of EGFP-LC3 dots compared with control cells (Figures 4(a) and 4(b)). Treating the cells with chloroquine (CQ), which suppresses lysosome function thereby inhibiting the degradation of autophagosomes, increased the number of LC3 dots (Figures 4(a) and 4(b)).

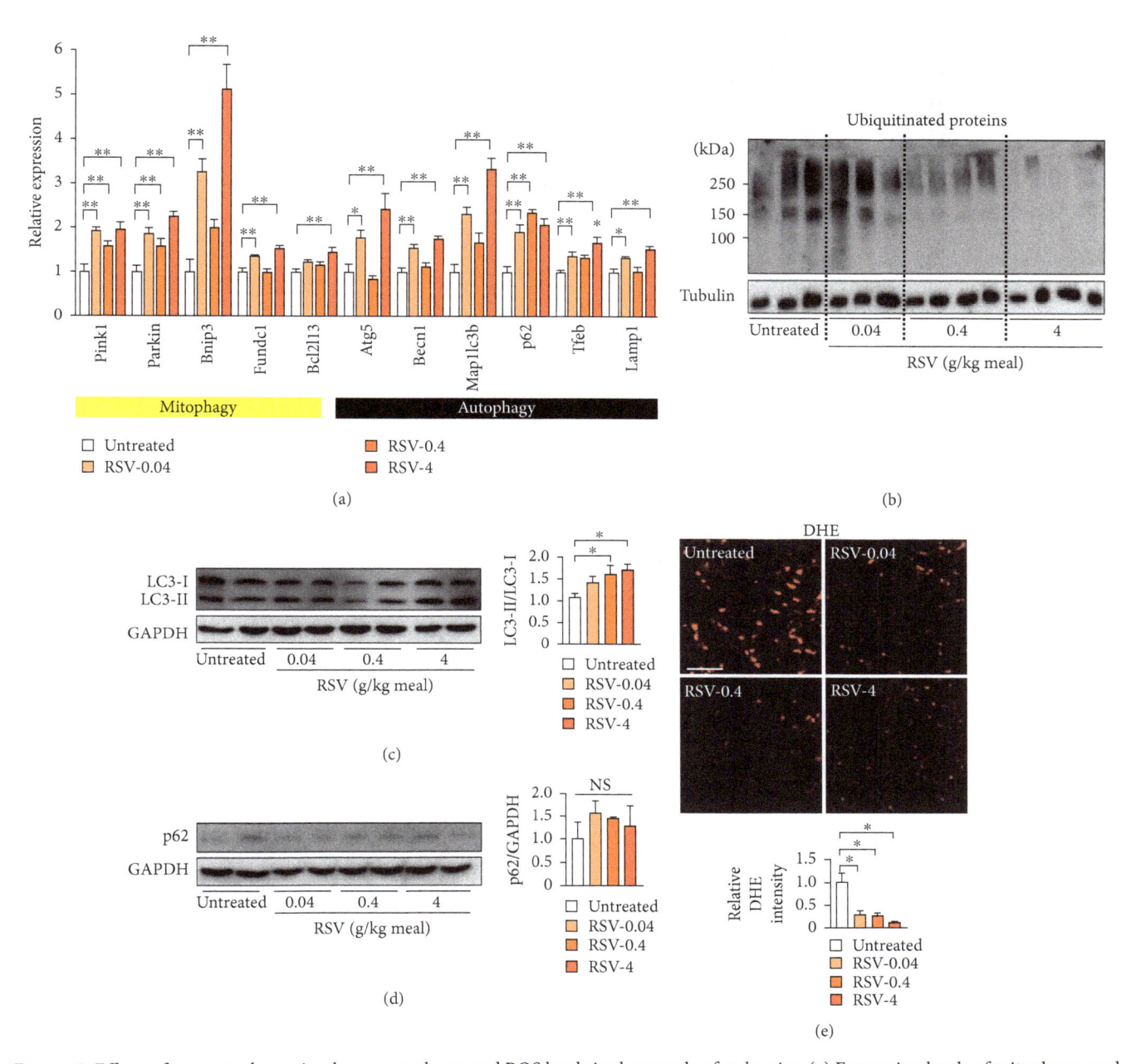

FIGURE 2: Effects of resveratrol on mitophagy, autophagy, and ROS levels in the muscle of *mdx* mice. (a) Expression levels of mitophagy- and autophagy-related genes in the quadriceps muscle from untreated *mdx* mice and *mdx* mice treated with resveratrol (RSV) at 0.04, 0.4, and 4 g/kg food (RSV-0.04, RSV-0.4, and RSV-4, respectively). $n = 4$ in each group. (b) Western blot analysis for ubiquitinated proteins in the quadriceps. (c) Representative Western blots for LC3 in the muscle tissue (left) and summary data of the LC3-II/LC3-I ratio (right). $n = 4$ in each group. (d) Representative Western blot for p62 in muscle (left) and summary data of p62 normalized to the GAPDH level (right). $n = 4$ in each group. (e) Representative dihydroethidium (DHE) staining in the quadriceps (upper) and summary data of DHE fluorescence intensity (lower). Six randomly selected images were captured in each muscle section, and 4 mice were analyzed in each group. Scale bar: 50 μm. $^*P < 0.05$; $^{**}P < 0.01$. NS: not significant.

In the presence of CQ, resveratrol treatment further increased the number of LC3 dots (Figures 4(a) and 4(b)). These observations indicated that resveratrol enhanced autophagosome formation.

To examine whether mitophagy was accelerated by resveratrol, we stained mitochondria with MitoTracker Red (MTR) and examined the colocalization of mitochondria with EGFP-LC3 dots. While there were few mitochondria-containing autophagosomes in the control cells, the administration of CQ significantly increased the number of EGFP-LC3 dots colocalized with mitochondria (Figure 4(a)). Thus, in the absence of CQ, mitochondria were constantly degraded by mitophagy in C2C12 cells (Figures 4(a) and 4(c)). Resveratrol significantly increased the number of EGFP-LC3 dots colocalized with mitochondria in the absence of CQ (Figures 4(a) and 4(c)). The highest numbers of EGFP-LC3

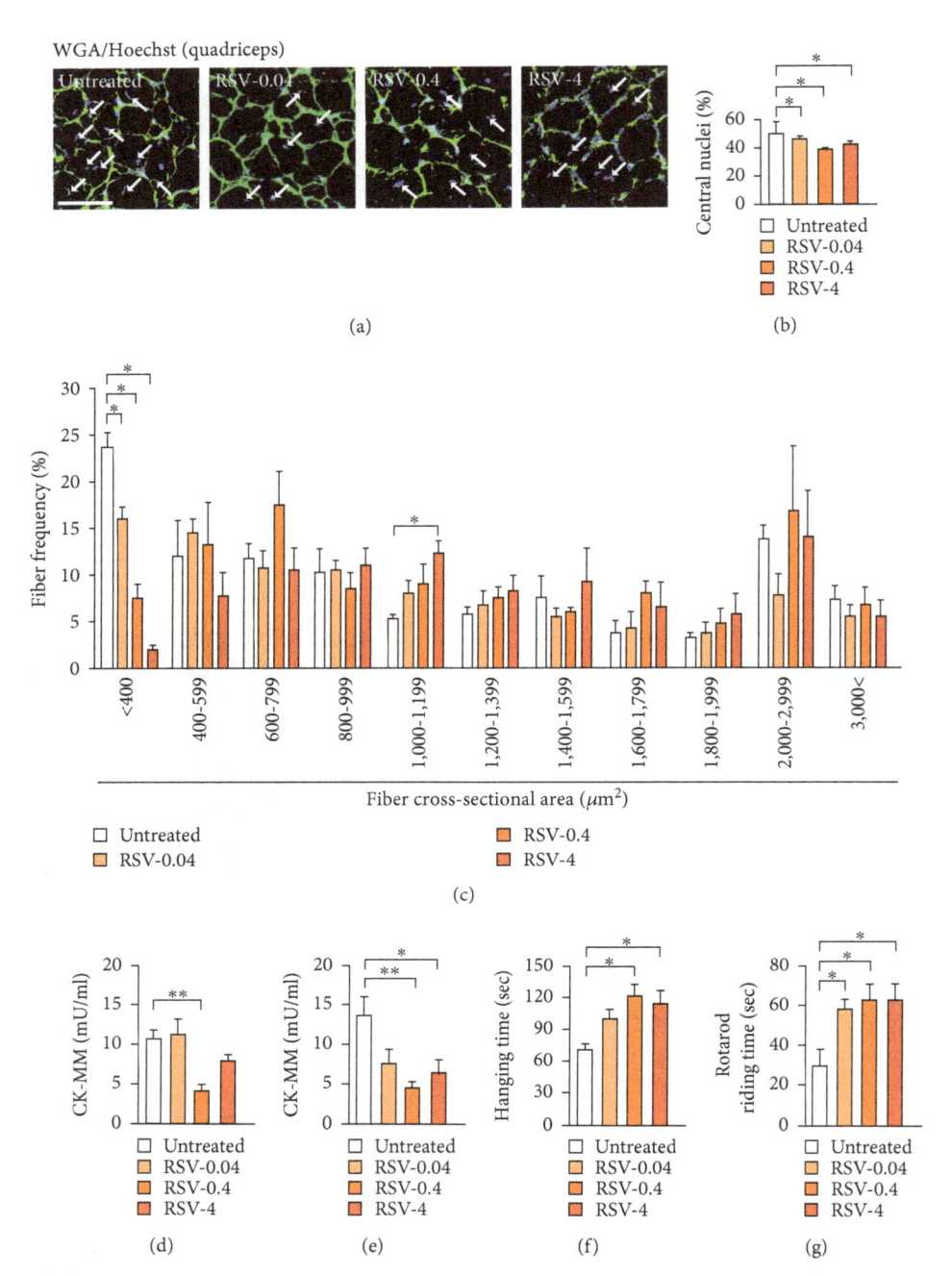

FIGURE 3: Resveratrol decreases muscular injury and improves muscle function in *mdx* mice. (a) Representative images of the quadriceps stained with FITC-conjugated wheat germ agglutinin (WGA, green) and Hoechst 33342 (blue) to detect myofiber membranes and nuclei, respectively. Muscle sections were obtained from untreated and resveratrol-treated *mdx* mice. Scale bar: 50 μm. (b) Percentage of myofibers with central nuclei in *mdx* mice. $n = 4$ in each group. (c) Cross-sectional area of myofibers in the quadriceps muscles in *mdx* mice. $n = 4$ mice per group. (d, e) Serum levels of the muscle isoform of creatine kinase (CK-MM) in the *mdx* mice at 23 (d) and 65 weeks old (e). $n = 5 - 6$ mice per group. (f) Hanging time assessed by the inverted hang test of *mdx* mice at 37 weeks of age. (g) Rotarod riding time in *mdx* mice at 40 weeks of age. $n = 4 - 5$ mice per group. $^*P < 0.05$; $^{**}P < 0.01$.

dots colocalized with mitochondria were detected in cells treated with resveratrol and CQ (Figures 4(a) and 4(c)). These results indicated that resveratrol promoted both autophagy and mitophagy.

3.5. *Mitochondrial ROS Reduction by Resveratrol-Induced Mitophagy.* Antimycin A (AA), an inhibitor of the electron transport chain in mitochondria, depolarizes mitochondria and increases the mitochondrial ROS levels. C2C12 cells were treated with AA, and the mitochondrial ROS levels were monitored with MitoSOX Red, a fluorescent indicator of mitochondrial superoxide levels. Treating the cells with AA increased the mitochondrial ROS levels, and this increase was significantly suppressed by resveratrol (Figure 5(a)).

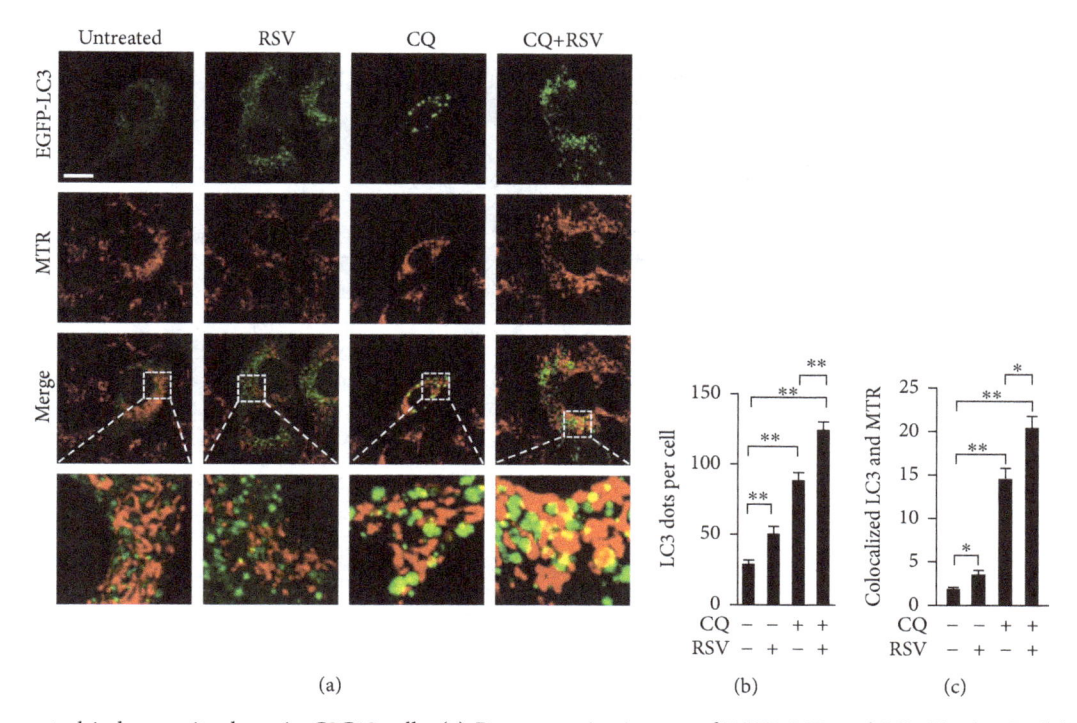

FIGURE 4: Resveratrol induces mitophagy in C2C12 cells. (a) Representative images of EGFP-LC3 and MitoTracker Red (MTR) and the merged images in C2C12 cells. Cells were cultured in the absence or presence of resveratrol (RSV, 30 $\mu$M), chloroquine (CQ, 50 $\mu$M), or RSV and CQ together, for 6 h. Yellow indicates EGFP-LC3 dots (green) colocalized with mitochondria (red). Scale bar: 10 $\mu$m. (b) Summary data of the number of LC3 dots per cell. (c) Summary data of the number of EGFP-LC3 dots colocalized with fragmented mitochondria per cell. In (c) and (d), the data were obtained from 40 randomly selected cells from 4 independent experiments. $^*P < 0.05$; $^{**}P < 0.01$.

The addition of CQ completely cancelled this effect of resveratrol on the ROS levels (Figure 5(a)), indicating that the autophagic flux was required for resveratrol's antioxidative effect.

Atg5, an E3 ubiquitin-like ligase, is involved in autophagic vesicle formation. Because Atg5 is deacetylated by SIRT1 [14], we examined the effect of Atg5 knockdown on resveratrol's function. Treating the C2C12 cells with *Atg5-siRNA* decreased the LC3-II/LC3-I ratio, indicating that the *Atg5-siRNA* inhibited autophagy (Figures 5(b) and 5(c)). MitoSOX Red staining showed that resveratrol failed to decrease the AA-induced ROS levels in cells treated with *Atg5-siRNA* (Figure 5(d)). Thus, disrupting autophagy/mitophagy with *Atg5-siRNA* inhibited resveratrol's antioxidative function.

Pink1 is indispensable for mitophagy [31]. Therefore, to inhibit mitophagy, C2C12 cells were treated with *Pink1-siRNA* (Figures 5(e)–5(g)). In the absence of AA, the knockdown of Pink1 alone altered the mitochondrial morphology and increased the size and area of mitochondria in C2C12 cells (Figures 5(e) and 5(f)), indicating that mitophagy was disrupted by the *Pink1-siRNA*. MitoSox Red staining showed that *Pink1-siRNA* completely cancelled resveratrol's antioxidative function against AA (Figure 5(g)). These findings together indicated that the induction of mitophagy by resveratrol reduced the number of damaged mitochondria and decreased the ROS levels.

## 4. Discussion

Membrane fragility due to dystrophin deficiency causes intracellular $Ca^{2+}$ dysregulation, resulting in mitochondrial dysfunction and ROS production [32]. Damaged mitochondria are a major source of cellular ROS and are selectively degraded by mitophagy, which decreases cellular ROS levels [17–19]. We showed that resveratrol induced mitophagy (Figure 4) and reduced the ROS levels in C2C12 cells in a mitophagy-dependent manner (Figure 5). *Pink1-siRNA* alone significantly increased the ROS levels in the absence of AA, indicating that mitophagy continuously contributes to the decrease in cellular ROS levels (Figure 5(g)). Although the knockdown efficiency by *Atg5-siRNA* was greater than that by *Pink1-siRNA* (Figures 5(b) and 5(e)), *Atg5-siRNA* alone could not increase the ROS levels in the absence of AA (Figure 5(d)). Atg5 is dispensable for the mitophagy occurring during erythroid maturation, and Atg5-independent mitophagy is found in various organs [33]. Thus, an Atg5-independent mitophagy pathway may contribute to decrease the ROS levels in C2C12 cells.

The expression of autophagy-related genes, i.e., *Atg12*, *Map1lc3b*, *Gabarapl1*, and *Bnip3*, was previously shown to be suppressed in *mdx* mice [28]. In this study, we found that other autophagy/mitophagy-related genes, i.e., *Pink1*, *parkin*, *Fundc1*, *Becn1*, *Atg5*, *p62*, and *Tfeb*, were downregulated in the quadriceps of *mdx* mice (Figure 1). TFEB is a master

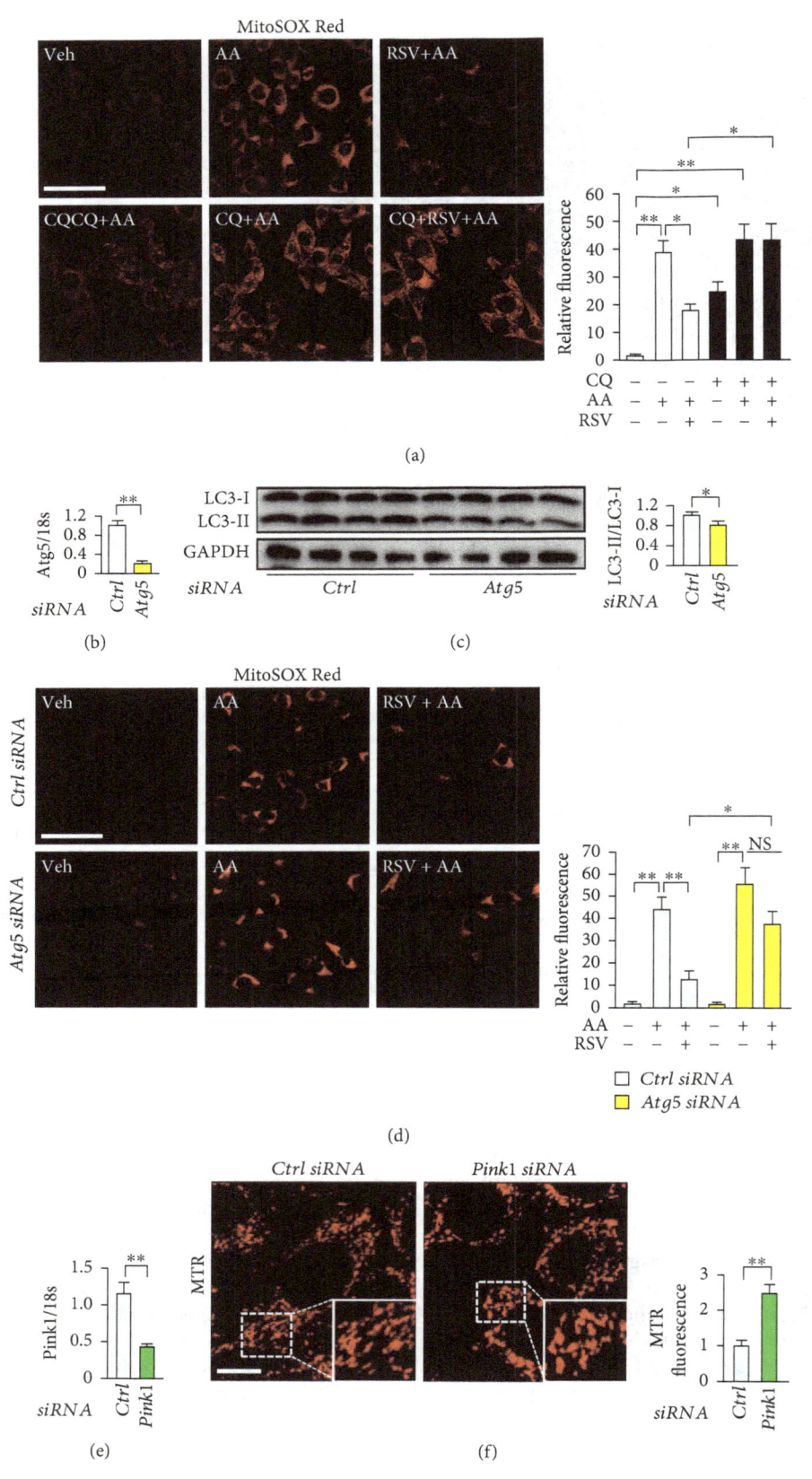

FIGURE 5: Continued.

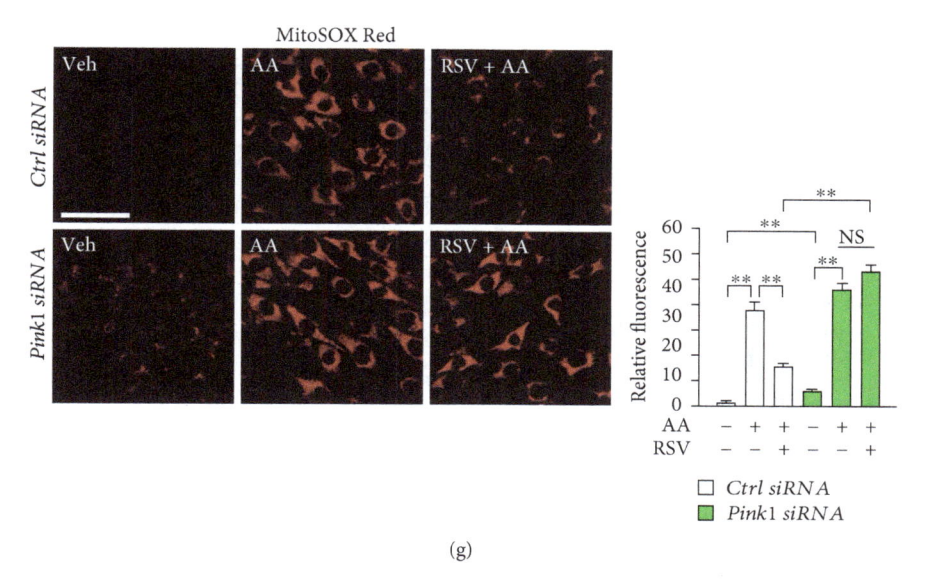

FIGURE 5: Resveratrol decreases ROS levels via mitophagy in C2C12 cells. (a) Representative images of MitoSOX Red fluorescence in C2C12 cells. Cells were cultured with vehicle (Veh), antimycin A (AA, 10 μM), or AA and resveratrol (RSV, 30 μM) in the presence or absence of chloroquine (CQ, 50 μM) for 24 h (left). Summary data of MitoSOX Red fluorescence intensity per cell (right). $n = 6$. Scale bar: 50 μm. (b) mRNA level of Atg5 normalized to 18s ribosomal RNA in C2C12 cells transfected with control (Ctrl) siRNA or siRNA against mouse Atg5. $n = 4$. (c) Representative Western blot for LC3 in cells transfected with control or Atg5 siRNA in C2C12 cells. $n = 4$. (d) Representative images of MitoSOX Red fluorescence in C2C12 cells transfected with control or Atg5 siRNA and treated with vehicle, AA, or AA and resveratrol for 24 h (left). Scale bar: 50 μm. Summary data of MitoSOX Red fluorescence intensity (right). $n = 6$. (e) Level of Pink1 mRNA in cells treated with control or Pink1 siRNA in C2C12 cells. $n = 4$. (f) Representative images of MitoTracker Red (MTR) fluorescence in C2C12 cells transfected with control or Pink1 siRNA (left). Scale bar: 10 μm. Summary data of MitoTracker Red fluorescence intensity (right). $n = 4$. (g) Representative images of MitoSOX Red fluorescence in C2C12 cells transfected with control or Pink1 siRNA and treated with vehicle, AA, or AA and resveratrol for 24 h (left). Scale bar: 50 μm. Summary data of MitoSOX Red fluorescence intensity per cell (right). $n = 6$. $^*P < 0.05$; $^{**}P < 0.01$. NS: not significant.

transcription factor for autophagy and lysosomal biogenesis [18]. Because the *Tfeb* mRNA levels were downregulated in *mdx* mice (Figure 1(a)), a decrease in TFEB level may downregulate autophagy/mitophagy-related genes in *mdx* mice. In addition, increase in the P-4EBP1 levels suggested the activation of mTORC1 in *mdx* mice (Figure 1(b)). mTORC1 is reported to downregulate autophagy-related genes in animal models of muscular dystrophies and DMD patients [28, 34]. Consistent with this finding, inhibiting mTORC1 by administering a low-protein diet or rapamycin ameliorates dystrophic muscle phenotypes [28, 34]. Since TFEB is phosphorylated and excluded from the nucleus by mTORC1, TFEB's inactivation by mTORC1 may also contribute to the downregulation of autophagy/mitophagy-related genes. We found that resveratrol restored the expression levels of autophagy/mitophagy machineries and autophagic flux (Figures 2(a)–2(c)). However, the phosphorylation levels of 4EBP1 were not changed by resveratrol (Supplemental Figure 1(b)), indicating that resveratrol did not affect the mTORC1 activity. FOXOs are known to positively regulate autophagy/mitophagy-related genes [19]. We previously showed that resveratrol decreases ROS levels in C2C12 cells by promoting the activation of FOXOs [35]. Because the knockdown of *Foxos*, i.e., *Foxo1*, *Foxo3a*, and *Foxo4*, by their *siRNAs* completely inhibited resveratrol's antioxidative function in AA-treated C2C12 cells [35], the activation of FOXOs by resveratrol may upregulate the autophagy/mitophagy-related genes and facilitate the autophagy/mitophagy flux in *mdx* mice. In addition, the upregulation of *Sirt1* mRNA by resveratrol in *mdx* mice (Supplemental Figure 1(a)) may have been caused by the activation of FOXOs, since FOXOs also induce *Sirt1* mRNA [4].

SIRT1 siRNA also inhibits resveratrol's antioxidative function in AA-treated C2C12 cells [35], suggesting that resveratrol activates FOXOs via SIRT1 activation in *mdx* mice. In addition, the deacetylation and activation of Atg5, Atg7, and LC3 by SIRT1 could be involved in the increased autophagy/mitophagy flux caused by resveratrol.

Recently, mitochondrial dysfunction and mitophagy insufficiency were shown to be involved in the pathogenesis of progeroid syndromes. Mitophagy disturbance worsens the phenotypes of xeroderma pigmentosum group A (XPA) deficiency and ataxia telangiectasia (AT), both of which are DNA repair disorders [36, 37]. DNA repair failure activates poly (ADP-ribose) polymerase 1 (PARP1), and then $NAD^+$ is depleted by the activated PARP1, thereby decreasing the activity of the $NAD^+$-dependent deacetylase SIRT1. Increased $NAD^+$ levels activate SIRT1's activity, and indeed, activating SIRT1 by adding nicotinamide riboside, an $NAD^+$ precursor, improves the mitochondria quality via mitophagy induction and retards the progression of the DNA repair disorders [36, 37]. Importantly, *mdx* mice have been shown to have increased PARP activities and low NAD$^+$ levels in their muscle tissues [38]. Because PARP is activated by DNA damage, *mdx* mice are expected to have enhanced levels of DNA damage, which may be derived from the disturbance of mitophagy. Replenishing the NAD$^+$ by administering nicotinamide riboside reduced the nuclear ADP-ribosylated protein levels and improved the muscle function and heart pathology in *mdx* mice [38]. Similar to XPA and AT, nicotinamide riboside may induce mitophagy and ameliorate phenotypes of dystrophin-deficient mice. Since nicotinamide riboside is much more costly than resveratrol, resveratrol has an economic advantage over nicotinamide riboside to treat DMD.

Stem cell depletion also plays a role in the progression of muscular dystrophies [3, 39]. SIRT1 induces the proliferation of myoblast cells, and muscle-specific SIRT1 knockout mice exhibit impaired muscle regeneration [40]. Because resveratrol appeared to decelerate the muscular turnover rate and to suppress excess muscle regeneration (Figures 3(a)–3(c)), resveratrol may preserve the number of muscle stem cells in DMD. However, resveratrol administration to *mdx* mice did not increase Pax7 expression levels (Supplemental Figure 2). Thus, the resveratrol's main function on the muscle of *mdx* mice seems to inhibit cell death and promote maturation of muscle cells by reducing oxidative stress.

In this study, we administered resveratrol to *mdx* mice at three doses to determine its optimal dose. The most effective dose of resveratrol was 0.4 g/kg food for muscle injury, function, and autophagic activity, although the administration of resveratrol at 0.04 g/kg or 4 g/kg was also effective. Because resveratrol is rather hydrophobic, it may accumulate in lipids such as cellular membranes and adipose tissues. For its clinical evaluation, the optimal dosage of resveratrol for treating muscular dystrophies needs to be determined. Our findings indicate that resveratrol would be effective for muscular dystrophy patients and may provide a combination therapy with other medicines such as glucocorticoids.

## Abbreviations

| | |
|---|---|
| AA: | Antimycin A |
| CK-MM: | Muscular isoform of creatine kinase |
| CQ: | Chloroquine |
| DHE: | Dihydroethidium |
| DMD: | Duchenne muscular dystrophy |
| 4EBP1: | Eukaryotic initiation factor 4E-binding protein 1 |
| FOXOs: | Forkhead box O transcription factors |
| Mitophagy: | Autophagy of damaged mitochondria |
| mTORC1: | Mechanistic target of rapamycin complex 1 |
| MTR: | MitoTracker Red |
| PGC-1$\alpha$: | Peroxisome proliferator-activated receptor gamma coactivator 1-alpha |
| Pink1: | PTEN-induced putative kinase 1 |
| ROS: | Reactive oxygen species |
| RSV: | Resveratrol |
| SOD2: | Superoxide dismutase 2 |
| WGA: | Wheat germ agglutinin |
| WT: | Wild type. |

## Authors' Contributions

Rio Sebori and Atsushi Kuno contributed equally to this work.

## Acknowledgments

This study was supported in part by the Japanese Society for the Promotion of Science Grants-in-Aid for Scientific Research (15K08312, 15K18992, 17K08600, 17K15582, 18K06965), grants from the Setsuro Fujii Memorial, the Osaka Foundation for Promotion of Fundamental Medical Research, the Osaka Medical Research Foundation for Intractable Diseases, Takeda Research Support, and donations from Dr. Hiroshige Kondo and Meisterbio Co. Ltd.

## Supplementary Materials

Supplemental Figure 1: effects of resveratrol on SIRT1 expression and P-4EBP1 levels of muscles from mdx mice. (a) SIRT1 mRNA levels analyzed by a qPCR method in the quadriceps, diaphragm, and tibialis anterior (TA) muscles from untreated and resveratrol- (RSV-) treated mdx mice. $n = 4$. (b) Representative Western blots (upper) and summary data (lower) for P-4EBP1 and total 4EBP1 in muscles from mdx mice. $n = 4$. *$P < 0.05$, NS: not significant. Supplemental Figure 2: effects of resveratrol on Pax7 mRNA expression levels of muscles from mdx mice. Pax7 mRNA levels analyzed by a qPCR method in the quadriceps and soleus muscles from untreated and resveratrol- (RSV-) treated mdx mice. $n = 4$. NS: not significant. Supplemental Figure 3: effects of resveratrol on P-AMPK and AMPK levels of quadriceps muscles from mdx mice. Western blots (upper) and summary data (lower) for P-AMPK and AMPK in muscles from mdx mice. $n = 4$. NS: not significant. Supplemental Table 1: primer sequences for quantitative PCR. Supplemental Table 2: antibodies used in the present study. *(Supplementary Materials)*

## References

[1] E. Mercuri and F. Muntoni, "Muscular dystrophies," *Lancet*, vol. 381, no. 9869, pp. 845–860, 2013.

[2] E. Matthews, R. Brassington, T. Kuntzer, F. Jichi, and A. Y. Manzur, "Corticosteroids for the treatment of Duchenne muscular dystrophy," *Cochrane Database of Systematic Reviews*, vol. 13, no. 5, article CD003725, 2016.

[3] X. Mu, Y. Tang, K. Takayama et al., "RhoA/ROCK inhibition improves the beneficial effects of glucocorticoid treatment in dystrophic muscle: implications for stem cell depletion,"

*Human Molecular Genetics*, vol. 26, no. 15, pp. 2813–2824, 2017.

[4] H. Pan and T. Finkel, "Key proteins and pathways that regulate lifespan," *The Journal of Biological Chemistry*, vol. 292, no. 16, pp. 6452–6460, 2017.

[5] K. T. Howitz, K. J. Bitterman, H. Y. Cohen et al., "Small molecule activators of sirtuins extend Saccharomyces cerevisiae lifespan," *Nature*, vol. 425, no. 6954, pp. 191–196, 2003.

[6] Y. S. Hori, A. Kuno, R. Hosoda et al., "Resveratrol ameliorates muscular pathology in the dystrophic mdx mouse, a model for Duchenne muscular dystrophy," *The Journal of Pharmacology and Experimental Therapeutics*, vol. 338, no. 3, pp. 784–794, 2011.

[7] A. Kuno, Y. S. Hori, R. Hosoda et al., "Resveratrol improves cardiomyopathy in dystrophin-deficient mice through SIRT1 protein-mediated modulation of p300 protein," *The Journal of Biological Chemistry*, vol. 288, no. 8, pp. 5963–5972, 2013.

[8] B. S. Gordon, D. C. Delgado Diaz, and M. C. Kostek, "Resveratrol decreases inflammation and increases utrophin gene expression in the mdx mouse model of Duchenne muscular dystrophy," *Clinical Nutrition*, vol. 32, no. 1, pp. 104–111, 2013.

[9] V. Ljubicic, M. Burt, J. A. Lunde, and B. J. Jasmin, "Resveratrol induces expression of the slow, oxidative phenotype in mdx mouse muscle together with enhanced activity of the SIRT1-PGC-1α axis," *American Journal of Physiology. Cell Physiology*, vol. 307, no. 1, pp. C66–C82, 2014.

[10] J. T. Selsby, K. J. Morine, K. Pendrak, E. R. Barton, and H. L. Sweeney, "Rescue of dystrophic skeletal muscle by PGC-1α involves a fast to slow fiber type shift in the mdx mouse," *PLoS One*, vol. 7, no. 1, article e30063, 2012.

[11] A. Chalkiadaki, M. Igarashi, A. S. Nasamu, J. Knezevic, and L. Guarente, "Muscle-specific SIRT1 gain-of-function increases slow-twitch fibers and ameliorates pathophysiology in a mouse model of Duchenne muscular dystrophy," *PLoS Genetics*, vol. 10, no. 7, article e1004490, 2014.

[12] M. Tanno, A. Kuno, T. Yano et al., "Induction of manganese superoxide dismutase by nuclear translocation and activation of SIRT1 promotes cell survival in chronic heart failure," *The Journal of Biological Chemistry*, vol. 285, no. 11, pp. 8375–8382, 2010.

[13] R. Huang, Y. Xu, W. Wan et al., "Deacetylation of nuclear LC3 drives autophagy initiation under starvation," *Molecular Cell*, vol. 57, no. 3, pp. 456–466, 2015.

[14] I. H. Lee, L. Cao, R. Mostoslavsky et al., "A role for the NAD-dependent deacetylase Sirt1 in the regulation of autophagy," *Proceedings of the National Academy of Sciences of the United States of America*, vol. 105, no. 9, pp. 3374–3379, 2008.

[15] G. Mariño, E. Morselli, M. V. Bennetzen et al., "Longevity-relevant regulation of autophagy at the level of the acetylproteome," *Autophagy*, vol. 7, no. 6, pp. 647–649, 2014.

[16] E. Morselli, G. Mariño, M. V. Bennetzen et al., "Spermidine and resveratrol induce autophagy by distinct pathways converging on the acetylproteome," *The Journal of Cell Biology*, vol. 192, no. 4, pp. 615–629, 2011.

[17] R. S. Balaban, S. Nemoto, and T. Finkel, "Mitochondria, oxidants, and aging," *Cell*, vol. 120, no. 4, pp. 483–495, 2005.

[18] L. Galluzzi, E. H. Baehrecke, A. Ballabio et al., "Molecular definitions of autophagy and related processes," *The EMBO Journal*, vol. 36, no. 13, pp. 1811–1836, 2017.

[19] J. Lee, S. Giordano, and J. Zhang, "Autophagy, mitochondria and oxidative stress: cross-talk and redox signalling," *The Biochemical Journal*, vol. 441, no. 2, pp. 523–540, 2012.

[20] T. N. Nguyen, B. S. Padman, and M. Lazarou, "Deciphering the molecular signals of PINK1/parkin mitophagy," *Trends in Cell Biology*, vol. 26, no. 10, pp. 733–744, 2016.

[21] E. Masiero, L. Agatea, C. Mammucari et al., "Autophagy is required to maintain muscle mass," *Cell Metabolism*, vol. 10, no. 6, pp. 507–515, 2009.

[22] N. Raben, V. Hill, L. Shea et al., "Suppression of autophagy in skeletal muscle uncovers the accumulation of ubiquitinated proteins and their potential role in muscle damage in Pompe disease," *Human Molecular Genetics*, vol. 17, no. 24, pp. 3897–3908, 2008.

[23] J. J. Wu, C. Quijano, E. Chen et al., "Mitochondrial dysfunction and oxidative stress mediate the physiological impairment induced by the disruption of autophagy," *Aging*, vol. 1, no. 4, pp. 425–437, 2009.

[24] J. C. Greene, A. J. Whitworth, I. Kuo, L. A. Andrews, M. B. Feany, and L. J. Pallanck, "Mitochondrial pathology and apoptotic muscle degeneration in Drosophila parkin mutants," *Proceedings of the National Academy of Sciences of the United States of America*, vol. 100, no. 7, pp. 4078–4083, 2003.

[25] J. Park, S. B. Lee, S. Lee et al., "Mitochondrial dysfunction in Drosophila PINK1 mutants is complemented by parkin," *Nature*, vol. 441, no. 7097, pp. 1157–1161, 2006.

[26] J. Dorrie, H. Gerauer, Y. Wachter, and S. J. Zunino, "Resveratrol induces extensive apoptosis by depolarizing mitochondrial membranes and activating caspase-9 in acute lymphoblastic leukemia cells," *Cancer Research*, vol. 61, no. 12, pp. 4731–4739, 2001.

[27] C. Sun, F. Zhang, X. Ge et al., "SIRT1 improves insulin sensitivity under insulin-resistant conditions by repressing PTP1B," *Cell Metabolism*, vol. 6, no. 4, pp. 307–319, 2007.

[28] C. de Palma, F. Morisi, S. Cheli et al., "Autophagy as a new therapeutic target in Duchenne muscular dystrophy," *Cell Death & Disease*, vol. 3, no. 11, article e418, 2012.

[29] J. P. White, A. N. Billin, M. E. Campbell, A. J. Russell, K. M. Huffman, and W. E. Kraus, "The AMPK/p27[Kip1] axis regulates autophagy/apoptosis decisions in aged skeletal muscle stem cells," *Stem Cell Reports*, vol. 11, no. 2, pp. 425–439, 2018.

[30] D. J. Klionsky, K. Abdelmohsen, A. Abe et al., "Guidelines for the use and interpretation of assays for monitoring autophagy (3rd edition)," *Autophagy*, vol. 12, no. 1, pp. 1–222, 2016.

[31] F. Koyano, K. Okatsu, H. Kosako et al., "Ubiquitin is phosphorylated by PINK1 to activate parkin," *Nature*, vol. 510, no. 7503, pp. 162–166, 2014.

[32] J. R. Terrill, H. G. Radley-Crabb, T. Iwasaki, F. A. Lemckert, P. G. Arthur, and M. D. Grounds, "Oxidative stress and pathology in muscular dystrophies: focus on protein thiol oxidation and dysferlinopathies," *The FEBS Journal*, vol. 280, no. 17, pp. 4149–4164, 2013.

[33] Y. Nishida, S. Arakawa, K. Fujitani et al., "Discovery of Atg5/Atg7-independent alternative macroautophagy," *Nature*, vol. 461, no. 7264, pp. 654–658, 2009.

[34] P. Grumati, L. Coletto, P. Sabatelli et al., "Autophagy is defective in collagen VI muscular dystrophies, and its reactivation rescues myofiber degeneration," *Nature Medicine*, vol. 16, no. 11, pp. 1313–1320, 2010.

[35] Y. S. Hori, A. Kuno, R. Hosoda, and Y. Horio, "Regulation of FOXOs and p53 by SIRT1 modulators under oxidative stress," *PLoS One*, vol. 8, no. 9, article e73875, 2013.

[36] E. F. Fang, M. Scheibye-Knudsen, L. E. Brace et al., "Defective mitophagy in XPA via PARP-1 hyperactivation and $NAD^+$/SIRT1 reduction," *Cell*, vol. 157, no. 4, pp. 882–896, 2014.

[37] E. F. Fang, H. Kassahun, D. L. Croteau et al., "$NAD^+$ replenishment improves lifespan and Healthspan in ataxia telangiectasia models via mitophagy and DNA repair," *Cell Metabolism*, vol. 24, no. 4, pp. 566–581, 2016.

[38] D. Ryu, H. Zhang, E. R. Ropelle et al., "$NAD^+$ repletion improves muscle function in muscular dystrophy and counters global PARylation," *Science Translational Medicine*, vol. 8, no. 361, article 361ra139, 2016.

[39] N. A. Dumont, Y. X. Wang, J. von Maltzahn et al., "Dystrophin expression in muscle stem cells regulates their polarity and asymmetric division," *Nature Medicine*, vol. 21, no. 12, pp. 1455–1463, 2015.

[40] J. G. Ryall, S. Dell'Orso, A. Derfoul et al., "The $NAD^+$-dependent SIRT1 deacetylase translates a metabolic switch into regulatory epigenetics in skeletal muscle stem cells," *Cell Stem Cell*, vol. 16, no. 2, pp. 171–183, 2015.

# Protective Effects of Aqueous Extracts of *Flos lonicerae Japonicae* against Hydroquinone-Induced Toxicity in Hepatic L02 Cells

Yanfang Gao,[1] Huanwen Tang,[2] Liang Xiong,[1] Lijun Zou,[1] Wenjuan Dai,[1,3] Hailong Liu,[1,3] and Gonghua Hu[1]

[1]*Department of Preventive Medicine, Gannan Medical University, 1 Yixueyuan Road, Ganzhou, 341000 Jiangxi, China*
[2]*Department of Environmental and Occupational Health, Dongguan Key Laboratory of Environmental Medicine, School of Public Health, Guangdong Medical University, Dongguan, 523808 Guangdong, China*
[3]*Department of Occupational Health and Toxicology, School of Public Health, Nanchang University, BaYi Road 461, Nanchang, 3300061 Jiangxi, China*

Correspondence should be addressed to Gonghua Hu; hgh0129@163.com

Academic Editor: Daniele Vergara

Hydroquinone (HQ) is widely used in food stuffs and is an occupational and environmental pollutant. Although the hepatotoxicity of HQ has been demonstrated both in vitro and in vivo, the prevention of HQ-induced hepatotoxicity has yet to be elucidated. In this study, we focused on the intervention effect of aqueous extracts of *Flos lonicerae Japonicae* (FLJ) on HQ-induced cytotoxicity. We demonstrated that HQ reduced cell viability in a concentration-dependent manner by administering 160 $\mu$mol/L HQ for 12 h as the positive control of cytotoxicity. The aqueous FLJ extracts significantly increased cell viability and decreased LDH release, ALT, and AST in a concentration-dependent manner compared with the corresponding HQ-treated groups in hepatic L02 cells. This result indicated that aqueous FLJ extracts could protect the cytotoxicity induced by HQ. HQ increased intracellular MDA and LPO and decreased the activities of GSH, GSH-Px, and SOD in hepatic L02 cells. In addition, aqueous FLJ extracts significantly suppressed HQ-stimulated oxidative damage. Moreover, HQ promoted DNA double-strand breaks (DSBs) and the level of 8-hydroxy-2′-deoxyguanosine and apoptosis. However, aqueous FLJ extracts reversed HQ-induced DNA damage and apoptosis in a concentration-dependent manner. Overall, our results demonstrated that the toxicity of HQ was mediated by intracellular oxidative stress, which activated DNA damage and apoptosis. The findings also proved that aqueous FLJ extracts exerted protective effects against HQ-induced cytotoxicity in hepatic L02 cells.

## 1. Introduction

Hydroquinone (HQ) is a ubiquitous environmental chemical in cosmetics, medicines, the environment, and human diet; HQ can be metabolized from benzene as potentially hematotoxic, genotoxic, and carcinogenic compounds [1]. Humans are exposed to HQ through various channels, including oral administration, inhalation, and through the skin [2, 3]. Although the effects of HQ exposure on human health have been extensively studied and reported, the actual mechanisms of such effects remain unclear. The involved mechanisms trigger oxidative stress, which causes DNA damage, mutation in cellular transformation, in vivo tumorigenesis, gene toxicity, and epigenetic changes [1, 4–8]. In our previous experiments, HQ induced apoptosis in hepatic L02 cells by changing the cellular redox status by reducing the cellular thiol level and increasing the cellular reactive oxygen species (ROS) level. In addition, ROS can lead to DNA damage by breaking DNA or producing lipid peroxidation in the membrane, thereby increasing the degrees of apoptosis and necrosis of L02 hepatocytes [1, 9–11]. These findings indicated that HQ can damage L02 hepatocytes through a series of oxidative stress reactions, so it is possible to use some antioxidants for reducing HQ toxicity.

*Flos lonicerae Japonicae* (FLJ) is the flower of *Lonicera japonica Thunb*, which is widely planted in China [12]. FLJ is used worldwide as a popular traditional herbal medicine with various pharmacological activities [13, 14]. In traditional Chinese medicine, FLJ is typically used to treat common colds and fevers. FLJ exerts various effects, such as antioxidant, anti-inflammatory, antihyperlipidemic, and anticancer [15–18]. In addition, FLJ protects cells against hydrogen peroxide-induced apoptosis by phosphorylating MAPKs and PI3K/Akt. FLJ is believed to dispel noxious heat from the blood and neutralize poisonous effects. FLJ significantly increases blood neutrophil activity and promotes neutrophil phagocytosis at appropriate concentrations [12]. Some investigators posited that the methanol extract of FLJ induces protective effects against rat hepatic injuries caused by carbon tetrachloride and aqueous extracts of FLJ flowers may act as therapeutic agents for inflammatory disease through the selective regulation of NF-$\kappa$B activation in rat liver [19]. FLJ is characterized by high biomass, easy cultivation, extensive competitive ability, wide geographic distribution, and strong resistance to environmental stresses, including bacterial, viral, and oxidative stresses [20]. However, the protective effects of aqueous FLJ extracts against HQ-induced cytotoxicity have not been demonstrated.

In this study, we examined the protective effects of aqueous FLJ extracts against HQ-induced cytotoxicity and their involvement in oxidative stress, DNA damage, and apoptosis. For this reason, we investigated the MDA and LPO levels as indexes of lipid peroxidation; the activities of SOD, GSH, and GSH-Px as antioxidant enzymes; DNA double-strand breaks (DSBs) and 8-hydroxy-2′-deoxyguanosine (8-OHdG) level as specific markers of oxidative damage of DNA; HQ-induced apoptosis; and the cytoprotective effect of aqueous FLJ extracts in hepatic L02 cells.

## 2. Materials and Methods

*2.1. Chemicals.* HQ, 3-(4,5-dimethylthiazol-2-yl)2,5-diphenyl-tetrazolium bromide (MTT), Hoechst33258, low-melting agarose (LMA), normal-melting agarose (NMA), 8-OHdG, deoxyguanosine (dG), and propidium iodide (PI) were purchased from Sigma (St. Louis, MO, USA). Fetal bovine serum (FBS) and RPMI-1640 medium were acquired from HyClone (Logan, UT, USA). Penicillin-streptomycin for the cell culture and trypsin were procured from Gibco/Invitrogen (Carlsbad, CA, USA). Other chemicals and reagents were of the highest analytical grade and bought from Sangon Biotech Co. Ltd. (Shanghai, China).

*2.2. Plant Material and Preparation of Extract.* The aqueous FLJ extracts were prepared by adopting the standard method used for treating patients with liver disease in traditional Chinese medicine. In brief, dried FLJ fruits (100 g) were boiled in 500 mL of distilled water for 3 h. The total extract was centrifuged at 5000 rcf for 30 min. The supernatant was filtered with filter paper, and the residue was further extracted twice under the same conditions. The filtrates were evaporated to dryness under vacuum and weighed. The final yield was 12.5% (w/w). The lyophilized extract was dissolved in distilled water to produce a final concentration of 100 g/mL FLJ extract, which was stored at −20°C.

*2.3. Cell Lines and Culture.* The immortalized human normal hepatocyte L02 cell line was provided by Dr. Zhixiong Zhuang (Shenzhen Center for Disease Control and Prevention, Guangdong, China). L02 cells were cultured in RPMI-1640 medium supplemented with 10% (v/v) heat-inactivated FBS and antibiotic supplement (100 U/mL of penicillin and 100 μg/mL of streptomycin) in a humidified incubator at 37°C under 95% air and 5% $CO_2$. After each specified treatment time of incubation, the cells were harvested for further analysis.

*2.4. Assessment of L02 Cell Viability.* The cell viability of L02 cells was determined by MTT assay. The cells were seeded into 96-well flat-bottomed plates overnight at 37°C under 5% $CO_2$ and immediately treated with HQ (5, 10, 20, 40, 80, and 160 μM) for 6, 12, 24, and 48 h; 160 μmol/L HQ was administered for 12 h as the positive control of oxidative damage. The samples were classified under five groups, namely, control (without HQ and FLJ), 160 μmol/L HQ, 0.25 g/mL FLJ + HQ, 0.50 g/mL FLJ + HQ, and 1.00 g/mL FLJ + HQ for 12 h. Subsequently, 20 μL of MTT solution (5 mg/mL) was added to the culture medium for 4 h before the end of the treatment time to allow the formation of formazan crystals. The supernatant was discarded, and 100 μL of cell lysis buffer (50% DMF, 20% SDS, pH 4.6–4.7) was added to dissolve the intracellular crystalline formazan products. After constant and gentle shaking for 10 min at room temperature, the absorbance was recorded at 570 nm. Cell viability was calculated using the equation cellular relative viability = $(OD_{treated\ wells} - OD_{blank})/(OD_{control\ wells} - OD_{blank})$.

*2.5. Measurement of Intracellular Cytotoxicity and Oxidative Damage.* The toxic effect of HQ in hepatic L02 cells was estimated in terms of LDH release and the activities of ALT and AST by using commercial kits (Nanjing Jiancheng Bioengineering Institute, China) in accordance with the manufacturer's protocol.

Commercial kits (Nanjing Jiancheng Bioengineering Institute, China) were utilized to assess the activities of SOD, GSH-Px, and GSH and the levels of MDA and LPO. The activities of SOD, GSH-Px, and GSH were expressed as units/milligram of protein. MDA and LPO contents were expressed as nmol/milligram of protein. The protein concentration was estimated using a BCA kit.

*2.6. DNA Damage Assay.* DNA damage was determined by the following assays. (i) DNA damage was evaluated using alkaline single-cell gel electrophoresis (comet assay). In brief, after 24 h of exposure, followed by washing with PBS, a $1.2 \times 10^5$ cell suspension was mixed with 0.8% LMA at 37°C and spread on a fully frosted microscope slide precoated with 0.65% NMA. After the agarose solidified, the slide was covered with another 75 μL of 0.8% LMA and then immersed in lysis solution (2.5 M NaCl, 100 mM $Na_2EDTA$, 10 mM Tris base, 1% Triton-X 100, 10% DMSO, pH 10) for 2 h at 4°C. The slides were placed in a gel-electrophoresis apparatus containing 300 mM NaOH and 1 mM $Na_2EDTA$ (pH 13) for 30 min at 4°C

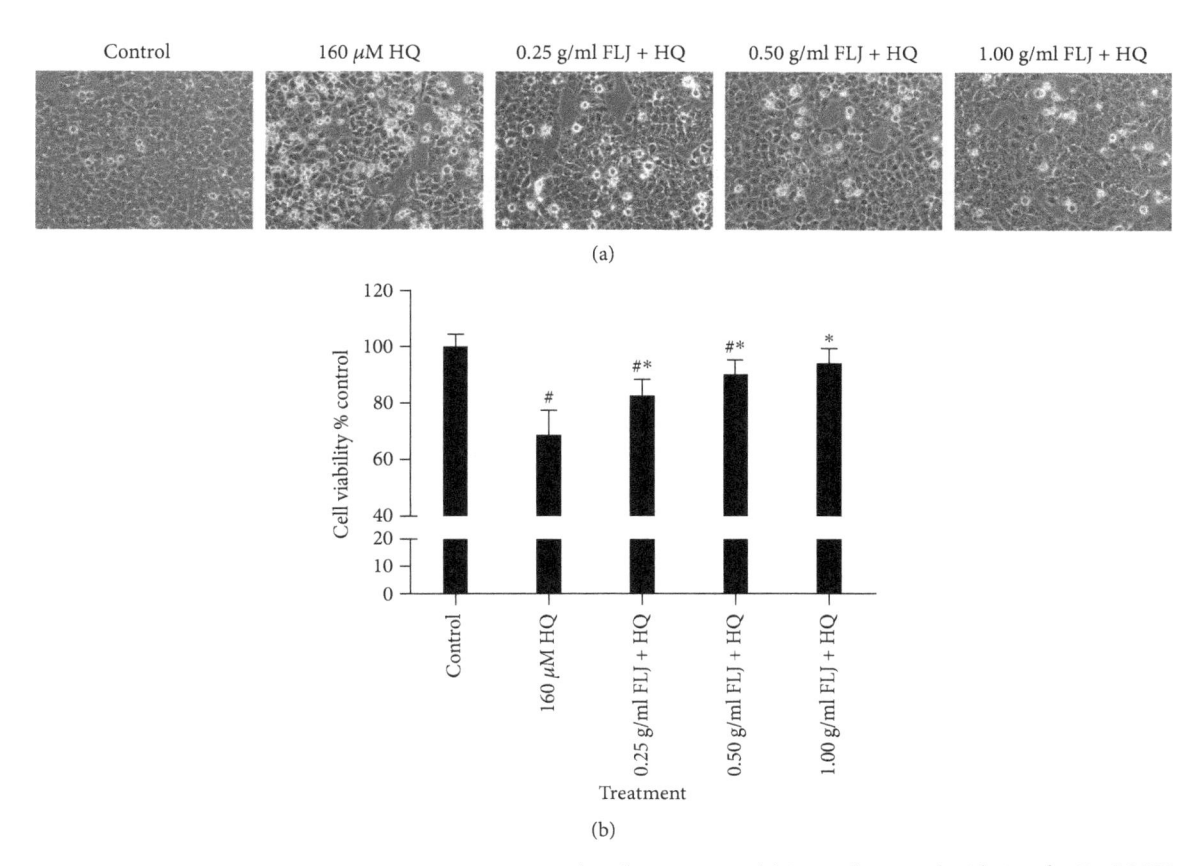

FIGURE 1: Protective effects of aqueous FLJ extracts against HQ-induced cytotoxicity (a) L02 cells treated with 0 and 160 μM HQ and 0.25, 0.50, and 1.00 g/mL FLJ + HQ for 12 h. Morphological features of L02 cells were observed by inverted microscope. (b) L02 cells treated with 0 and 160 μM HQ and 0.25, 0.50, and 1.00 g/mL FLJ + HQ for 12 h. Cell proliferation was detected by using the MTT proliferating reagent. All data were representative of at least three independent experiments. $^{\#}P < 0.05$ compared with the control. $^{*}P < 0.05$ compared with 160 μM HQ treatment.

to allow DNA unwinding and alkali labile damage. Electrophoresis was performed at 25 V (300 mA) and 4°C for 20 min. Subsequently, all slides were washed three times with a neutralizing buffer (0.4 M Tris, pH 7.5) for 5 min each time and stained with 80 μL of PI (5 μg/mL). A total of 100 randomly chosen cells (comets) were visually scored using a fluorescence microscope (Nikon Eclips TE-S, Japan) equipped with an excitation filter of 515–560 nm and a barrier filter of 590 nm. The "tail length and tail moment" of each comet were calculated using casp-1.2.2 analysis software. (ii) 8-OHdG was evaluated by high-performance liquid chromatography with electrochemical detection (HPLC-ECD) [21]. In brief, genomic DNA was extracted using a Genomic DNA Purification Kit in 500 μg hydrolyzed DNA samples through the nuclease P1 and alkaline phosphatase hydrolysis of DNA. The samples were then filtered through 0.22 μm nylon filters. 8-OHdG and dG levels were measured by using HPLC-ECD and HPLC with variable wavelength detector (HPLC-UV) systems as previously described. About 100 μL of final hydrolysates was analyzed by HPLC-ECD with reverse phase-C18 (RP-C18) analytical column as the column. The mobile phase consisted of 50 mM $KH_2PO_4$ buffer (pH 5.5 and containing 10% methanol). The separations were performed at a flow rate of 1 mL/min. The amount of 8-OHdG in DNA was calculated as the number of 8-OHdG molecules/$10^6$ unmodified dG molecules.

2.7. Apoptosis Assay. Apoptosis was assessed by flow cytometry using PI assay. In brief, the cells were incubated in RPMI 1640 with 10% FBS. The cells consisted of the control group, the group treated with 160 μM HQ, and the group coincubated in the absence or presence of FLJ (0.25, 0.50, and 1.00 g/mL) for 12 h. Subsequently, the cells were washed with cold PBS and incubated with PI (0.5 μg) for 20 min at room temperature in the dark. The cells were immediately analyzed on a FACSCanto II flow cytometer (Becton and Dickinson, San Jose, CA, USA), and data from 10,000 events were obtained. The results were expressed as the percentage of cells labeled with PI (apoptotic).

2.8. Statistical Analysis. All data are presented as the mean ± standard deviation (SD). Statistical evaluation of data analysis was performed using SPSS 16.0 for Windows. The differences between the mean values of multiple groups were analyzed by one-way ANOVA, followed by SNK test. $P < 0.05$ was considered statistically significant.

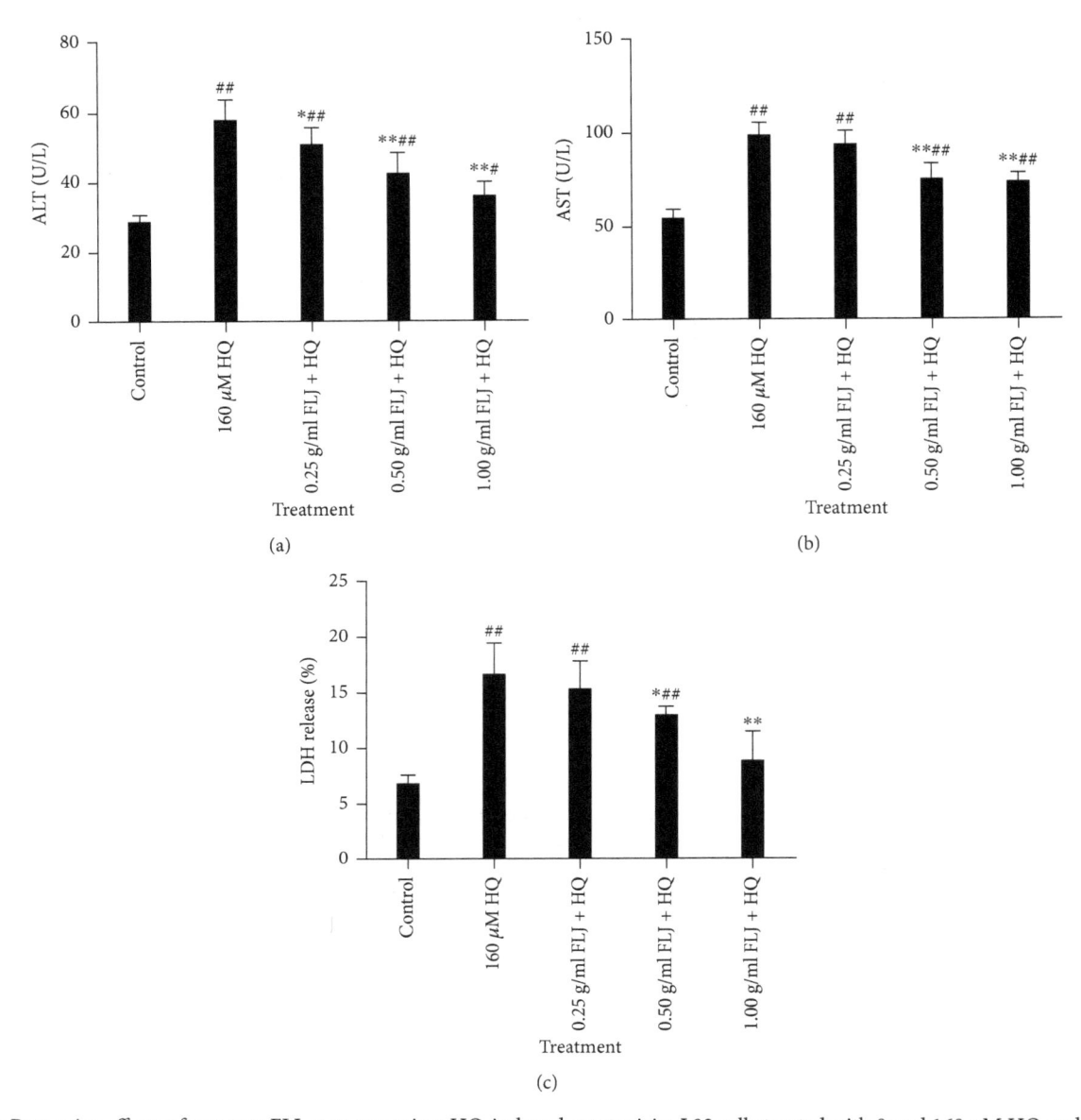

FIGURE 2: Protective effects of aqueous FLJ extracts against HQ-induced cytotoxicity L02 cells treated with 0 and 160 μM HQ and 0.25, 0.50, and 1.00 g/mL FLJ + HQ for 12 h. LDH release and the activities of aminotransferase (ALT) and aspartate aminotransferase (AST) were detected by using commercial kits. All data were representative of at least three independent experiments. $^{\#}P < 0.05$ compared with the control. $^{*}P < 0.05$ compared with 160 μM HQ treatment.

## 3. Results

### 3.1. Protective Effects of Aqueous FLJ Extracts against HQ-Induced Cytotoxicity.
As the first step in determining the necessary HQ concentration to induce cytotoxicity, the viability of L02 cells was assessed by MTT assay. HQ induced a concentration- and time-dependent reduction in L02 cell viability compared with untreated control cells. We administered 160 μmol/L HQ for 12 h as the positive control of cytotoxicity. After 12 h of coexposure to aqueous FLJ extracts at the specified concentrations, the cell significantly increased compared with that in the corresponding HQ-treated groups (Figures 1(a) and 1(b)). These results revealed that aqueous FLJ extracts intervened the HQ-induced cytotoxicity.

To evaluate the protective effects of aqueous FLJ extracts against HQ, we examined cytotoxicity in terms of the LDH release and the ALT and AST levels. The results showed that aqueous FLJ extracts significantly downregulated LDH release, ALT, and AST in a concentration-dependent manner compared with the corresponding HQ-treated groups (Figure 2). Thus, aqueous FLJ extracts could reverse HQ-induced cytotoxicity.

### 3.2. Protective Effects of Aqueous FLJ Extracts against HQ-Induced Cellular Oxidative Damage.
Oxidative stress can mediate apoptosis in various cell models and is considered an important apoptotic signal. Therefore, we investigated whether intracellular oxidative damage is involved in HQ-

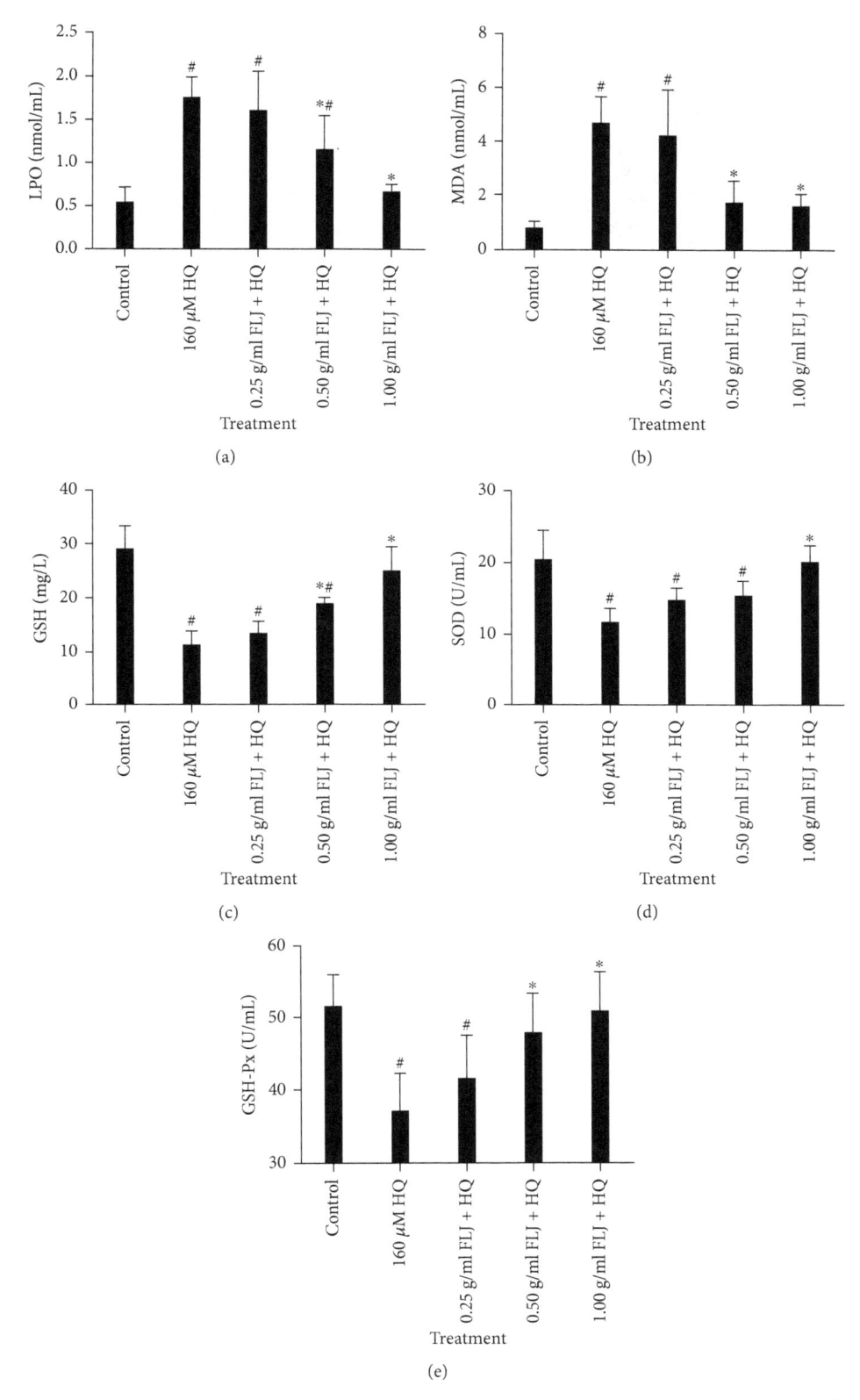

Figure 3: Protective effects of aqueous FLJ extracts against HQ-induced oxidative damage L02 cells treated with 0 and 160 μM HQ and 0.25, 0.50, and 1.00 g/mL FLJ + HQ for 12 h. MDA and LPO levels as the indexes of lipid peroxidation and the SOD, GSH, and GSH-Px activities as antioxidant enzymes were detected by using commercial kits for oxidative stress. All data were representative of at least three independent experiments. $^{\#}P < 0.05$ compared with the control. $^{*}P < 0.05$ compared with 160 μM HQ treatment.

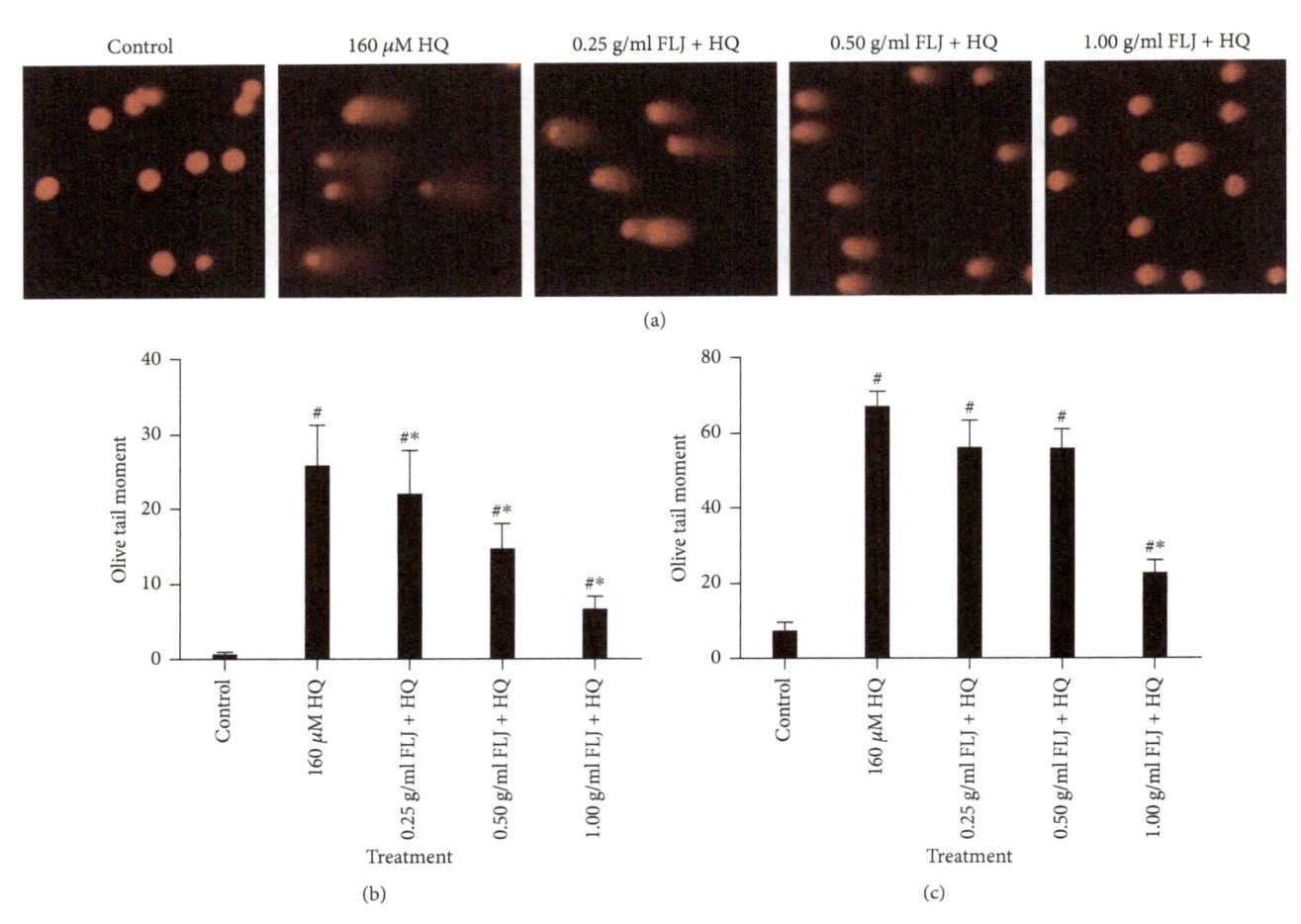

FIGURE 4: Protective effects of aqueous FLJ extracts against HQ-induced DNA damage (a, b) L02 cells treated with 0 and 160 $\mu$M HQ and 0.25, 0.50, and 1.00 g/mL FLJ + HQ for 12 h. DNA damage was evaluated by alkaline single-cell gel electrophoresis (comet assay). (c) L02 cells treated with 0 and 160 $\mu$M HQ and 0.25, 0.50, and 1.00 g/mL FLJ + HQ for 12 h. 8-Hydroxy-2′-deoxyguanosine (8-OHdG) level as a specific marker of oxidative damage of DNA. 8-OHdG was evaluated using high-performance liquid chromatography with electrochemical detection (HPLC-ECD). All data were representative of at least three independent experiments. #$P < 0.05$ compared with the control. *$P < 0.05$ compared with 160 $\mu$M HQ treatment.

induced cell death and found that HQ exposure increased MDA and LPO production, but aqueous FLJ extracts reduced the production of MDA and LPO in a concentration-dependent manner. In addition, aqueous FLJ extracts reversed the production of antioxidant enzymes (such as GSH, SOD, and GSH-Px), indicating that aqueous FLJ extracts exerted a protective effect against HQ-induced oxidative stress in L02 cells (Figure 3).

### 3.3. Protective Effects of Aqueous FLJ Extracts against HQ-Induced DNA Damage.

To elucidate the protective effects of aqueous FLJ extracts against HQ-induced DNA damage, L02 cells were exposed to various treatments: 0 and 160 $\mu$M HQ and 0.25, 0.5, and 1.0 g/mL FLJ + HQ for 12 h. DNA damage was assessed. We performed comet assay, which is a commonly used indicator of genomic instability and genotoxic exposure. As shown in Figure 4(a), HQ induced a marked increase in DNA damage. However, aqueous FLJ extracts significantly decreased DNA damage compared with the corresponding HQ-treated groups as measured in terms of the tail moment and tail length. To further validate our findings regarding the protective effects of aqueous FLJ extracts against HQ-induced DNA damage, we used HLPC-ECD to measure the level of 8-OHdG, which is a widely used marker of oxidative DNA damage. The results revealed a concentration-dependent decrease in 8-OHdG levels in the samples treated with aqueous FLJ extracts (Figure 4(b)), implying that aqueous FLJ extracts exerted a protective effect against HQ-induced DNA damage.

### 3.4. Protective Effects of Aqueous FLJ Extracts against HQ-Induced Apoptosis.

Given that apoptosis is regulated by oxidative stress and DNA damage, the apoptosis rate in the L02 cells was also investigated by flow cytometry using PI assay. The apoptosis rate significantly decreased in the HQ-intoxicated L02 cells compared with that in the control. To further evaluate whether the protective effect of aqueous FLJ extracts on the L02 cells involves cell apoptosis, the L02 cells were incubated with aqueous FLJ extracts and HQ, and the apoptosis rate was then assessed. The results revealed that the apoptosis rate of the group exposed to both aqueous FLJ extracts and HQ was lower than that of the HQ group,

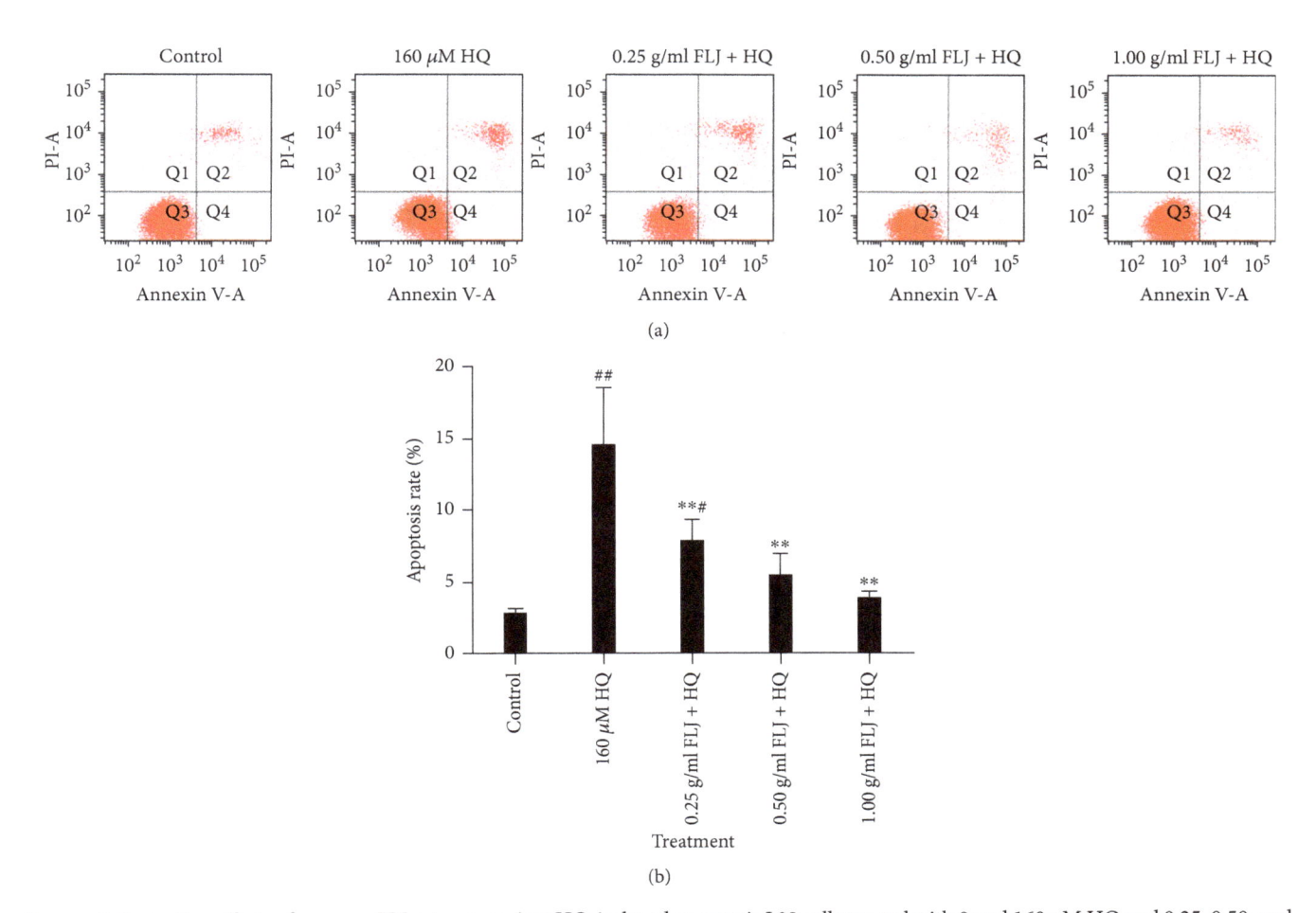

FIGURE 5: Protective effects of aqueous FLJ extracts against HQ-induced apoptosis L02 cells treated with 0 and 160 $\mu$M HQ and 0.25, 0.50, and 1.00 g/mL FLJ + HQ for 12 h. Apoptosis was assessed by flow cytometry using propidium iodide (PI) assay. All data were representative of at least three independent experiments. $^{\#}P < 0.05$ compared with the control. $^{*}P < 0.05$ compared with 160 $\mu$M HQ treatment.

indicating that aqueous FLJ extracts played an important role in reducing apoptosis in HQ-exposed L02 cells (Figure 5).

## 4. Discussion

HQ is a well-known toxicant of liver and induces many effects on the hepatic system. Although many studies have explored HQ, the actual mechanisms underlying these effects remain poorly understood. FLJ, which is also known as *Jinyinhua* or *Japanese honeysuckle*, is the dried flower bud or open flower of *Lonicera japonica Thunb* [22]. It is one of the most popular traditional Chinese medicines, and it has been applied in the healthcare of China and other East Asian countries for a long time. It has been proven to display antioxidant, antiviral, anticarcinogenic, anti-inflammatory, analgesic, antipyretic, and antimicrobial functions [23]. This study focused on the protective effects of aqueous FLJ extracts against HQ-induced cytotoxicity.

Serum ALT and AST levels are used as biochemical markers of liver damage, as the membrane destruction of hepatocytes releases hepatic enzymes, such as ALT and AST, into blood circulation [24]. The water extracts of FLJ containing 20% chlorogenic acid are protective against alcohol-induced chemical liver injury in mice [12, 24], which was similar to the protection conferred by aqueous FLJ extracts to the L02 hepatic cell model induced by HQ.

Oxidative stress plays a critical role in the development of drug-induced liver damage [25] and in HQ-induced toxicity. Intercellular ROS may cause detrimental alterations in cell membranes, DNA, and other cellular structures. ROS are critical intermediates under normal physiological conditions that contribute to pathophysiological events in liver injury. Intracellular oxidative stress balance is crucial in maintaining normal cellular function in response to exogenous and endogenous factors [26]. MDA is a key marker of lipid peroxidation. The progression of liver damage was correlated with oxidative stress, as confirmed by MDA measurements [27]. FLJ polyphenol extracts were protective against the lipid peroxidation of erythrocyte and lipid membranes [28]. Consistent with the literature, the levels of MDA and LPO in our study were significantly higher in the HQ-treated group than in the controls. The levels of MDA and LPO concentrations were lower in aqueous FLJ extracts than in the HQ group.

The primary antioxidant defense systems include SOD and the GSH redox cycle [29]. SOD, which is produced from the mitochondrion electron transfer chain and scavenges superoxide anions, is an important antioxidant enzyme. SOD can transform superoxide anions into $H_2O_2$ and catalase and then continuously detoxify them to $H_2O$ [30]. The GSH redox cycle, which mainly includes GSH and GPx, modulates the redox-mediated responses of hepatic cells induced by external or intracellular stimulation. GSH is the main nonenzymatic regulator of intracellular redox homeostasis. GSH directly scavenges hydroxyl radicals and is a cofactor in detoxifying hydrogen peroxide, lipid peroxides, and alkyl peroxides. GPx, which is a selenocysteine-containing enzyme, reduces lipid hydroperoxides to their corresponding alcohols and hydrogen peroxide to water in the liver. Therefore, enhancing the hepatic antioxidant system capacity may be an effective therapeutic strategy for alleviating and treating liver damage [25]. Our study showed that the cytotoxicity of HQ is mediated by intracellular oxidative stress and confirmed the protective effects exerted by aqueous FLJ extracts as an antioxidant against HQ-induced cytotoxicity in L02 cells.

DNA damage is caused by multiple factors, including oxidative stress, vitamin B12 deficiency, and ischemia-reperfusion injury [31]. The ratio of 8-OHdG/dG is correlated with the severity of oxidative stress. The high activity of an antioxidant enzyme may be a compensatory regulation in response to increased oxidative stress. In our previous study, we demonstrated that DNA damage is related to HQ-induced hepatotoxicity [11]. ROS causes oxidative stress, which leads to DNA damage. Cells have evolved elaborate mechanisms to respond to DNA damage, at the core of which is the signaling pathway known as the DNA damage checkpoint [32]. This pathway initiates many aspects of the DNA damage response (DDR), including activation of DNA repair and induction of apoptosis [33, 34]. DNA damage is an early event in DDR. Thus, we examined DNA damage through the comet assay and 8-OHdG level, which is a specific marker of oxidative damage of DNA. In different kinds of cells, HQ induces the production of superoxides and hydroperoxides, which are implicated in the initiation and promotion stages of apoptosis and DNA damage. In the present study, although HQ induced DNA damage and apoptosis, we observed a concentration-dependent reverse in 8-OHdG levels, DDBs in the nucleus, and apoptosis in the group treated with aqueous FLJ extracts.

In conclusion, the results of this study strongly suggested that aqueous FLJ extracts played a role in protecting against HQ-induced cytotoxicity. On the basis of our results, we suggest a possible mechanism involved in the protective effects of aqueous FLJ extracts against HQ-induced cytotoxicity. HQ induced MDA and LPO formation, which activated antioxidant enzyme production. Intracellular oxidative stress balance was disturbed, thereby inducing DNA damage and apoptosis and promoting HQ-induced cytotoxicity. However, aqueous FLJ extracts could increase the activation of the antioxidant, which may reduce DNA damage and apoptosis and exert a protective effect against HQ-induced cytotoxicity. The molecular mechanism could further elucidate the antioxidant role of aqueous FLJ extracts.

## Abbreviations

| | |
|---|---|
| DDR: | DNA-damage response |
| HQ: | Hydroquinone |
| MDA: | Malondialdehyde |
| LPO: | Lipid peroxidation |
| ROS: | Reactive oxygen species |
| FLJ: | Flos lonicerae Japonicae |
| LDH: | Lactic dehydrogenase |
| ALT: | Alanine transaminase |
| AST: | Aspartate aminotransferase |
| PI: | Propidium iodide |
| HPLC-EC: | High-performance liquid chromatography with electrochemical |
| DSBs: | DNA double-strand breaks |
| 8-OHdG: | 8-hydroxy-2′-deoxyguanosine. |

## Authors' Contributions

Yanfang Gao, Huanwen Tang, and Liang Xiong contributed equally to this work.

## Acknowledgments

This work was supported by grants from National Natural Science Foundation of China (81460504, 81260434, 81273116, and 81060241), the Traditional Chinese Medicine Scientific Research Project funded by the Health Department of Jiangxi Province (2009A074), the Science and Technology Project funded by the Education Department of Jiangxi Province (GJJ13689, GJJ170862), the Natural Science Foundation of Jiangxi Province (20132BAB205071, 20114BAB205041, and 20142BAB215014), and supported by Doctor Initial Funding for Gannan Medical University (QD201302 and QD201701).

## References

[1] S. Chen, H. Liang, G. Hu et al., "Differently expressed long noncoding RNAs and mRNAs in TK6 cells exposed to low dose hydroquinone," *Oncotarget*, vol. 8, no. 56, pp. 95554–95567, 2017.

[2] C. B. Hebeda, F. J. Pinedo, S. M. Bolonheis et al., "Intracellular mechanisms of hydroquinone toxicity on endotoxin-activated neutrophils," *Archives of Toxicology*, vol. 86, no. 11, pp. 1773–1781, 2012.

[3] S. H. Inayat-Hussain, H. A. Ibrahim, E. L. Siew et al., "Modulation of the benzene metabolite hydroquinone induced toxicity: evidence for an important role of *fau*," *Chemico-Biological Interactions*, vol. 184, no. 1-2, pp. 310–312, 2010.

[4] F. J. Enguita and A. L. Leitao, "Hydroquinone: environmental pollution, toxicity, and microbial answers," *BioMed Research International*, vol. 2013, Article ID 542168, 14 pages, 2013.

[4] F. J. Enguita and A. L. Leitao, "Hydroquinone: environmental pollution, toxicity, and microbial answers," *BioMed Research International*, vol. 2013, Article ID 542168, 14 pages, 2013.

[5] Z. Li, C. Wang, J. Zhu et al., "The possible role of liver kinase B1 in hydroquinone-induced toxicity of murine fetal liver and bone marrow hematopoietic stem cells," *Environmental Toxicology*, vol. 31, no. 7, pp. 830–841, 2016.

[6] J. Liu, Q. Yuan, X. Ling et al., "PARP-1 may be involved in hydroquinone-induced apoptosis by poly ADP-ribosylation of ZO-2," *Molecular Medicine Reports*, vol. 16, no. 6, pp. 8076–8084, 2017.

[7] J. Zhu, H. Wang, S. Yang et al., "Comparison of toxicity of benzene metabolite hydroquinone in hematopoietic stem cells derived from murine embryonic yolk sac and adult bone marrow," *PLoS One*, vol. 8, no. 8, article e71153, 2013.

[8] H. Bahadar, F. Maqbool, S. Mostafalou et al., "The molecular mechanisms of liver and islets of Langerhans toxicity by benzene and its metabolite hydroquinone *in vivo* and *in vitro*," *Toxicology Mechanisms and Methods*, vol. 25, no. 8, pp. 628–636, 2015.

[9] G. H. Hu, Z. X. Zhuang, H. Y. Huang, L. Yu, J. H. Yuan, and L. Q. Yang, "Effect of hydroquinone on expression of ubiquitin-ligating enzyme Rad18 in human L-02 hepatic cells," *Zhonghua Lao Dong Wei Sheng Zhi Ye Bing Za Zhi*, vol. 27, no. 4, pp. 222–225, 2009.

[10] G. H. Hu, Z. X. Zhuang, H. Y. Huang, L. Yu, L. Q. Yang, and W. D. Ji, "Relationship between polymerase eta expression and DNA damage-tolerance in human hepatic cells by hydroquinone," *Zhonghua Yu Fang Yi Xue Za Zhi*, vol. 43, no. 1, pp. 56–60, 2009.

[11] G. Hu, H. Huang, L. Yang et al., "Down-regulation of Polη expression leads to increased DNA damage, apoptosis and enhanced S phase arrest in L-02 cells exposed to hydroquinone," *Toxicology Letters*, vol. 214, no. 2, pp. 209–217, 2012.

[12] Y. Nam, J. M. Lee, Y. Wang, H. S. Ha, and U. D. Sohn, "The effect of *Flos Lonicerae Japonicae* extract on gastro-intestinal motility function," *Journal of Ethnopharmacology*, vol. 179, pp. 280–290, 2016.

[13] Y. P. Guo, L. G. Lin, and Y. T. Wang, "Chemistry and pharmacology of the herb pair Flos Lonicerae japonicae-Forsythiae fructus," *Chinese Medicine*, vol. 10, no. 1, p. 16, 2015.

[14] W. Zhou, A. Yin, J. Shan, S. Wang, B. Cai, and L. Di, "Study on the rationality for antiviral activity of *Flos Lonicerae Japonicae*-Fructus Forsythiae herb couple preparations improved by chito-oligosaccharide via integral pharmacokinetics," *Molecules*, vol. 22, no. 4, p. 654, 2017.

[15] Y. Yang, L. Wang, Y. Wu et al., "On-line monitoring of extraction process of Flos Lonicerae Japonicae using near infrared spectroscopy combined with synergy interval PLS and genetic algorithm," *Spectrochimica Acta Part A: Molecular and Biomolecular Spectroscopy*, vol. 182, pp. 73–80, 2017.

[16] W. Zhou, J. Shan, S. Wang, B. Cai, and L. di, "Transepithelial transport of phenolic acids in *Flos Lonicerae Japonicae* in intestinal Caco-2 cell monolayers," *Food & Function*, vol. 6, no. 9, pp. 3072–3080, 2015.

[17] S. T. Kao, C. J. Liu, and C. C. Yeh, "Protective and immunomodulatory effect of flos *Lonicerae japonicae* by augmenting IL-10 expression in a murine model of acute lung inflammation," *Journal of Ethnopharmacology*, vol. 168, pp. 108–115, 2015.

[18] Y. Li, W. Cai, X. Weng et al., "Lonicerae Japonicae Flos and Lonicerae Flos: a systematic pharmacology review," *Evidence-Based Complementary and Alternative Medicine*, vol. 2015, Article ID 905063, 16 pages, 2015.

[19] B. C. Y. Cheng, H. Yu, H. Guo et al., "A herbal formula comprising Rosae Multiflorae Fructus and Lonicerae Japonicae Flos, attenuates collagen-induced arthritis and inhibits TLR4 signalling in rats," *Scientific Reports*, vol. 6, no. 1, article 20042, 2016.

[20] J. W. Kang, N. Yun, H. J. Han, J. Y. Kim, J. Y. Kim, and S. M. Lee, "Protective effect of *Flos Lonicerae* against experimental gastric ulcers in rats: mechanisms of antioxidant and anti-inflammatory action," *Evidence-Based Complementary and Alternative Medicine*, vol. 2014, Article ID 596920, 11 pages, 2014.

[21] Y. Kaya, A. Çebi, N. Söylemez, H. Demir, H. H. Alp, and E. Bakan, "Correlations between oxidative DNA damage, oxidative stress and coenzyme Q10 in patients with coronary artery disease," *International Journal of Medical Sciences*, vol. 9, no. 8, pp. 621–626, 2012.

[22] J. Li, Y. Wang, J. Xue, P. Wang, and S. Shi, "Dietary exposure risk assessment of flonicamid and its effect on constituents after application in *Lonicerae Japonicae Flos*," *Chemical and Pharmaceutical Bulletin*, vol. 66, no. 6, pp. 608–611, 2018.

[23] Z. Hu, C. Lausted, H. Yoo et al., "Quantitative liver-specific protein fingerprint in blood: a signature for hepatotoxicity," *Theranostics*, vol. 4, no. 2, pp. 215–228, 2014.

[24] S. E. Fomenko, N. F. Kushnerova, V. G. Sprygin, and T. V. Momot, "Hepatorotective activity of honeysuckle fruit extract in carbon tetrachloride intoxicated rats," *Eksperimental'naia i Klinicheskaia Farmakologiia*, vol. 77, no. 10, pp. 26–30, 2014.

[25] J. Arauz, E. Ramos-Tovar, and P. Muriel, "Redox state and methods to evaluate oxidative stress in liver damage: from bench to bedside," *Annals of Hepatology*, vol. 15, no. 2, pp. 160–173, 2016.

[26] C. Wang, J. D. Harwood, and Q. Zhang, "Oxidative stress and DNA damage in common carp (*Cyprinus carpio*) exposed to the herbicide mesotrione," *Chemosphere*, vol. 193, pp. 1080–1086, 2018.

[27] S. Ullah, Z. Li, Z. Hasan, S. U. Khan, and S. Fahad, "Malathion induced oxidative stress leads to histopathological and biochemical toxicity in the liver of rohu (*Labeo rohita*, Hamilton) at acute concentration," *Ecotoxicology and Environmental Safety*, vol. 161, pp. 270–280, 2018.

[28] D. Bonarska-Kujawa, H. Pruchnik, S. Cyboran, R. Żyłka, J. Oszmiański, and H. Kleszczyńska, "Biophysical mechanism of the protective effect of blue honeysuckle (*Lonicera caerulea* L. var. kamtschatica Sevast.) polyphenols extracts against lipid peroxidation of erythrocyte and lipid membranes," *The Journal of Membrane Biology*, vol. 247, no. 7, pp. 611–625, 2014.

[29] T. Fukai and M. Ushio-Fukai, "Superoxide dismutases: role in redox signaling, vascular function, and diseases," *Antioxidants & Redox Signaling*, vol. 15, no. 6, pp. 1583–1606, 2011.

[30] Y. Nojima, K. Ito, H. Ono et al., "Superoxide dismutases, SOD1 and SOD2, play a distinct role in the fat body during pupation in silkworm *Bombyx mori*," *PLoS One*, vol. 10, no. 2, article e0116007, 2015.

[31] Y. Nakabeppu, K. Sakumi, K. Sakamoto, D. Tsuchimoto, T. Tsuzuki, and Y. Nakatsu, "Mutagenesis and carcinogenesis caused by the oxidation of nucleic acids," *Biological Chemistry*, vol. 387, no. 4, pp. 373–379, 2006.

[32] C. Park, S. Hong, S. Shin et al., "Activation of the Nrf2/HO-1 signaling pathway contributes to the protective effects of *Sargassum serratifolium* extract against oxidative stress-induced DNA damage and apoptosis in SW1353 human chondrocytes," *International Journal of Environmental Research and Public Health*, vol. 15, no. 6, article 1173, 2018.

[33] S. Khan, A. Zafar, and I. Naseem, "Copper-redox cycling by coumarin-di(2-picolyl)amine hybrid molecule leads to ROS-mediated DNA damage and apoptosis: a mechanism for cancer chemoprevention," *Chemico-Biological Interactions*, vol. 290, pp. 64–76, 2018.

[34] D. Liu, R. Li, X. Guo et al., "DNA damage regulated autophagy modulator 1 recovers the function of apoptosis-stimulating of p53 protein 2 on inducing apoptotic cell death in Huh7.5 cells," *Oncology Letters*, vol. 15, no. 6, pp. 9333–9338, 2018.

# Multifunctional Phytocompounds in *Cotoneaster* Fruits: Phytochemical Profiling, Cellular Safety, Anti-Inflammatory and Antioxidant Effects in Chemical and Human Plasma Models *In Vitro*

Agnieszka Kicel,[1] Joanna Kolodziejczyk-Czepas,[2] Aleksandra Owczarek,[1] Magdalena Rutkowska,[1] Anna Wajs-Bonikowska,[3] Sebastian Granica,[4] Pawel Nowak,[2] and Monika A. Olszewska[1]

[1]*Department of Pharmacognosy, Faculty of Pharmacy, Medical University of Lodz, 1 Muszynskiego, 90-151 Lodz, Poland*
[2]*Department of General Biochemistry, Faculty of Biology and Environmental Protection, University of Lodz, Pomorska 141/143, 90-236 Lodz, Poland*
[3]*Institute of General Food Chemistry, Faculty of Biotechnology and Food Sciences, Lodz University of Technology, 4/10 Stefanowskiego, 90-924 Lodz, Poland*
[4]*Department of Pharmacognosy and Molecular Basis of Phytotherapy, Faculty of Pharmacy, Medical University of Warsaw, 1 Banacha, 02-097 Warsaw, Poland*

Correspondence should be addressed to Agnieszka Kicel; agnieszka.kicel@umed.lodz.pl

Academic Editor: Felipe L. de Oliveira

The work presents the results of an investigation into the molecular background of the activity of *Cotoneaster* fruits, providing a detailed description of their phytochemical composition and some of the mechanisms of their anti-inflammatory and antioxidant effects. GS-FID-MS and UHPLC-PDA-ESI-MS$^3$ methods were applied to identify the potentially health-beneficial constituents of lipophilic and hydrophilic fractions, leading to the identification of fourteen unsaturated fatty acids (with dominant linoleic acid, 375.4–1690.2 mg/100 g dw), three phytosterols (with dominant $\beta$-sitosterol, 132.2–463.3 mg/100 g), two triterpenoid acids (10.9–54.5 mg/100 g), and twenty-six polyphenols (26.0–43.5 mg GAE/g dw). The most promising polyphenolic fractions exhibited dose-dependent anti-inflammatory activity in *in vitro* tests of lipoxygenase ($IC_{50}$ in the range of 7.7–24.9 $\mu$g/U) and hyaluronidase ($IC_{50}$ in the range of 16.4–29.3 $\mu$g/U) inhibition. They were also demonstrated to be a source of effective antioxidants, both in *in vitro* chemical tests (DPPH, FRAP, and TBARS) and in a biological model, in which at *in vivo*-relevant levels (1–5 $\mu$g/mL) they normalized/enhanced the nonenzymatic antioxidant capacity of human plasma and efficiently protected protein and lipid components of plasma against peroxynitrite-induced oxidative/nitrative damage. Moreover, the investigated extracts did not exhibit cytotoxicity towards human PMBCs. Among the nine *Cotoneaster* species tested, *C. hjelmqvistii*, *C. zabelii*, *C. splendens*, and *C. bullatus* possess the highest bioactive potential and might be recommended as dietary and functional food products.

## 1. Introduction

Edible fruits are widely recognized as a valuable source of structurally diverse phytochemicals with a broad spectrum of health-promoting properties. Decreased cholesterol levels, lower blood pressure, better mental health, and protection against cancer are only a few of the many benefits associated with the regular intake of fruit products, as indicated by numerous epidemiological studies [1]. Among the different fruit-bearing families, the Rosaceae seems to be of special importance. With over 3000 species, the family provides numerous types and varieties of fruits, some of which, such

FIGURE 1: The fruits of *C. bullatus* (a) and *C. splendens* (b).

as apples, pears, strawberries and cherries, have great economic and dietary importance, and are frequently and willingly consumed due to their excellent flavors and proven nutritional value [2]. Many other taxa (e.g., *Aronia* sp., *Sorbus* sp., *Pyracantha* sp., and *Prunus spinosa* L.) produce fruits, that while less attractive in taste and appearance, are, nonetheless, distinguished by especially high quantities of bioactive constituents, which makes them perfect candidates for more specialized food applications, for example, as functional food products or food additives [3–6].

The chemical diversity of health-beneficial phytochemicals contained in rosaceous plant materials is immense and ranges from highly lipophilic to strongly polar constituents. Unsaturated fatty acids of almond oil, the cholesterol-regulating phytosterols of *Prunus africana* (Hook.f.) Kalkman, and the pentacyclic triterpenes, ubiquitous throughout the Rosaceae, with proven anti-inflammatory activity are some examples of the possible structures from the hydrophobic end of the spectrum [7, 8]. On the other hand, the hydrophilic fractions often contain an abundance of highly-valued polyphenol antioxidants belonging to numerous chemical classes, such as flavonoids, phenolic acids, and tannins. The bioactive potential of Rosaceae fruits is, therefore, associated not with a single fraction but rather is an effect of the presence of a range of phytochemicals.

The genus *Cotoneaster* Medikus is one of the largest genera of the Rosaceae family (subfamily Spiraeoideae, tribe Pyreae) comprising about 500 species of shrubs or small trees. Its members are native to the Palearctic region (temperate Asia, Europe, north Africa) but are often cultivated throughout Europe as ornamental plants due to their decorative bright red fruits (Figure 1). The center of diversity of the taxon are the mountains of southwestern China and the Himalayas [9, 10], where the fruits have been used for culinary purposes by the local communities. The nutritional value of the fruits as a source of vitamins and minerals has been confirmed [11, 12] and additional beneficial health effects of the fruit consumption have been also reported in the traditional medicine for the treatment of diabetes mellitus, cardiovascular diseases, nasal hemorrhage, excessive menstruation, fever, and cough [9, 10]. The phytochemical research on the subject is scarce, but the available data indicate the tendency of the fruits to accumulate a wide range of active metabolites. In particular, the fruits of *Cotoneaster pannosus* Franch. are a source of linoleic acid, those of *Cotoneaster microphylla* Wall ex Lindl contain pentacyclic triterpenoids, and the polyphenolic fractions of *C. pannosus* and *Cotoneaster integerrimus* Medik. fruits are rich in epicatechin, shikimic acid, and chlorogenic acid [9, 11, 12]. However, broader generalization of their properties is troublesome, and the possible wider application of the fruits, for example, as functional food products, is hindered by a lack of systematic studies. Similarly limited is the information on the activity of *Cotoneaster* fruits. Preliminary studies have been performed on the fruits of *C. integerrimus* and *C. pannosus* with regard to their antioxidant, anticholinesterase, antityrosinase, antiamylase, and antiglucosidase properties, and their free radical-scavenging potential was proven to be the most promising [9, 12]. Still, the research was carried out using only simple *in vitro* chemical tests and did not cover *in vivo*-relevant antioxidant mechanisms.

The aim of this study was, therefore, to provide a more detailed insight into the chemical composition and activity of *Cotoneaster* fruits. To this end, the fruits from nine species of *Cotoneaster* cultivated in Poland were analyzed for a range of lipophilic and hydrophilic (polyphenolic) constituents with acknowledged health-promoting properties using a combination of chromatographic and spectroscopic methods (GC-FID-MS, UHPLC-PDA-ESI-MS$^3$, and UV-Vis spectrophotometry). The most promising polyphenolic fractions were then subjected to an analysis of antioxidant activity comprising eight complementary *in vitro* tests (both chemical and biological plasma models) covering some of the mechanisms crucial for reducing the level of oxidative damage in the human organism, that is, scavenging of free radicals, enhancement of the nonenzymatic antioxidant capacity of blood plasma, and protection of its lipid and protein components against oxidative/nitrative changes. Additionally, the inhibitory effects of the fruit extracts on the proinflammatory enzymes, that is, lipoxygenase and hyaluronidase, were also measured. Finally, the cellular safety of the extracts was evaluated in cytotoxicity tests employing human peripheral blood mononuclear cells (PMBCs).

## 2. Materials and Methods

*2.1. Plant Material.* The fruit samples of nine selected *Cotoneaster* Medik. species, that is, *C. lucidus* Schltdl. (AR), *C. divaricatus* Rehder et E.H. Wilson (BG), *C. horizontalis* Decne. (BG), *C. nanshan* Mottet (BG), *C. hjelmqvistii* Flinck et B. Hylmö (BG), *C. dielsianus* E. Pritz. (BG), *C. splendens* Flinck et B. Hylmö (BG), *C. bullatus* Bois (BG), and *C. zabelii* C.K. Schneid. (BG) were collected in September 2013, in the Botanical Garden (BG; 51°45′N 19°24′E) in Lodz (Poland) and in the Arboretum (AR; 51°49′N 19°53′E), Forestry Experimental Station of Warsaw University of Life Sciences (SGGW) in Rogow (Poland). The voucher specimens were deposited in the Herbarium of the Department of Pharmacognosy, Medical University of Lodz (Poland). The raw materials were powdered with an electric grinder, sieved through a 0.315 mm sieve, and stored in airtight containers until use.

*2.2. General.* Reagents and standards of analytical or HPLC grade such as 2,2-diphenyl-1-picrylhydrazyl (DPPH), 2,4,6-tris-(2-pyridyl)-s-triazine (TPTZ), 2,2′-azobis-(2-amidinopropane)-dihydrochloride (AAPH), linoleic acid, 2-thiobarbituric acid, Tween® 40, 5,5′-dithiobis-(2-nitrobenzoic acid) (DNTB), xylenol orange disodium salt, Histopaque®-1077 medium *N,O-bis*-(trimethylsilyl)-trifluoroacetamide with 1% 1-trimethylchlorosilane (BSTFA + TMCS), boron trifluoride, bovine testis hyaluronidase, lipoxygenase from soybean, reference standards of fatty acid methyl esters (FAMEs), ethyl oleate, 5-$\alpha$-cholesterol, ($\pm$)-6-hydroxy-2,5,7,8-tetramethylchromane-2-carboxylic acid (Trolox®), butylated hydroxyanisole (BHA), 2,6-di-*tert*-butyl-4-methylphenol (BHT), gallic acid monohydrate, quercetin dehydrate, chlorogenic acid hemihydrate (5-*O*-caffeoylquinic acid), 3-*O*- and 4-*O*-caffeoylquinic acids, hyperoside semihydrate, isoquercitrin, rutin trihydrate, procyanidins B-2 and C-1, (−)-epicatechin, and indomethacin were purchased from Sigma-Aldrich (St. Louis, MO, USA). The standards of quercetin 3-*O*-$\beta$-D-(2″-*O*-$\beta$-D-xylosyl)-galactoside and quercitrin (quercetin 3-*O*-$\alpha$-L-rhamnoside) have previously been isolated in our laboratory from *C. bullatus* and *C. zabelii* leaves with at least 95% HPLC purity (unpublished results). A ($Ca^{2+}$ and $Mg^{2+}$)-free phosphate buffered saline (PBS) was purchased from Biomed (Lublin, Poland). Peroxynitrite was synthesized according to Pryor et al. [13]. Anti-3-nitrotyrosine polyclonal antibody, biotin-conjugated secondary antibody, and streptavidin/HRP were purchased from Abcam (Cambridge, UK). HPLC grade solvents such as acetonitrile and formic acid were from Avantor Performance Materials (Gliwice, Poland). For chemical tests, the samples were incubated at a constant temperature using a BD 23 incubator (BINDER, Tuttlingen, Germany) and measured using a UV-1601 Rayleigh spectrophotometer (Beijing, China). Activity tests in blood plasma models and enzyme inhibitory assays were performed using 96-well plates and monitored using a SPECTROStar Nano microplate reader (BMG LABTECH, Ortenberg, Germany).

*2.3. Phytochemical Profiling*

*2.3.1. Extraction and Derivatization of Lipophilic Phytochemicals.* The fruit samples (7.0 g) were exhaustively extracted in a Soxhlet apparatus with chloroform (150 mL, 24 h), to give lipid extracts (288–467 mg dw), which were then subjected to quantification of lipophilic compounds. Fatty acids were assayed as fatty acid methyl esters (FAMEs) prepared according to a method described earlier [14]. Phytosterols and triterpenes were assayed after their transformation to trimethylsilyl ethers (TMSs) according to Thanh et al. [15]. The FAME and TMS mixtures were independently analyzed by GC-FID-MS.

*2.3.2. GC-FID-MS Analysis.* The analyses of lipophilic fractions were performed on a Trace GC Ultra instrument coupled with a DSQII mass spectrometer (Thermo Electron, Waltham, MA, USA) and a MS-FID splitter (SGE Analytical Science, Trajan Scientific Americas, Austin, TX, USA). The applied mass range was 33–550 amu, ion source-heating was 200°C, and ionization energy was 70 eV. The conditions for FAMEs were as follows: capillary column: TG-WaxMS (30 m × 0.25 mm i.d., film thickness 0.25 $\mu$m; Thermo Fisher Scientific, Waltham, MA, USA); temperature program: 3–30 min: 50–240°C at 4°C/min; and injector and detector temperatures: 250°C and 260°C, respectively. The conditions for TMSs were as follows: capillary column: HP-5 (30 m × 0.25 mm i.d., film thickness 0.25 $\mu$m; Agilent Technologies, Santa Clara, CA, USA); temperature program: 1–15 min: 100–250°C, at 10°C/min; 15–30 min: 250–300°C, at 4°C/min; and injector and detector temperatures: 310°C and 300°C, respectively. In all cases, the carrier gas was helium (constant pressure: 300 kPa). The lipophilic analytes were identified by comparison of their MS profiles with those stored in the libraries NIST 2012 and Wiley Registry of Mass Spectral Data (10th and 11th eds). Retention times ($t_R$) of FAMEs were also compared with those of the commercial FAME mixture. The analyte levels were expressed as mg/100 g fruit dry weight (dw), calculated using the internal standards of ethyl oleate and 5-$\alpha$-cholesterol (for the fatty acids as well as phytosterols and triterpenoids, respectively) and it was recalculated to the content in the plant material taking into account the extraction yield.

*2.3.3. Extraction of Polyphenolic Compounds.* The fruit samples (100–500 mg) were first defatted by preextraction with chloroform (20 mL, 15 min; the chloroform extracts were discarded), then refluxed for 30 min with 30 mL of 70% ($v/v$) aqueous methanol, and twice for 15 min with 20 mL of the same solvent. The combined extracts were diluted with the extractant to 100 mL. Each sample was extracted in triplicate to give the test extracts, which were analyzed for their total phenolic contents (TPCs) and antioxidant activity in chemical models. For UHPLC analyses and antioxidant activity evaluation in the human plasma models, the test extracts were evaporated *in vacuo* and lyophilized using an Alpha 1-2/LDplus freeze dryer (Christ, Osterode am Harz, Germany) before weighing.

*2.3.4. UHPLC-PDA-ESI-$MS^3$ Analysis.* Metabolite profiling was performed on an UltiMate 3000 RS UHPLC system (Dionex, Dreieich, Germany) with PDA detector scanning in the wavelength range of 220–450 nm and an amaZon SL ion trap mass spectrometer with ESI interface (Bruker Daltonics, Bremen, Germany). Separations were carried out on a Kinetex XB-C18 column (150 × 2.1 mm, 1.7 $\mu$m;

Phenomenex Inc., Torrance, CA, USA). The mobile phase consisted of solvent A (water-formic acid, 100 : 0.1, $v/v$) and solvent B (acetonitrile-formic acid, 100 : 0.1, $v/v$) with the following elution profile: 0–45 min, 6–26% ($v/v$) B; 45–55 min, 26–95% B; 55–60 min, 95% B; and 60–63 min, 95–6% B. The flow rate was 0.3 mL/min. The column temperature was 25°C. Before injections, samples of dry extracts (15 mg) were dissolved in 1.5 mL of 70% aqueous methanol, filtered through PTFE syringe filters (25 mm, 0.2 $\mu$m, Vitrum, Czech Republic) and injected (3 $\mu$L) into the UHPLC system. UV-Vis spectra were recorded over a range of 200–600 nm, and chromatograms were acquired at 280, 325, and 350 nm. The LC eluate was introduced directly into the ESI interface without splitting and analyzed in a negative ion mode using a scan from $m/z$ 70 to 2200. The $MS^2$ and $MS^3$ fragmentations were obtained in Auto MS/MS mode for the most abundant ions at the time. The nebulizer pressure was 40 psi, dry gas flow was 9 L/min, dry temperature was 300°C, and capillary voltage was 4.5 kV.

*2.3.5. Determination of Total Phenolic Content (TPC).* The TPC levels were determined according to the Folin-Ciocalteu method as described previously [16]. The results were expressed as mg of gallic acid equivalents (GAE) per g of dry weight of the plant material (mg GAE/g dw).

*2.4. Lipoxygenase (LOX) and Hyaluronidase (HYAL) Inhibition Tests.* The ability of the fruit extracts to inhibit lipoxygenase (LOX) and hyaluronidase (HYAL) was evaluated according to the method optimized earlier [17]. The results of both tests were expressed as $IC_{50}$ values ($\mu$g/mL) from concentration-inhibition curves.

*2.5. Antioxidant Activity in Chemical Models.* The DPPH free-radical scavenging activity was determined according to a previously optimized method [16] and expressed as normalized $EC_{50}$ values calculated from concentration-inhibition curves. The FRAP (ferric reducing antioxidant power) was determined according to [16] and expressed in $\mu$mol of ferrous ions ($Fe^{2+}$) produced by 1 g of the dry extract or standard, which was calculated from the calibration curve of ferrous sulfate. The ability of the extracts to inhibit AAPH-induced peroxidation of linoleic acid was assayed as described previously [18] with peroxidation monitored by quantification of thiobarbituric acid-reactive substances (TBARS) according to a previously optimized method [19], and the antioxidant activity was expressed as $IC_{50}$ values calculated from concentration-inhibition curves. Additionally, the activity parameters in all of the assays were also expressed as $\mu$mol Trolox® equivalents (TE) per g of dry weight of the plant material ($\mu$mol TE/g dw).

*2.6. Antioxidant Activity in Human Plasma Models*

*2.6.1. Isolation of Blood Plasma and Sample Preparation.* Blood (buffy coat units) from eight healthy volunteers, received from the Regional Centre of Blood Donation and Blood Treatment in Lodz (Poland), was centrifuged to obtain plasma [20]. All experiments were approved by the committee on the Ethics of Research at the Medical University of Lodz RNN/347/17/KE. Plasma samples, diluted with 0.01 M Tris/HCl pH 7.4 (1 : 4 $v/v$), were preincubated for 15 min at 37°C with the examined extracts, added to the final concentration range of 1–50 $\mu$g/mL, and then exposed to 100 or 150 $\mu$M peroxynitrite ($ONOO^-$). Control samples were prepared with plasma untreated with the extracts and/or peroxynitrite. To eliminate the possibility of direct interactions of the extracts with plasma proteins and lipids, several experiments with blood plasma and the extracts only (without adding $ONOO^-$) were also performed and no prooxidative effect was found.

*2.6.2. Determination of 3-Nitrotyrosine and Thiols in Human Plasma Proteins.* The peroxynitrite-induced protein damage in blood plasma was determined by the use of 3-nitrotyrosine and protein thiol levels (–SH) as biomarkers of oxidative stress. Immunodetection of 3-nitrotyrosine-containing proteins by the competitive ELISA (C-ELISA) method in plasma samples (control or antioxidants and 100 $\mu$M $ONOO^-$-treated plasma) was performed according to [20]. The nitrofibrinogen (3NT-Fg, at a concentration of 0.5 $\mu$g/mL and 3–6 mol nitrotyrosine/mol protein) was prepared for use in the standard curve. The concentrations of nitrated proteins that inhibit antinitrotyrosine antibody binding were estimated from the standard curve and are expressed as the 3NT-Fg equivalents (in nmol/mg of plasma protein). The concentration of free thiol groups (–SH) in plasma samples (control or antioxidants and 100 $\mu$M $ONOO^-$-treated plasma) was measured spectrophotometrically according to Ellman's method [20]. The free thiol group concentration was calculated from the standard curve of glutathione (GSH) and expressed as umol/mL of plasma.

*2.6.3. Determination of Lipid Hydroperoxides and TBARS in Human Blood Plasma.* The peroxynitrite-induced lipid peroxidation in blood plasma was determined spectrophotometrically by evaluation of the level of lipid hydroperoxides and TBARS. The concentration of hydroperoxides in plasma samples (control or antioxidants and 100 $\mu$M $ONOO^-$-treated plasma) was determined by a ferric-xylenol orange (FOX-1) protocol with a later modification [20]. The amount of lipid hydroperoxides was calculated from the standard curve of hydrogen peroxide and expressed in nmol/mg of plasma proteins. Determination of TBARS in plasma samples (control or antioxidants and 100 $\mu$M $ONOO^-$-treated plasma) was performed according to [20]. The TBARS values were expressed in $\mu$mol TBARS/mL of plasma.

*2.6.4. Ferric Reducing Ability of Human Blood Plasma (FRAP).* The influence of the extracts on the nonenzymatic antioxidant status of plasma was conducted by measurements of their ability to reduce ferric ions ($Fe^{3+}$) to ferrous ions ($Fe^{2+}$). The experiments were performed according to Benzie and Strain [21] and modified by Kolodziejczyk-Czepas et al. [20]. The FRAP values of plasma samples (control or antioxidants and 150 $\mu$M $ONOO^-$-treated plasma) were expressed in mM $Fe^{2+}$ in plasma as calculated from the calibration curve of ferrous sulphate.

*2.7. Cellular Safety Testing.* The cytotoxicity of the examined extracts was conducted in an experimental system of peripheral blood mononuclear cells (PBMCs). PBMCs were isolated from fresh human blood using the Histopaque®-1077 medium, according to a procedure described in our previous work [19]. Then, the cells ($1 \times 10^6$ PBMCs/mL, suspended in PBS) were incubated with *Cotoneaster* fruit extracts at the final concentrations of 5, 25, and 50 μg/mL. Measurements of cell viability were executed after two, four, and six hours of incubation (at 37°C) in a routine dye excluding test, based on a staining with 0.4% Trypan blue. The procedure was carried out according to the manufacturer's protocol using a microchip-type automatic cell counter Bio-Rad (Hercules, CA, USA).

*2.8. Statistical and Data Analysis.* The statistical analysis was performed using STATISTICA 13Pl software for Windows (StatSoft Inc., Krakow, Poland). The results were reported as means ± standard deviation (SD) or ±standard error (SE) for the indicated number of experiments. The significance of differences between the samples and controls were analyzed by one-way ANOVA, followed by the post hoc Tukey's test for multiple comparison. A level of $p < 0.05$ was accepted as statistically significant.

## 3. Results and Discussion

*3.1. GC-FID-MS Analysis of Fatty Acids.* The fatty acid profiles of the lipophilic fractions in the chloroform extracts of the *Cotoneaster* fruits were determined by GC-FID-MS analysis of methyl ester derivatives (FAMEs). As shown in Table 1 and Figure 2, fourteen fatty acids were identified, including saturated, mono-, and polyunsaturated acids with chain lengths ranging from 6 to 22 carbon atoms. Their total content (TFA) varied among the *Cotoneaster* species from 902.5 to 2683.8 mg/100 g of fruit dry weight (dw) with the highest levels noted for *C. zabelii* (2683.8 mg/100 g dw) and *C. splendens* (2024.1 mg/100 g dw). All analyzed fruits contained primarily poly- and monounsaturated acids, constituting 41.6–66.8% and 18.6–29.6% of TFA, respectively. The major component in each sample was linoleic acid C18 : 2 $\Delta^{9,12}$, the sole representative of the polyunsaturated acids. Its content varied among species from 375.4 to 1690.2 mg/100 g fruit dw with the highest amounts (above 10 mg/g dw) recorded for the fruits of *C. zabelii, C. splendens, C. hjelmqvistii,* and *C. horizontalis*. Relatively high levels of oleic acid C18 : 1 $\Delta^9$, a monounsaturated acid, were also noted, especially for the *C. zabelii* and *C. splendens* (649.7 and 473.7 mg/100 g dw, respectively). Regarding saturated acids, they accounted for only 12.3–28.8% of TFA. The highest content of this group was observed in the fruits of *C. zabelii, C. splendens,* and *C. nanshan,* with palmitic acid C16 : 0 being the dominant compound (226.5, 212.6 and 168.7 mg/100 g dw, respectively).

The present work is the first comparison of several *Cotoneaster* fruits in terms of their fatty acid profile. Despite some quantitative differences observed between the investigated fruits, a high level of consistency can be noticed in the qualitative composition of this fraction. The results are in accordance with previous reports for the fruits of *C. pannosus* from Italy, as well as the branches of *C. horizontalis* Decke. of Egyptian origin and the seeds of *C. bullatus, C. dielsianus, C. francheti* Bois, *C. moupinensis* Franch., and *C. simonsii* Baker cultivated in Germany, in which linoleic and palmitic acids were also detected as the major fatty acid components [9, 22, 23].

The unsaturated fatty acids are known factors associated with the prevention of various chronic and acute diseases, such as cardiovascular diseases, osteoporosis, immune disorders, and cancer [7]. Linoleic acid, the representative of the omega-6 fatty acid family (essential fatty acids (EFA)), is considered a vital constituent of a healthy human diet, due to its contribution to cholesterol metabolism (regulation of plasma total cholesterol and low-density lipoprotein cholesterol levels and HDL-LDL ratio) and its association with a lower risk of atherosclerosis [24]. Main sources of this compound are plant oils, derived, inter alia, from the seeds of safflower, sunflower, grape, pumpkin, and corn. The available literature data [25, 26] indicate that whole fruits of some Rosaceae members, such as *Crataegus monogyna* Jacq., *Prunus spinosa* L., and *Rubus ulmifolius* Schott., might be considered as abundant in linoleic acid, constituting over 10% of their lipophilic fraction [26]. Our present results indicate that the analyzed *Cotoneaster* fruits also deserve more attention as rich sources of this compound.

*3.2. GC-FID-MS Analysis of Phytosterols and Triterpenoids.* Apart from fatty acids, three phytosterols (campesterol, β-sitosterol, and stigmasterol) and four triterpenes (α- and β-amyrins, ursolic and oleanolic acids) were identified in the chloroform extracts of the *Cotoneaster* fruits, based on GC-FID-MS analysis of their trimethylsilyl ether derivatives (TMSs). As reported in Table 2 and Figure 2, the total content of sterols and triterpenoids, depending on the tested species, was in the range of 154.6–515.6 mg/100 g of fruit (dw) with the highest levels observed for *C. splendens* (515.6 mg/100 g dw) and *C. nanshan* (438.0 mg/100 g dw). The dominant compound in all samples was β-sitosterol, with the levels ranging from 132.2 to 463.3 mg/100 g dw (76.5–89.3% of the total sterols and triterpenes). The highest content of β-sitosterol was observed for the fruit of *C. splendens* (463.3 mg/100 g dw) followed by those of *C. nanshan* (391.3 mg/100 g dw) and *C. horizontalis* (316.3 mg/100 g dw). Other individual components were observed at much lower concentrations, reaching at most 42 mg/100 g dw.

Regarding the phytosterol and triterpenoid profile, the present results are generally similar to the data obtained previously for different organs of *Cotoneaster* species, although some differences can be noticed in relative proportions of particular compounds. Among the sterols and triterpenoids identified earlier for the *C. horizontalis* branches collected in Egypt, α-amyrin was the dominant compound, constituting 14.4% of the total lipophilic constituents, followed by β-sitosterol (8.5%) and stigmasterol (1.1%) [23]. The ursolic acid was isolated previously from *C. simonsii* twigs [27], *C. racemiflora* Desf. twigs [28], and *C. microphylla* fruits [11], but the present work is the first to describe its quantitative levels in the *Cotoneaster* plants.

TABLE 1: Content of fatty acids (mg/100 g dw) in the *Cotoneaster* fruits.[a]

| Fruit sample | 6:0 | 8:0 | 12:0 | 14:0 | 15:0 | 16:0 | 17:0 | 16:1 $\Delta^9$ | 18:0 | 18:1 $\Delta^9$ | 20:0 | 18:2 $\Delta^{9,12}$ | 20:1 $\Delta^{11}$ | 22:0 |
|---|---|---|---|---|---|---|---|---|---|---|---|---|---|---|
| *C. lucidus* | 3.41 ±0.10[B] | 2.35 ±0.10[F] | 2.77 ±0.01[F] | 5.97 ±0.30[E] | nd | 126.25 ±5.23[A] | tr | 8.96 ±0.51[E] | 92.98 ±5.12[F] | 258.05 ±12.11[A,B] | 6.02 ±0.20[A] | 375.35 ±18.01[A] | tr | 18.77 ±0.80[B] |
| *C. divaricatus* | 2.24 ±0.11[A] | 0.61 ±0.03[A,B] | 0.82 ±0.05[A] | 2.86 ±0.15[A] | tr | 136.65 ±6.20[A] | 0.82 ±0.01[B] | 4.90 ±0.21[C] | 63.23 ±2.45[E] | 262.09 ±10.09[B] | 6.73 ±0.31[A] | 566.60 ±25.03[B] | 2.04 ±0.08[B] | 7.55 ±0.22[A] |
| *C. horizontalis* | tr | 0.69 ±0.01[B] | 2.28 ±0.10[E] | 4.80 ±0.25[D] | tr | 174.10 ±5.40[B] | 1.37 ±0.05[C] | 8.68 ±0.43[E] | 38.38 ±2.10[B,C] | 294.05 ±13.01[B,C] | 13.25 ±0.60[C] | 1012.83 ±45.02[D] | tr | 26.05 ±1.00[C] |
| *C. nanshan* | nd | 0.65 ±0.01[B] | 1.73 ±0.10[C,D] | 4.33 ±0.20[C,D] | 0.65 ±0.01[A] | 168.68 ±6.40[B] | 2.16 ±0.12[D] | 8.22 ±0.38[D,E] | 87.58 ±3.54[F] | 384.72 ±18.12[D] | 17.30 ±0.75[D] | 736.79 ±30.01[C] | 3.89 ±0.20[D] | 19.25 ±0.95[B] |
| *C. hjelmqvistii* | tr | 1.12 ±0.04[C] | 1.96 ±0.10[D] | 4.48 ±0.32[D] | nd | 174.55 ±8.00[B] | 0.56 ±0.03[A] | 3.36 ±0.16[A,B] | 41.19 ±2.05[D] | 335.10 ±14.10[C] | 17.09 ±0.71[D] | 1216.27 ±50.01[E] | 0.56 ±0.02[A] | 24.38 ±1.05[C] |
| *C. dielsianus* | 6.11 ±0.20[C] | 1.77 ±0.03[E] | 1.58 ±0.08[C] | 4.14 ±0.25[B,C,D] | 0.59 ±0.02[A] | 177.81 ±6.43[B] | 3.15 ±0.16[E] | 4.14 ±0.19[B,C] | 37.85 ±1.04[B,C] | 273.21 ±15.02[B] | 16.76 ±0.80[D] | 643.22 ±15.15[B,C] | 0.79 ±0.03[A] | 19.91 ±0.55[B] |
| *C. splendens* | tr | 1.55 ±0.04[D] | 2.80 ±0.15[F] | 5.59 ±0.32[E] | nd | 212.60 ±11.00[C] | 2.18 ±0.11[D] | 8.39 ±0.50[D,E] | 32.95 ±1.14[A,B] | 473.70 ±20.01[E] | 18.34 ±0.61[D] | 1225.89 ±30.12[E] | 2.49 ±0.11[C] | 37.30 ±1.85[E] |
| *C. bullatus* | tr | 0.51 ±0.01[A] | 1.53 ±0.04[C] | 3.73 ±0.22[B,C] | 0.51 ±0.03[A] | 120.49 ±5.20[A] | 2.37 ±0.10[D] | 2.54 ±0.12[A] | 29.83 ±1.10[A] | 215.22 ±10.00[A] | 10.85 ±0.55[B] | 677.53 ±16.15[C] | 4.07 ±0.15[D] | 30.33 ±1.10[D] |
| *C. zabelii* | 3.25 ±0.11[B] | 1.44 ±0.05[D] | 1.08 ±0.05[B] | 3.61 ±0.15[B] | nd | 226.45 ±5.40[C] | 1.44 ±0.06[C] | 7.58 ±0.35[D] | 53.09 ±2.70[D] | 649.73 ±25.05[F] | 30.34 ±1.32[E] | 1690.23 ±55.01[F] | 5.78 ±0.21[E] | 9.75 ±0.20[A] |

[a]Values presented as means ± SD calculated per dw of the plant material ($n = 3$); tr—trace, the content less than 0.5 mg/100 g dw; nd—not detected; different capital letters within the same row indicate significant differences at $\alpha = 0.05$ in HSD Tukey's test; 6:0—caproic acid, 8:0—caprylic acid, 12:0—lauric acid, 14:0—myristic acid, 15:0—pentadecylic acid, 16:0—palmitic acid, 17:0—margaric acid, 16:1 $\Delta^9$—palmitoleic acid, 18:0—stearic acid, 18:1 $\Delta^9$—oleic acid, 20:0—arachidic acid, 18:2 $\Delta^{9,12}$—linoleic acid, 20:1 $\Delta^{11}$—eicosenoic acid and 22:0—behenic acid.

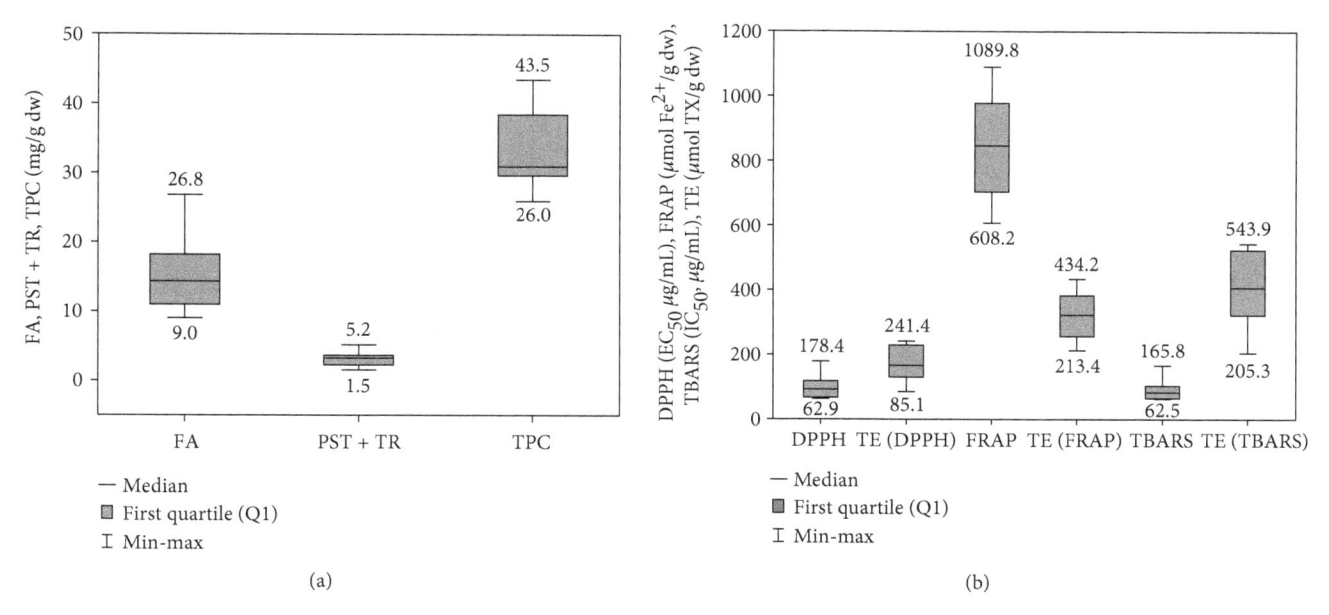

FIGURE 2: Variability of the measured quantitative and activity parameters among the investigated *Cotoneaster* fruits. (a) FA, total fatty acids; PS + TR, sum of phytosterols and tritrepenes; TPC, total phenolic content, expressed in gallic acid equivalents (GAE). (b) DPPH, radical scavenging activity expressed as $EC_{50}$ value; FRAP, ferric reducing antioxidant power; TBARS, inhibition of linoleic acid peroxidation; TE, Trolox® equivalent antioxidant activity.

TABLE 2: Content of phytosterols and triterpenes (mg/100 g dw) in the *Cotoneaster* fruits.[a]

| Fruit sample | Campesterol | β-Sitosterol | Stigmasterol | β-Amyrin | α-Amyrin | Ursolic acid | Oleanolic acid |
|---|---|---|---|---|---|---|---|
| *C. lucidus* | 6.83 ± 0.30[C] | 195.31 ± 5.31[B] | nd | nd | 1.05 ± 0.05[A] | 6.61 ± 0.30[B] | 15.52 ± 0.53[D] |
| *C. divaricatus* | 9.06 ± 0.31[E] | 132.19 ± 4.23[A] | nd | nd | 2.48 ± 0.07[B] | 2.21 ± 0.04[A] | 8.65 ± 0.32[A,B] |
| *C. horizontalis* | 6.04 ± 0.22[B,C] | 316.31 ± 15.03[D] | nd | nd | 0.88 ± 0.02[A] | 25.45 ± 1.10[F] | 17.24 ± 0.50[E] |
| *C. nanshan* | 8.94 ± 0.40[E] | 391.26 ± 17.02[E] | nd | nd | 5.26 ± 0.21[C] | 6.04 ± 0.22[B] | 26.52 ± 1.05[F] |
| *C. hjelmqvistii* | 4.31 ± 0.12[A] | 211.99 ± 10.13[B] | nd | 1.17 ± 0.05[A] | 14.37 ± 0.61[F] | 27.03 ± 0.98[F] | 18.41 ± 0.50[E] |
| *C. dielsianus* | 5.38 ± 0.21[B] | 181.96 ± 5.22[B] | tr | 2.12 ± 0.10[B] | 6.32 ± 0.24[D] | 10.49 ± 0.35[C] | 7.30 ± 0.18[A] |
| *C. splendens* | 13.11 ± 0.56[F] | 463.26 ± 15.10[F] | nd | nd | 8.79 ± 0.30[E] | 13.42 ± 0.45[D] | 17.05 ± 0.45[D,E] |
| *C. bullatus* | 7.98 ± 0.31[D] | 274.47 ± 12.15[C] | 2.70 ± 0.07[B] | 0.88 ± 0.04[A] | 14.15 ± 0.50[F] | 41.45 ± 1.50[G] | 13.05 ± 0.52[C] |
| *C. zabelii* | 6.77 ± 0.30[C] | 273.25 ± 10.22[C] | 1.00 ± 0.01[A] | nd | 14.89 ± 0.22[F] | 20.70 ± 1.03[E] | 9.27 ± 0.36[B] |

[a]Values presented as means ± SD calculated per dw of the plant material ($n = 3$); tr—trace, the content less than 0.5 mg/100 g dw; nd—not detected; different capital letters within the same row indicate significant differences at $\alpha = 0.05$ in HSD Tukey's test.

On the other hand, betulinic acid, reported earlier for *C. microphylla* fruits [11], was not detected during the present study in any fruit sample.

Phytosterols (β-sitosterol, stigmasterol, and their analogues) are important dietary components which help regulate serum lipid profile, reduce total- and LDL-cholesterol levels, and increase HDL/LDL ratio. In addition, plant sterols possess anticancer, anti-inflammatory, and moderate antioxidant activities [29]. For instance, β-sitosterol, the most abundant plant sterol in the human diet, displays significant effects on reducing the symptoms of benign prostatic hyperplasia and prostate cancer. Moreover, this compound has been associated with antidiabetic, immunomodulatory, and analgesic properties [30]. Phytosterols are found abundantly in nonpolar fractions of plants, and their daily consumption is estimated in the range of 200–400 mg with the main dietary sources being vegetable oils, nuts, cereal products, vegetables, fruits, and berries [30]. They are also known to be present in abundance in the fruits derived from numerous genera of Rosaceae, including *Prunus*, *Crataegus*, and *Rosa* [25]. In the lipid fraction of rosaceous fruits, β-sitosterol was often identified as the predominant lipophilic compound, constituting usually more than 60% of the total sterols. As the daily intake of phytosterols (1.5–2.4 g) required for beneficial health effects, especially for cardiovascular and antiatherogenic protection, is usually higher than consumed with the common diet [30], dietary supplementation is a rational solution, and new plant sources of these biomolecules, such as the *Cotoneaster* fruits, offer promise in this aspect.

*3.3. Polyphenolic Profiling of Fruit Extracts.* LC-MS analysis of the hydrophilic (70% aqueous methanolic) extracts of the *Cotoneaster* fruits revealed the presence of a number of

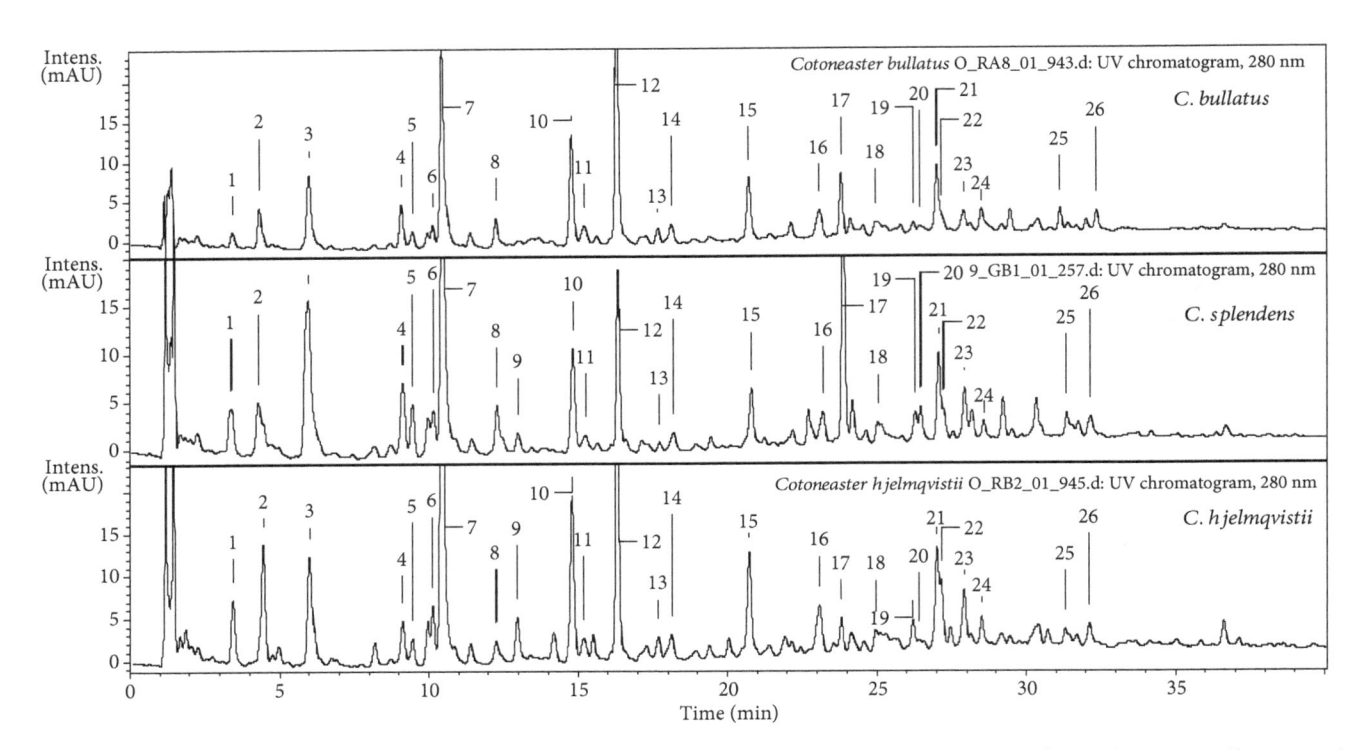

FIGURE 3: Representative UHPLC-UV chromatograms of the *C. bullatus*, *C. splendens*, and *C. hjelmqvistii* fruit polar extracts ($\lambda = 280$ nm). The peak numbers refer to those applied in Table 3.

polyphenols (UHPLC peaks 1–26, Figure 3, Table 3) that were fully or tentatively identified by comparison of their chromatographic behavior and ESI-MS$^3$ fragmentation pattern with authentic standards or literature values. Three major groups of polyphenols were recognized, including phenolic acids (3, 7, and 8) and their derivatives (1, 4, 5, and 11), flavan-3-ols including proanthocyanidins (9, 10, 12–16, 18, and 24), and flavonoids (17, 20, 21–23, 25, and 26). The recorded UHPLC fingerprints (Table 3) indicate that the phenolic profiles of all nine *Cotoneaster* fruits were qualitatively similar. However, noticeable differences were found in the proportions of individual polyphenols, which allowed the subgroups of species to be distinguished depending on the prevalent phenolic class. A distinctive feature of most *Cotoneaster* samples, especially *C. divaricatus*, *C. horizontalis*, and *C. nanshan*, was the predominance of phenolic acid derivatives (1, 3–5, 7, 8, and 11), mainly caffeoylquinic acids, with the dominant peak being chlorogenic acid (7). On the other hand, *C. zabelii*, *C. bullatus*, and *C. hjelmqvistii* contained relatively high amounts of flavan-3-ols and proanthocyanidins (9, 10, 12–16, 18, and 24), with dominating (−)-epicatechin (12). The contribution of flavonoids (17, 20, 21–23, 25, and 26) to the overall phenolic fraction was generally the lowest, but *C. splendens* was distinguished by a particularly large proportion of quercetin 3-(2″-xylosyl)-galactoside (17), and *C. dielsianus* contained a relatively higher level of hyperoside (21).

This report is the first comprehensive study of the LC-MS characteristics of the *Cotoneaster* fruits; the previous studies on *C. integerrimus* and *C. pannosus* have focused only on a selected aspect (HPLC-PDA) of their polyphenolic profiles [9, 12]. In contrast to the present results, the occurrence of low-molecular phenolic acids, including shikimic, p-coumaric, and benzoic acids, has been previously reported, and this phenomenon may be explained by the individual attributes of the tested samples or by differences in the methodology employed for the structural identification. On the other hand, the reported high level of (−)-epicatechin in the fruits of *C. integerrimus* [12] indicates its similarity to those of *C. zabelii* and *C. bullatus* analyzed in the present study.

The total phenolic content (TPC) of the 70% aqueous methanolic extracts of the *Cotoneaster* fruits was determined by the Folin-Ciocalteu photometric assay, commonly used to estimate phenolic metabolites as gallic acid equivalents (GAE). As shown in Table 4 and Figure 2, the TPC values in the analyzed fruits varied from 26.0 to 43.5 mg GAE/g of fruit dw. The highest phenolic content was found for the fruits of *C. hjelmqvistii* and *C. zabelii* (43.5 and 43.0 mg/g dw, respectively), followed by those of *C. splendens* and *C. bullatus* (38.5 and 37.3 mg/g dw, respectively). The level of phenolics in these species is comparable with those observed for other Rosaceae fruits reported in the literature as rich sources of natural polyphenols, for example, *Aronia melanocarpa* (Michx.) Elliott (34.4–78.5 mg GAE/g dw; [3]) and *Sorbus* species (22.4–29.8 mg GAE/g dw; [16]).

The presence of polyphenolic compounds in fruits and vegetables is strongly linked with the beneficial effects of these food products for human health, and the influence of polyphenols on closely intertwined processes of inflammation and oxidative stress is recognized as the most feasible mode of this action. As free radical scavengers, metal chelators, prooxidant and proinflammatory enzyme inhibitors, and modifiers of cell signaling pathways, polyphenols are

TABLE 3: UHPLC-PDA-ESI-MS$^3$ data of polyphenols identified in the polar extracts from *Cotoneaster* fruits.

| Number | Compounds | $t_R$ | UV | (M-H)$^-$ m/z | MS/MS m/z (% base peak) | CL | CDV | CHR | CN | CH %$^b$ | CDL | CS | CB | CZ |
|---|---|---|---|---|---|---|---|---|---|---|---|---|---|---|
| 1 | Vanillic acid-hexoside | 3.5 | 250, 290 | 329 | MS$^2$: 167 (100); 123 (2); 107 (4) | 3.8 | 3.1 | 3.4 | 3.4 | 2.5 | 1.4 | 2.3 | 0.9 | 1.7 |
| 2 | Unidentified | 4.4 | 250, 295 | 255 | MS$^2$: 165 (23) | 33.9 | 3.1 | 2.1 | 16.1 | 5.7 | 5.0 | 2.6 | 3.1 | 3.2 |
| 3 | 3-O-Caffeoylquinic acid | 6.0 | 294, 325 | 353 | MS$^2$: 191 (100); 179 (47); 135 (6) | 1.8 | 10.4 | 5.3 | 15.1 | 5.1 | 5.1 | 8.1 | 5.7 | 1.9 |
| 4 | 3-O-p-Coumaroylquinic acid | 9.4 | 285, 310 | 337 | MS$^2$: 163 (100); 119 (10) | 2.1 | 5.0 | 4.0 | 2.3 | 1.9 | 2.4 | 3.7 | 3.3 | 2.5 |
| 5 | Caffeic acid hexoside | 9.8 | 290, 323 | 341 | MS$^2$: 179 (100); 135 (10) | 2.3 | 8.2 | 2.7 | 5.1 | 1.2 | 0.6 | 2.5 | 1.0 | 2.8 |
| 6 | Unidentified | 10.0 | 285, 323 | 439 | MS$^2$: 391 (100); 338 (17); 243 (10); 195 (55) | 3.9 | 1.9 | 4.3 | 4.8 | 2.3 | 1.2 | 2.0 | 1.3 | 1.3 |
| 7 | 5-O-Caffeoylquinic acid (chlorogenic acid)$^a$ | 10.4 | 294, 325 | 353 | MS$^2$: 191 (100); 179 (6) | 29.4 | 28.3 | 29.5 | 26.0 | 23.5 | 23.0 | 17.9 | 17.3 | 10.8 |
| 8 | 4-O-Caffeoylquinic acid | 10.9 | 294, 325 | 353 | MS$^2$: 191 (21); 179 (47); 173 (100) | 1.1 | 4.1 | 2.4 | 5.1 | 1.0 | 1.4 | 2.4 | 2.1 | 2.0 |
| 9 | Procyanidin B-type dimer | 13.7 | 280 | 577 | MS$^2$: 451 (30); 425 (100); 407 (55); 289 (10) MS$^3$ (425): 407 (80); 273 (13) | 0.7 | 1.6 | 0.5 | 1.2 | 1.1 | 1.2 | 0.9 | nd | 0.7 |
| 10 | Procyanidin B-2$^a$ | 14.9 | 280 | 577 | MS$^2$: 451 (25); 425 (100); 407 (62); 289 (14); MS$^3$ (425): 407 (95); 273 (9) | 0.3 | 2.5 | 4.6 | 0.8 | 8.0 | 6.2 | 5.5 | 9.3 | 10.2 |
| 11 | 5-O-p-Coumaroylquinic acid | 15.7 | 285, 310 | 337 | MS$^2$: 191 (100); 163 (7) | 11.7 | 3.1 | 1.2 | 3.6 | 1.2 | 0.6 | 0.8 | 1.4 | 1.9 |
| 12 | (−)-Epicatechin$^a$ | 16.4 | 280 | 289 | MS$^2$: 245 (100); 205 (28) | 2.4 | 4.4 | 8.5 | 2.1 | 15.8 | 12.3 | 9.7 | 18.6 | 34.4 |
| 13 | Procyanidin B-type dimer | 17.3 | 280 | 577 | MS$^2$: 451 (25); 425 (100); 407 (45); 289 (6); MS$^3$ (425): 407 (75); 273 (9) | nd | nd | nd | nd | 0.4 | nd | 0.6 | 0.9 | 1.2 |
| 14 | Procyanidin B-type tetramer | 18.3 | 280 | 1153 | MS$^2$: 1027 (15); 863 (80); 739 (15); 501 (05); 491 (58); 289 (100) | nd | nd | nd | nd | 1.0 | nd | 0.6 | 1.1 | 1.4 |

TABLE 3: Continued.

| Number | Compounds | $t_R$ | UV | (M-H)$^-$ m/z | MS/MS m/z (% base peak) | CL | CDV | CHR | CN | CH %[b] | CDL | CS | CB | CZ |
|---|---|---|---|---|---|---|---|---|---|---|---|---|---|---|
| 15 | Procyanidin C-1[a] | 20.6 | 280 | 865 | MS$^2$: 847 (19); 739 (77); 713 (51); 695 (100); 577 (26); MS$^3$ (713): 695 (100); 561 (30); 543 (31); 425 (32); 407 (36) | 0.5 | 1.9 | 3.2 | nd | 5.3 | 3.7 | 3.3 | 5.6 | 7.3 |
| 16 | Procyanidin B-type tetramer | 23.3 | 280 | 1153 | MS$^2$: 863 (90); 739 (10); 501 (65); 491 (62); 289 (100) | nd | nd | 2.1 | nd | 2.7 | 2.2 | 2.0 | 2.7 | 3.5 |
| 17 | Quercetin 3-O-β-D-(2″-O-β-D-xylosyl)galactoside[a] | 23.9 | 268, 355 | 595 | MS$^2$: 463 (10); 445 (14); 300 (85); MS$^3$ (463): 343 (62); 301 (100) | nd | nd | 4.0 | nd | 2.0 | 2.6 | 16.1 | 5.9 | nd |
| 18 | Epicatechin derivative | 26.2 | 280 | 739 | MS$^2$: 587 (100); 451 (19); 339 (40); 289 (35) | nd | 2.2 | 2.0 | nd | 1.1 | nd | 1.5 | 1.2 | nd |
| 19 | Unidentified | 26.3 | 280 | 451 | MS$^2$: 341 (100); 217 (8) | nd | 2.5 | 2.6 | 2.4 | 1.5 | 2.0 | 2.0 | 1.2 | 1.2 |
| 20 | Quercetin rhamnoside-hexoside | 26.7 | 275, 350 | 609 | MS$^2$: 301 (100) | 0.6 | 0.4 | 1.4 | nd | 0.7 | 3.2 | 2.3 | 0.9 | 1.8 |
| 21 | Quercetin 3-O-β-D-galactoside (hyperoside)[a] | 27.1 | 265, 355 | 463 | MS$^2$: 301 (100) | 2.5 | 5.0 | 4.9 | 5.5 | 5.5 | 9.5 | 5.2 | 6.6 | 2.4 |
| 22 | Quercetin 3-O-β-D-(6″-O-α-L-Rhamnosyl)glucoside (rutin)[a] | 27.3 | 260, 355 | 609 | MS$^2$: 301 (100) | 0.8 | 2.5 | 2.6 | 2.8 | 3.8 | 2.5 | 2.0 | nd | 2.2 |
| 23 | Quercetin 3-O-β-D-glucoside (isoquercitrin)[a] | 28.0 | 265, 355 | 463 | MS$^2$: 301 (100) | 1.6 | 3.1 | 2.4 | 2.4 | 3.5 | 2.5 | 3.3 | 2.6 | 3.2 |
| 24 | Procyanidin B-type dimer | 28.6 | 280 | 577 | MS$^2$: 425 (100); 407 (52); 289 (18) | 0.6 | 1.6 | 2.0 | 1.0 | 1.6 | 2.0 | 1.6 | 2.0 | 2.4 |
| 25 | Quercetin rhamnoside-hexoside | 31.3 | 276, 350 | 609 | MS$^2$: 301 (100) | nd | 2.2 | 2.2 | nd | 0.4 | 4.1 | nd | 2.9 | nd |
| 26 | Quercetin 3-O-α-L-rhamnoside (quercitrin)[a] | 32.4 | 276, 350 | 447 | MS$^2$: 301 (100) | nd | 2.8 | 2.2 | nd | 1.4 | 5.1 | 1.8 | 2.6 | nd |

[a]Identified with the corresponding standards; [b]relative contribution based on peak area on the UHPLC chromatograms ($\lambda$ = 280 nm) recorded at the extract concentration of 10 mg/mL and injection volume of 3 μL; nd—not detected; the values are means ($n$ = 3); with RSD ≤ 5%. CL, *C. lucidus*; CDV, *C. divaricatus*; CHR, *C. horizontalis*; CN, *C. nanshan*; CH, *C. hjelmqvistii*; CDL, *C. dielsianus*; CS, *C. splendens*; CB, *C. bullatus*; CZ, *C. zabelii*.

TABLE 4: Total phenolic content (TPC) and antioxidant activity (DPPH, FRAP, and TBARS tests) of the *Cotoneaster* fruits and standard antioxidants.

| Fruit sample/ standard | TPC[a] (mg GAE/g) | Radical scavenging activity DPPH[b] | | Reducing power[c] | | LA-peroxidation TBARS[d] | |
|---|---|---|---|---|---|---|---|
| | | $EC_{50}$ ($\mu$g/mL) | TE ($\mu$mol TE/g) | FRAP (mmol $Fe^{2+}$/g) | TE ($\mu$mol TE/g) | $IC_{50}$ ($\mu$g/mL) | TE ($\mu$mol TE/g) |
| *C. lucidus* | 28.70 ± 1.01[B] | 123.41 ± 1.70[E] | 122.75 ± 1.69[C] | 0.70 ± 0.01[B] | 257.22 ± 4.96[B,C] | 108.70 ± 4.11[F] | 314.84 ± 6.03[C] |
| *C. divaricatus* | 29.71 ± 0.91[B] | 91.47 ± 2.01[C] | 165.58 ± 3.62[D] | 0.76 ± 0.01[C] | 281.61 ± 4.43[C] | 83.16 ± 0.58[D] | 406.94 ± 1.43[D] |
| *C. horizontalis* | 30.50 ± 0.72[B] | 93.32 ± 1.90[C] | 162.38 ± 3.31[D] | 0.85 ± 0.01[D] | 322.75 ± 4.06[D] | 84.89 ± 2.11[D] | 401.23 ± 5.03[D] |
| *C. nanshan* | 26.02 ± 0.74[A] | 178.35 ± 2.81[F] | 84.91 ± 1.33[B] | 0.61 ± 0.01[A] | 213.41 ± 4.42[A] | 165.76 ± 3.74[G] | 205.30 ± 2.33[B] |
| *C. hjelmqvistii* | 43.50 ± 1.21[D] | 64.51 ± 0.84[B] | 234.84 ± 2.91[E,F] | 1.05 ± 0.02[F] | 414.38 ± 11.14[F,G] | 62.96 ± 1.10[C] | 532.92 ± 4.63[E,F] |
| *C. dielsianus* | 31.02 ± 1.02[B] | 117.10 ± 2.40[D] | 129.37 ± 2.65[C] | 0.67 ± 0.03[B] | 240.90 ± 13.83[A,B] | 103.72 ± 2.58[E] | 322.66 ± 3.98[C] |
| *C. splendens* | 38.51 ± 0.81[C] | 67.15 ± 1.80[B] | 225.49 ± 6.04[E] | 0.98 ± 0.01[E] | 383.06 ± 6.24[E,F] | 66.21 ± 2.94[C] | 518.18 ± 11.79[E] |
| *C. bullatus* | 37.31 ± 0.80[C] | 66.31 ± 1.70[B] | 228.54 ± 5.86[E] | 0.97 ± 0.01[E] | 378.87 ± 2.90[E] | 64.99 ± 1.55[C] | 523.90 ± 6.30[E,F] |
| *C. zabelii* | 43.02 ± 1.11[D] | 62.93 ± 1.91[B] | 240.93 ± 7.28[F] | 1.09 ± 0.04[G] | 434.27 ± 20.50[G] | 62.54 ± 1.32[C] | 543.86 ± 5.76[F] |
| QU | — | 1.70 ± 0.11[A] | 8.96 ± 0.58[A] | 31.20 ± 0.98[K] | 11878.15 ± 15.20[J] | 1.85 ± 0.12[A] | 18.37 ± 1.69[A] |
| BHA | — | 2.90 ± 0.15[A] | 5.24 ± 0.27[A] | 16.14 ± 0.77[I] | 7726.31 ± 10.52[H] | 3.16 ± 0.22[A] | 10.76 ± 1.06[A] |
| BHT | — | 6.50 ± 0.13[A] | 2.34 ± 0.05[A] | 18.89 ± 0.45[J] | 9247.66 ± 12.30[I] | 9.31 ± 0.16[B] | 3.64 ± 0.09[A] |
| TX | — | 3.80 ± 0.20[A] | — | 9.34 ± 0.35[H] | — | 8.47 ± 0.45[B] | — |

[a–d]Results expressed as means ± SD calculated per dw of the plant material ($n = 3$); different capital letters within the same row indicate significant differences at $\alpha = 0.05$ in HSD Tukey's test. [a]Total phenolic content (TPC), expressed in gallic acid equivalents (GAE). [b]Scavenging efficiency in the DPPH test, the amount of the plant materials or standards required for 50% reduction of the initial DPPH concentration expressed as $EC_{50}$, effective concentration. [c]Ferric reducing antioxidant power. [d]Ability to inhibit linoleic acid (LA) peroxidation monitored by TBARS test and expressed as $IC_{50}$, concentration of plant materials or standards needed to decrease the LA-peroxidation by 50%; TE, Trolox® equivalent antioxidant activity. Standards: QU, quercetin; BHA, butylated hydroxyanisole; BHT, 2,6-di-*tert*-butyl-4-methylphenol; TX, Trolox®.

effective agents preventing damages related to the oxidative stress and inflammation implicated in the etiology and progression of numerous chronic diseases, including cardiovascular diseases, diabetes mellitus, neurodegenerative disorders, and cancer [31–33]. The occurrence of polyphenolic compounds in the investigated fruits might thus largely define their bioactivity, especially that *Cotoneaster*-derived polyphenols have been previously linked with strong antioxidant capacity in our earlier study regarding the leaves [34].

*3.4. Biological Activity.* The above presented phytochemical studies proved that fruits of *Cotoneaster* species are indeed a rich source of diverse phytochemicals with a wide spectrum of recognized biological properties. However, based on the results of the quantitative studies, the polyphenolic fraction with the highest content would appear to have the greatest beneficial health effects of the fruits in a human organism. Thus, further studies were focused on providing a more detailed insight into potential mechanisms of the activity of the hydrophilic components, that is, their anti-inflammatory and antioxidant effects.

*3.4.1. Inhibitory Effects on Two Enzymes Involved in Inflammation.* Inflammation is a complex process that constitutes a part of the immune system defense against harmful stimuli, but may lead to negative effects if uncontrolled. The inflammatory response is regulated by numerous enzymes and mediators and thus can be intercepted at different points, and several of these key enzymes, including lipoxygenases (LOX) and hyaluronidases (HYAL), are most often used to determine the anti-inflammatory potential of natural products [35]. LOX catalyze the dioxygenation of arachidonic acid to form hydroperoxides, the first step in the biosynthesis of several proinflammatory mediators [36]. HYAL, on the other hand, are highly specific hydrolases that degrade hyaluronic acid, an important component of the extracellular matrix, thus increasing the permeability of the tissues and facilitating the spread of inflammation [37]. Our present findings indicate that all fruit extracts inhibit the activity of LOX and HYAL in a dose-dependent manner (Table 5). The strongest inhibitory effect towards LOX was demonstrated by the leaf extracts of *C. hjelmqvistii* and *C. zabelii* ($IC_{50} = 7.70$ and $9.97\,\mu$g/U, respectively), while the activity of HYAL was most strongly hindered by the leaf extract of *C. lucidus* ($IC_{50} = 16.44\,\mu$g/U). The activity of the extracts was weaker in comparison to indomethacin ($IC_{50} = 1.89\,\mu$g/U for LOX and $5.60\,\mu$g/U for HYAL), but after recalculating the results to adjust for the actual polyphenol content (which gives $IC_{50}$ values in the range of $0.33$–$0.77\,\mu$g/U for LOX and $0.47$–$1.93\,\mu$g/U for HYAL inhibition), the activity of the extracts looks quite advantageous in comparison to the positive standard. The anti-inflammatory potential of *Cotoneaster* polyphenols is further confirmed by the high activity of (−)-epicatechin, quercetin, and chlorogenic acid, the main constituents of the investigated leaf extracts.

*3.4.2. Antioxidant Activity in Chemical Models.* The basic antioxidant mechanism of *Cotoneaster* polyphenols was verified in chemical models using three complementary *in vitro* assays: DPPH and FRAP tests, two of the most frequently

TABLE 5: Inhibitory effects of Cotoneaster fruit extracts and standards towards lipoxygenase (LOX) and hyaluronidase (HYAL).

| Fruit sample/standard | LOX | | HYAL | |
|---|---|---|---|---|
| | $IC_{50}^a$ ($\mu$g/mL) | $IC_{50}^b$ ($\mu$g/U) | $IC_{50}^a$ ($\mu$g/mL) | $IC_{50}^b$ ($\mu$g/U) |
| C. lucidus | $487.75 \pm 6.57^F$ | $13.29 \pm 0.18^F$ | $25.65 \pm 0.95^C$ | $16.44 \pm 0.61^C$ |
| C. divaricatus | $479.98 \pm 12.79^F$ | $13.08 \pm 0.35^F$ | $34.22 \pm 1.48^D$ | $21.93 \pm 0.95^D$ |
| C. horizontalis | $421.85 \pm 5.78^E$ | $11.50 \pm 0.16^E$ | $40.51 \pm 2.11^{E,F,G}$ | $25.97 \pm 1.35^{E,F,G}$ |
| C. nanshan | $626.16 \pm 5.04^H$ | $17.07 \pm 0.14^H$ | $45.64 \pm 0.76^G$ | $29.25 \pm 0.49^G$ |
| C. hjelmqvistii | $290 \pm 2.75^C$ | $7.70 \pm 0.07^C$ | $44.44 \pm 1.72^{F,G}$ | $28.48 \pm 1.10^{F,G}$ |
| C. dielsianus | $914.97 \pm 2.15^J$ | $24.94 \pm 0.06^J$ | $35.07 \pm 2.60^{D,E}$ | $22.48 \pm 1.66^{D,E}$ |
| C. splendens | $734.25 \pm 5.86^I$ | $20.01 \pm 0.16^I$ | $34.36 \pm 0.11^D$ | $22.03 \pm 0.07^D$ |
| C. bullatus | $585.43 \pm 16.14^G$ | $15.96 \pm 0.44^G$ | $39.04 \pm 0.82^{D,E,F}$ | $25.03 \pm 0.53^{D,E,F}$ |
| C. zabelii | $375.87 \pm 9.89^D$ | $9.97 \pm 0.26^D$ | $33.33 \pm 2.12^D$ | $21.37 \pm 1.36^D$ |
| QU | $69.60 \pm 2.62^A$ | $2.46 \pm 0.01^A$ | $21.04 \pm 1.03^C$ | $13.87 \pm 0.06^C$ |
| ECA | $124.38 \pm 1.56^B$ | $3.39 \pm 0.04^B$ | $18.51 \pm 0.50^B$ | $11.87 \pm 0.32^B$ |
| CHA | $151.71 \pm 7.52^B$ | $4.14 \pm 0.21^B$ | $20.35 \pm 0.36^B$ | $13.05 \pm 0.23^B$ |
| IND | $90.12 \pm 0.40^A$ | $1.89 \pm 0.10^A$ | $8.61 \pm 0.22^A$ | $5.60 \pm 0.07^A$ |

Results expressed as means ± SD calculated per dry weight (dw) of the extracts; different capital letters within the same row indicate significant differences at $\alpha = 0.05$ in HSD Tukey's test. Standards: QU, quercetin; ECA, (−)-epicatechin; CHA, chlorogenic acid; IND, indomethacin. Ability to inhibit lipoxygenase (LOX) and hyaluronidase (HYAL) calculated as the amount of analyte needed for 50% inhibition of enzyme activity was expressed as follows: $^a\mu$g of the dry extracts or standards/mL of the enzyme solution and $^b\mu$g of the extracts/enzyme units (U).

employed SET (single electron transfer) type methods, and the inhibition of AAPH-induced linoleic acid peroxidation test (monitored by TBARS assay), a more physiologically relevant system which involves the HAT (hydrogen atom transfer) mechanism. In all of the applied tests, the investigated fruits displayed concentration-dependent activity with the capacity parameters (expressed in $\mu$mol TE/g dw) of a similar order of magnitude, which shows that Cotoneaster antioxidants can effectively act via both basic mechanisms. The highest activity in comparison to the natural (quercetin) and synthetic standards (BHA and BHT) were observed in the FRAP and TBARS assays for all fruits (Table 4 and Figure 2). In all tests, the fruits of C. zabelii, C. hjelmqvistii, C. bullatus, and C. splendens, indicated in the present study as the richest sources of polyphenols, displayed the highest antioxidant efficiency, with the activity parameters varying in the narrow range of 225.5–240.9 $\mu$mol TE/g dw (DPPH), 378.9–434.3 $\mu$mol TE/g (FRAP), and 518.2–543.9 $\mu$mol TE/g (TBARS), respectively. Interestingly, these were the species that also exhibited the relatively largest proportions of proanthocyanidins/flavan-3-ols (C. zabelii, C. bullatus, C. splendens) or quercetin 3-(2″-xylosyl)-glucoside (C. hjelmqvistii), which suggest that these polyphenols play a significant role in the activity of fruits. Additionally, the close connection between the phenolic levels and antioxidant parameters was also evidenced by statistically significant linear correlations between TPCs and the results of the DPPH ($|r| = 0.9352$, $p < 0.001$), FRAP ($|r| = 0.9491$, $p < 0.001$), and TBARS ($|r| = 0.9116$, $p < 0.001$) tests.

*3.4.3. Protective Effects on Human Plasma Components Exposed to Oxidative Stress.* To provide a more detailed insight into the antioxidant effects of Cotoneaster polyphenols, the four most promising species (C. zabelii, C. bullatus, C. splendens, and C. hjelmqvistii) were selected for further studies in a biological model. Since according to traditional application and our present results, Cotoneaster fruits appear to be promising sources of phytochemicals with properties especially advantageous for the circulatory system (i.e., linoleic acid and $\beta$-sitosterol), a human plasma model was selected to evaluate their additional benefits for cardiovascular health, this time mediated by polyphenols. This approach allowed for the *in vitro* monitoring of the protective effects of the extracts towards human plasma components under oxidative stress conditions. The peroxynitrite ($ONOO^-$) used for inducing oxidative stress is a known *in vivo*-operating oxidant, responsible for structural changes in plasma proteins and lipids and implicated in numerous oxidative stress-related disorders [38]. The concentrations of $ONOO^-$ (100 and 150 $\mu$M) selected for the study enabled quantitative measurements of the resulting modifications in plasma components, but may be also regarded as physiologically-relevant as they can be reached *in vivo* in local compartments, for example, during a serious inflammation of blood vessels [39].

The addition of $ONOO^-$ to the plasma samples resulted in an overall decrease ($p < 0.001$) in the nonenzymatic antioxidant capacity of the plasma, measured as the FRAP parameter, and in oxidative and nitrative alterations of its protein and lipid components, which was evidenced by a significant increase ($p < 0.001$) in lipid peroxidation biomarkers (lipid hydroperoxides and TBARS), a noticeable rise ($p < 0.001$) in 3-nitrotyrosine level (marker of protein nitration), and a decrease ($p < 0.001$) in the level of thiol groups (marker of protein oxidation). On the other hand, in the plasma samples incubated with $ONOO^-$ in the presence of Cotoneaster extracts (1–50 $\mu$g/mL), the extent of oxidative/nitrative damage to both proteins and lipids was noticeably limited ($p < 0.05$), regardless of the tested species and the

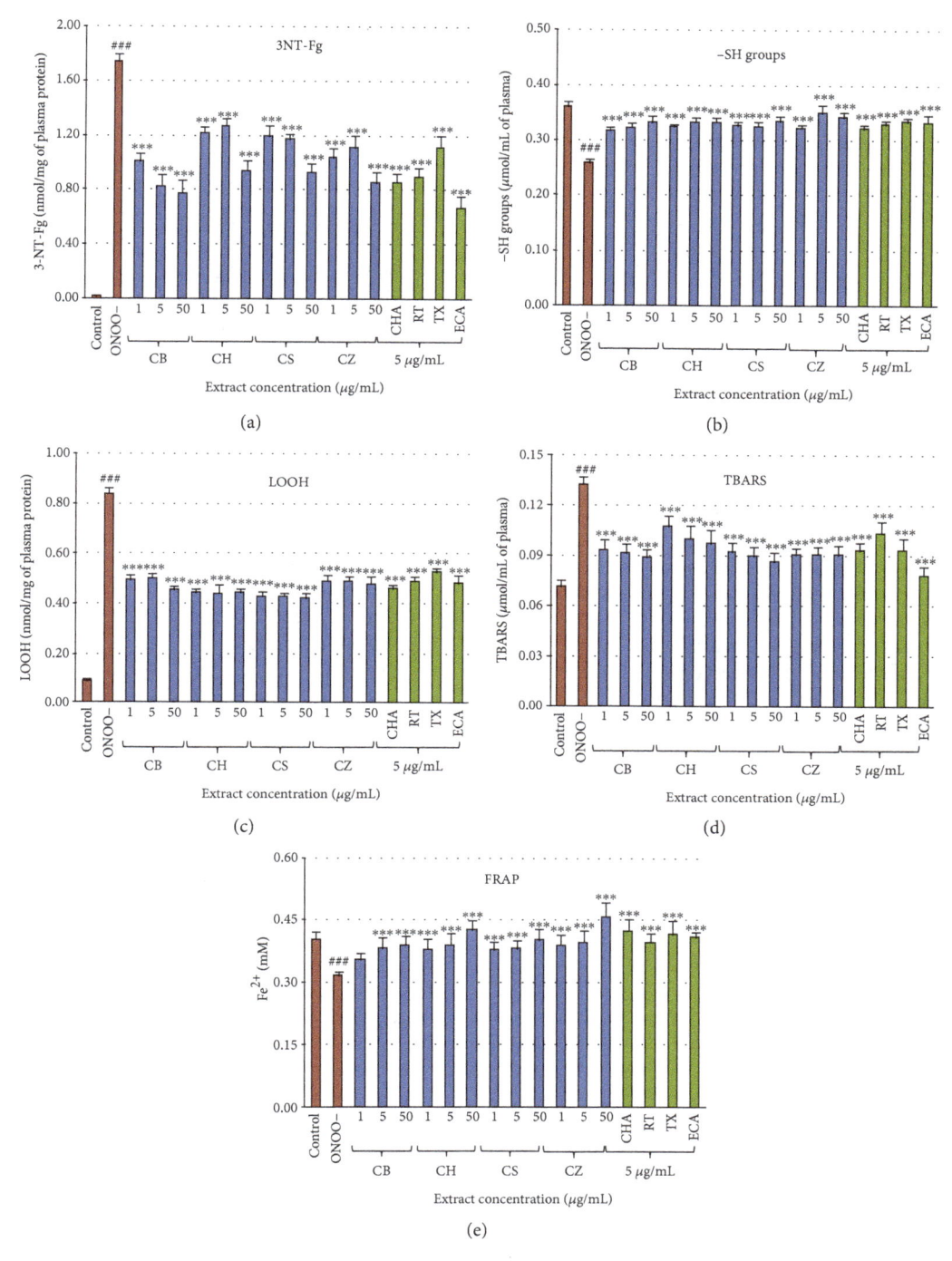

FIGURE 4: Effects of the *Cotoneaster* fruit extracts on human plasma exposed to oxidative stress: (a) effects on the nitration of tyrosine residues in plasma proteins and formation of 3-nitrotyrosine (3-NT-Fg); (b) effects on the oxidation of free thiol groups (-SH); effects on the peroxidation of plasma lipids including (c) formation of lipid hydroperoxides (LOOH), and (d) thiobarbituric acid-reactive substances (TBARS); (e) effects on ferric reducing ability of blood plasma (FRAP). Results expressed as means ± SE ($n = 8$) for repeated measures: $^{\#\#\#}p < 0.001$, for ONOO$^-$-treated plasma (without the extracts) versus control plasma, and $^{***}p < 0.001$ for plasma treated with ONOO$^-$ in the presence of the investigated extracts (1–50 μg/mL) or the standards (5 μg/mL). CB, *C. bullatus*; CH, *C. hjelmqvistii*; CS, *C. splendens*; CZ, *C. zabelii*. Standards: CHA, chlorogenic acid; RT, rutin; TX, Trolox®; ECA, (−)-epicatechin.

extract concentration. As shown in Figures 4(a) and 4(b), even at the lowest concentrations of 1 μg/mL, the extracts were able to reduce tyrosine nitration by about 29–42% and thiol group oxidation by about 24–26%, while at the concentration of 50 μg/mL the effectiveness rose to 46–55% and 29–32%, respectively. Moreover, as demonstrated in Figures 4(c) and 4(d), all fruit samples inhibited the generation of plasma lipid hydroperoxides by 40–50% and reduced

TBARS levels by 19–35%. All extract-treated samples, apart from those fortified with 1 μg/mL of *C. bullatus* extract, demonstrated a statistically significant ($p < 0.001$) improvement in the nonenzymatic antioxidant capacity of blood plasma of up to 44% in comparison to the samples not protected by the extracts (Figure 4(e)). In most cases, little difference was observed in the activity between the tested fruits; however, the inhibition of tyrosine nitration assay found *C. bullatus* and *C. zabelii* displaying stronger activity than the other two extracts at all concentrations tested ($p < 0.05$). A dose dependency was noticeable for *C. bullatus* and *C. splendens* in antinitrative activity (Figure 4(a)) and for most *Cotoneaster* species in the TBARS test, with the exception of *C. zabelii* (Figure 4(d)). Some significant correlations were also found, between the TPCs and the activity parameters. The most prominent was the relationship for the FRAP assay ($|r| = 0.7587, p < 0.01$). In the tests for protein protection, the correlation between the percentage inhibition of tyrosine nitration and phenolic level was stronger ($|r| = 0.6774, p < 0.05$) than the analogous relationship for the reduction of thiol group oxidation ($|r| = 0.4885, p < 0.05$). Contrastingly, the correlations in the lipid peroxidation assays were not statistically significant ($p > 0.05$).

The effectiveness of the extracts was further supported by the fact that in all of the tests, the observed antioxidant effects of the fruit extracts at the corresponding concentration levels (5 μg/mL) were similar or higher to that of Trolox®, a synthetic analog of vitamin E often used as a positive standard in antioxidant studies. Moreover, the significant activity of rutin, chlorogenic acid, and, especially, (−)-epicatechin confirm the important role of polyphenols in the capacity of the extracts.

The wide range of the extract concentrations tested (1–50 μg/mL) was in accordance with the general practice of *in vitro* studies [20] and allowed for the study of different interactions in the system. Additionally, the lower levels (1–5 μg/mL) might be considered physiologically-relevant as they correspond to the levels of phenolics attainable *in vivo* after consumption of polyphenol-rich plant materials. For example, according to the accumulated research [40, 41], the maximal achievable concentration of plant phenolics in blood plasma can reach up to 5–10 μM, which generally corresponds to less than 5 μg/mL. Taking into account the TPC levels evaluated for *Cotoneaster* fruits in the present study and the extraction efficiency (15–30%, depending on the species), the levels of phenolics corresponding to the applied extract concentration of 1–5 μg/mL are about 0.13–1.25 μg/mL: well within the obtainable plasma range. This suggests that the protective activity of the *Cotoneaster* extracts towards $ONOO^-$-induced changes observed *in vitro* may translate to their positive *in vivo* effects.

The harmful influence of $ONOO^-$ is often associated with serious pathological consequences in many organs and systems of the human body. The nitration/oxidation of biomolecules such as enzymes, receptors, lipoproteins, fatty acids, or nucleic acids changes their function and may impair cellular signalization pathways, induce inflammatory responses, or even promote cell apoptosis [38, 39]. In the case of the circulatory system, the negative effects of $ONOO^-$ result in a higher risk of cardiovascular disorders, such as stroke, myocardial infarction, or chronic heart failure [38], and are connected with the direct modifications of plasma proteins and lipids. For instance, the formation of 3-nitrotyrosine in fibrinogen might contribute to prothrombotic events in the blood coagulation cascade and fibrinolysis process [42], while thiol oxidation in platelet proteins leads to the inhibition of platelet function [43]. Additionally, oxidation of low-molecular-weight thiols, such as reduced glutathione, diminishes the endogenous antioxidant capacity of plasma and primes further oxidative damage in the system [38]. Similarly, lipid peroxidation initiated by $ONOO^-$ may propagate platelet aggregation [44], while peroxynitrite-modified LDL binds with high affinity to macrophage scavenger receptors leading to foam cell formation, which represent a key early event in atherogenesis [38, 45]. The prevention of these processes partially explains the beneficial effects of *Cotoneaster* fruits reported by traditional medicine and might be regarded as a good strategy in prophylaxis of various cardiovascular complaints.

*3.5. Cellular Safety.* Due to its long tradition of consumption and application in folk medicine, the *Cotoneaster* fruits might be regarded as nontoxic. However, in the case of the concentrated extracts, a more detailed evaluation of their safety is required. Therefore, the next step of our research was a viability test on PMBCs which assessed the cytotoxicity of the extracts. After two, four, and six-hour incubation periods with the plant extracts at concentrations of 5, 25, and 50 μg/mL, the viability of the extract-treated cells constituted 97.3–101.7% of that of the control (non-treated cells) and no statistically significant differences were found ($p > 0.05$) between the two values (Figure 5). These findings suggest that the *Cotoneaster* extracts do not have cytotoxic effects at these concentrations.

## 4. Conclusion

The current paper presents the first comprehensive phytochemical and activity study of *Cotoneaster* fruits. The fruits were found to possess distinct lipophilic and phenolic profiles, significant antioxidant activity in both chemical and biological models, noticeable inhibitory effects on the proinflammatory enzymes, and cellular safety. Hence, *Cotoneaster* fruits appear to be promising candidates for the production of pharma- and nutraceuticals associated with preventing and treating oxidative stress and inflammatory-related chronic diseases; they may also contribute to a balanced and varied diet comprising food rich in bioactive compounds. Furthermore, the protective effects against $ONOO^-$-induced modifications in the plasma components, demonstrated by the polyphenolic fractions from the fruits of *C. hjelmqvistii*, *C. zabelii*, *C. splendens*, and *C. bullatus* at *in vivo*-relevant levels, may be considered as a molecular basis for the beneficial effects of *Cotoneaster* fruits within the cardiovascular system reported by traditional medicine. The biological activity demonstrated in the present study might therefore be a starting point of more extensive investigation on the nutritional

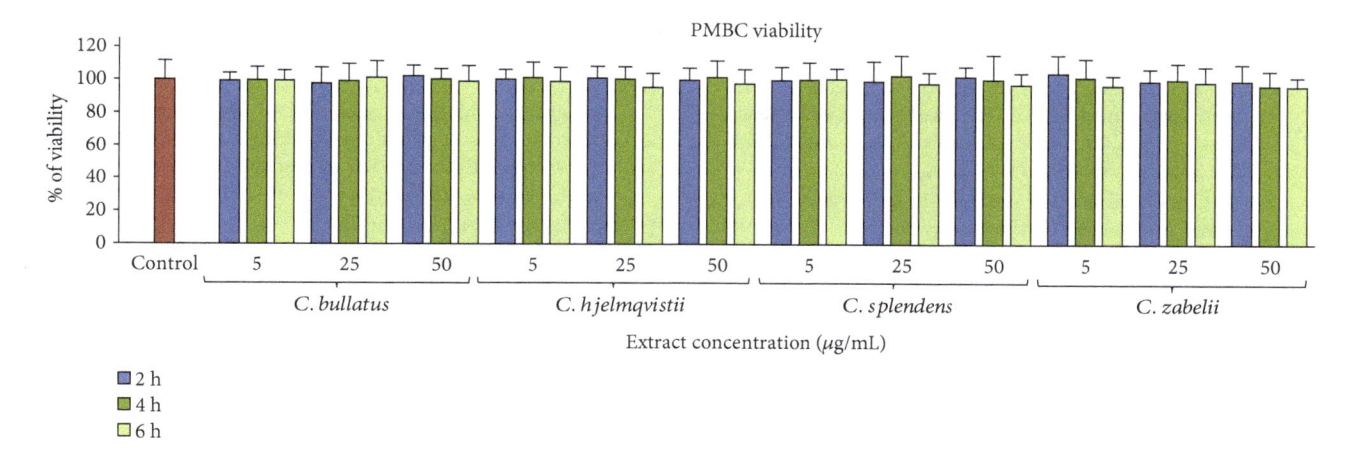

FIGURE 5: Viability of peripheral blood mononuclear cells (PMBCs) after 2, 4, and 6 h incubation with the *Cotoneaster* fruit extracts at 5, 25, and 50 $\mu$g/mL. Results are presented as means ± SD ($n = 14$).

value and bioactivity of *Cotoneaster* fruits, including their effects in *in vivo* systems.

## Acknowledgments

This work was financially supported by the University of Lodz, Poland through Grant no. 506/1136 (the synthesis of peroxynitrite, ONOO$^-$), the Medical University of Lodz, Poland through Grant no. 503/3-022-01/503-31-001 (the phytochemical and anti-inflammatory activity studies), and the National Science Centre, Poland through Grant no. 2017/01/X/NZ7/00520 (all other costs). The authors would like to thank the staff of the Botanical Garden in Lodz and the Forestry Experimental Station of the Warsaw University of Life Sciences in Rogow for providing and authenticating the plant material.

## References

[1] H. Boeing, A. Bechthold, A. Bub et al., "Critical review: vegetables and fruit in the prevention of chronic diseases," *European Journal of Nutrition*, vol. 51, no. 6, pp. 637–663, 2012.

[2] O. Ogah, C. S. Watkins, B. E. Ubi, and N. C. Oraguzie, "Phenolic compounds in Rosaceae fruit and nut crops," *Journal of Agricultural and Food Chemistry*, vol. 62, no. 39, pp. 9369–9386, 2014.

[3] P. N. Denev, C. G. Kratchanov, M. Ciz, A. Lojek, and M. G. Kratchanova, "Bioavailability and antioxidant activity of black chokeberry (*Aronia melanocarpa*) polyphenols: in vitro and in vivo evidences and possible mechanisms of action: a review," *Comprehensive Reviews in Food Science and Food Safety*, vol. 11, no. 5, pp. 471–489, 2012.

[4] R. Raudonis, L. Raudonė, K. Gaivelytė, P. Viškelis, and V. Janulis, "Phenolic and antioxidant profiles of rowan (*Sorbus* L.) fruits," *Natural Product Research*, vol. 28, no. 16, pp. 1231–1240, 2014.

[5] C. F. Zhao, S. Li, S. J. Li, G. H. Song, L. J. Yu, and H. Zhang, "Extraction optimization approach to improve accessibility of functional fraction based on combination of total polyphenol, chromatographic profiling and antioxidant activity evaluation: *Pyracantha fortuneana* fruit as an example," *Journal of Functional Foods*, vol. 5, no. 2, pp. 715–728, 2013.

[6] R. Pinacho, R. Y. Cavero, I. Astiasarán, D. Ansorena, and M. I. Calvo, "Phenolic compounds of blackthorn (*Prunus spinosa* L.) and influence of in vitro digestion on their antioxidant capacity," *Journal of Functional Foods*, vol. 19, pp. 49–62, 2015.

[7] J. M. Lee, H. Lee, S. B. Kang, and W. J. Park, "Fatty acid desaturases, polyunsaturated fatty acid regulation, and biotechnological advances," *Nutrients*, vol. 8, no. 1, p. 23, 2016.

[8] Z. Ovesná, A. Vachálková, K. Horváthová, and D. Tóthová, "Pentacyclic triterpenoic acids: new chemoprotective compounds. Minireview," *Neoplasma*, vol. 51, no. 5, pp. 327–333, 2004.

[9] F. Les, V. López, G. Caprioli et al., "Chemical constituents, radical scavenging activity and enzyme inhibitory capacity of fruits from *Cotoneaster pannosus* Franch.," *Food & Function*, vol. 8, no. 5, pp. 1775–1784, 2017.

[10] G. Zengin, A. Uysal, E. Gunes, and A. Aktumsek, "Survey of phytochemical composition and biological effects of three extracts from a wild plant (*Cotoneaster nummularia* Fisch. et Mey.): a potential source for functional food ingredients and drug formulations," *PLoS One*, vol. 9, no. 11, article e113527, 2014.

[11] G. Bisht, "Chemical constituents from the fruits of *Cotoneaster microphylla* Wall ex Lindl," *Asian Journal of Chemistry*, vol. 7, pp. 455-456, 1995.

[12] A. Uysal, G. Zengin, A. Mollica et al., "Chemical and biological insights on *Cotoneaster integerrimus*: a new (−)-epicatechin source for food and medicinal applications," *Phytomedicine*, vol. 23, no. 10, pp. 979–988, 2016.

[13] W. A. Pryor, R. Cueto, X. Jin et al., "A practical method for preparing peroxynitrite solutions of low ionic strength and free of hydrogen peroxide," *Free Radical Biology & Medicine*, vol. 18, no. 1, pp. 75–83, 1995.

[14] J. Nazaruk, A. Wajs-Bonikowska, and R. Bonikowski, "Components and antioxidant activity of fruits of *Cirsium palustre* and *C. rivulare*," *Chemistry of Natural Compounds*, vol. 48, pp. 9-10, 2012.

[15] T. T. Thanh, M. F. Vergnes, J. Kaloustian, T. F. El-Moselhy, M. J. Amiot-Carlin, and H. Portugal, "Effect of storage and heating on phytosterol concentrations in vegetable oils deter-

mined by GC/MS," *Journal of the Science of Food and Agriculture*, vol. 86, no. 2, pp. 220–225, 2006.

[16] M. A. Olszewska and P. Michel, "Antioxidant activity of inflorescences, leaves and fruits of three *Sorbus* species in relation to their polyphenolic composition," *Natural Product Research*, vol. 23, no. 16, pp. 1507–1521, 2009.

[17] P. Michel, A. Owczarek, M. Matczak et al., "Metabolite profiling of eastern teaberry (*Gaultheria procumbens* L.) lipophilic leaf extracts with hyaluronidase and lipoxygenase inhibitory activity," *Molecules*, vol. 22, no. 3, p. 412, 2017.

[18] M. A. Olszewska, A. Presler, and P. Michel, "Profiling of phenolic compounds and antioxidant activity of dry extracts from the selected *Sorbus* species," *Molecules*, vol. 17, no. 3, pp. 3093–3113, 2012.

[19] M. Matczak, A. Marchelak, P. Michel et al., "*Sorbus domestica* L. leaf extracts as functional products: phytochemical profiling, cellular safety, pro-inflammatory enzymes inhibition and protective effects against oxidative stress in vitro," *Journal of Functional Foods*, vol. 40, pp. 207–218, 2018.

[20] J. Kolodziejczyk-Czepas, P. Nowak, B. Wachowicz et al., "Antioxidant efficacy of *Kalanchoe daigremontiana* bufadienolide-rich fraction in blood plasma in vitro," *Pharmaceutical Biology*, vol. 54, no. 12, pp. 3182–3188, 2016.

[21] I. F. F. Benzie and J. J. Strain, "The ferric reducing ability of plasma (FRAP) as a measure of "antioxidant power": the FRAP assay," *Analytical Biochemistry*, vol. 239, no. 1, pp. 70–76, 1996.

[22] B. Matthaus and M. M. Özcan, "Fatty acid, tocopherol and squalene contents of Rosaceae seed oils," *Botanical Studies*, vol. 55, no. 1, p. 48, 2014.

[23] S. A. Mohamed, N. M. Sokkar, O. El-Gindi, Z. Y. Ali, and I. A. Alfishawy, "Phytoconstituents investigation, anti-diabetic and anti-dyslipidemic activities of *Cotoneaster horizontalis* Decne cultivated in Egypt," *Life Science Journal*, vol. 9, pp. 394–403, 2012.

[24] J. Lunn and H. E. Theobald, "The health effects of dietary unsaturated fatty acids," *Nutrition Bulletin*, vol. 31, no. 3, pp. 178–224, 2006.

[25] F. Anwar, R. Przybylski, M. Rudzinska, E. Gruczynska, and J. Bain, "Fatty acid, tocopherol and sterol compositions of Canadian prairie fruit seed lipids," *Journal of the American Oil Chemists' Society*, vol. 85, no. 10, pp. 953–959, 2008.

[26] P. Morales, I. C. F. R. Ferreira, A. M. Carvalho et al., "Wild edible fruits as a potential source of phytochemicals with capacity to inhibit lipid peroxidation," *European Journal of Lipid Science and Technology*, vol. 115, no. 2, pp. 176–185, 2013.

[27] E. Palme, A. R. Bilia, and I. Morelli, "Flavonols and isoflavones from *Cotoneaster simonsii*," *Phytochemistry*, vol. 42, no. 3, pp. 903–905, 1996.

[28] S. Khan, N. Riaz, Aziz-Ur-Rehman, N. Riaz, N. Afza, and A. Malik, "Isolation studies on *Cotoneaster racemiflora*," *Journal of the Chemical Society of Pakistan*, vol. 29, pp. 620–623, 2007.

[29] P. J. H. Jones and S. S. AbuMweis, "Phytosterols as functional food ingredients: linkages to cardiovascular disease and cancer," *Current Opinion in Clinical Nutrition and Metabolic Care*, vol. 12, no. 2, pp. 147–151, 2009.

[30] S. Saeidnia, A. Manayi, A. R. Gohari, and M. Abdollahi, "The story of beta-sitosterol—a review," *European Journal of Medicinal Plants*, vol. 4, no. 5, pp. 590–609, 2014.

[31] D. Vauzour, A. Rodriguez-Mateos, G. Corona, M. J. Oruna-Concha, and J. P. E. Spencer, "Polyphenols and human health: prevention of disease and mechanisms of action," *Nutrients*, vol. 2, no. 11, pp. 1106–1131, 2010.

[32] M. Locatelli, G. Zengin, A. Uysal et al., "Multicomponent pattern and biological activities of seven *Asphodeline* taxa: potential sources of natural-functional ingredients for bioactive formulations," *Journal of Enzyme Inhibition and Medicinal Chemistry*, vol. 32, no. 1, pp. 60–67, 2017.

[33] G. Zengin, S. Nithiyanantham, M. Locatelli et al., "Screening of in vitro antioxidant and enzyme inhibitory activities of different extracts from two uninvestigated wild plants: *Centranthus longiflorus* subsp. longiflorus and *Cerinthe minor* subsp. Auriculata," *European Journal of Integrative Medicine*, vol. 8, no. 3, pp. 286–292, 2016.

[34] A. Kicel, P. Michel, A. Owczarek et al., "Phenolic profile and antioxidant potential of leaves from selected *Cotoneaster* Medik. species," *Molecules*, vol. 21, no. 6, p. 688, 2016.

[35] R. Medzhitov, "Origin and physiological roles of inflammation," *Nature*, vol. 454, no. 7203, pp. 428–435, 2008.

[36] S. T. Prigge, J. C. Boyington, M. Faig, K. S. Doctor, B. J. Gaffney, and L. M. Amzel, "Structure and mechanism of lipoxygenases," *Biochimie*, vol. 79, no. 11, pp. 629–636, 1997.

[37] N. S. El-Safory, A. E. Fazary, and C. K. Lee, "Hyaluronidases, a group of glycosidases: current and future perspectives," *Carbohydrate Polymers*, vol. 81, no. 2, pp. 165–181, 2010.

[38] P. Pacher, J. S. Beckman, and L. Liaudet, "Nitric oxide and peroxynitrite in health and disease," *Physiological Reviews*, vol. 87, no. 1, pp. 315–424, 2007.

[39] C. Szabó, H. Ischiropoulos, and R. Radi, "Peroxynitrite: biochemistry, pathophysiology and development of therapeutics," *Nature Reviews Drug Discovery*, vol. 6, no. 8, pp. 662–680, 2007.

[40] S. Baba, N. Osakabe, M. Natsume, Y. Muto, T. Takizawa, and J. Terao, "In vivo comparison of the bioavailability of (+)-catechin, (−)-epicatechin and their mixture in orally administered rats," *Journal of Nutrition*, vol. 131, no. 11, pp. 2885–2891, 2001.

[41] C. Burak, V. Brüll, P. Langguth et al., "Higher plasma quercetin levels following oral administration of an onion skin extract compared with pure quercetin dihydrate in humans," *European Journal of Nutrition*, vol. 56, no. 1, pp. 343–353, 2017.

[42] J. Kolodziejczyk-Czepas, M. B. Ponczek, and P. Nowak, "Peroxynitrite and fibrinolytic system—the effects of peroxynitrite on t-PA-induced plasmin activity," *International Journal of Biological Macromolecules*, vol. 81, pp. 212–219, 2015.

[43] P. Nowak, B. Olas, E. Bald, R. Głowacki, and B. Wachowicz, "Peroxynitrite-induced changes of thiol groups in human blood platelets," *Platelets*, vol. 14, no. 6, pp. 375–379, 2003.

[44] C. Calzada, E. Vericel, and M. Lagarde, "Low concentrations of lipid hydroperoxides prime human platelet aggregation specifically via cyclo-oxygenase activation," *Biochemical Journal*, vol. 325, no. 2, pp. 495–500, 1997.

[45] N. Hogg, V. M. Darley-Usmar, A. Graham, and S. Moncada, "Peroxynitrite and atherosclerosis," *Biochemical Society Transactions*, vol. 21, no. 2, pp. 358–362, 1993.

# Photobiomodulation at Multiple Wavelengths Differentially Modulates Oxidative Stress *In Vitro* and *In Vivo*

Katia Rupel,[1] Luisa Zupin,[1] Andrea Colliva,[2] Anselmo Kamada,[3] Augusto Poropat,[1] Giulia Ottaviani,[1] Margherita Gobbo,[1] Lidia Fanfoni,[1] Rossella Gratton,[4] Massimo Santoro,[5] Roberto Di Lenarda,[1] Matteo Biasotto,[1] and Serena Zacchigna[1,2]

[1]Department of Medical, Surgical and Health Sciences, University of Trieste, 34127 Trieste, Italy
[2]Cardiovascular Biology Laboratory, International Centre for Genetic Engineering and Biotechnology (ICGEB), 34149 Trieste, Italy
[3]Department of Genetics, Federal University of Pernambuco, Recife, Brazil
[4]Institute for Maternal and Child Health, IRCCS "Burlo Garofolo", 34137 Trieste, Italy
[5]Laboratory of Angiogenesis and Redox Metabolism, Department of Biology, University of Padua, 35131 Padova, Italy

Correspondence should be addressed to Serena Zacchigna; zacchign@icgeb.org

Academic Editor: Rodrigo Franco

Photobiomodulation (PBM) is emerging as an effective strategy for the management of multiple inflammatory conditions, including oral mucositis (OM) in cancer patients who receive chemotherapy or radiotherapy. Still, the poor understanding of the mechanisms by which the light interacts with biological tissues and the heterogeneity of light sources and protocols employed worldwide significantly limits its applicability. Reactive oxygen species (ROS) are massively generated during the early phases of OM and play a major role in the pathogenesis of inflammation in general. Here, we report the results of a clinical and experimental study, aimed at evaluating the effect of laser light at different wavelengths on oxidative stress *in vivo* in oncologic patients suffering from OM and *in vitro* in two cell types abundantly present within the inflamed oral mucosa, neutrophil polymorphonuclear (PMN) granulocytes, and keratinocytes. In addition to standard ROS detection methods, we exploited a roGFP2-Orp1 genetically encoded sensor, allowing specific, quantitative, and dynamic imaging of redox events in living cells in response to oxidative stress and PBM. We found that the various wavelengths differentially modulate ROS production. In particular, the 660 nm laser light increases ROS production when applied either before or after an oxidative stimulus. In contrast, the 970 nm laser light exerted a moderate antioxidant activity both in the saliva of OM patients and in both cell types. The most marked reduction in the levels of ROS was detected in cells exposed either to the 800 nm laser light or to the combination of the three wavelengths. Overall, our study demonstrates that PBM exerts different effects on the redox state of both PMNs and keratinocytes depending on the used wavelength and prompts the validation of a multiwavelength protocol in the clinical settings.

## 1. Introduction

Chemotherapy (CT) and radiant therapy (RT) are two of the most common and widely used treatments for both solid and haematologic tumors. While advances are constantly made to improve their specificity towards cancer cells in order to limit toxic effects on surrounding tissues, both therapies often lead to debilitating side effects, such as nausea, vomiting, diarrhea, and mucositis/dermatitis. Oral mucositis (OM) is a severe inflammation of the oral and oropharyngeal mucosa, leading to the development of extensive erythema and ulcerations. OM importantly limits the quality of life of patients, as it causes oral pain, dysgeusia, dysphagia, incapacity of autonomous nutrition with the consequent need of parenteral alimentation, and narcotic analgesia. In severe cases, the anticancer therapy has to be either reduced or stopped, with a significant negative impact on overall patient prognosis [1, 2]. The prevalence of OM raises up to 90–100% in patients receiving RT for head-neck cancer malignancies, CT, or high-dose myeloablative chemotherapy for hematopoietic stem cell transplantation [3]. While the etiology and the pathogenic mechanisms of OM have not been fully elucidated yet, several studies have reported a major role of DNA damage and reactive oxygen species (ROS) during the early phases of oral and intestinal mucositis [4–7]. Thus, either limiting ROS generation or increasing their detoxification during CT/RT stands as a rational and promising strategy to prevent OM onset and/or reduce its severity [8–11].

Various approaches have been proposed to manage OM, by either reducing its severity or preventing its onset, including cryotherapy, growth factors, and anti-inflammatory drugs, with very minimal efficacy [12, 13]. Laser therapy, also known as photobiomodulation (PBM), is emerging as an effective intervention in the management of OM [14], and it was recently added to the guidelines issued by both the Mucositis Study Group of the Multinational Association of Supportive Care in Cancer and the International Society of Oral Oncology for the prevention of OM in oncologic patients [15]. Following these guidelines, we have successfully applied PBM to foster the healing of OM lesions in both adult [16–18] and pediatric [19] patients. Yet the exact mechanism by which light at certain wavelengths promotes mucosal healing, also reducing pain and inflammation, is not clearly understood [20, 21].

PBM is based on the assumption that red (600–700 nm) and near-infrared (NIR, 770–1200 nm) light at low irradiance excites specific chromophores, such as the cytochrome c oxidase, essentially located inside the mitochondria, which represents the main source for intracellular ROS generation. In vitro studies have so far provided conflicting results on the effect of PBM on ROS balance. While PBM induces a modest and dose-dependent increase in ROS production in normal cell lines, it appears to reduce ROS levels in cells previously exposed to oxidative stress [22, 23]. Whether these effects of PBM can be reproduced on various cell types and to what extent the various wavelengths are differentially able to modulate ROS production still remain open questions.

Here, we report the results of a clinical and experimental study, aimed at evaluating the effect of PBM on oxidative stress *in vivo* in oncologic patients suffering from CT/RT-induced OM and *in vitro* in keratinocytes and neutrophil polymorphonuclear (PMN) granulocytes. In addition, we exploited the roGFP2-Orp1 sensor to monitor redox changes in response to oxidative stress and laser light at different wavelengths in immortalized human keratinocytes.

## 2. Materials and Methods

*2.1. Laser Device.* A gallium arsenide (GaAs) + indium gallium aluminum arsenide phosphide (InGaAlAsP) diode laser device (class IV, K-Laser Cube series, K-Laser d.o.o., Sežana, Slovenia) was used to perform all the experiments. The laser device is associated with a programmable scanner conveniently designed to provide uniform irradiation to different multiwell plates (12, 24, and 96-well plates). Plate covers were removed during irradiation, and the emission tip was held perpendicular above the cells. The emitted light completely covered the irradiated field of each culture plate, and power was adapted to the spot size to provide the desired irradiance and fluence using an optical power meter (LaserPoint Plus+, Via Burona, 51-20090 Vimodrone, Milan, Italy). The device is able to provide 660 nm, 800 nm, and 970 nm wavelength laser light in different combinations of power and energy.

*2.2. Clinical Study.* The study was performed according to the ethical standards of the 1975 Declaration of Helsinki (7th revision, 2013) upon approval by the local ethical committee. All subjects enrolled in the study signed an informed consent to participate. A total of 10 patients affected by OM were enrolled at the Oral Medicine and Pathology Department, Ospedale Maggiore, Trieste, Italy according to the following inclusion criteria:

(a) Age between 40 and 95 years

(b) Diagnosis of a solid or haematologic malignancy undergoing CT and/or RT

(c) Presence of OM of grade 2 or 3 (CTC) related to ongoing oncological therapies

(d) Availability to undergo PBM for four consecutive days (T0, T1, T2, T3, and one follow-up recall, CTRL)

The exclusion criteria are the following:

(a) OM previously treated by PBM

(b) Application of topical oral medication

(c) Systemic antioxidant therapy

A schematic representation of the study design is shown in Figure 1(a). During the first visit on day 0 (T0), experienced clinicians registered information about patients' clinical history and scored OM severity according to the common toxicity criteria (CTC) scale (WHO, 1976), considering ulceration and erythema distribution and size. A 0–10 visual analogue scale (VAS) was employed to quantify subjective parameters, including pain, difficulties in swallowing, speaking, and chewing. The patients were treated with PBM at T0, T1, T2, and T3 using a previously optimized protocol ($\lambda$ 970 nm, 200 mW/cm$^2$, 6 J/cm$^2$, in continuous wave) [18, 24]. Both patients and operators wore protective glasses during the treatment to avoid possible eye damage. PBM was provided using a rotatory motion all over the oral cavity to cover both ulcerated and healthy areas, keeping a 3 cm distance between the laser probe and the tissue. Irradiation time was calculated considering a mean oral mucosa surface area of 215 cm$^2$ [25].

Unstimulated saliva samples were collected for 5 minutes before and after each PBM session (T0–T4) and once on day 5 (CTRL). Patients were asked to fast for at least 2 hours before sampling, avoiding tooth brushing, excess alcohol intake, and physical activity since the previous evening. Patients rinsed the mouth with water for 1 minute and then spit saliva for 5 minutes in a sterile collection tube. Samples were stored at −20°C until analysis.

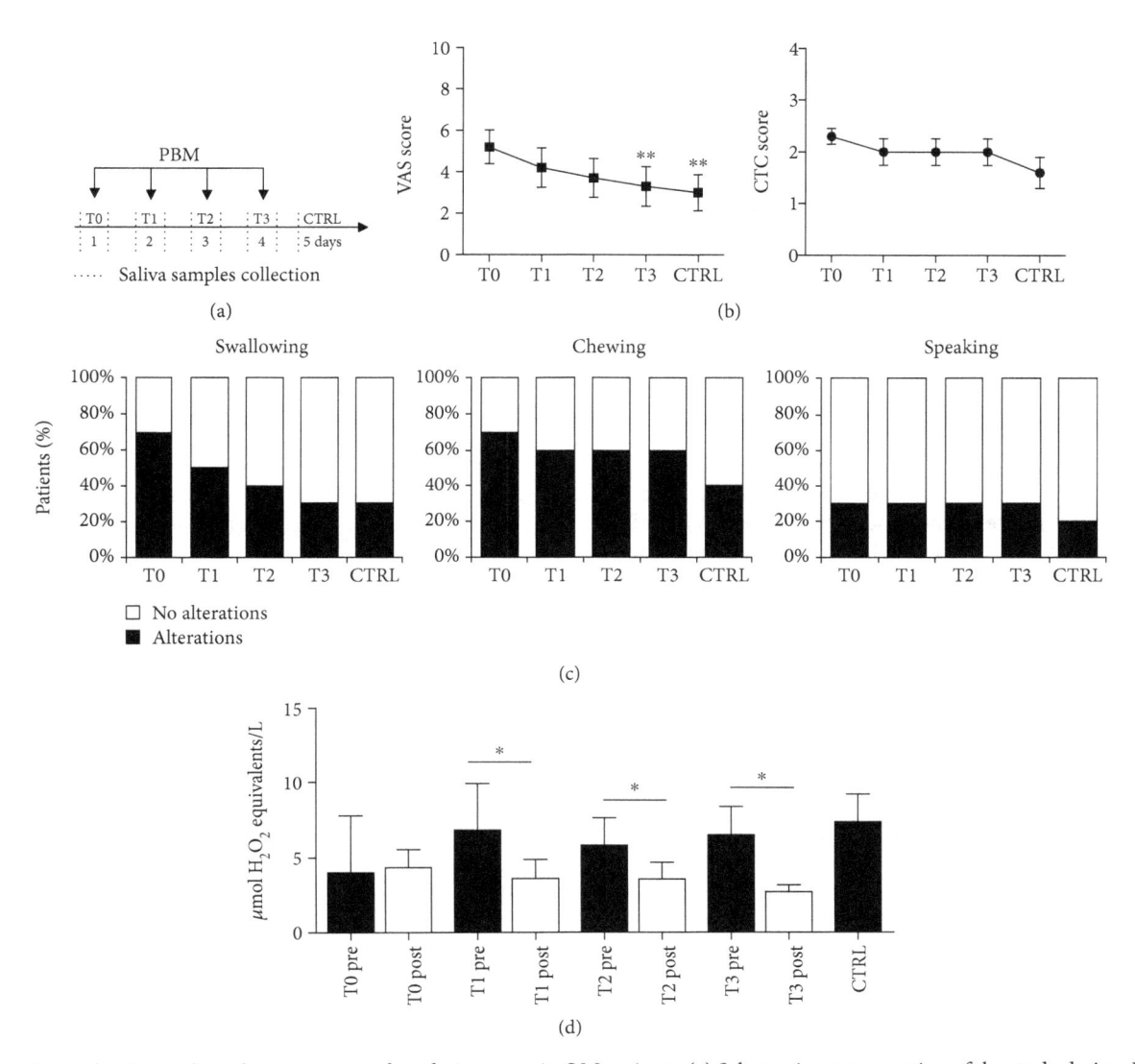

FIGURE 1: Effect of PBM on clinical parameters and oxidative stress in OM patients. (a) Schematic representation of the study design. Enrolled patients were treated with PBM for 4 consecutive days followed by a control session on day 5. Saliva samples were collected on treatment days before and after each PBM session and once on day 5. (b) Evaluation of OM severity by VAS score (left panel) and CTC score (right panel) over time. Both parameters significantly decreased over time (Friedman's test $p = 0.0003$ for VAS and $p = 0.0034$ for CTC). **Dunn's multiple comparison test $p < 0.001$ compared to T0. (c) Percentage of patients indicating the presence (black bars) or absence (white bars) of either pain or functional alterations in swallowing, chewing, or speaking over time. (d) TOS levels in the patient saliva at the indicated time points. Data are the means ± SD. *Mann–Whitney $U$ test $p < 0.05$.

### 2.3. TOS Assay

Oxidative stress status was measured in saliva at each time point using the total oxidant status (TOS) method [26]. This method is based on the principle that oxidant species present in the saliva oxidize the ferrous ion-o-dianisidine complex to ferric ion, creating a coloured complex with xylenol orange in an acidic medium. In each well of a 96 multiwell plate, centrifuged saliva (35 μL) was added to 225 μL of reagent 1 (xylenol orange 150 μM, NaCl 140 mM, and glycerol 1.35 M in 25 mM $H_2SO_4$ solution, pH 1.75). Subsequently, 11 μL of reagent 2 (ferrous ion 5 mM and o-dianisidine 10 mM in 25 mM $H_2SO_4$ solution) was added. The solutions were gently mixed for 5 minutes, and then the absorbance was measured with a multiwell reader spectrophotometer (Glomax multi+ detection system, Promega, Italy) at $OD_{560}$. The absorbance measured before mixing reagent 1 and reagent 2 was used as a sample blank. The assay was calibrated using standard aqueous solutions of hydrogen peroxide ($H_2O_2$). TOS values are reported as the mean of 4 measurements ± standard deviation and are expressed as micromolar hydrogen peroxide equivalent per liter (μmol $H_2O_2$ Equiv/L).

### 2.4. Cells

Human keratinocytes (HaCaT) were maintained in DMEM culture medium supplemented with 10% fetal bovine serum, 100 U/mL penicillin/streptomycin, and 2 mM glutamine (Euroclone, Pero, Milan, Italy). Cells were seeded one day prior to the experiment (10,000 cells/well in 96 multiwell plates and 50,000 cells/well in 24 multiwell plates), avoiding the use of cells over the 10th passage.

*2.5. Evaluation of Intracellular ROS Production and Kinetics in PMN.* PMN isolation from the venous whole blood of 5 healthy volunteers was performed using a Ficoll gradient (Ficoll-Paque Plus, GE Healthcare) through a centrifugation of $400 \times g$ for 30 minutes. Erythrocytes were lysed with a buffered ammonium chloride solution (ammonium [Tris (1.7 mM)-NH4Cl (16 mM)]). The PMNs were washed in HANKS balanced salt solution ($150 \times g$ for 10 minutes) and seeded at a concentration of $10^6$ cells/mL in RPMI-1640 medium (Sigma-Aldrich) in 96-well plates (100 μL in each well).

After isolation, PMNs were either stimulated with 100 ng/mL LPS (lipopolysaccharide from *Escherichia coli* 055:B5) for 15 minutes or not and then exposed to laser light at either 970 nm ($\lambda$ 970 nm, irradiance 200 mW/cm$^2$, fluence 6 J/cm$^2$) or 660 nm ($\lambda$ 660 nm, irradiance 50 mW/cm$^2$, fluence 3 J/cm$^2$). The cells were then incubated for $10'$ with $2'$ $7'$ dichlorofluorescein diacetate (DCFH-DA, Sigma-Aldrich, Saint Louis, Missouri, USA), and OD$_{529}$ emission was measured using a spectrophotometer (Envision plate reader, PerkinElmer, Waltham, Massachusetts, USA) for 2 hours (1 read/minute). ROS levels are reported as the mean of 4 measurements ± standard deviation.

*2.6. Real-Time Quantification of Oxidative Stress Using Fluorescent Protein-Based Redox Probes in Keratinocytes.* HaCaT cells were seeded three days before the experiment in 8-well Ibidi μ-slides (100,000 cells/well). The following day, cells were transfected using lipofectamine 2000 with a plasmid expressing roGFP2-Orp1 [27]. After 48 hours, the cells were treated with different PBM protocols [24] and imaged at a Leica LSM 810 confocal microscope.

The following laser protocols were applied, either individually or in combination: (i) $\lambda$ 660 nm, irradiance 50 mW/cm$^2$, fluence 3 J/cm$^2$, continuous wave; (ii) $\lambda$ 800 nm, irradiance 200 mW/cm$^2$, fluence 6 J/cm$^2$, continuous wave; and (iii) $\lambda$ 970 nm, irradiance 200 mW/cm$^2$, fluence 6 J/cm$^2$, continuous wave. An oxidative stimulus (0.5 mM $H_2O_2$) was added either 10 seconds prior to the laser light or immediately after the laser light. Cells were imaged for 240 seconds (12 frames every 20 seconds). The ratio between the fluorescence excited at 408 nm and 488 nm was used as a surrogate indicator of the redox status of the cells, with an expected increase in the 408/488 nm ratio in the presence of oxidative stress. Images were analyzed using the ImageJ software. ROS levels are reported as the mean of 4 measurements ± standard deviation.

*2.7. Keratinocyte Viability and Oxidative Stress Assay following 5-FU.* 5-FU 0.1 mg/mL (F6627, Sigma-Aldrich, Saint Louis, Missouri, USA) was added to HaCaT cells for 18 hours, followed by PBM using combined laser wavelengths ($\lambda$ 660 nm, irradiance 50 mW/cm$^2$, fluence 3 J/cm$^2$; $\lambda$ 800 nm, irradiance 200 mW/cm$^2$, fluence 6 J/cm$^2$; and $\lambda$ 970 nm, irradiance 200 mW/cm$^2$, fluence 6 J/cm$^2$). After additional 24 hours, the MTT assay was performed following the manufacturer's instructions (Trevigen, Gaithersburg, Maryland, USA). Results are expressed as % OD$_{570}$ absorbance relative to untreated cells. ROS production was measured 30 minutes after PBM using the cell-permeant $2',7'$-dichlorodihydrofluorescein diacetate ($H_2$DCFDA) dye (D399, Invitrogen, Thermo Fisher Scientific, Waltham, Massachusetts, USA), and the obtained fluorescence was normalized on living cells. The results are reported as the mean of 8 measurements ± standard deviation.

*2.8. Gene Expression Analysis.* Total RNA was extracted 30 minutes after PBM using Eurogold Trifast reagent (EMR507100, Euroclone, Pero, Milan, Italy) following the manufacturer's instructions and retrotranscribed using the High-Capacity cDNA Reverse transcription kit (Thermo Fisher Scientific, Waltham, Massachusetts, USA). TaqmanTM probes for HMOX1 (hs00167309_m1) and SOD2 (hs01110250_m1) genes and β-actin (as calibrator and reference, ACTB: Hs99999903_m1) were used to quantify the relative mRNAs using an Applied Biosystems 7900HT Fast Real-Time PCR System (Thermo Fisher Scientific, Waltham, Massachusetts, USA) platform. Data were analyzed using the Relative Quantification manager software (Thermo Fisher Scientific, Waltham, Massachusetts, USA) using untreated cells as reference. The experiment was performed in three replicates including duplicate wells, and the results are expressed as the mean ± standard deviation.

*2.9. Statistical Analysis.* The Prism 6.0 software (GraphPad Software, La Jolla, California, USA) was used to perform statistical analysis. All statistical assessments were two-sided, and a $p$ value < 0.05 was used for the rejection of the null hypothesis. Friedman's test was employed to test the significance of changes over time in VAS and CTC scores, while Dunn's multiple comparison test was used as post hoc to compare each pair of time points. Pearson's nonparametric correlation test was used to evaluate the relation between TOS levels and VAS or CTC scores at each time point. The Mann–Whitney $U$ test was employed to evaluate differences between treated and untreated samples. The Mann–Whitney $U$ test was performed to determine gene expression differences among samples. Linear regression analysis was employed to evaluate differences in PMN ROS levels among groups and over time. Two-way ANOVA was applied to determine the significance of differences between the curves representing oxidative status in keratinocytes with genetically encoded redox probes treated with $H_2O_2$ with and without PBM.

## 3. Results

*3.1. PBM Improves Clinical Parameters in OM Patients and Determines a Transient Reduction in the Oxidant Status of the Saliva.* Baseline demographic and clinical characteristics of the 10 patients suffering from OM who completed the study are reported in Table 1. PBM was performed daily for 4 consecutive days (T0 to T3, Figure 1(a)) and well tolerated by all patients, without any adverse event. Effectiveness of

Table 1: Selected baseline characteristics of the patients enrolled in the study.

| Patient | Age | Gender | Malignancy | Anticancer therapy | VAS score at T0 | CTC score at T0 |
|---|---|---|---|---|---|---|
| 1 | 58 | M | Gastrointestinal | CT | 2 | 2 |
| 2 | 58 | F | Breast | CT | 2 | 2 |
| 3 | 62 | M | Head neck | CT and RT | 8 | 2 |
| 4 | 92 | F | Head neck | RT | 8 | 2 |
| 5 | 74 | M | Haematological | CT | 7 | 3 |
| 6 | 56 | M | Head neck | CT | 5 | 2 |
| 7 | 65 | M | Gastrointestinal | CT | 6 | 3 |
| 8 | 44 | M | Haematological | CT | 2 | 2 |
| 9 | 69 | M | Head neck | CT and RT | 4 | 2 |
| 10 | 71 | M | Head neck | RT | 8 | 3 |

PBM was confirmed by the progressive decrease of both the VAS and the CTC scores over time, as reported in Figure 1(b) (Friedman's test $p = 0.0003$ for VAS and $p = 0.0034$ for CTC). The number of patients reporting discomfort in subjective parameters (swallowing, chewing and speaking) also decreased over time following PBM (Figure 1(c)).

In the same patients, we analyzed the total oxidant status (TOS) in the saliva before and after each PBM session, as well as the day following the last session (CTRL). TOS level was significantly correlated to the VAS score at T3 and T4 (Pearson's nonparametric correlation test $p < 0.05$), but not to the CTC score. While the oxidative stress did not change after PBM at T0, starting from T1, ROS levels decreased promptly after each PBM session (Mann–Whitney $U$ test $p < 0.01$ for T1, T2, and T3). Despite this constant trend, ROS levels increased again during the following 24 hours, as well as at day 5 (CTRL). Thus, PBM transiently reduces oxidative stress after treatment, but this effect lasts less than 24 hours.

### 3.2. PBM at Different Wavelengths Modulate ROS Production in Both Stimulated and Unstimulated PMNs.
As PMNs are known to be a major source of ROS in inflammatory conditions, including OM, we investigated the potential of PBM, set at two different wavelengths commonly employed in the clinics (970 nm and 660 nm), to modulate ROS production in unstimulated and LPS-stimulated PMNs. In unstimulated PMNs (Figure 2(a)), the two wavelengths had opposite effects. While the 970 nm protocol significantly reduced ROS levels (linear regression $p = 0.023$ and $p < 0.0001$ after 90 and 120 minutes, respectively), the 660 nm protocol increased ROS levels at 2 hours after PBM (linear regression $p < 0.0001$). As expected, stimulation of PMNs with LPS resulted in increased ROS production in all experimental groups (Figure 2(b)). In line with the results obtained with unstimulated PMNs, we observed a reduction in ROS levels upon irradiation with the 970 nm protocol, whereas the 660 nm did not exert any effect.

### 3.3. Dynamic Real-Time Detection of PBM-Induced Redox Changes in Keratinocytes Using Genetically Encoded Fluorescent Sensors.
Compared to standard ROS detection methods, based on the use of chemical indicators, genetically encoded probes, exploiting redox-sensitive fluorescent proteins, allow specific, quantitative, and dynamic imaging of redox events in living cells [27]. To monitor redox changes in keratinocytes upon PBM, we employed a genetically encoded probe based on the redox-active green fluorescent protein 2 (roGFP2) fused to Orp1, a highly sensitive thiol peroxidase that is oxidized by $H_2O_2$ [28]. By introducing the roGFP2-Orp1 probe in keratinocytes, we could monitor any change in the intracellular redox state upon exposure to PBM either in the absence or in the presence of $H_2O_2$-induced oxidative stress. This time we compared the effect of the three wavelengths (660 nm, 800 nm, and 970 nm) that have been so far considered for the treatment of OM and other inflammatory conditions.

First, we determined whether PBM induces any change in the redox state of cells at baseline and found that any of the three tested laser protocols, and not even their combination, exert a significant effect (Figures 3(a)–3(d)). As expected, the addition of $H_2O_2$ determined a rapid increase in ROS levels (Figure 3(e)). Whereas the 660 nm protocol did not induce any change in ROS production ($p = $ ns), the 970 nm protocol and, even more evident, the 800 nm protocol significantly reduced ROS levels ($p < 0.01$ and $0.001$, respectively; Figures 3(a)–3(c)). Of notice, the simultaneous delivery of the three wavelengths was also very effective in reducing the oxidative status upon exposure to $H_2O_2$ ($p < 0.01$; Figure 3(d)). We also tested whether PBM was able to modulate ROS levels when applied to cells previously exposed to $H_2O_2$. In agreement with the results obtained in PMNs, the 660 nm protocol further increased ROS levels ($p < 0.01$; Figure 3(f)). In contrast, both the 800 nm and the combined protocols significantly reduced ROS production also in this condition ($p < 0.01$ and $p < 0.05$, respectively; Figures 3(g)–3(i)).

### 3.4. PBM Improves Keratinocyte Resistance to Oxidative Stress Induced by 5-Fluorouracil In Vitro.
Based on the promising results obtained using the multiwavelength PBM protocol, we wanted to confirm its actual capacity to reduce oxidative stress in keratinocytes treated with 5-fluorouracil (5-FU) treatment, partially mimicking the scenario of OM patients. Indeed, 5-FU is widely used in patients affected by different solid tumors, including five patients enrolled in

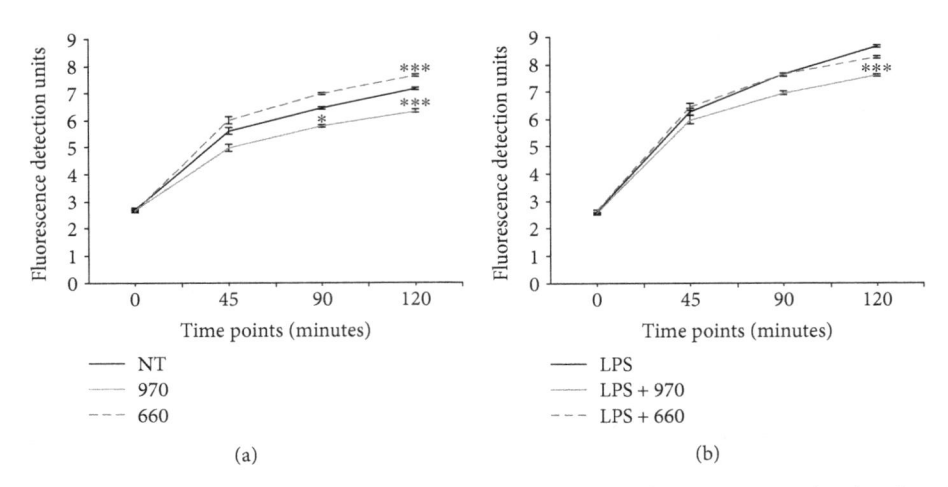

FIGURE 2: Effect of PBM on intracellular ROS production in unstimulated PMNs and in PMNs stimulated with LPS. (a) Monitoring of fluorescence detection units at $OD_{529}$ over time in unstimulated PMNs. NT: not treated, 970: treated with 970 nm laser light, and 660: treated with 660 nm laser light. *Linear regression analysis $p < 0.05$ compared to NT. ***Linear regression analysis $p < 0.0001$ compared to NT. (b) Monitoring of fluorescence detection units at $OD_{529}$ over time in PMNs stimulated with LPS. NT: not treated, 970: treated with 970 nm laser light, and 660: treated with 660 nm laser light. ***Linear regression analysis $p < 0.0001$ compared to NT.

our study. 5-FU acts as an inducing cell apoptosis through several mechanisms, including the generation of mitochondrial oxidative stress [29]. We treated human keratinocytes with 5-FU for 18 hours and found a significant rate of cell death compared to untreated cells (Mann–Whitney $U$ test $p < 0.001$), as represented in Figure 4(a). When the same cells were exposed to PBM after 5-FU treatment, cell viability was significantly higher (Mann–Whitney $U$ test $p < 0.001$).

Besides cell death, 5-FU also induced ROS production in human keratinocytes (Mann–Whitney $U$ test $p = 0.02$), as assessed using the $H_2DCFDA$ dye, while PBM significantly lowered the level of intracellular ROS in both 5-FU treated and untreated cells (Mann–Whitney $U$ test $p = 0.048$ and $p = 0.015$, respectively; Figure 4(b)).

We also assessed the expression levels of HMOX1 and SOD2, two enzymes playing a key role in the cellular response to oxidative stress, and consistently found that 5-FU upregulated both enzymes as expected, whereas PBM markedly decreased their expression (Mann–Whitney $U$ test $p = 0.03$ for both genes), as shown in Figures 4(c) and 4(d). Similar to what was observed for ROS production, exposure of the cells to PBM resulted in a significant downregulation of HMOX1 and SOD2, even in cells treated with 5-FU (Mann–Whitney $U$ test $p = 0.03$ for both genes, Figures 4(c) and 4(d)).

## 4. Discussion

Despite increasing evidence of its effectiveness in the treatment of OM and other inflammatory conditions, PBM still does not stand as a universally recognized and accepted therapy, essentially because of the variety of light sources and irradiation protocols employed in different trials in terms of wavelength (usually in the range of 600–1100 nm), irradiance, and fluence. Therefore, a major effort is needed to understand whether the different wavelengths exert specific biological effects and to establish the optimal parameters to be used to achieve the best therapeutic activity in each condition.

This work contributes to establishing the capacity of laser light to modulate oxidative stress, which is crucial in OM onset and progression. We report our clinical experience on cancer patients suffering from OM induced by either CT or RT. These patients were treated with a protocol optimized on the base of our previous clinical and experimental work [18, 24], using 970 nm laser light. Consistent with previous studies, this treatment was effective in improving all clinical parameters, starting from the fifth day of PBM. We then moved to assess the levels of ROS in the saliva of the same patients and observed a marked antioxidant activity exerted by each PBM session. However, this drop in ROS level was transient and not maintained during the 24 hours of posttreatment.

To better understand the cellular mechanism and the dynamics underlying this effect, with the ultimate goal of optimizing our clinical protocol, we wanted to assess the effect of multiple wavelengths of laser light on two different cell types, reasonably representing the major sources of ROS in OM, PMNs, and keratinocytes.

First, we evaluated the effect of red (660 nm) and near-infrared (970 nm) laser light on PMNs, either in resting conditions or upon stimulation with LPS and observed an opposite response. While the 970 nm light reduced ROS production in both conditions, the 660 nm light increased ROS levels in unstimulated PMNs and did not exert any effect in LPS-stimulated cells. Our result is also in accordance with the work by Cerdeira et al. [30], in which neutrophil irradiation with 660 nm laser light resulted in a significant increase in their respiratory burst, associated with intra- and extracellular superoxide radical production and improved fungicidal activity against *Candida albicans*. Our data add an important piece of evidence, showing that different wavelengths exert specific and even opposite effects on ROS production. While a definitive explanation for these differences is still missing, it

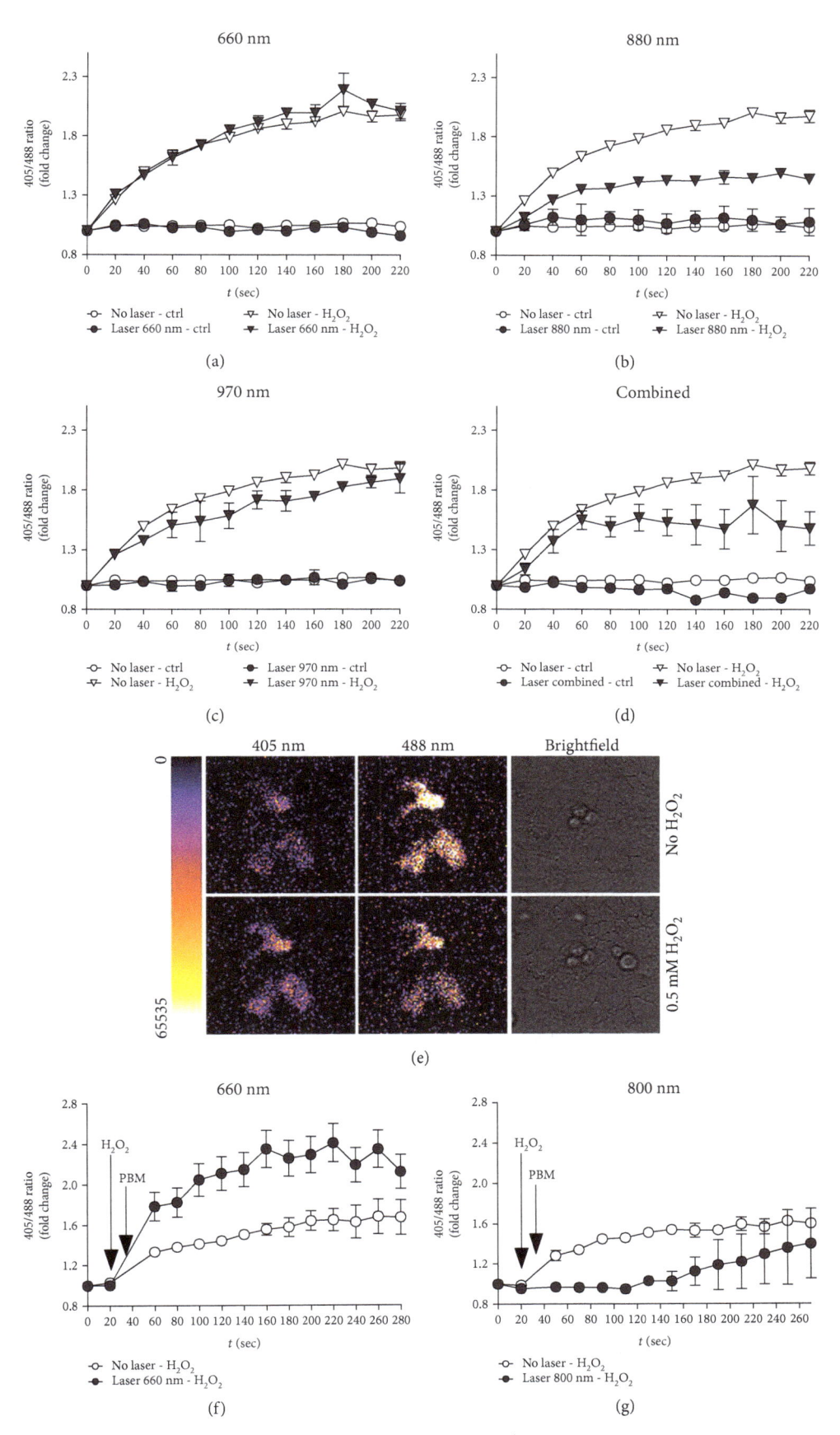

FIGURE 3: Continued.

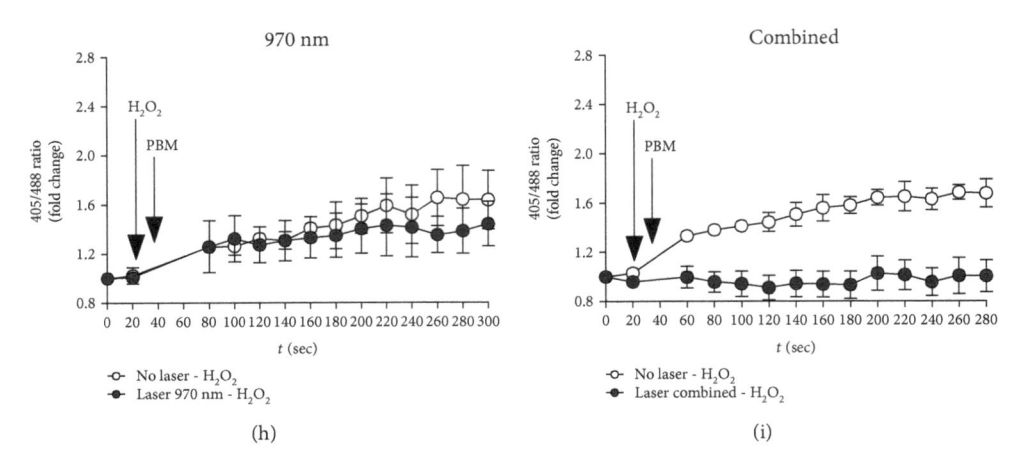

(h) 970 nm — plot of 405/488 ratio (fold change) vs t (sec), arrows at H$_2$O$_2$ and PBM; curves: No laser - H$_2$O$_2$ (open circles), Laser 970 nm - H$_2$O$_2$ (filled circles).

(i) Combined — plot of 405/488 ratio (fold change) vs t (sec), arrows at H$_2$O$_2$ and PBM; curves: No laser - H$_2$O$_2$ (open circles), Laser combined - H$_2$O$_2$ (filled circles).

FIGURE 3: Real-time evaluation of the effect of PBM on the redox status in HaCaT cells using genetically encoded fluorescent sensors. (a–d) Cells were treated with PBM at the indicated wavelength (660 nm in (a), 800 nm in (b), and 970 nm in (c) and the combination of the three wavelengths in (d)) and subsequently exposed to 0.5 mM H$_2$O$_2$. Measurement of fluorescence started immediately after exposure to oxidative stress. Data are the means ± SD. Signals recorded in treated cells and in cells treated only with PBM are also plotted. (e) Representative images of the fluorescence intensity at 405 nm (left) and 488 nm (center) and transmitted light (right) of the same cells at baseline (upper raw) and upon treatment with 0.5 mM H$_2$O$_2$ (lower raw). (f–i) Cells were first treated with 0.5 mM H$_2$O$_2$ and subsequently exposed to PBM at the indicated wavelength (660 nm in (a), 800 nm in (b), and 970 nm in (c) and the combination of the three wavelengths in (d)). Measurement of fluorescence started 20 seconds prior to the exposure to oxidative stress. Data are the means ± SD.

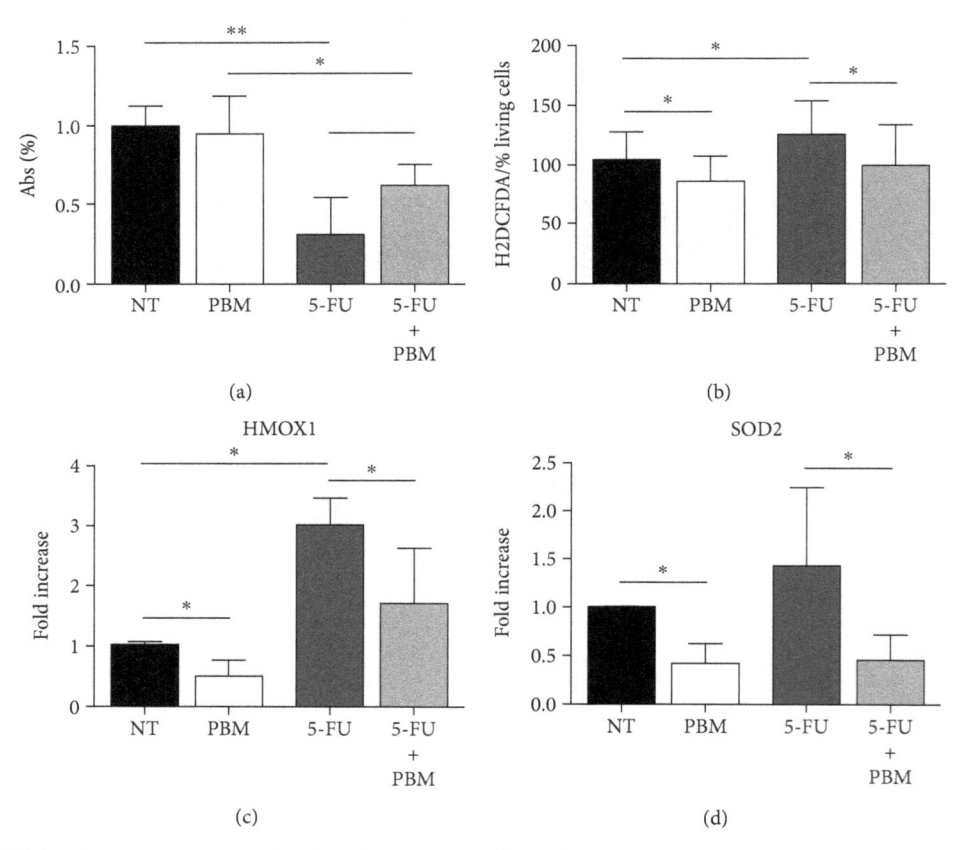

(a) Abs (%) bar chart: NT, PBM, 5-FU, 5-FU + PBM. ** between NT and 5-FU; * between PBM and 5-FU, and between 5-FU and 5-FU+PBM.

(b) H2DCFDA/% living cells bar chart: NT, PBM, 5-FU, 5-FU + PBM. * between NT and PBM; * between NT and 5-FU; * between 5-FU and 5-FU+PBM.

(c) HMOX1 Fold increase bar chart: NT, PBM, 5-FU, 5-FU + PBM. * between NT and PBM; * between NT and 5-FU; * between 5-FU and 5-FU+PBM.

(d) SOD2 Fold increase bar chart: NT, PBM, 5-FU, 5-FU + PBM. * between NT and PBM; * between 5-FU and 5-FU+PBM.

FIGURE 4: Effect of PBM on keratinocyte survival and oxidative stress induced by exposure to 5-FU (0.1 mg/ml). (a) MTT assay showing the percentage of living cells in the absence of any treatment (not treated, NT) and after exposure to either PBM, 5-FU, or their combination. (b) ROS production (normalized to the percentage of living cells) in the absence of any treatment (not treated, NT) and after exposure to either PBM, 5-FU, or their combination. (c) Levels of HMOX1 gene expression in the absence of any treatment (not treated, NT) and after exposure to either PBM, 5-FU, or their combination. (d) Levels of SOD2 gene expression in the absence of any treatment (not treated, NT) and after exposure to either PBM, 5-FU, or their combination. *Mann–Whitney $U$ test $p < 0.05$; **Mann–Whitney $U$ test $p < 0.01$. Data are the means ± SD.

might depend on the fact that intracellular chromophores, and in particular the cytochrome c oxidase, which is considered the main target of PBM, change their absorption spectra depending on their oxidation state [31].

This implies that PBM cannot be considered a single therapeutic entity and that the optimal wavelength can be chosen not only on the base of the depth of tissue penetration but also considering the biological activity exerted on the irradiated cells. This is particularly relevant considering that neutrophils of patients undergoing CT usually present a defective capacity in ROS generation, with reduced production of IL-1$\beta$ and overall impaired antimicrobial function [32–34]. Thus, the use of a laser wavelength able to stimulate ROS production in PMNs may represent a powerful tool to boost their ability to respond to infections in immunocompromised patients.

To understand whether the same effects are also exerted on keratinocytes, we exploited genetically encoded redox biosensors to monitor ROS production in real time in response to PBM, this time also testing an additional wavelength (800 nm), which is being progressively used to reach inflamed tissues in depth, as this wavelength is poorly absorbed by water. We confirmed that the 660 nm laser light increases ROS production when applied either before or after an oxidative stimulus. More importantly, we found that the 970 nm laser light exerted a moderate antioxidant activity, whereas a striking reduction in the levels of ROS was detected in cells exposed to either the 800 nm laser light or the combination of the three different wavelengths.

Since this multiwavelength PBM protocol could represent a promising therapeutic tool to be introduced into the clinics, we wanted to confirm its actual capacity to reduce oxidative stress in keratinocytes treated with 5-FU treatment, partially mimicking the scenario of OM patients. We found that this combined PBM protocol reduced ROS levels and ROS-induced genes in both untreated and 5-FU-treated cells.

These results differ partially from those obtained using single wavelengths, in which PBM seems to decrease ROS levels in stressed or injured tissues, while increasing them in healthy ones [35]. For instance, Tatmatsu-Rocha et al. showed that the 904 nm laser light reduces oxidative stress markers in the wounded skin of diabetic mice, whereas it increased them in irradiated controls [36]. In primary cortical neurons, PBM at 810 nm increased ROS levels in basal conditions, but it reduced ROS when the neurons were treated with oxidant reagents [23]. The same wavelength was reported to lead to ROS increment in healthy murine embryonic fibroblast through the activation of nuclear factor kappa B (NF-$\kappa$B) [37]. Additional variability could depend on the timing at which ROS levels are assessed following PBM. While in healthy cells our combined PBM protocol did not induce any significant change in ROS levels during the first 5 minutes, as assessed using the fluorescent sensors, it was effective in reducing both ROS generation and ROS-induced gene expression after 18 hours, as determined by the H2DCFDA assay and real-time PCR.

These findings suggest that a combined multiwavelength irradiation protocol may be considered, to reach tissues located at different depths and exploit the different characteristics of single wavelengths, i.e., increasing ROS levels in neutrophils by the 660 nm laser light and reducing the oxidative stress in keratinocytes by the 800 and 970 nm light. Future studies using both cellular and animal models will further confirm the rationale and efficacy of this approach.

## 5. Conclusions

Overall, our study demonstrates that PBM exerts different effects on the redox state of both PMNs and keratinocytes, depending on the wavelength, and prompts the validation of a multiwavelength protocol in the clinical settings.

## Authors' Contributions

Katia Rupel and Luisa Zupin contributed equally to this work.

## Acknowledgments

This work was supported by the intramural grant of the International Centre for Genetic Engineering and Biotechnology and grant AIRC IG 2016 19032 to SZ.

## References

[1] S. T. Sonis, "Mucositis: the impact, biology and therapeutic opportunities of oral mucositis," *Oral Oncology*, vol. 45, no. 12, pp. 1015–1020, 2009.

[2] S. T. Sonis, L. S. Elting, D. Keefe et al., "Perspectives on cancer therapy-induced mucosal injury: pathogenesis, measurement, epidemiology, and consequences for patients," *Cancer*, vol. 100, no. S9, pp. 1995–2025, 2004.

[3] A. Trotti, L. A. Bellm, J. B. Epstein et al., "Mucositis incidence, severity and associated outcomes in patients with head and neck cancer receiving radiotherapy with or without chemotherapy: a systematic literature review," *Radiotherapy and Oncology*, vol. 66, no. 3, pp. 253–262, 2003.

[4] R. W. Moss, "Do antioxidants interfere with radiation therapy for cancer?," *Integrative Cancer Therapies*, vol. 6, no. 3, pp. 281–292, 2007.

[5] F. Yoshino, A. Yoshida, A. Nakajima, S. Wada-Takahashi, S. S. Takahashi, and M. C. Lee, "Alteration of the redox state with reactive oxygen species for 5-fluorouracil-induced oral mucositis in hamsters," *PLoS One*, vol. 8, no. 12, article e82834, 2013.

[6] K. Rtibi, S. Selmi, D. Grami, M. Amri, H. Sebai, and L. Marzouki, "Contribution of oxidative stress in acute intestinal mucositis induced by 5 fluorouracil (5-FU) and its prodrug capecitabine in rats," *Toxicology Mechanisms and Methods*, vol. 28, no. 4, pp. 262–267, 2018.

[7] E. G. Russi, J. E. Raber-Durlacher, and S. T. Sonis, "Local and systemic pathogenesis and consequences of regimen-induced

inflammatory responses in patients with head and neck cancer receiving chemoradiation," *Mediators of Inflammation*, vol. 2014, Article ID 518261, 14 pages, 2014.

[8] K. I. Block, A. C. Koch, M. N. Mead, P. K. Tothy, R. A. Newman, and C. Gyllenhaal, "Impact of antioxidant supplementation on chemotherapeutic efficacy: a systematic review of the evidence from randomized controlled trials," *Cancer Treatment Reviews*, vol. 33, no. 5, pp. 407–418, 2007.

[9] P. Urbain, A. Raynor, H. Bertz, C. Lambert, and H. K. Biesalski, "Role of antioxidants in buccal mucosa cells and plasma on the incidence and severity of oral mucositis after allogeneic haematopoietic cell transplantation," *Support Care Cancer*, vol. 20, no. 8, pp. 1831–1838, 2012.

[10] L. de Freitas Cuba, F. G. Salum, K. Cherubini, and M. A. de Figueiredo, "Antioxidant agents: a future alternative approach in the prevention and treatment of radiation-induced oral mucositis?," *Alternative Therapies in Health and Medicine*, vol. 21, no. 2, pp. 36–41, 2015.

[11] Z. Shen, J. Wang, Q. Huang et al., "Genetic modification to induce CXCR2 overexpression in mesenchymal stem cells enhances treatment benefits in radiation-induced oral mucositis," *Cell Death & Disease*, vol. 9, no. 2, p. 229, 2018.

[12] S. Kanuga, "Cryotherapy and keratinocyte growth factor may be beneficial in preventing oral mucositis in patients with cancer, and sucralfate is effective in reducing its severity," *Journal of the American Dental Association (1939)*, vol. 144, no. 8, pp. 928-929, 2013.

[13] R. V. Lalla, G. B. Gordon, M. Schubert et al., "A randomized, double-blind, placebo-controlled trial of misoprostol for oral mucositis secondary to high-dose chemotherapy," *Supportive Care in Cancer*, vol. 20, no. 8, pp. 1797–1804, 2012.

[14] S. Elad, P. Arany, R. J. Bensadoun, J. B. Epstein, A. Barasch, and J. Raber-Durlacher, "Photobiomodulation therapy in the management of oral mucositis: search for the optimal clinical treatment parameters," *Support Care Cancer*, vol. 26, no. 10, pp. 3319–3321, 2018.

[15] R. V. Lalla, J. Bowen, A. Barasch et al., "MASCC/ISOO clinical practice guidelines for the management of mucositis secondary to cancer therapy," *Cancer*, vol. 120, no. 10, pp. 1453–1461, 2014.

[16] M. Gobbo, G. Ottaviani, G. Mustacchi, R. di Lenarda, and M. Biasotto, "Acneiform rash due to epidermal growth factor receptor inhibitors: high-level laser therapy as an innovative approach," *Lasers in Medical Science*, vol. 27, no. 5, pp. 1085–1090, 2012.

[17] M. Gobbo, G. Ottaviani, G. Perinetti et al., "Evaluation of nutritional status in head and neck radio-treated patients affected by oral mucositis: efficacy of class IV laser therapy," *Support Care Cancer*, vol. 22, no. 7, pp. 1851–1856, 2014.

[18] G. Ottaviani, M. Gobbo, M. Sturnega et al., "Effect of class IV laser therapy on chemotherapy-induced oral mucositis: a clinical and experimental study," *The American Journal of Pathology*, vol. 183, no. 6, pp. 1747–1757, 2013.

[19] M. Chermetz, M. Gobbo, L. Ronfani et al., "Class IV laser therapy as treatment for chemotherapy-induced oral mucositis in onco-haematological paediatric patients: a prospective study," *International Journal of Paediatric Dentistry*, vol. 24, no. 6, pp. 441–449, 2014.

[20] M. R. Hamblin, "Mechanisms and mitochondrial redox signaling in photobiomodulation," *Photochemistry and Photobiology*, vol. 94, no. 2, pp. 199–212, 2018.

[21] L. F. de Freitas and M. R. Hamblin, "Proposed mechanisms of photobiomodulation or low-level light therapy," *IEEE Journal of Selected Topics in Quantum Electronics*, vol. 22, no. 3, pp. 348–364, 2016.

[22] S. K. Sharma, G. B. Kharkwal, M. Sajo et al., "Dose response effects of 810 nm laser light on mouse primary cortical neurons," *Lasers in Surgery and Medicine*, vol. 43, no. 8, pp. 851–859, 2011.

[23] Y.-Y. Huang, K. Nagata, C. E. Tedford, T. McCarthy, and M. R. Hamblin, "Low-level laser therapy (LLLT) reduces oxidative stress in primary cortical neurons in vitro," *Journal of Biophotonics*, vol. 6, 2013.

[24] G. Ottaviani, V. Martinelli, K. Rupel et al., "Laser therapy inhibits tumor growth in mice by promoting immune surveillance and vessel normalization," *eBioMedicine*, vol. 11, pp. 165–172, 2016.

[25] L. M. C. Collins and C. Dawes, "The surface area of the adult human mouth and thickness of the salivary film covering the teeth and oral mucosa," *Journal of Dental Research*, vol. 66, no. 8, pp. 1300–1302, 1987.

[26] O. Erel, "A new automated colorimetric method for measuring total oxidant status," *Clinical Biochemistry*, vol. 38, no. 12, pp. 1103–1111, 2005.

[27] E. Panieri, C. Millia, and M. M. Santoro, "Real-time quantification of subcellular $H_2O_2$ and glutathione redox potential in living cardiovascular tissues," *Free Radical Biology & Medicine*, vol. 109, pp. 189–200, 2017.

[28] B. Morgan, M. C. Sobotta, and T. P. Dick, "Measuring $E_{GSH}$ and $H_2O_2$ with roGFP2-based redox probes," *Free Radical Biology & Medicine*, vol. 51, no. 11, pp. 1943–1951, 2011.

[29] P. M. Hwang, F. Bunz, J. Yu et al., "Ferredoxin reductase affects p53-dependent, 5-fluorouracil–induced apoptosis in colorectal cancer cells," *Nature Medicine*, vol. 7, no. 10, pp. 1111–1117, 2001.

[30] C. D. Cerdeira, M. R. P. Lima Brigagão, M. L. Carli et al., "Low-level laser therapy stimulates the oxidative burst in human neutrophils and increases their fungicidal capacity," *Journal of Biophotonics*, vol. 9, no. 11-12, pp. 1180–1188, 2016.

[31] M. G. Mason, P. Nicholls, and C. E. Cooper, "Re-evaluation of the near infrared spectra of mitochondrial cytochrome *c* oxidase: implications for non invasive in vivo monitoring of tissues," *Biochimica et Biophysica Acta (BBA) - Bioenergetics*, vol. 1837, no. 11, pp. 1882–1891, 2014.

[32] M. Lejeune, E. Sariban, B. Cantinieaux, A. Ferster, C. Devalck, and P. Fondu, "Granulocyte functions in children with cancer are differentially sensitive to the toxic effect of chemotherapy," *Pediatric Research*, vol. 39, no. 5, pp. 835–842, 1996.

[33] E. Latz, T. S. Xiao, and A. Stutz, "Activation and regulation of the inflammasomes," *Nature Reviews Immunology*, vol. 13, no. 6, pp. 397–411, 2013.

[34] V. A. K. Rathinam, S. K. Vanaja, and K. A. Fitzgerald, "Regulation of inflammasome signaling," *Nature Immunology*, vol. 13, no. 4, pp. 333–342, 2012.

[35] H. Chung, T. Dai, S. K. Sharma, Y. Y. Huang, J. D. Carroll, and M. R. Hamblin, "The nuts and bolts of low-level laser (light) therapy," *Annals of Biomedical Engineering*, vol. 40, no. 2, pp. 516–533, 2012.

[36] J. C. Tatmatsu-Rocha, C. Ferraresi, M. R. Hamblin et al., "Low-level laser therapy (904nm) can increase collagen and reduce oxidative and nitrosative stress in diabetic wounded mouse skin," *Journal of Photochemistry and Photobiology B: Biology*, vol. 164, pp. 96–102, 2016.

[37] A. C.-H. Chen, P. R. Arany, Y.-Y. Huang et al., "Low-level laser therapy activates NF-$k$B via generation of reactive oxygen species in mouse embryonic fibroblasts," *PLoS One*, vol. 6, no. 7, article e22453, 2011.

# Permissions

All chapters in this book were first published in OMCL, by Hindawi Publishing Corporation; hereby published with permission under the Creative Commons Attribution License or equivalent. Every chapter published in this book has been scrutinized by our experts. Their significance has been extensively debated. The topics covered herein carry significant findings which will fuel the growth of the discipline. They may even be implemented as practical applications or may be referred to as a beginning point for another development.

The contributors of this book come from diverse backgrounds, making this book a truly international effort. This book will bring forth new frontiers with its revolutionizing research information and detailed analysis of the nascent developments around the world.

We would like to thank all the contributing authors for lending their expertise to make the book truly unique. They have played a crucial role in the development of this book. Without their invaluable contributions this book wouldn't have been possible. They have made vital efforts to compile up to date information on the varied aspects of this subject to make this book a valuable addition to the collection of many professionals and students.

This book was conceptualized with the vision of imparting up-to-date information and advanced data in this field. To ensure the same, a matchless editorial board was set up. Every individual on the board went through rigorous rounds of assessment to prove their worth. After which they invested a large part of their time researching and compiling the most relevant data for our readers.

The editorial board has been involved in producing this book since its inception. They have spent rigorous hours researching and exploring the diverse topics which have resulted in the successful publishing of this book. They have passed on their knowledge of decades through this book. To expedite this challenging task, the publisher supported the team at every step. A small team of assistant editors was also appointed to further simplify the editing procedure and attain best results for the readers.

Apart from the editorial board, the designing team has also invested a significant amount of their time in understanding the subject and creating the most relevant covers. They scrutinized every image to scout for the most suitable representation of the subject and create an appropriate cover for the book.

The publishing team has been an ardent support to the editorial, designing and production team. Their endless efforts to recruit the best for this project, has resulted in the accomplishment of this book. They are a veteran in the field of academics and their pool of knowledge is as vast as their experience in printing. Their expertise and guidance has proved useful at every step. Their uncompromising quality standards have made this book an exceptional effort. Their encouragement from time to time has been an inspiration for everyone.

The publisher and the editorial board hope that this book will prove to be a valuable piece of knowledge for researchers, students, practitioners and scholars across the globe.

# List of Contributors

**Li Lu, Yating Qin, Chen Chen and Xiaomei Guo**
Department of Cardiology, Tongji Hospital, Tongji Medical College, Huazhong University of Science and Technology, Wuhan 430030, China

**Hannah K. Drescher, Fabienne Schumacher, Teresa Schenker, Christian Trautwein, Konrad L. Streetz and Daniela C. Kroy**
Department of Internal Medicine III, University Hospital, RWTH, Aachen, Germany

**Maike Baues and Twan Lammers**
Institute for Experimental Molecular Imaging, University Hospital, RWTH, Aachen, Germany

**Thomas Hieronymus**
Institute for Biomedical Engineering-Cell Biology, University Hospital, RWTH, Aachen, Germany

**Lingling Zhang**
Translational Medicine Center, Honghui Hospital, Xi'an Jiaotong University, Xi'an 710054, China

**Jianzong Chen**
Traditional Chinese Medicine Department, Xijing Hospital, Fourth Military Medical University, Xi'an 710032, China

**A-Reum Ryu**
Department of Medical Science, Soonchunhyang University, 22 Soonchunhyang-ro, Asan, Chungnam, Republic of Korea

**Do Hyun Kim**
Department of Biology, The College of Wooster, 1189 Beall Ave, Wooster, OH, USA

**Eunjoo Kim**
Companion Diagnostics and Medical Technology Research Group, Daegu Gyeongbuk Institute of Science and Technology (DGIST), Daegu, Republic of Korea

**Mi Young Lee**
Department of Medical Science, Soonchunhyang University, 22 Soonchunhyang-ro, Asan, Chungnam, Republic of Korea
Department of Medical Biotechnology, Soonchunhyang University, 22 Soonchunhyang-ro, Asan, Chungnam, Republic of Korea

**Liang Liu and Zhiyong Gong**
Key Laboratory for Deep Processing of Major Grain and Oil, Ministry of Education, Wuhan 430023, China
College of Food Science and Engineering, Wuhan Polytechnic University, Wuhan 430023, China
Hubei Key Laboratory for Processing and Transformation of Agricultural Products (Wuhan Polytechnic University), Wuhan 430023, China

**Qinling Hu, Huihui Wu and Xiujing Wang**
Key Laboratory for Deep Processing of Major Grain and Oil, Ministry of Education, Wuhan 430023, China
College of Food Science and Engineering, Wuhan Polytechnic University, Wuhan 430023, China

**Chao Gao**
National Institute for Nutrition and Health, Chinese Center for Disease Control and Prevention, Beijing 100050, China

**Guoxun Chen**
Department of Nutrition, University of Tennessee at Knoxville, Knoxville 37996, USA

**Ping Yao**
Department of Nutrition and Food Hygiene, School of Public Health, Tongji Medical College, Huazhong University of Science and Technology, Wuhan 430030, China

**O. V. Yakovleva, A. R. Ziganshina, A. N. Arslanova, A. V. Yakovlev, N. N. Khaertdinov, G. K. Ziyatdinova and G. F. Sitdikova**
Kazan Federal University, Kazan 420008, Russia

**S. A. Dmitrieva and F. V. Minibayeva**
Kazan Institute of Biochemistry and Biophysics, FRC Kazan Scientific Center of RAS, Kazan 420011, Russia

**R. A. Giniatullin**
Kazan Federal University, Kazan 420008, Russia
A.I. Virtanen Institute, University of Eastern Finland, Kuopio 70211, Finland

**Bruno Alexandre Quadros Gomes and João Paulo Bastos Silva**
Neuroscience and Cell Biology Graduate Program, Institute of Biological Sciences, Federal University of Pará, Belém, Pará, Brazil

**Camila Fernanda Rodrigues Romeiro and Pricila Rodrigues Gonçalves**
Faculty of Pharmacy, Institute of Health Sciences, Federal University of Pará, Belém, Pará, Brazil

**Sávio Monteiro dos Santos, Caroline Azulay Rodrigues, Joni Tetsuo Sakai, Paulo Fernando Santos Mendes and Everton Luiz Pompeu Varela**
Pharmaceutical Sciences Graduate Program, Institute of Health Sciences, Federal University of Pará, Belém, Pará, Brazil

**Marta Chagas Monteiro**
Neuroscience and Cell Biology Graduate Program, Institute of Biological Sciences, Federal University of Pará, Belém, Pará, Brazil
Faculty of Pharmacy, Institute of Health Sciences, Federal University of Pará, Belém, Pará, Brazil
Pharmaceutical Sciences Graduate Program, Institute of Health Sciences, Federal University of Pará, Belém, Pará, Brazil

**Md Mizanur Rahman, Allal Ouhtit and Haissam Abou-Saleh**
Department of Biological and Environmental Sciences, College of Arts and Sciences, Qatar University, Doha, Qatar

**Amina El Jamali**
Division of Nephrology, Department of Medicine, University of Texas Health Science Center at San Antonio, 7703 Floyd Curl Drive, Texas 78229-3900, USA

**Ganesh V. Halade**
Division of Cardiovascular Disease, Department of Medicine, University of Alabama at Birmingham, Birmingham, Alabama 35294, USA

**Gianfranco Pintus**
Department of Biomedical Sciences, College of Health Sciences, Qatar University, Doha, Qatar
Department of Biomedical Sciences, College of Medicine, University of Sassari, 07100 Sassari, Italy
Biomedical Research Center, Qatar University, Doha, Qatar

**Wenjun Han, Changli Wang, Jinjun Bian, Lulong Bo and Xiaoming Deng**
Faculty of Anesthesiology, Changhai Hospital, Naval Medical University, Shanghai 200433, China

**Xiongwei Yu**
Faculty of Anesthesiology, Changhai Hospital, Naval Medical University, Shanghai 200433, China
Department of Anesthesiology, 285th Hospital of the CPLA, Handan 056001, China

**Daming Sui**
Faculty of Anesthesiology, Changhai Hospital, Naval Medical University, Shanghai 200433, China
Department of Anesthesiology, Chengdu Military General Hospital, Chengdu 610083, China

**Barbara Carpita, Dario Muti and Liliana Dell'Osso**
Department of Clinical and Experimental Medicine, University of Pisa, Pisa 55100, Italy

**Marta Loureiro**
Laboratory of Cardiovascular Proteomics, Centro Nacional de Investigaciones Cardiovasculares (CNIC), 28029 Madrid, Spain

**María Gómez-Serrano**
Laboratory of Cardiovascular Proteomics, Centro Nacional de Investigaciones Cardiovasculares (CNIC), 28029 Madrid, Spain
Centro de Investigación Biomédica en Red sobre Enfermedades Cardiovasculares (CIBERCV), Spain

**Emilio Camafeita**
Centro de Investigación Biomédica en Red sobre Enfermedades Cardiovasculares (CIBERCV), Spain
Proteomics Unit, Centro Nacional de Investigaciones Cardiovasculares (CNIC), 28029 Madrid, Spain

**Belén Peral**
Instituto de Investigaciones Biomédicas Alberto Sols (IIBM), Consejo Superior de Investigaciones Científicas and Universidad Autónoma de Madrid (CSIC-UAM), 28029 Madrid, Spain

**Rio Sebori, Atsushi Kuno, Ryusuke Hosoda, Takashi Hayashi and Yoshiyuki Horio**
Department of Pharmacology, Sapporo Medical University School of Medicine, S1, W 17, Chu-ouku, Sapporo 060-8556, Japan

**Yanfang Gao, Liang Xiong, Lijun Zou and Gonghua Hu**
Department of Preventive Medicine, Gannan Medical University, 1 Yixueyuan Road, Ganzhou, 341000 Jiangxi, China

**Huanwen Tang**
Department of Environmental and Occupational Health, Dongguan Key Laboratory of Environmental Medicine, School of Public Health, Guangdong Medical University, Dongguan, 523808 Guangdong, China

**Wenjuan Dai and Hailong Liu**
Department of Preventive Medicine, Gannan Medical University, 1 Yixueyuan Road, Ganzhou, 341000 Jiangxi, China
Department of Occupational Health and Toxicology, School of Public Health, Nanchang University, BaYi Road 461, Nanchang, 3300061 Jiangxi, China

## List of Contributors

**Agnieszka Kicel, Aleksandra Owczarek, Magdalena Rutkowska and Monika A. Olszewska**
Department of Pharmacognosy, Faculty of Pharmacy, Medical University of Lodz, 1 Muszynskiego, 90-151 Lodz, Poland

**Joanna Kolodziejczyk-Czepas and Pawel Nowak**
Department of General Biochemistry, Faculty of Biology and Environmental Protection, University of Lodz, Pomorska 141/143, 90-236 Lodz, Poland

**Anna Wajs-Bonikowska**
Institute of General Food Chemistry, Faculty of Biotechnology and Food Sciences, Lodz University of Technology, 4/10 Stefanowskiego, 90-924 Lodz, Poland

**Sebastian Granica**
Department of Pharmacognosy and Molecular Basis of Phytotherapy, Faculty of Pharmacy, Medical University of Warsaw, 1 Banacha, 02-097 Warsaw, Poland

**Katia Rupel, Luisa Zupin, Augusto Poropat, Giulia Ottaviani, Margherita Gobbo, Lidia Fanfoni, Roberto Di Lenarda and Matteo Biasotto**
Department of Medical, Surgical and Health Sciences, University of Trieste, 34127 Trieste, Italy

**Andrea Colliva**
Cardiovascular Biology Laboratory, International Centre for Genetic Engineering and Biotechnology (ICGEB), 34149 Trieste, Italy

**Serena Zacchigna**
Department of Medical, Surgical and Health Sciences, University of Trieste, 34127 Trieste, Italy
Cardiovascular Biology Laboratory, International Centre for Genetic Engineering and Biotechnology (ICGEB), 34149 Trieste, Italy

**Anselmo Kamada**
Department of Genetics, Federal University of Pernambuco, Recife, Brazil

**Rossella Gratton**
Institute for Maternal and Child Health, IRCCS "Burlo Garofolo", 34137 Trieste, Italy

**Massimo Santoro**
Laboratory of Angiogenesis and Redox Metabolism, Department of Biology, University of Padua, 35131 Padova, Italy

# Index

**A**
Alzheimer's Disease, 26-28, 32, 37, 55, 74, 78, 82-87, 124, 132, 137-138
Antioxidants, 3, 5, 37-38, 42-43, 46, 68-70, 72, 74, 77, 85, 97, 107, 133, 138, 157, 165, 167-168, 170, 177, 191-192
Astroglia, 29, 33
Atherosclerosis, 1-3, 5-11, 21, 25, 30-31, 39, 43, 47, 50, 55, 171, 182
Autism Spectrum Disorders, 109, 114, 116

**B**
Body Mass Index (BMI), 124
Bone Marrow, 12-14, 18-19, 21-23, 89-92, 94-96, 165
Bone Marrow Transplantation, 19
Bone Remodeling, 89-90, 94

**C**
Cardiovascular Disease, 8, 10, 31, 41-43, 47, 60, 89, 118, 123, 132, 182
Cellular Longevity, 36, 40, 72, 82
Chronic, 1-2, 9, 12-14, 19-22, 24, 27, 30, 37, 39, 41-50, 60, 67, 78-79, 86-87, 107, 110, 115-116, 125, 155, 171, 177, 180-181
Chronic Obstructive Pulmonary Disease, 41, 44, 47-48
Colon Carcinoma, 20

**D**
Drug Delivery System, 2
Dystrophin, 144, 151, 154-156

**E**
Endothelial Cells, 8-10, 27, 30-31, 33, 35, 39, 72, 80, 128
Endothelium Dysfunction, 3
Exosomes, 41-48
Extracellular Vesicles, 41, 46-47

**F**
Fatty Acid Composition, 49, 51, 54-55, 59-60
Fibrosis, 7, 10, 12-14, 16, 18, 21-22, 24-25, 35-36, 60, 106, 108, 144-145

**H**
Heme Oxygenase, 5, 37, 79, 81, 99-100, 107-108
Hemin, 99-106, 108
Hepatic Inflammation, 18, 50, 55, 58
Hepatocellular Carcinoma, 12, 22-24
Homocysteine, 9, 61-63, 65, 67-73, 87
Hydroquinone-induced Toxicity, 165

**I**
Immunohistochemical Staining, 20-21

Inflammatory Damage, 31, 39
Inflammatory Disease, 31, 34-35, 39
Inflammatory Response, 5, 22, 42, 58, 79, 111-112, 116, 177
Ischemia, 10, 26-27, 29-30, 33, 36, 38-39, 75, 79, 86, 124, 132, 134, 136-139, 164

**L**
Lipid, 1-2, 5-6, 8-10, 13-14, 17, 24-25, 27, 30-31, 34-35, 37, 39, 43, 47, 49-50, 52-53, 55, 57-63, 66-68, 71, 77, 79, 81, 83-84, 86, 96, 101, 114, 123, 125-126, 157-158, 161, 163-165, 167-170, 173, 178-180, 182
Lipid Dysbolism, 5
Lipid Metabolism, 1, 6, 8, 10, 14, 24, 30, 39, 49-50, 53, 57-59
Lipid Profiles, 5
Lipopolysaccharides, 3, 7, 47
Lipoprotein, 3, 5, 7-10, 30, 35, 49, 51, 55, 86, 171
Liver Cell, 12-14, 18, 21
Liver Fibrosis, 12, 22, 25, 106, 108
Lung Cancer, 22, 24, 41-48

**M**
Mammary Hyperplasia, 20, 24
Maternal Diabetes, 109, 112, 114, 116
Metabolic Syndrome, 13-14, 21, 23, 55, 60, 72
Microglial Activation, 28-29, 31, 77, 79, 83, 110
Mitochondrial Apoptotic Pathway, 31, 34
Mitochondrial Dysfunction, 29, 37-38, 75, 77, 116, 118, 123-126, 136-138, 145, 153, 155
Mitoproteomics, 118, 121, 123-127, 129, 132
Monocyte, 5, 10, 25, 54

**N**
Neurodegenerative Diseases, 31, 38, 40, 72, 74, 80, 82, 84-85, 124, 138
Neuroinflammation, 27, 29, 31, 33, 36, 38, 40, 62, 66, 68, 79-80, 86-87
Neuroprotective Mechanism, 81
Neurotoxicity, 28-29, 32-33, 38, 61-62, 68, 70-73, 79, 86-88
Nonalcoholic Fatty Liver Disease, 14, 23-24, 49, 59, 125, 137
Nonalcoholic Steatohepatitis, 12, 14, 18, 23, 60, 125

**O**
Obesity, 14, 59-60, 89-90, 94, 96-97, 109-117, 123-125, 133, 135, 137
Obstructive Pulmonary Disease, 41, 44
Oral Cancer, 41-42, 44-46, 48
Oxidative Medicine, 9, 36, 38, 40, 72, 82, 84
Oxidative Stress, 1-3, 5, 9, 12-14, 18, 22, 26-33, 35-40, 42, 44, 47, 49, 55, 58-60, 62, 66-73, 75, 77-78, 80, 82-84, 86-87, 90, 97, 104-107, 109-113, 116, 118-119, 134, 144, 154-158, 170, 178-180, 190-191

## P
Paeonia Suffruticosa, 1, 8-9
Paeonol, 1-3, 5-11
Parkinson's Disease, 26, 29, 33, 36-38, 124, 135, 137
Photobiomodulation, 183-184, 192
Polygonum Multiflorum, 26, 35-36, 39
Prenatal Hyperhomocysteinemia, 61, 70-71
Proinflammatory Cytokines, 31, 59, 89, 94, 99-100, 105-106, 110, 115
Proinflammatory Immune Response, 16, 20, 22
Proinflammatory Status, 50, 58

## R
Redox Homeostasis, 58, 72, 141, 164
Resveratrol, 26-27, 36, 74-88, 144-151, 153-155

## Rheumatoid Arthritis, 11, 90
Rhizome, 26, 36

## S
Sepsis, 99-107
Steatohepatitis, 12-14, 16-25, 58, 60, 125, 138

## T
Tetrahydroxystilbene Glucoside, 26, 36-38, 40
Thrombophilia, 42-43
Thrombosis, 2, 6, 10, 39, 46-47, 72, 136
Tumor Necrosis Factor, 10, 28, 31, 35, 51, 60, 79, 90

## V
Vascular Endothelial Growth Factor, 30

CPSIA information can be obtained
at www.ICGtesting.com
Printed in the USA
BVHW091611160820
586552BV00002B/13